Smart Nanomaterials Technology

Series Editors

Azamal Husen, Wolaita Sodo University, Wolaita, Ethiopia

Mohammad Jawaid, Department of Chemical and Petroleum Engineering, United Arab Emirates University, Al Ain, United Arab Emirates

Nanotechnology is a rapidly growing scientific field and has attracted a great interest over the last few years because of its abundant applications in different fields like biology, physics and chemistry. This science deals with the production of minute particles called nanomaterials having dimensions between 1 and 100 nm which may serve as building blocks for various physical and biological systems. On the other hand, there is the class of smart materials where the material that can stimuli by external factors and results a new kind of functional properties. The combination of these two classes forms a new class of smart nanomaterials, which produces unique functional material properties and a great opportunity to larger span of application. Smart nanomaterials have been employed by researchers to use it effectively in agricultural production, soil improvement, disease management, energy and environment, medical science, pharmaceuticals, engineering, food, animal husbandry and forestry sectors.

This book series in Smart Nanomaterials Technology aims to comprehensively cover topics in the fabrication, synthesis and application of these materials for applications in the following fields:

- Energy Systems—Renewable energy, energy storage (supercapacitors and electrochemical cells), hydrogen storage, photocatalytic water splitting for hydrogen production
- Biomedical—controlled release of drugs, treatment of various diseases, biosensors,
- Agricultural—agricultural production, soil improvement, disease management, animal feed, egg, milk and meat production/processing,
- Forestry—wood preservation, protection, disease management
- Environment—wastewater treatment, separation of hazardous contaminants from wastewater, indoor air filters.

Ayan Chatterjee · Tanmay Sarkar · Slim Smaoui
Editors

Nanofuel: The Future of Sports Nutrition

Boosting Performance with Nanotech Nutrients

Springer

Editors
Ayan Chatterjee
School of Health Science and Technology
Medhavi Skills University
Singtam, Sikkim, India

Slim Smaoui
Center of Biotechnology of Sfax
University of Sfax
Sfax, Tunisia

Tanmay Sarkar
Department of Food Processing
Technology, Malda Polytechnic
West Bengal State Council of Technical
Education, Government of West Bengal
Malda, West Bengal, India

ISSN 3004-8273 ISSN 3004-8281 (electronic)
Smart Nanomaterials Technology
ISBN 978-981-96-5470-3 ISBN 978-981-96-5471-0 (eBook)
https://doi.org/10.1007/978-981-96-5471-0

This Springer imprint is published by the registered company Springer Nature Singapore Pte Ltd.
The registered company address is: 152 Beach Road, #21-01/04 Gateway East, Singapore 189721, Singapore

To my beloved mother, Arati Sarkar, and father, Bholanath Sarkar, and my sister, Tanusree Sarkar, whose unwavering love, support, and encouragement have shaped my journey.

—Dr. Tanmay Sarkar

To my beloved mother, Baisakhi Chatterjee, and my father, Ashis Chatterjee, and my wife, Parna Chatterjee.

—Dr. Ayan Chatterjee

This book is dedicated to the memory of my deceased Father Mohamed Nejib Smaoui, to my beloved mother Raoudha, my wife Hajer and my children Mohamed Yassine and Yassmine, whose unwavering love, support, and encouragement have shaped my life.

—Dr. Slim Smaoui

And to God, for the strength and guidance throughout every step of the way.

Preface

This book, *Nanofuel: The Future of Sports Nutrition*, is where two groundbreaking areas—sports nutrition and nanotechnology—come together. People around the world are always looking for ways to help athletes perform at their best, and this has led to the search for new ideas that go beyond the usual ways of eating and fueling the body. Nanotechnology is changing the game in fields like medicine, biochemistry, and food science, and it's ready to do the same for sports nutrition. It promises to bring more accuracy, efficiency, and better results than ever before.

The book explores how nanotechnology can be used to meet the special nutritional needs of athletes and active people. It takes a close look at how nano-enhanced products, advanced ways of delivering nutrients, and new methods can improve how the body absorbs and uses nutrients. Whether it's helping with recovery, staying hydrated, boosting endurance, or building strength, nanotechnology is paving the way for a future where sports nutrition is more tailored, precise, and effective. From improving recovery and hydration to boosting endurance and strength, nanotechnology leads the way to a future where sports nutrition is more customized, precise, and effective than ever before.

The book is designed for researchers, professionals, and students; this book covers important topics like nano-encapsulation, improving nutrient absorption, and how nanomaterials can enhance performance. Each chapter blends scientific depth with practical insights, offering a clear understanding of the topic while encouraging new ideas and real-world applications.

Exploring *Nanofuel: The Future of Sports Nutrition* is a step forward in advancing athlete health and performance. We hope this book not only educates but also sparks new innovations and teamwork in this exciting area of study.

Singtam, India — Dr. Ayan Chatterjee
Malda, India — Dr. Tanmay Sarkar
Sfax, Tunisia — Dr. Slim Smaoui

Contents

Editors and Contributors

About the Editors

Dr. Ayan Chatterjee is currently working as an associate professor and associate dean of School of Health Science and Technology, Medhavi Skills University, Sikkim, India. He also has teaching experience in human physiology, biochemistry, and allied subjects at undergraduate, postgraduate, and diploma levels for 12 years. He also served as a member of Editorial Board of three UGC CARE-listed journals and reviewer of three Scopus-indexed and three UGC CARE-listed journals. Dr. Ayan also served as an editor of five books of different publication houses (Elsevier, Springer Nature, IGI Global USA).

Dr. Tanmay Sarkar is currently working as a lecturer in the Government of West Bengal. He has received his Bachelor of Technology, Master of Technology, and Ph.D. in Engineering from Jadavpur University, Kolkata, India. With nine years of experience in both teaching and research, his areas of interest include bioactive components from natural sources, ethnic foods, mathematical modeling, and phytochemicals. He is an editor for several SCI, SCIE, and Scopus-indexed journals and has edited 12 books with Elsevier, Springer, and Wiley. Dr. Sarkar has been classified among the top 2% of global highly cited researchers published by Stanford University, USA.

Dr. Slim Smaoui is currently working as an associate professor of engineering at the Centre of Biotechnology of Sfax (CBS)—Tunisia. According to August 2024 data update for "Updated science-wide author databases of standardized citation indicators" Ioannidis list Elsevier and Stanford University, Dr. Slim Smaoui is in the top 2% of scientists worldwide. Dr. Smaoui's research interests are centered on the development of new healthier meat products; use of agro-food industries by-products as potential sources to obtain valuable and bioactive compounds, and incorporation of bioactive compounds and development of functional foods. Dr. Slim Smaoui is the author or co-author of several papers (Scopus indexed) published in international peer-reviewed journals in food science.

Contributors

Suad Ajab Institute of Public Health, College of Medicine and Health Sciences, United Arab Emirates University, Al Ain, Abu Dhabi, United Arab Emirates

Faiza Ameen British School of Leadership and Management, Glasgow Caledonian University, London Campus, London, England

K. A. Anju Department of Food Technology, Saintgits College of Engineering, Kottayam, Kerala, India

Akramul Ansary NETES Institute of Pharmaceutical Science, NEMCARE Group of Institutions, Mirza, Kamrup, Assam, India

Emmanuel Achumboro Apanga Ho Polyclinic, Ghana Health Service, Ho, Volta Region, Ghana

Mahenderan Appukutty School of Sports Science and Recreation, Universiti Tecknologi MARA, Shah Alam, Malaysia

M. K. Aravind Department of Physical Education and Sports, Pondicherry University, Puducherry, India

N. K. Ashif Ali Department of Physical Education and Sports, Pondicherry University, Puducherry, India

Debarati Basu Goenka College of Pharmacy, Lachhmangarh (Sikar), Rajasthan, India

Abhilash Bharadwaj Department of Pharmacognosy, NETES Institute of Pharmaceutical Science, NEMCARE Group of Institution, Shantipur, Assam, India

Anirbandeep Bose Department of Pharmaceutical Technology, School of Health and Medical Sciences, Adamas University, Barasat, Kolkata, West Bengal, India

Ayan Chatterjee School of Health Science and Technology, Medhavi Skills University, Topakhani, Sikkim, India

Nafih Cherappurath Department of Physical Education, Amal College of Advanced Studies, Nilambur, India

Elsa Cherian Department of Food Technology, Saintgits College of Engineering, Kottayam, Kerala, India

Kirti Chhabra Department of Physiotherapy, RPIIMS, Karnal, Haryana, India

Deijy Choudhury NETES Institute of Pharmaceutical Science, NEMCARE Group of Institutions, Mirza, Kamrup, Assam, India

Anindita Das Department of Physical Education Pedagogy, Lakshmibai National Institute of Physical Education, Gwalior, Madhya Pradesh, India

Pulen Das Lakshmibai National Institute of Physical Education NERC Guwahati, Guwahati, India

Soma Das Department of Pharmaceutical Technology, School of Health and Medical Sciences, Adamas University, Barasat, Kolkata, West Bengal, India

Panchali Dasgupta Child Fund India, Kolkata, India

Suchandra Datta Department of Home Science, University of Calcutta, Kolkata, West Bengal, India

Babita Deka NETES Institute of Pharmaceutical Science, NEMCARE Group of Institutions, Mirza, Kamrup, Assam, India

Juti Rani Devi NETES Institute of Pharmaceutical Science, NEMCARE Group of Institutions, Mirza, Kamrup, Assam, India

Bajrang Dhundura Healthcare Australia, Osborne Park, Australia

Tarun Kumar Dua Department of Pharmaceutical Technology, University of North Bengal, Raja Rammohunpur, Darjeeling, West Bengal, India

Koushik Nandan Dutta Department of Pharmacognosy, NETES Institute of Pharmaceutical Science, NEMCARE Group of Institution, Shantipur, Assam, India

Masilamani Elayaraja Department of Physical Education and Sports, Pondicherry University, Puducherry, India

Arunava Ghosh Bengal School of Technology, Hooghly, West Bengal, India

Gnanaratnam Giritharan Nevada Center for Reproductive Medicine, Reno, NV, USA

Bitu Gogoi NETES Institute of Pharmaceutical Science, NEMCARE Group of Institutions, Mirza, Kamrup, Assam, India

Rajdip Goswami Department of Pharmaceutical Technology, School of Health and Medical Sciences, Adamas University, Barasat, Kolkata, West Bengal, India

Karuppasamy Govindasamy Department of Sports, Recreation and Wellness, Symbiosis International (Deemed University), Hyderabad Campus, Modallaguda (V), Nandigama (M), Rangareddy, Telangana, India

Prakash Kishore Hazam Marine Research Station, Institute of Cellular and Organismic Biology, Academia Sinica, Jiaushi, Ilan, Taiwan

Darshana Hazarika Department of Community Health Nursing, NEMCARE Institute of Nursing Sciences, NEMACRE Group of Institutions, Santipur, Assam, India

Rukshana Irani Department of Food and Nutrition, Raidighi College, Raidighi, West Bengal, India

Beenish Israr Faculty of Food Nutrition and Home Sciences, Institute of Home Science, University of Agriculture, Faisalabad, Pakistan

Nizwa Itrat Faculty of Food Nutrition and Home Sciences, Institute of Home Science, University of Agriculture, Faisalabad, Pakistan

Dilshith A. Kabeer Department of Physical Education and Sports, Pondicherry University, Puducherry, India

Debajit Karmakar Department of Physical Education Pedagogy, Lakshmibai National Institute of Physical Education, Gwalior, Madhya Pradesh, India

Neha Kashyap SAHS, Galgotias University, Greater Noida, UP, India

Ketheeswaran Science Unit, Allied Health Sciences, University of Jaffna, Jaffna, Sri Lanka

Waseem Khalid Department of Organic Chemistry, Faculty of Chemical Sciences and Technologies, University of Castilla La Mancha, Ciudad Real, Spain

Rameer Thaslim Khan Department of Food Science and Technology, Pondicherry University, Kalapet, Puducherry, India

Soumik Laha Department of Pharmaceutical Technology, School of Health and Medical Sciences, Adamas University, Barasat, Kolkata, West Bengal, India

G. Manibharathi Department of Physical Education and Sports, Pondicherry University, Puducherry, India

Sammra Maqsood National Institute of Food Science and Technology, University of Agriculture, Faisalabad, Pakistan

Dua Munawwar Department of Nutrition and Dietetics, The University of Faisalabad, Faisalabad, Pakistan

P. Muneer Department of Physical Education, University of Calicut, Thenhipalam, India

Bijina Naushad Applied Food Safety and Quality Management, University of Greenwich, London, England

Mohammed Navaf Department of Food Science and Technology, Pondicherry University, Puducherry, India

Bhaskar Pal Department of Pharmacology, Charaktala College of Pharmacy, Charaktala, Mothabari, Malda, West Bengal, India

Koulla Parpa Faculty of Sports and Exercise Science, UCLan University of Cyprus, Pyla, Cyprus

Pooja Patowary NETES Institute of Pharmaceutical Science, NEMCARE Group of Institutions, Mirza, Kamrup, Assam, India

Pollobi Porasar Department of Pharmacognosy, NETES Institute of Pharmaceutical Science, NEMCARE Group of Institution, Shantipur, Assam, India

Borem Siri Reddy Conestoga College Institute of Technology, Kitchener, ON, Canada

Rosamma Department of Food Technology, Saintgits College of Engineering, Kottayam, Kerala, India

Arjit Prosad Roy DmbH Institute of Medical Science, Dadpur, Puinan, Hooghly, West Bengal, India

Sukanta Roy ABS Clinical Solutions, Garia, Kolkata, India

Kushal Roychoudhuri Department of Pharmaceutical Technology, Charaktala College of Pharmacy, Charaktala, Mothabari, Malda, West Bengal, India

A. K. Safad Department of Physical Education and Sports, Pondicherry University, Puducherry, India

Medha Samineni University of Houston-Clear Lake, Houston, TX, USA

Saikat Santra Department of Pharmaceutics, JRSET College of Pharmacy, Panchpota, Chakdah, Nadia, West Bengal, India

Rabiya Saroosh Department of Nutrition and Dietetics, The University of Faisalabad, Faisalabad, Pakistan

Javeria Shabbir Department of Nutrition and Dietetics, The University of Faisalabad, Faisalabad, Pakistan

Vinitha Shankar Department of Chemical Engineering, AC Tech Campus, Anna University, Chennai, India

K. K. Shareef Department of Physical Education and Sports, Pondicherry University, Puducherry, India

Dipjyoti Sharma Department of Pharmacognosy, NETES Institute of Pharmaceutical Science, NEMCARE Group of Institution, Shantipur, Assam, India

Ashutosh Singh Department of Physiotherapy, Callingwood Physical Therapy and Sports Injury Clinic, ProActive Physical Therapy and Sports Injury Clinic, Edmonton, Canada

Shikha Singh Jyotiraophule Subharti College of Physiotherapy, Subharti University, Meerut, India

Shobana Srinivasan Department of Chemical Engineering, AC Tech Campus, Anna University, Chennai, India

Valentina Stefanica Department of Physical Education and Sport, Faculty of Sciences, Physical Education and Informatics, National University of Science and Technology Politehnica Bucharest, Pitesti University Center, Pitesti, Romania

Kappat V. Sunooj Department of Food Science and Technology, Pondicherry University, Puducherry, India

Aadesh Upadhyay Project Lead-Oncology, One Clinical Research, Perth, WA, Australia

Sivakumar Venkatachalam Department of Chemical Engineering, AC Tech Campus, Anna University, Chennai, India

W. Vinu Department of Physical Education and Sports, Pondicherry University, Puducherry, India

Wong Tin Wui Non-Destructive Biomedical and Pharmaceutical Research Centre, Smart Manufacturing Research Institute/Faculty of Pharmacy, Universiti Teknologi MARA, Puncak Alam, Selangor, Malaysia

Introduction to Nanotechnology in Sports Science and Nutrition

Pulen Das and Emmanuel Achumboro Apanga

Abstract **Sports nutrition** is a specialized field within nutrition that focuses on the dietary needs of athletes and active individuals. To maximize athletic performance, improve recovery, and promote general health, it includes the strategic planning and execution of nutrient intake. **Nanotechnology** aids in making sports equipment lighter, durable, and more flexible. Nanotechnology could be especially advantageous in altering physical characteristics that involve size, shape, volume, color, weight, and temperature to develop novel sports products. Nanoscience is a branch of material science. Sports nutrition is changing quickly due to technological breakthroughs and a better understanding of human physiology. Nanotechnology has been a game-changer among these developments, resulting in the creation of "**nano fuels**." Sports cannot exist without technology due to the constant advancement of new technologies. The sports sector is improving daily by utilizing several sources. Around the globe, the sports industry additionally collaborates with the smallest things termed 'nanotechnology'. Currentchapter examines concept of nanofuels, their possible advantages for athletes, their working mechanics, and the difficulties and a moral dilemma associated with their application in sports and also covers the various applications of nanomaterials in sports industry and recent research on sports nutrition.

Keywords Sports nutrition · Nanotechnology · Nano fuel · Nanomaterial · Supplements of sports nutrition

---

Although Nanotech Innovation of Tomorrow is Small-Scale, It's Kind of a Big Deal

---

P. Das (✉)
Lakshmibai National Institute of Physical Education NERC Guwahati, Guwahati, India
e-mail: pulendas86@gmail.com

E. A. Apanga
Ho Polyclinic, Ghana Health Service, Volta Region, Ho, Ghana

A. Chatterjee et al. (eds.), *Nanofuel: The Future of Sports Nutrition*, Smart Nanomaterials Technology, https://doi.org/10.1007/978-981-96-5471-0_1

1 Introduction

1.1 Overview of Sports Nutrition

The research investigation examining how to apply nutrition to enhance athletic performance has been termed sports nutrition. To assist athletes excel in their sport, it involves comprehending the right diet, hydration, nutrients, and supplements. Employing nutrition to support all aspects of athletic performance is research and practice of sports nutrition. Educating one on right diets, nutrition, hydration techniques, and supplements will help one excel in their sport.

The potential that athletes need different nutritional intakes than non-athletes distinguish **sportsnutrition** from general nutrition. Sports nutrition, along with exercise, may enhance an athlete's performance.

Sports nutrition is an interdisciplinary field that combines nutrition knowledge with exercise science and research on the human body. It also considers psychological aspects of nutrition, such as eating disorders and other psychiatric conditions that may affect athletes.

Here Are Some Tips for Sports Nutrition:

- **Carbohydrates**: Carbohydrates are the primary fuel source for athletes, providing glucose for energy. Athletes should consume 45–65% of their calories from them.Yogurt, milk, whole grains, fruits, and vegetables are all good sources of carbohydrates.
- **Eating before an event**: Eat carbohydrate-rich food 3–4 h preceding an event.Eat digestible foods instead of fried or high-fat ones.Within an hour of the event, restrict sugary drinks and snacks.
- **Supplements**: Some supplements like creatine, can help improve muscle strength and speed. However, supplements are not FDA-tested for safety and effectiveness and can cause side effects like bloating, diarrhea, and low blood pressure.

1.2 Overview of Nanotechnology

Nanotechnology produces materials at the molecular level. Its advanced qualities could strengthen numerous industries and improve material [1]. The sports sector has benefited from the introduction of nanotechnology by increasing efficiency and lowering the weight of equipment. Nanotechnology's improved durability, potency, and lightweight construction make athletes more comfortable, safer, less prone to injuries, and more nimble.

In recent years, nanotechnology has become a buzzword in the world of sports nutrition. With its promise of delivering enhanced performance and improved recovery, it has sparked the interest of athletes and fitness enthusiasts alike.

Nanotechnology regulates and controls materials and devices smaller than 100 nm. This multidisciplinary approach includes "applied physics, materials science, chemistry, biology, surface science, robotics, engineering, electrical engineering, and biomedical engineering". At this scale, properties of matter are established, and there are very few discrepancies between scientific domains. Nanotechnology (often abbreviated to "nanotech") is investigation of manipulating matter at atomic and molecular scale. Primary objective of nanotechnology is to produce materials, equipment, or other structures with at least one dimension within 1–100 nm [2].

However, nanotechnology goes beyond only reducing the size of what occurs at eye level. Objects might display peculiar characteristics that differ from where they appear on macroscopic scale when they are altered on nanometric scale. Physical, chemical, optical, or mechanical composition can all change as a result of a change in surface area. Compared to their life-sized equivalents, materials can become more robust, conductive, or durable.

Example: Football utilizes nanotechnology in a variety of ways, such as:

- **Shin pads**: Shin pads are frequently composed of strong, lightweight nanostructured polymers.
- **Footballs**: Footballs utilize nano clay elements in their linings for helping retain pressure.
- **Kit fabrics**: Nanoparticles (NP) of titanium and silver (Ag) have been employed for developing fabrics that are oil-repellent, waterproof, antibacterial, anti-odor, and anti-stain.

The scale of **nanomaterials** and their high surface area-to-volume ratio provide them with unique properties and behaviors. In general, they could be characterized by having dimensions ranging from 1 to 100 nm. NP, nanotubes, nanofibers, and nanocomposites are all members of nanomaterial family. Variations in electronic, optical, and magnetic properties might result from these materials' unique characteristics, including high surface areas and quantum effects.

Synthesis of nanomaterials through a bottom-up method that builds materials from atomic or molecular components, or a top-down method that breaks down larger materials. From improving electronics and catalysis to advancing energy and medical technology, nanomaterials have a wide range of applications. As institutions and regulatory agencies deal with the safety and ethical ramifications of nanotechnology, its application presents safety and environmental issues that require careful investigation into possible toxicity and long-term impacts.

Types of Nanotechnology

Based on the sequence that it has been developed or media in which it operates, nanotechnology could be categorized into four primary fields:

- **Descending (top-down)**: By reducing existing structures and systems to nanoscale from atomic levels to100nm novel technologies could be established.

- **Ascending (bottom-up)**: An atom or molecule serves as one of the fundamental building components of nanometric structure that nanotechnologists begin to operate with.
- **Dry**: Nanotechnology applied to water-incompatible semiconductors and metals is distinctive.
- **Wet**: addresses biological systems that occur in aquatic environments, including cells and water-dependent activities.

Nano Fuels can significantly improve sports nutrition by enhancing nutrient delivery, bioavailability, and overall athletic performance.Materials designed at the nanoscale to maximize energy release and nutrient absorption are known as **nanofuels.** These can comprise nutrients that have been encapsulated, nanoemulsions, and NP that are intended to enhance the effectiveness and delivery of bioactive substances. These developments may result in better performance indicators, quicker recuperation times, and more efficient fuelling plans for athletes.

Mechanisms of Action

1. **Enhanced Absorption**: Nanotechnology allows for the creation of nano-sized carriers that encapsulate nutrients, which can enhance their solubility and bioavailability. For instance, curcumin, known for its anti-inflammatory properties, has low bioavailability when consumed in its natural form. However, when encapsulated in NP, its absorption rate significantly increases, making it more effective for athletes.
2. **Targeted Delivery**: Nutrients may be released precisely where they are required by engineering NP that target specific tissues or organs. This targeted approach can improve the efficacy of supplements like creatine, ensuring that it reaches muscle cells more effectively than conventional forms.
3. **Sustained Release**: Nanofuels can be designed for controlled and sustained nutrient release. This feature can help athletes maintain energy levels during prolonged activity without the need for frequent re-fueling, optimizing endurance performance.
4. **Functionalization**: By modifying the surface properties of NP, researchers can enhance their interaction with biological systems. This functionalization can improve how nutrients are absorbed and utilized within the body, providing a more holistic approach to sports nutrition.

2 Applications of Nano Fuels Technology in Sports

The technology of Nano fuel is used in sports in a number of ways that helps athletes build nutritional values, which includes equipment, medicine, and clothing [3]:

- **Equipment**: Sports equipment may prove made lighter, more durable, and more effective with the addition of nanomaterials. For instance, carbon nanotubes (CNTs) are 6 times lighter than steel and 100 times stronger than steel, they are employed in bike frames, golf clubs, and tennis rackets. Graphene, silica NPs, and nanoclays are further nanomaterials utilized in sports equipment.

- **Medicine**: Nanotechnology can be used in sports medicine for medical interventions like tissue regeneration, drug therapies, and implantation. It can also be used for supporting elements like imaging, strappings, and tapings.
- **Clothing:** Nanotechnology can be used in clothing to make it more water-repellent, quick-drying, and environmentally friendly. For example, Sun Dry Swim uses NP to create a water-repellent surface on swimwear fabric.

2.1 Equipments

Sports equipment including "fly-fishing rods, tennis/badminton racquets, racing bicycles, golf balls, baseball bats, archery arrows, skis, and hockey sticks" are among many that have improved performance since nanotechnology has been introduced to the sports industry.

Equipment for "golf, kayaking, bowling, watercraft, road racing, skiing, tennis/badminton, and archery" employs **nanomaterials** involving CNTs, fullerenes, nanoclay, silica NPs, carbon nanofibers, carbon nanonickel, carbon NP to improve qualities consisting of repulsion, strength, durability, scratch resistance, increased stiffness, decreased weight, and chipping resistance.

Nanotechnology in the sports industry is essential, and the stakes are undeniably elevated for athletes. Nanotechnology has introduced breathable materials into sporting apparel. Socks are enhanced by capturing microorganisms and smell due to the incorporation of AgNP in the material. Materials could be made stronger, lighter, resilient, more reactive, more permeable, or stronger electrical conductors, among other qualities, via nanotechnology.

In addition to improving resistance to bacterial growth, stains, and wrinkles, **nanoscale** additions or surface treatments for textiles may enhance lightweight ballistic energy deflection in personal body armor. On eyeglasses, computer and camera screens, windows, and other surfaces, **nanoscalefilms** can make them electrically conductive, water and residue-repellent, antireflective, self-cleaning, UV or infrared radiation-resistant, antifog, antimicrobial, and scratch-resistant. Health monitoring, solar energy capture, and movement-based energy harvesting are all made possible by the development of washable, long-lasting "smart fabrics" that are integrated with flexible nanoscale sensors and electronics.

Significant fuel savings could result from **light weighting** of vehicles including cars, trucks, airplanes, boats, and spacecraft. Nanoscale additions in polymer composite materials are"employed in "baseball bats, tennis rackets, bicycles, motorcycle helmets, car parts, luggage, and power tool housings to make them lightweight, rigid, resilient, and long-lasting". CNT"sheets are now being produced for application in airplanes of future. For instance, they are ideal for applications involving heat control and electromagnetic shielding due to their conductivity and low weight.

By adding carbon nanotubes filled with graphite to the row, sports equipment beefed up **rackets** made of metal and wood weigh too much less than their predecessors. However, the racquets made with nanotechnology are more durable and lighter for handling, and the ball that provides them the term "Nanotechnology in Sports Equipment" also has nanotechnology.

Nanotechnology in sports equipment additionally assists in the development of a range of NP coatings on **balls** to change their inflating capabilities and reduce the requirement for ball replacement since air cannot escape from the center of the ball.

Every second matters in professional cycling when it comes to winning. By adding carbon nanotubes to the carbon fiber resin, the process of making bicycles has changed. Bicycles have become stronger and lighter due to the introduction of nanotechnology in sports equipment, which enables cyclists to ride more effectively.

Comparing Traditional versus Nano-Enhanced Equipment

There are significant advantages when employing nanomaterials when compared to conventional sporting equipment. Although robust, traditional equipment often lacks nano-enhanced materials' durability and wear resistance. Due to materials including carbon nanotubes, nano-enhanced equipment is incredibly robust and durable, greatly lowering the probability of breakage and damage during intense athletic activity. A further significant factor is weight. An athlete's performance may be affected by the additional weight of traditional equipment. Due to improved materials, nano-enhanced equipment is lighter, allowing athletes to move more efficiently and without fatigue, improving agility and performance.

Nano-enhanced equipment excels in sports, where performance is crucial factor. It provides better performance qualities in variety of sports, including increased power, control, accuracy, and aerodynamics. Traditional equipment couldn't perform additionally due to its limitations regarding design and materials. Comfort and safety are essential for sports. Novel impact-absorbing materials in nano-enhanced safety gear like helmets and padding improve comfort and safety, whereas standard gear may only offer limited benefits. Another significant benefit of nano-enhanced equipment is customization, that enables it to be customized to fulfill the specifications and preferences of certain athletes. Traditional equipment, conversely, is typically mass-produced and offers minimal personalizing options.

Longevity is another advantage of nano-enhanced equipment. The lifespan of traditional equipment may be reduced by wear and tear, however, sportsmen may conserve money by utilizing nano-enhanced equipment, that typically offers longer durability and improved performance. However, safety and environmental issues related to nanomaterials must be taken into account. Although nano-enhanced equipment has several advantages, careful application, and comprehensive safety evaluations are essential for protecting environment and athletes. When considered, sports equipment has evolved significantly due to the application of nanomaterials, providing athletes with improved performance, increased safety, and a more enjoyable athletic experience.

In the future, nanomaterials in sports apparel will advance. Nanomaterials will improve sports performance as nanocomposites increase precision, power, speed, and endurance. As customization becomes prevalent athletes would be able to customize their equipment to fulfill their unique requirements.

Recyclable and biodegradable NP might reduce sports equipment's environmental effects and promote **sustainability**. As NP continues to enhance impact absorption in padding and helmets, safety will become more essential. Future sporting equipment will be built with **durability and innovative designs** in mind, and athletes will be able to get real-time performance insights through data integration and smart devices. Further developments will be fueled by a stronger regulatory environment and cooperative research, which will help athletes and promote nanotechnology across a range of industries.

2.2 Medicines

The medical tools, information, and treatment that physicians now have access to are already being expanded by nanotechnology. Nanomedicine applies nanotechnology to biomedicine, providing precise solutions for disease prevention, diagnosis, and treatment.

Nanotechnology enhances athlete and human health while making sports safer. Faster, more intensive medical assistance is provided via nanotechnology. Early disease detection and innovative therapy for both novel and existing conditions will result from molecular disease detection.

For example, General Electric is focused on material performance. They are developing self-assembled block polymers, nanotubes, nanorods, nanoceramics, and nanostructured metal alloys.

Global Data, an analytics and consulting firm, discovered several emerging nanotech trends in patient care. Target specification is process of attaching NP to medications or synthetic vesicles termed liposomes basically, encapsulated water droplets made for a particular application to target particular cells and tissues. According to Global Data, this enables medicine to target sick or malignant "cells of interest" directly while avoiding healthy ones.

Another potential application is controlled drug release, which would enable healthcare professionals to regulate the release of a medication or medicinal substance as a response to a trigger. This might be induced externally by stimuli including heat, light, or ultrasound, or internally by a shift in the tissue as it develops around a tumor.

Below are some examples of **recent advances** in this area:

In addition to being therapeutically researched as possible treatments for cancer and other illnesses, **gold nanoparticles** have been modified for use in commercial applications as probes for the detection of certain nucleic acid sequences.

Improved therapy success rates, customized treatment options, and earlier diagnosis are being made possible by **improved imaging and diagnostic instruments** made feasible by nanotechnology.

Researchers are exploring the application of nanotechnology for the detection and treatment of **atherosclerosis**, characterized by plaque deposition in arteries. Researchers developed a nanoparticle that replicates HDL (high-density lipoprotein), the body's beneficial cholesterol that facilitates plaque elimination.

Advanced solid-state nanopore materials may facilitate development of innovative gene sequencing methods enabling single-molecule detection at low costs at high speeds, requiring minimal sample preparation and apparatus.

Researchers in nanotechnology have developed several medicines that utilize NPs to encapsulate or facilitate targeted delivery of medication to cancer cells, thereby minimizing risk of damaging healthy tissue. This could significantly decrease the adverse effects of chemotherapy and transform approach physicians employ in cancer treatment.

The field of nanotechnology for regenerative medicine is exploring several application areas, that include bone and neural tissue engineering. For instance, innovative materials may be developed to replicate the crystalline mineral composition of human bone or employed as a dental restorative resin. Researchers are exploring strategies for nurturing complex tissues to ultimately generate human organs for donation. Initial investigations suggest that neurons proliferate effectively on conductive graphene surfaces; hence, researchers are further exploring the application of graphene nanoribbons to facilitate spinal cord repair.

Researchers in nanomedicine are exploring if nanotechnology may improve vaccinations, involving administration of needle-free vaccines. Researchers are attempting to develop universal vaccine scaffold for annual flu shot that would encompass a broader range of strains while minimizing the resources required for annual development.

2.3 *Clothing*

Nanotechnology is used in **sports clothing** to improve athletes' performance, comfort, [4] and safety. Nanofibers, or nanotechnology applied to clothing, contribute to the raindrops, the way your windbreaker rolls and then adheres to it, and the unexpectedly odor-resistant athleisure clothes that are stacked in the corner of your bedroom weeks ago.

Whether woven into the fabric or sprayed on its surface, silica NP contributes to keeping us dry in water-repellent clothing and beneath umbrellas. AgNP which frequently appears in socks and T-shirts, have antibacterial qualities that eliminate odorous bacteria and reduce the frequency of washing cycles. Copper is added to combination to provide a protective layer that, when heated or exposed to sunlight, breaks down food and dirt. Copper-silica NP goes one step further and actively targets and alters stench-causing chemicals to chemically deodorize.

Some examples of how nanotechnology is used in sports clothing include:

- **Temperature regulation**: Nanofibers and NP can assist athletes in maintaining a cool temperature in hot conditions and warm in cold environments.
- **UV protection**: NP can protect athletes from harmful UV radiation.
- **Antimicrobial activity**: NP can reduce bacteria and odor.
- **Water repellency**: Nano-coatings can make surfaces water-repellent to keep athletes dry.
- **Thermoregulation**: NP can convert body heat into infrared radiation to promote blood flow and muscle oxygenation.
- **Electronics**: Electronic sensors and GPS can help athletes in emergencies.
- **Doping detection**: Nanosensors can improve the accuracy of detecting doping substances.

Some **companies** that use nanotechnology in their products include:

- **Polartec**: Has partnered with Nike, Under Armour, and Salomon.
- **Celliant**: Has partnered with Reebok and Adidas.
- **Finetex Inc.** Created Polartec_NeoShell, breathable and waterproof sportswear.
- **Reebok**: Designed cross-training shoes that balance breathability, durability, and flexibility.

Nanotechnology has been discovered to provide antibacterial, odor-controlling, UV-protective, water-repellent, wrinkle-resistant, antistatic, and strength-enhancing qualities in clothes (Table 1). Stadium and gym floor coverings are one of the major NT employs in clothing, improving durability and cleanliness of the flooring while additionally having an impact on athlete's performance and safety [5].

Below are a few case studies. Gore-Tex brand manufactures apparel for motorcyclists and vehicle racers, featuring jackets made from waterproof polymer membranes embedded with nanoscale carbon particles that resist static charge. NanoTex brand manufactures stain-resistant Nano-Care clothing utilizing nanotechnology. They

Table 1 Properties of sportswear and shoes incorporated with nanotechnology

Main properties of sportswear and shoes incorporated with nanotechnology	Benefits for sports
Water-Proof	Swimsuit, diving, canoeing, sport shoes, sailing
Comfort	All types of sportswear
Enhanced blood circulation and recovery of muscles	Therapeutic knee bands, elbow bands, back-belts esp. for weightlifting
Electronic textiles	Mountaineering, running and fencing clothes, referee gadgets
Self-Cleaning	Mountaineering tents
Protection from cold and heat	Skiing, snowboard, diving, mountaineering, canoeing, and cycling clothes
UV protection	Clothing for outdoor sports
Antibacterial	All types of sportswear and shoes

apply a layer of polymer to the fibers prior to weaving them into fabric, this maintains the fabric's delicate texture and feel while remaining stain-resistant despite numerous washings. They grant licenses to other companies, such as clothing behemoths Old Navy, Nike, and The Gap. Schoeller Textiles AG (Switzerland) has developed a nano-based technology for the manufacturing of garments that optimally combine comfort, air permeability, wind and water resistance, and self-cleaning capabilities for extreme cold weather sports including mountaineering and skiing.

2.4 Applications of Nanotechnology in Different Sports

Nanotechnology has dramatically enhanced sporting equipment across various sports:

- **Tennis:** Nanomaterials in club heads and racket frames increase strength and decrease weight, which improves shot control and distance. Current study indicates that **Wilson**, a manufacturer of equipment, has produced tennis racquets that are up to 22% more powerful and twice as stable as traditional racquets. In a game that is already incredibly fast-paced, this improvement in the racquet's performance potential may result in significant speed increases. Wilson is researching cutting-edge methods to enhance the current generation of tennis balls in addition to racquets. Given their research, they have been able to develop balls that retain their bounce for considerably longer than those that are already on the market. This is achieved by using clay particles to bond microscopic butyl rubber balls. An inner layer of the ball is subsequently coated with this mixture, forming an impermeable yet flexible barrier that prolongs gas's stay inside the tennis. Figure 1 provides a graphic representation of the results of this novel airtight barrier.
- **Football**: Football athletes' shin pads are frequently composed of nano-structured polymers due to their lightweight nature and enhanced strength, which allows for thin layers to provide enough protection against shin kicks. Football linings also use nano clay materials as barrier materials to assist in preserving ball's pressure

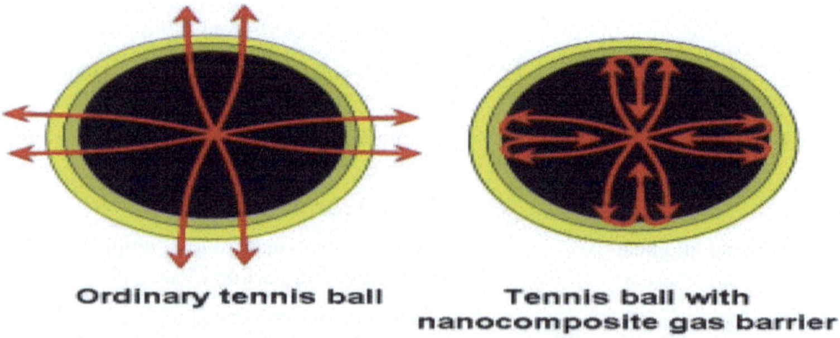

Fig. 1 Ordinary tennis ball (left). New Wilson tennis ball coated with gas barrier (right)

for a longer time. Furthermore, football kit manufacturers have introduced fabrics that are naturally antibacterial, waterproof, oil repellent, anti-odor, and anti-stain, with the help of added NP ofAg and titanium.

- **Golf**: Nanotechnology has significantly improved equipment and performance in the golf industry. Manufacturers have been filling imperfections in club shaft materials with NP, improving the uniformity of the material that makes up the shaft and thereby enhancing your swing. This integration of nanotechnology in golf clubs has revolutionized the game, giving golfers better control and precision in their shots. Moreover, golf balls have also benefited from nanotechnology. Through the utilization of nanoscale technologies, chemicals, and physical properties in the ball's material, golf balls can now redistribute their weight on the fly. This innovation allows golfers to achieve slight changes in direction, ensuring their shots reach the intended target accurately.
- **Cycling**: Bicycle frames with nanomaterials improve strength and aerodynamics, increasing cycling efficiency. Nanotechnology is used in cycling to make bicycle frames lighter, thinner, and more durable. Carbon nanotubes are a nanomaterial that is used to achieve this. Nanotechnology is a field that can be used to change the physical properties of materials, such as their size, shape, weight, and color. It can also make materials stronger, more reactive, and better electrical conductors.
- **Hockey**: Nanomaterial-based sticks provide more strength and accuracy, and safety equipment increases safety. Nanotechnology is used in hockey sticks to improve their strength, durability, and feel. **Carbon nanotubes**: These are used in N12 Technologies' NanoStitch technology to make hockey sticks stronger and more durable. Carbon nanotubes are one of the smallest and toughest materials in the universe and are 50 times stronger than spider silk. They can:
  - Give a lighter, thinner laminate durability, fatigue tolerance, and impact resistance.
  - Assist a stick in maintaining its "snap" and feel that players desire.
- **Swimming**: Nanotechnology is used in competitive swimwear to reduce water drag, which can help swimmers achieve faster times.
- **Skiing and Snowboarding**: Nanotechnology is used in skiing and snowboarding equipment to improve performance and durability:
  - **Nanomaterial coatings**: These coatings can enhance the performance and durability of skiing and snowboarding equipment.
  - **Improved strength and stiffness**: Nanomaterials can improve the strength and stiffness of skis.
  - **Reduced weight**: Nanomaterials can help reduce the weight of skis.
- **Athletic Footwear**: Nanotechnology can be used in athletic footwear to improve the shoe's performance, comfort, and safety:
  - **Water, dirt, and stain protection**: Nanotechnology products can protect shoes from water, dirt, and stains without affecting the shoe's look, feel, color, or breathability.

  - **Improved comfort and safety**: Nanotechnology can help athletes perform better and enjoy a more successful sports experience.
  - **Enhanced durability**: Nanomaterials can make equipment stronger and more durable. For example, carbon nanotubes can reinforce the frame of tennis rackets, reducing the risk of deformation.
  - **Reduced weight**: Nanotechnology can help make equipment lighter. For example, nanomaterials can be used in footballs to make them more resistant to wear and tear while maintaining a lightweight profile.

- **Sports Balls**: Performance, durability, and aerodynamics are all improved with nanomaterial coatings. Sports balls may employ nanotechnology for a variety of uses that consist of:

  - **Longer play**: Nanotechnology can be used to extend the life of a ball by making it harder for air to escape. For example, some tennis balls have a clay nanoparticle coating on the inner core that acts as a sealant. For twice as long as a typical ball, this enables the ball to maintain its pressure and bounce.
  - **Barrier material**: Footballs can have nanoclay linings added to them to serve as a barrier while maintaining the ball's internal pressure constant.
  - **Antimicrobial treatment**: Clothing for sports fibers might be coated with AgNP to prevent odor-causing bacteria and fungi from growing.
  - **Smart foam**: It is possible to measure the impact of each hit by inserting smart foam into a football helmet.Electrical signals are produced when the foam is squeezed and sent to a trainer or coach.

- **Safety Equipment**: Nanomaterials have been employed in helmets along with other protective gear for additional safety and impact resistance.
- **Sports Cars**: Additionally, nanotechnology has been employed in car racing. Sports cars equipped with nanotechnology have less friction, that amplifies engine power and improves fuel efficiency. In addition to applying new additives that increase fuel efficiency, cars will have stronger and lighter engine blocks and frames. The development of materials that are ten times stronger than steel but only a tenth of its weight claims to improve the speed and efficiency of transportation. Bicycles are made via solid and light carbon NP, that enhances their power and decrease their weight.

For athletes in a variety of sports, these applications lead to improved performance, durability, and safety.

3 Benefits of Nanotechnology and Nanofuels for Sports Nutrition and Performance

Many of the benefits of nanotechnology, that greatly broadens the scope of materials science, are based on the capacity to modify material structures at extraordinarily small scales to achieve particular properties. Among numerous additional properties, nanotechnology could effectively make materials stronger, lighter, more resilient, more reactive, more sieve-like, or better electrical conductors. There are currently many well-known commercial products utilizing nanoscale materials and technologies that are employed daily:

- Nanoscale surface treatments or additions to fabric allow them to resist bacterial growth, stains, and wrinkles, or they could offer lightweight ballistic energy deflection in personal body armor.
- Windows, computer and camera screens, eyeglasses, and other surfaces can be coated with clear nanoscale films to make them electrically conductive, antireflective, self-cleaning, UV or infrared light resistant, antifog, antimicrobial, and scratch-resistant.
- Nanoscale materials have begun to enable washable, durable "smart fabrics" with flexible nanoscale sensors and electronics that may monitor health, absorb solar energy, and then convert that energy into movement-based energy.
- **Lightweighting** "automobiles, trucks, airplanes, boats, and spaceships" could lead to significant fuel savings. Nanoscale additions in polymer composite materials have been employed in "baseball bats, tennis rackets, bicycles, motorcycle helmets, car parts, luggage, and power tool housings" to make them "lightweight, stiff, durable, and long-lasting". CNT sheets for next-generation airplanes are now being manufactured. For instance, given their conductivity and low weight, they are perfect for applications involving heat management and electromagnetic shielding [6].
- High-quality household products, which include antibacterial cleansers, environmental sensors, air purifiers, filters, degreasers, and stain removers, in addition to specialty paints and sealing solutions, including self-cleaning house paints that are resistant to dirt and stains, have been manufactured employing **nano-engineered materials**.
- **Nanoscale materials** are additionally being integrated into numerous personal care products to improve performance. Nanoscale zinc oxide and titanium dioxide have been applied in sunscreen for many years to provide sun protection without being visible on the skin.

Potential Benefits of Nano Fuels in Sports Nutrition

1. **Improved Performance**: With enhanced nutrient delivery, athletes may experience significant improvements in performance metrics, such as speed, strength, and endurance. The faster absorption rates can lead to quicker energy replenishment and improved recovery times.

2. **Customization**: Nanofuels may be modified to satisfy each athlete's unique nutritional requirements. Personalized nutrition plans that incorporate nanotechnology could provide athletes with optimized fueling strategies based on their unique physiological requirements and performance goals.
3. **Convenience**: Nanofuels can be incorporated into various formats, including gels, powders, or even bars, making it easier for athletes to consume essential nutrients on the go. The reduced volume of consumption necessary for effective fueling can also lead to improved gastrointestinal comfort during exercise.
4. **Reduction of Side Effects**: Nanofuels could decrease potential negative effects of traditional supplements, that includes nutrient overload or gastrointestinal distress, by improving targeted delivery of nutrients.
5. **Improved nutrient absorption**: NP can increase the absorption of nutrients and moisture.
6. **Reduced lactic acid**: Nanotechnology can help reduce lactic acid levels during exercise.
7. **Improved antioxidant activity**: Nano-supplements can increase antioxidant activity, which can help with exercise performance.
8. **Reduced inflammation**: Nano-supplements can reduce inflammation, which can help with exercise performance.
9. **Improved mitochondrial function**: Nano-supplements can improve mitochondrial function and quality, which can help with exercise performance.
10. **Improved energy substrate balance**: Nano-supplements can improve energy substrate balance, which can help with exercise performance.
11. **Increased iron content**: Nano-supplements can increase serum and liver iron content, which can help with exercise performance. Nanotechnology can also be used to create NP from essential nutrients, which can help athletes optimize their nutrient intake.

4 Literature Review

Sports medicine nanotechnology [7] is the broad application of several nanoscale advancements to medical applications about sports injury.

Composite nanomaterials based on aluminum have been employed in making sports facilities and equipment especially cars, bicycles, and structures lighter, stronger, and safer. This investigation on new nanocomposites utilizing functionally graded materials [8] and nano microcrystalline aluminum alloys [9] could prove extremely beneficial in the research and development of these aluminum-based nanocomposites.

Certain sports facilities and equipment may prove significantly improved by nanomaterials. However, because nanotechnology is very costly and difficult to use, there aren't many sports products with nanotechnology enhancements on the market, despite the many benefits that have been suggested [10].

Although nanomaterials in various sports have greatly benefited the sports community, they may also have detrimental effects on organisms and the environment, threatening human health [11].

5 Recent Research on Supplements of Sports Nutrition

Employing nutrition to support all aspects of athletic performance is the research and practice of sports nutrition. Educating them on proper foods, nutrition, hydration techniques, and supplements may assist to excel in your sport.

The fact that athletes may require different nutrient intakes than non-athletes is a significant element that sets sports nutrition apart from general nutrition. When combined with other elements of athletic development, that include training, sports nutrition may considerably enhance an athlete's performance.

Irrespective of sport, most athletes may benefit from equal treatment of sports nutrition advice. Athletes should consume nutrient-rich foods from 5 basic food groups for conducting high-intensity training or competition:

- Vegetables
- Low-fat dairy or dairy alternatives
- Lean proteins
- Grains
- Fruits

To optimize the nutritional content of the foods selected, they should generally be minimally processed. It should also steer away from excessive sodium and minimize additional preservatives.

High-performance eating plan objectives:

- To increase consumption of good fats, utilize oils including olive oil as an alternative to butter and margarine.
- To ensure a healthy omega-3 intake regularly consist of "wild-caught fatty fish" incorporating salmon in daily diet. If don't consume fish, look for "plant-based omega-3 s" including "chia seeds and walnuts".
- Since these nutrients are often deficient in traditional diets, emphasize foods that are abundant in "potassium, fiber, calcium, and vitamin D".
- Make at least half of your grains whole and give whole grains priority over refined ones.
- Increase consumption of dairy substitutes or low-fat dairy products.
- Add a range of lean proteins, including fish, poultry, eggs, soy products, chicken, pork, and meat substitutes.
- A broad spectrum of fruits and vegetables that are high in essential vitamins, minerals, and phytonutrients should be included.

5.1 Macronutrients

The essential elements of food that provide body requiresto survive are **proteins, carbohydrates, and fats**. When eating to promote high-performance exercise, it's particularly crucial to consume the proper mix of macronutrients.

Proteins: Building blocks of body are proteins, which are composed of individual amino acids. It promotes the development of bones, teeth, skin, and muscle. Although it is usually obtained from animal sources, protein is additionally obtained among certain plant-based diets. Most of the time, leaner proteins are the better alternative.

Including:

- Fish
- Eggs and egg whites
- Low-fat cheese and cottage cheese
- Tofu and other high-protein meat substitutes
- Legumes
- Lean ground turkey
- Lean cuts of pork or beef
- Some grains, such as quinoa
- Chicken breast

For stimulating muscle growth protein is essential assisting in the healing process after exercise. This is because it facilitates the process of muscle protein synthesis, essentially means that new muscle develops.

Carbohydrates: Body's preferred energy source is carbohydrates. It supports all of our daily activities, including breathing, eating, exercising, and thinking.

Attempt to obtain at least half of the carbohydrates consumed daily from sources high in fiber, including:

- Low sugar cereals
- Sweet potatoes
- Rice cakes
- Quinoa
- Oats
- Non-starchy vegetables like broccoli, brussels sprouts, green beans, carrots, and cucumbers
- Brown rice
- Fruit
- Whole grain pasta
- Legumes
- Whole grain bread

Simpler starches, including "white rice, white potatoes, pasta, and occasionally sweets and desserts", can provide the other half.

The optimal time to ingest simple carbohydrates is during workout to top-up or refill glycogen reserve. Sports can have a significant impact on the amount of carbohydrates consumed daily.

For instance, amount of carbohydrates required by an ultra-marathon runner and an Olympic weightlifter will differ significantly.

Fat: Fats serve as the body's primary or secondary energy source for the majority of humans. They will supply a greater percentage of daily energy needs in some situations, such as in athletes who have undergone **keto-adaptation**.

The distinctive feature of fats is that they contain nine calories per gram, while proteins and carbohydrates only include four. Apart from supplying energy, fats additionally assist with hormonessynthesis, operate as structural elements of cell membranes, which promote metabolic processes, among others. In addition to being a good source of calories, fats may assist with the promotion of post-exercise recovery and support hormones connected to sports. It has been demonstrated that omega-3 fatty acids, in particular, offer anti-inflammatory qualities that aid in athletes' recovery following intense training.

Generally, it is recommended to prioritize the consumption of fats from beneficial sources that include:

- Seeds (sunflower, chia, hemp, flax)
- Olive oil
- Fatty fish
- Avocado
- Nuts and nut butter
- Whole eggs

Following "proteins and carbohydrates, fats" contribute remaining proportion of caloric intake. This constitutes 20–35% of caloric intake.

5.2 Timing

The timing of meal consumption or nutrient intake concerning training or competition is a crucial element in optimizing sports nutrition. Meals Following high-intensity exercise, scheduling your training or competition around it may promote improved muscle growth, tissue regeneration, and mood enhancement.

"International Society of Sports Nutrition (ISSN)" Trusted Source recommends eating a meal with "20–40 g of protein" every "3–4 h" throughout day for maximum muscle protein synthesis. Since protein synthesis is increased in the two hours following an intense workout session, it is particularly crucial to consume protein during this time.

After vigorous activity, you additionally require refueling with carbohydrates. Following 30 min of working out, consider eating 30–60 g of a basic carbohydrate source. ISSN advises optimizing energy levels during training or events spanning longer than 60 min, some endurance athletes should consume 30–60 g of carbohydrates every hour.

5.3 *Hydration Needs*

One important aspect of sports nutrition that might affect performance is hydration state. Sweating is one way for the body to cool itself since it causes fluid and electrolyte loss during activity.

Maintaining fluids and electrolytes is necessary during prolonged high-intensity exercise to avoid mild to potentially severe dehydration. Because fluids and electrolytes can be rapidly lost in hot weather, athletes that train or compete in hot settings must pay special attention to their level of hydration.

It's recommended to take preventative measures and consume at least "16 oz(0.5 L) of drinks" within "2 h" prior to vigorous training sessions or competition to assist prevent dehydration.

For proper fluid balance during a vigorous training session, athletes should drink 6–8 oz of fluid every 15 min. Weighing before and after exercising is an effective method for determining amount of fluid to consume. 16 oz (0.5 L) of fluid is lost for every pound (0.45 kg) lost.

To prevent dehydration, it's particularly critical for restoring electrolytes during and after prolonged, intense physical activity, should replenish electrolytes by consuming foods high in potassium and salt and sports drinks. Some people decide to prepare their own sports drinks considering most of them don't contain enough electrolytes.

Furthermore, a lot of companies produce electrolyte tablets that one may consume with water to fulfill electrolytes that are required to remain hydrated.

5.4 *Snacks*

Countless snack options can replenish energy reserves without making one feel full or sluggish. Ideal snack is simple to make and offers a decent macronutrient ratio.

Here are several pre- and post-training snack options:

- A piece of fruit and a low-fat cheese stick
- Chia pudding
- A protein bar
- A protein smoothie
- A salad with some kind of protein, such as chicken

- A sandwich on whole grain bread (such as turkey, egg salad, tuna, peanut butter and jelly, or low-fat grilled cheese)
- Trail mix
- Pita chips and bell peppers dipped in hummus
- Low-sugar cereal and your choice of milk mixed with protein powder
- Hard-boiled eggs on toast
- Greek yogurt with granola
- Apple slices with nut butter

Avoid high-fat snacks right before working out since they may leave one feeling less satisfied and digest more quickly. It is particularly crucial to have a snack that contains a healthy amount of carbohydrates and protein after working out to promote muscle protein synthesis and replenish glycogen levels.

6 Supplements for Sports Nutrition

Predominant component of an athlete's diet should generally be whole foods. They contribute to providing an appropriate balance of nutrients, energy, and other bioactive compounds in food that are typically lacking in supplements. However, as athletes frequently require more nutrients than the average person, supplements can be taken to make up for any dietary deficiencies.

These are the best supplements with scientific support that sports dietitians frequently suggest.

6.1 Protein Powders

Isolated versions of several proteins, including "whey, egg white, pea, brown rice, and soy", can be identified in protein powders. Consuming a substantial amount of protein remains simple and convenient with protein powders, that typically include "10–25 g of protein in each scoop". According to research, consuming a protein supplement prior to exercise could assist with lean body mass gains and recovery.

For instance, some people consider slightly increasing the protein content of their oats by adding protein powder.

6.2 Carbohydrate Gels and Powders

Supplementing with carbohydrates may assist in maintaining energy levels, especially if one participates in endurance sports that last longer than an hour. Each serving of these concentrated types of carbohydrates typically contains roughly 25 g of simple carbohydrates, and some also contain vitamins or caffeine. They are available as powder or gel. It isn't necessary to combine gels with water. During an exercise session longer than an hour, many long-distance endurance athletes will attempt to take 25 g of carbohydrates per 30–45 min utilizing carbohydrate energy gel.

Though some athletes prefer gels to avoid consuming too much fluid during training or competition, as this could cause digestive distress, sports drinks frequently contain enough carbohydrates to sustain energy levels.

6.3 Vitamins and Minerals

To address potential nutritional shortages, numerous athletes prefer "high-quality multivitamin" that encompasses "all vital vitamins and minerals". For most people, this is probably an intelligent choice since taking a multivitamin may have more advantages than disadvantages.

Vitamin D is one nutrient that athletes frequently take supplements of, particularly in winter months when sun exposure is lower. Supplementation is often recommended as insufficient vitamin D levels may adversely affect athletic performance. To determine if one is lacking in a specific vitamin or mineral, it is recommended to visit a qualified healthcare practitioner and undergo laboratory testing.

6.4 Caffeine

Caffeine is effective in improving strength and endurance in a variety of sports, including weightlifting, running, jumping, and throwing. Caffeine's brain-stimulating effects may reduce the perceived exertion level. To enhance their performance, several athletes prefer to have a strong cup of coffee prior to training, whereas others utilize pre-workout supplements that include artificial caffeine.

Caffeine, however, isn't for everyone and can be harmful in excess, particularly for people under the age of 18, must begin with a low dosage and progressively raise it as long as the body is capable of handling it.

6.5 Fish Oil

Consuming omega-3 fat supplements, including fish oil, could assist athletes perform better and recuperate after intense training. To obtain omega-3 s from diet by consuming foods including soybeans, nuts, flax and chia seeds, and fatty fish. However, it might be an excellent choice to supplement with fish oil if you don't frequently eat these meals. For vegetarians and vegans, there are additional plant-based omega-3 supplements available.

6.6 Creatine

A substance that the body makes from amino acids is termed creatine. Additionally, fish and meat contain trace levels of it. When engaging in brief, high-intensity exercises, assists in generating energy. Research indicates that regular consumption of "5 g of creatine monohydrate", the most prevalent form, improves power and strength output during resistance exercise, potentially improving athletic performance.

Since creatine's effects are mild in comparison to those of other substances, the majority of sporting federations do not consider it to be a prohibited substance. Given their affordability, accessibility, and the wealth of data supporting them, creatine supplements might be beneficial for certain athletes.

6.7 Beta-Alanine

Another amino acid-based compound "Beta-alanine" present in animal products including "chicken and beef".

Carnosine, a compound that reduces acidity in active muscles during high-intensity exercise, is synthesized in body with beta-alanine.

The most prominent advantage when consuming beta-alanine supplements is an increase in performance during 1–10-min high-intensity workouts. For instance, swimmers and runners who run short to medium distances might benefit.

A few preferred the lower end of the spectrum to prevent paraesthesia, that causes tingling in the extremities. Beta-alanine is an excellent supplement option for athletes since, similar to creatine, it is typically safe for most people and isn't listed as a prohibited chemical.

7 Role of Sports Nutritionists

Sports nutritionists are in charge of following the most recent research and putting science-based nutrition plans for athletes into practice. At the highest level, licensed dietitians with expertise in sports nutrition are often in charge of managing and implementing these programs.

CSSD (certified specialist in sports dietetics), is a prestigious qualification that they frequently possess. These experts help players learn about proper hydration, nutritional intake, and supplementation when necessary, among other areas of nutrition that affect athletic performance.

A day in life of sports nutritionist includes:

- Meetings with athletes for developing personalized programs.
- Organizing sessions for group education.
- Strategizing with team management.
- Being available during meals pre- and post-training helps athletes make food choices.
- Ensuring athletes are fed properly so they can compete at their highest level.

In conclusion, sports nutritionists often assist athletes with medical concerns about nutrition, food allergies, intolerances, and any disordered eating or eating disorders they may be experiencing. They often do this in collaboration with psychotherapists.

8 Challenges and Ethical Considerations of Nano Fuel

Although nano fuels have a lot of potential advantages, a few issues need to be resolved before they are widely used in sports nutrition:

Safety and Regulation: Since the long-term impacts of NP on human health are yet unclear, safety is a worry. Regulations must be established to ensure that athletes can safely consume nanofuels.

Research and Development: For confirmingnanofuelefficacyin diverse sports populations, further thorough research is required. To evaluate the performance advantages and possible hazards of their use, clinical trials had to be carried out.

Ethical Implications: The application of cutting-edge technologies in sports nutrition presents moral dilemmas related to access and equity. The potential for the benefits of nanofuels to worsen already-existing inequalities in sports by favoring those who can afford these cutting-edge nutritional solutions exists.

Misleading Claims and Market Saturation: Like any new technology, there is a chance that the market will become saturated with goods that convey false information. For athletes to make educated dietary choices, clear labeling and education will be crucial.

9 Summary

In conclusion, through better nutrient delivery and absorption, nano fuels hold promise as a new area of sports nutrition that could increase athletic performance. It is crucial that stakeholders, including academics, dietitians, athletes, and regulatory agencies, collaborate to address the difficulties and moral dilemmas related to the use of nanotechnology in this field as studies continue to reveal its advantages and applications. By doing so, we can ensure that nano fuels contribute to a fair and effective approach to sports nutrition, thereby enabling athletes to reach their top performance.

A dynamic and developing field with great potential for athletes, producers, and the sports community is the incorporation of nanomaterials into sporting goods. These innovative materials have already made significant progress in enhancing the overall performance, reducing weight, and increasing strength of equipment in a range of sports. These innovative materials have already made significant progress in increasing equipment's strength, reducing its weight, and enhancing its overall performance across a range of sports. With specialized equipment, environmentally friendly procedures, and stricter safety regulations at the forefront, the future is expected to bring about even more significant changes. Nanomaterials will enable athletes to attain new heights of performance as they continue to push the limits of human potential.

However, we must approach the coming years with a dual attitude of responsible innovation and environmental stewardship. It becomes imperative to resolve the safety, health, and regulatory concerns associated with nanomaterials. To protect the environment and athletes, biodegradable and sustainable nanomaterials must be developed. Scientists, engineers, athletes, and regulatory bodies need to collaborate to properly integrate nanomaterials into sports equipment that assist athletes succeed in an environment-friendly and performance-driven manner.

References

1. Da Costa LP (2020) Engineered nanomaterials in the sports industry. Handb Nanomaterials Manuf Appl 309–320. https://doi.org/10.1016/B978-0-12-821381-0.00014-4
2. Nanotechnology in sports equipment: The game changer (n.d.) Retrieved 16 Jan 2024, from https://www.nanowerk.com/spotlight/spotid=30661.php#google_vignette
3. Nanotechnology in sports equipment: The game changer (n.d.) Retrieved 16 Jan 2024, from https://www.nanowerk.com/spotlight/spotid=30661.php
4. Amer Sports (2004) Amer Sports. 3 Dec 2008
5. Shishoo R (2005) Textiles in sport. Elsevier, Cambridge: Woodhead Publishing10. Yetisen AK, et al. (2016) Nanotechnology in textiles. ACS nano 10(3):3042–3068. https://doi.org/10.1021/acsnano.5b08176
6. Harifi T, Montazer M (2017) Application of nanotechnology in sports clothing and flooring for enhanced sports activities, performance, efficiency, and comfort: a review. J Ind Text 46(5):1147–1169. https://doi.org/10.1177/1528083715601512
7. Bayram C, Işıklı C, Denkbaş EB (2014) Nanotechnology in sports medicine. In: Doral M, Karlsson J (eds) Sports injuries. Springer, Berlin, pp 1–9. https://doi.org/10.1007/978-3-642-36801-1_243-1
8. Zhao HE, Shen F (2012) The applied research of nanophase materials in sports engineering. Adv Mater Res 496:126–129. https://doi.org/10.4028/www.scientific.net/AMR.496.126
9. Miteva A (2014) An overview of the functionally graded materials. Mach Technol Mater 8(3):13–16. https://stumejournals.com/journals/mtm/2014/3/13
10. Petrova A, Stefanov G, Miteva A (2020) Some properties of the nanozone in nano-microcrystalline ribbons of aluminium alloys. Comptesrendus de l'Académiebulgare des Sciences 73(10):1434–1442. Available from: https://doi.org/10.7546/CRABS.2020.10.13

11. Ćibo M, Šator A, Kazlagić A, Omanović-Mikličanin E (2019) Application and impact of nanotechnology in sport. In: Brka M, Omanović-Mikličanin E, Karić L, Falan V, Toroman A (eds) 30th Sci-Experts Conf Agric Food Ind. Springer, Cham, pp 349–362. https://doi.org/10.1007/978-3-030-40049-1

Nanocarriers: Revolutionizing Nutrient Delivery

Rukshana Irani and Suchandra Datta

Abstract The popularity of nanocarriers is increasing at a higher pace due to their potential benefits. The application of nanocarriers in food and nutrient delivery is an area of profound interest. When nanocarriers are administered into the biological system, their bioavailability and effectiveness increase manifold. Improving solubility, stability, controlled release, site-specific delivery, and biodistribution increases the utilisation of nutrients or drugs. Among these nanocarriers, liposomes, nanoemulsions, and solid lipid nanoparticles (SLNs) are most commonly used for nutrient delivery. Their physical and chemical properties control the mechanism of interaction of nanocarriers with nutrients. Nanocarriers in nutrient delivery aim to enhance health and wellness, treat nutritional deficiencies, and support medical therapies. The role of nanocarriers in sports nutrition is quite remarkable. They enhance exercise performance and endurance amongst athletes by meeting their nutritional needs, which are significantly higher than an average individual. More clinical trials on humans are needed, which will help to understand the effectiveness of nanocarriers and promote the development of advanced formulations.

Keywords Nanocarriers · Liposomes · Bioavailability · Drug delivery · Nutrition

1 Introduction

Innovations have been the driving force behind the progress of human civilisation. Nanotechnology is one such invention which has been used in almost all disciplines. It includes material science, electronics, photonics, supramolecular assembly, drug delivery, agriculture, food industry etc. [1]. Nanocarriers have been in huge demand due to their multiple benefits. Nano-sized particles have been found to exert their

R. Irani (✉)
Department of Food and Nutrition, Raidighi College, Raidighi, West Bengal, India
e-mail: iranirukshana1@gmail.com

S. Datta
Department of Home Science, University of Calcutta, Kolkata, West Bengal, India

A. Chatterjee et al. (eds.), *Nanofuel: The Future of Sports Nutrition*, Smart Nanomaterials Technology, https://doi.org/10.1007/978-981-96-5471-0_2

impact at various levels i.e., cellular, molecular, and even atomic levels. In this present era, an area of science and technology that does not exploit the abilities of nanotechnology can rarely be found [2].

Nanomedicine is a branch of medicine that utilises the potential of nanocarriers and nanotechnology. Nanotechnology has triggered the production of various nanomaterials, including liposomes, micelles, polymeric nanoparticles, and albumin-based formulations [3].

These encapsulated drugs have multiple properties, such as their expansive surface area, extended circulation time within the bloodstream, distinct structural properties, lower toxicities, increased bioavailability, better stability, and site-specific controlled release. The demand for nanocarriers has increased substantially. It is highly effective in the treatment of diseases and pathological conditions than the conventional forms of therapy. Nanotechnology that employs nanocarriers has a particle size of <1000 nm. The high surface area-to-volume ratio of nanocarriers helps in the treatment of diseases with optimized efficacy and reduced collateral effects. The usage of these tiny structures in the diagnosis of illnesses is due to their biocompatibility and easier removal from the system. The use of nanocarriers in the delivery of nutraceuticals is amongst the fastest emerging fields of nanotechnology which has opened new avenues in the improvement of health and well-being [4].

The application of nanotechnology in the food industry is one of the solutions for improving the functional properties of food products. Nanostructured materials (e.g. nanoparticles, nanocomposites or nanoemulsions) represent a novel technology for food enrichment with bioactive compounds, vitamins, minerals and antioxidants. Moreover, these nanomaterials can be engineered in a way that flavour, fragrance and other active elements are encapsulated in such a manner that not only protects them from adverse conditions such as temperature, and light but also controls their release during the processing and consumption of food, thus improving food quality, safety and retention time [5].

Nanotechnology has brought revolution to the food industry by ensuring food safety, quality, nutritional improvement, and sustainability. The development in the application of nanomaterials with their unique characteristics in physical, chemical, and biological properties has provided food scientists and food technologists with a wide range of innovative strategies to address existing challenges while meeting the never-ending demands of consumers [6].

Several studies have also been carried out to understand the role of nanocarriers in sports nutrition. Administration of nutritional supplements through nanocarriers has increased their efficacy and thus enhanced endurance and exercise performance (Fig. 1). Nutritional supplements administered through traditional mode have decreased bioavailability and impart lesser ergogenic activity. Nutritional nanocomposites have better solubility with higher dissipation rate, bioavailability, absorption etc. The chemical and physical changes on the surface of the nano compounds lead to various effects, like targeted activities, increased half-life, controlled release, and decreased toxicity [5].

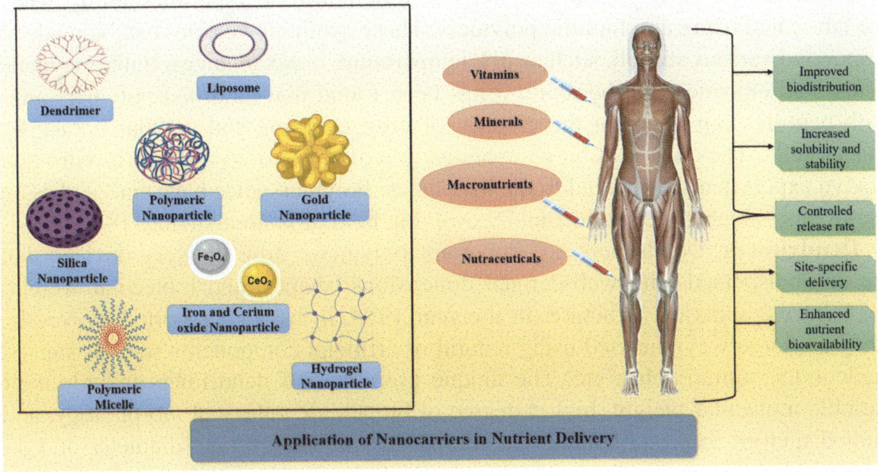

Fig. 1 Nanocarriers and their application in nutrient delivery

2 Types of Nanocarriers in Nutrient Delivery

Nano-carriers are developed with a diameter ranging from 1 to 1000 nm. They are known to make drugs safer and more effective for therapy. Nano-delivery systems enhance drug administration and biodistribution. The immunogenicity and side effects are also reduced [4]. The drugs in the nanocarriers can be delivered actively or passively to the targeted site. Nanocarriers enable the delivery of lesser doses in a more controlled manner with fewer side effects in comparison to medical therapy. They can also overcome major drawbacks associated with traditional approaches to pharmacology, i.e., non-specific distribution, speedy clearance rate, unpredictable drug delivery, and poor bioavailability [7].

Researchers gave many proposals about using nanoparticles to encapsulate nutrients and bioactive compounds. Depending on different modes of nutrient or drug delivery, nanoparticles can be classified into three major categories i.e., polymeric nanoparticles, inorganic nanoparticles, and lipid-based nanoparticles. These have been discussed in detail below.

2.1 Polymeric Nanoparticle

Polymersomes: They are unilamellar drug delivery systems engineered at the nanoscale. It exhibits remarkable colloidal stability and advantages. It includes limited permeation of aqueous fluids due to their enhanced bilayer thickness, higher encapsulation capacity and improved adaptability. While liposome-based drug carriers are widely used, polymersomes demonstrate superior stability. They are fabricated using amphiphilic polymers. These synthetic vesicles can be made to respond to various stimuli, such as pH, temperature, redox changes, light, magnetic fields, and enzymes. Furthermore, it has been found that modified polymersomes with peptides can augment the efficacy of drug targeting and cellular uptake by cancer cells. They also exhibit comparatively reduced toxicity in both in vitro and in vivo experiments compared to the liposomes. Polymersomes have emerged as an useful and valuable tool in the rapidly evolving field of nanomedicines [8].

Dendrimers: Dendrimers are important polymeric drug delivery devices that are monodispersed with well-defined dimensions, shapes, and molecular weight. They have a spherical structure. In the year 1978, it was first reported by Vogtle. They are largely synthesised using natural or artificial components such as sugars, nucleotides, amino acids, etc. The unique properties of dendrimers include their variable molecular weight, higher degree of branching, spherical morphology, and monodispersed macromolecular structure. They have an average diameter of 1.4–1.45 nm, which makes them suitable particularly for drug delivery. Nowadays, dendrimers are extensively used in various biomedical fields, like gene delivery, immunology, magnetic resonance imaging, and the delivery of antiviral, antibacterial, and anticancer agents. Generally, dendrimers consist mainly of smaller units like dendrons which are capable of enhancing the dissolution and bioavailability of their hydrophobic drugs encapsulated within their intra-molecular cavities and are delivered to their peripheral surface functional groups. The dendrons are branched tree-like structures consisting of three distinct sections, mainly the core, interior, and periphery. This free space in the interior of dendrons can be utilised for entrapment of drug molecules, which would allow solubilisation, controlled release, and targeting of drugs. The advantages of dendrimers include their uniform particle size, the ability to attach a wide range of targeting agents to their high-density peripheral functional groups, and the polyvalency of their end-groups, which increases the ability to bind to various types of receptors [9].

Polymeric Micelles: Polymer micelles are widely applied for the targeted delivery of hydrophobic anticancer drugs. These micellar structures are also used in diagnostics, gene delivery, and imaging. These allow better in vivo biodistribution of the drug. This micellar structure can be designed to be responsive to different stimuli such as pH, temperature, light, and ultrasound. It is done in a way that triggers drug release by controlled dissociation of the block copolymers. Polymeric micelles are typically smaller than 100 nm. They are slowly cleared by renal secretion while exhibiting the enhanced permeability and retention effect, which enables their accumulation in tumour tissue. These amphiphilic block copolymers can make various nanostructures, that include micelles, nanospheres, nanocapsules, and polymersomes. The properties of polymeric micelles as drug delivery systems are far better than other nanocarriers, especially because of their small size, facile synthesis and sterilisation, and excellent solubility [10].

Nanosphere: The role of nanospheres as therapeutic molecule carriers has been extensively studied. They are spherical with a solid structure. In these matrix-type nanodevices, drugs can be incorporated either uniformly into the core or onto the surface of the nanospheres, which allows diffusion of the drug into the target tissue. The other benefits of nanospheres include encapsulation of high drug payloads especially in the case of drugs with poor solubility. As a result, the possibility for sustained and targeted release of the therapeutic molecule increases. However, compared to nanorods, nanospheres have relatively low intracellular uptake with relatively rapid drug release, which indicates that the shape of the nanocarrier can impact the intracellular delivery of drugs [11].

Nanogels: These belong to the category of advanced nanoparticles which are used in drug delivery. It falls under the group of three-dimensional hydrogels. Its size varies from 1 to 1000 nm. Its stability, versatility, and ability to adapt to the changes in the external environment determine its utility. Nanogels are made up of hydrophilic swellable polymeric networks. These networks are cross-linked by covalent bonds, hydrogen bonds, or van der Waals forces. The inherent biocompatible characteristics of nanogels make them useful for both therapeutic interventions and imaging techniques. The sensitivity of nanogels towards specific stimuli can be optimised by careful selection of the polymers and crosslinkers. These distinct properties make nanogels more effective as drug-delivery systems than the other alternative nanomaterials [12].

2.2 Inorganic Nanoparticles

Silica nanoparticles (SiNPs): Silica nanoparticles have come out as a potentially beneficial drug delivery medium because of the desirable characteristics they possess, including biocompatibility, high surface area, and ease of functionality. These particles exist in several configurations i.e., nonporous, mesoporous, hollow mesoporous, and core–shell silica. Among these, mesoporous silica nanoparticles are particularly favoured for their adjustable pore sizes and volumes, biocompatibility, ease of functionality, high drug-loading capacity, antimicrobial drug delivery capability and minimising drug-related side effects. That's why mesoporous silica nanoparticles have been proposed to enhance drug delivery performances [13].

Quantum dots (QDs): Quantum dots are nanomaterials that have zero dimensions and a size between 1 and 20 nm. Besides, the radius of such QDs is equal to the Bohr exciton radius. These QDs can be characterised by broad spectra of absorption with high brilliance and persistence; thus, they can be found in various applications because of the quantum confinement effect. Graphene quantum dots (GQDs) are emerging as an important class of carbon-based nanomaterials, finding their applications in tissue engineering, stem cell research, and drug delivery. Compared to selenium, tellurium, or metal sulfide-based QDs, GQDs possess lower toxicity and better biocompatibility. Both doped and pure GQDs can be fabricated using bottom-up and top-down approaches, following the green synthesis protocol. QDs have been demonstrated to have long-lasting, brightly luminescent absorption as well as emission spectra, enabling them to be applied directly for fluorescent assays for several diseases as well as for monitoring drug metabolism by the body. Despite their tremendous potential, the clinical translation of the use of QDs has proved problematic because of concerns regarding their toxicity [14].

Iron oxide nanoparticles (IONPs): Iron oxide nanoparticles are another important nanocarriers. The application of the drug through iron oxide nanoparticles leads to their transport to the target site by the use of a magnet which can enhance the therapeutic efficacy of the drug under different physiological conditions. The application of magnetic nanoparticles depends on responsiveness to magnetic fields, along with their biocompatibility and the ease of surface modification for attaching drug molecules. These nanoparticles also adsorb and release drugs in a controlled manner. Also, it has been shown in a study that cationic peptides from bee venom like lasioglossin which are known for their antimicrobial properties can be loaded using iron oxide nanoparticles [15].

Gold nanoparticles (AuNPs): Gold nanoparticles are one of the emerging drug delivery platforms. The surface of AuNPs can be conjugated with various therapeutic ligands, which enhances their drug-delivery efficiency. The use of AuNPs in photothermal, antiangiogenic, and radiotherapy techniques is very effective. God nanoparticles are biocompatible with lower cytotoxicity and favourable optical properties. These characteristics make AuNPs suitable for cancer treatment. They are also found to have toxic effects within the biological system which is considered a limitation of its usage. They are absorbed rapidly because of their size and stability. They are found to combine with other materials easily [16].

Nanocomposite: Nanocomposites are an innovation greatly used in the field of material science. The unique characteristic of nanocomposites is the integration of nanoscale fillers into a matrix structure. The fillers consist of particulate, fibrous, or fragmental entities, which are enveloped and bound by the matrix as discrete units. Nanocomposites encompass a wide spectrum of materials, i.e., three-dimensional metal matrix composites, two-dimensional lamellar composites, single-dimensional nanowires and zero-dimensional core shells. Nanocomposites have been used in many fields, including in the making of thin film capacitors for computer chips, solid polymer electrolytes for batteries, parts for car engines and fuel tanks, and food packaging materials. Nanocomposite has various advantages such as biocompatibility, controlled drug release, targeted delivery of drugs, and enhanced stability of drugs which makes them a boon for healthcare [17].

2.3 Lipid-Based Nanoparticles

Liposome: Liposome nanoparticles are spherical vesicle-like structures that are formed when certain lipids are suspended in an aqueous medium. They have two major components i.e., an inner core and a phospholipid bilayer membrane around it. They can be made up of a single or multiple lipid bilayers surrounding the inner aqueous core. The phospholipid bilayer that encapsulates the aqueous solution can be used for the delivery of both hydrophilic and hydrophobic therapeutic agents. Hydrophilic drugs are encapsulated within the inner core, while the hydrophobic environment of the bilayer membrane is preferred by the hydrophobic drugs. The liposomes serve as a transport system for different bioactive substances. Drugs or medications are enclosed inside their double-layered phospholipid structure, making their delivery to particular cells or tissues in the body more precise and accurate. Liposome enhances the effectiveness of drugs and reduces undesirable side effects. Some studies reported traditional liposomes to have some issues like instability, insufficient drug loading, rapid drug release, and short blood circulation time. However, smart liposomes are found to be extremely responsive to various external and internal stimuli, such as pH changes, enzyme transformations, light, ultrasound, microwaves etc. Liposomes can encapsulate medicines, vaccines, cosmetics, nutraceuticals etc. and can also be conjugated with antibodies or ligands to improve their target specificity [18].

Niosomes: Niosomes belong to a distinct class of vesicular drug delivery systems. They are characterized by a spherical, bi-layered structure. This structure is formed by the self-assembly of non-ionic surfactants in an aqueous environment. Cholesterol is a predominant component of the membrane of the niosome, which contributes to its stability and functionality.

The size of niosomes is influenced by many factors, like their cholesterol content, the hydrophilic-lipophilic balance (HLB) value of the surfactant, and the techniques applied to reduce their size. Niosomes are often favoured more than liposomes because of their improved stability and cost-effective process of preparation. Several studies indicate that drug delivery using niosomes can significantly enhance the pharmacological efficacy of the drug. Niosomes extend their circulation time in the biological system, protecting them from adverse conditions, and allowing targeted delivery to specific cells [19].

Solid Lipid Nanoparticles (SLNs): Solid lipid nanoparticles (SLNs) were first introduced in the year 1991 as another alternative carrier system for drug delivery. They are composed of solid lipids. These nanoparticles have gained significant popularity as colloidal drug carriers. They are used for intravenous applications, due to their remarkable potential as effective particulate delivery systems. SLNs are submicron colloidal carriers with sizes ranging from 50 to 1000 nm. They have been constructed from physiological lipids dispersed in water or an aqueous surfactant solution. They have distinct characteristics, such as smaller size, larger surface area, high drug loading capacity, and interactions at the interface of phases. The drawbacks associated with the fluid state of oil droplets are mitigated by replacing liquid lipids with solid lipids, allowing for the formation of solid lipid nanoparticles. The process involved in its preparation is notably simpler than that of other biopolymeric nanoparticles. The advantages of SLNs include controlled or targeted drug release, biocompatibility, stability, scalability, higher drug loading and easy sterilization [20].

Nanoemulsion: Nanoemulsions are characterized by their smaller droplet sizes, ranging from 10 to 100 nm. They are also of high potential as drug-delivery systems. These liquid emulsions can be made by dispersing one liquid phase into another, leading to the formation of droplets that are stabilised by using surfactants, like peptides, proteins, phospholipids etc. They can also regulate controlled drug delivery. They can encapsulate and transport drugs to the site of action without degradation and thus improve bioavailability. Although, some adverse effects have been observed and were linked to parenteral administration of their high doses. The administration of nanoemulsions orally is considered safe. Moreover, the combination of nanoemulsions with surfactants and co-surfactants has been highly effective in the delivery of ocular drugs. The success of nanoemulsion-based drug delivery systems in clinical application depends on various factors, like the pharmacological profile of the drug, the methods of fabrication, the functionalization procedure, and the opted route of administration [21].

3 Mechanisms of Nanocarrier-Nutrient Interaction

Nanotechnology has altered the world and aided numerous fields, such as nutrition, food processing, agriculture and the pharmaceutical industry. The use of nanoscience in nutrition studies has quickly grown over the last few decades, generating a considerable desire for the application of micronutrient delivery systems in humans [22].

There are various techniques present in nanotechnology to prepare nanocarriers, including nano-composite formation, nanostructuration, nanofibre and nano-emulsification. They are employed to encapsulate the materials in small forms to transport the bioactive substances more effectively in the human system [23]. Over the past few years, lipid nanoemulsion has gained considerable attention in the food sector. Studies have found that the nanoemulsion has various advantageous properties such as enhanced absorption, biocompatibility, reduced degradation, stability, and safety of nutritional components, and organic extracts, particularly lipophilic bioactive compounds such as some vitamins and nutraceuticals [24].

Nanocapsules can release bioactive substances via several mechanisms, including diffusion, dissolution, erosion, swelling, osmosis, degradation, and fragmentation. For instance, the diffusion process involves the movement of a bioactive substance from the interior to the exterior of an encapsulation system [25]. The movement is controlled by the solubility of the material in the contained system and the diffusion coefficient of the material through the capsule material. This release of bioactive compounds is determined by several factors which include the molecular weight, polarity and vapour pressure of the bioactive molecule and the polarity, physical state and intermolecular forces of the encapsulating system [26].

For the lipid nanoemulsion, the size and interfacial properties of oil droplets are crucial factors in the digestibility of lipid droplets and the release of bioactive components within the gastrointestinal tract. Salivary salts can influence the stability and ionic strength of this nanoemulsion [27]. Saliva mucin either bridges or depletes flocculation, leading to the aggregation of nanodroplets. In the stomach, the oil droplets form lumps because of the low pH and high ionic strength. Gastric lipase hydrolyses lipid droplets in the stomach, whereas pancreatic lipase largely hydrolyses them in the small intestine. In the gastrointestinal fluids, bile salts displace emulsifier molecules from the oil–water interface and facilitate enzyme interaction.

During lipolysis, free fatty acids and monoacylglycerols accumulate on the surface of the oil surface. Due to their elevated surface activity, long-chain free fatty acids and monoglycerides can inhibit lipolysis by displacing lipase enzymes. Bile salts dissolve digestion products in mixed micelles on the surface of the oil droplet, facilitating complete digestion. Calcium ions can form insoluble soaps with long-chain free fatty acids, effectively removing them from droplet surfaces. Ultimately, the breakdown products of fat and fat-soluble compounds are dissolved in micelles or single-layered structures made of phospholipids, which attach to the inner lining of the intestines [28].

Research has shown that large particles and high concentrations of wall components can impact the absorption of encapsulated materials in the gastrointestinal tract. The larger size and smaller surface area of macro and microparticles may lead to reduced nutrient absorption within the gastrointestinal tract. Prolonged dissolution time for larger particles can delay the release and absorption of encapsulated nutrients as they pass through the digestive system. The high concentration of wall components in encapsulated particles may also interfere with particle degradability, affecting the release and availability of the encapsulated nutrients for absorption [29].

Studies have also found that liposomes are an efficient technique to encapsulate and shield vitamin C from oxygen and heat-induced deterioration. Liposomal vitamin C has higher stability and bioavailability than traditional formulations, which improves absorption and antioxidant activity [30]. Encapsulated bioactive constituents (e.g., flavonoids and vitamins) can be formulated with polymeric nanomaterials for protected nutrient delivery. A study has stated that polymeric nanoparticles made from PLGA [Poly (lactic-co-glycolic acid)] can improve the bioavailability of iron supplements. The properties of these nanoparticles ensure that iron is released gradually and absorbed more efficiently in the gastrointestinal tract, addressing issues of iron deficiency anaemia [31].

Nowadays, the production of supplements mostly involves encapsulation techniques where the desirable probiotics and other bioactive particles are directed into the human body with the help of Zn and Fe nano-structured capsules. The use of nano-encapsulated probiotics in foods offers several advantages in protecting the probiotics and greater efficacy in delivery to the intestine. It has been observed that nanoparticles in food supplements are more active than commonly used supplements because they respond more effectively with human cells due to their special size. Nano-capsules are also adapted as carriers for antioxidants, essential oils, flavours, coenzyme Q10, vitamins, phytochemicals, and minerals to improve their bioavailability in the human system. The encapsulation of polyphenols with nanoparticles may prevent any oxidative actions and provide them with an acceptable taste [32].

The role of nanotechnology is also assumed to improve the characteristics of bioactive particles in spices and herbs by enhancing their bioavailability, water solubility, and antioxidant properties, enabling the active ingredients to dissolve homogeneously. The nano-materials are considered to increase the bioavailability of some important phytochemicals, for example, genistein and curcumin. Moreover, nano nutraceuticals are also available as dietary supplements, herbal products, and bioactive particles in nano-formulations [33].

4 Applications of Nanocarriers in Nutrient Delivery

The awareness regarding health and well-being has increased throughout the globe. As a result, people are looking for new strategies for the delivery of nutrients and bioactive compounds. In this context, using nanotechnology to encapsulate nutrients has been found to be an effective tool for promoting absorption and increasing nutrient bioavailability. Numerous studies have indicated that encapsulation of nutrients promotes controlled release, solubility, and targeted delivery of drugs to specific tissues. Nanocarriers safeguard the nutrients from degradation during digestion and maximise their utilisation in the body. This technique is useful in improving the bioavailability of poorly absorbable nutrients such as vitamins, minerals and some bioactive substances present in foods [34].

The revolutionary role of nanotechnology has been outlined below.

4.1 Nanocarriers Help to Improve the Stability of Nutrients

Nanotechnology shows promising improvements in increasing the stability of nutrients by protecting them from several environmental factors such as heat, light, and oxygen. Many research studies indicated that traditional nutrient delivery methods were incapable of preserving the stability of sensitive nutrients, resulting in their deterioration during food processing and storage [1]. Nanocarriers like liposomes, nanoemulsions, polymeric nanoparticles etc. provide a favourable condition for sensitive nutrients, thus preventing their degradation and assuring their efficacy until consumption. For instance, when encapsulating vitamins and antioxidants, liposomes, a lipid bilayer nanocarrier, can protect these compounds from oxidative stress, thermal decomposition, and light degradation. Similarly, nanoemulsions stabilize lipophilic nutrients by dispersing them in a nanoscale emulsion and decreasing their exposure to degrading environments. Polymeric nanoparticles are also found to provide extended stability with controlled-release mechanisms. Hence, they protect the nutrients from environmental degradation and gradually release them into the body [35].

4.2 Nanocarriers Enhance Solubility and absorption of Nutrients

Nanotechnology greatly enhances the solubility and bioavailability of nutrients, particularly nutrients with low water solubility. Indeed, several important types of nutrient molecules (vitamins and antioxidants) are highly hydrophobic and therefore have limited solubility that prevents their absorption in the gastrointestinal tract. Nanoscale delivery systems such as nanoemulsions or nanocrystals can drastically increase the surface area per unit volume of these nutrients thus enhancing their solubility and hence their absorption efficiency. For example, nanoemulsions can be used to reduce lipophilic nutrient droplet size, thereby allowing them to be more easily taken up by the absorptive cells of our digestive system. Similarly, Polymer nanoparticles could be engineered for enhanced hydrophobic substrate-solvent interactions i.e. improved targeted delivery within the aqueous environment of our gastrointestinal tract. This increased solubility enhances the concentration of bioavailable moiety, allowing increased uptake i.e., they allow more efficient utilisation of consumed nutrients [36].

4.3 Nanocarriers Help in Targeted Delivery and Controlled Release of Nutrients

Nanotechnology also has the potential for the targeted transport and controlled release of nutrients, which can increase their therapeutic effects and reduce side effects. One such approach is a targeted delivery strategy that consists of engineering nanocarriers which specifically attach to certain cells or tissues, enabling targeted delivery of nutrients to desired sites [37]. Surface modifications and ligand-receptor interactions are the mechanisms that can regulate this process of nutrient delivery. The controlled release systems regulate the rate of nutrient release leading to a sustained effect over a while. For example, polymeric nanoparticles can be designed in a manner that they release their contents in response to certain stimuli in the environment, such as changes in pH or temperature etc. [38].

5 Nanocarriers in Sports Nutrition

Nanocarriers, especially liposomes, play a critical role in improving the stability and bioavailability of nutrients. This technology can be highly advantageous in sports nutrition. Degradation of certain compounds during digestion and absorption is the key challenge in manufacturing nutritional supplements and fortified foods, which causes loss of their nutritional value. Amongst microencapsulation techniques, liposomal technology has gained popularity due to its application in food or nutrient delivery. It helps to encapsulate a wide range of nutritional components, including essential oils, antioxidants, amino acids, bioactive principles, vitamins, and minerals. Initially, it was developed for pharmaceutical delivery, but it has also proven to be highly beneficial to the food industry.

Vitamin C is very important for enhancing endurance and exercise performance. As we know, Vitamin C or ascorbic acid cannot be synthesized within the human body but regulates multiple cellular functions as well as its integrity. The sources of Vitamin C are easily accessible but modern lifestyle at times causes its deficiency.

In that situation, improving vitamin C status becomes indispensable for optimizing the performance of athletes. Encapsulation of vitamin C using liposomal technology can be highly efficacious as it enhances the bioavailability. Studies have indicated that liposomal vitamin C can cause higher spikes in plasma levels compared to standard oral supplementation. The incidence of iron deficiency is also very common in athletes. Studies indicate liposomal iron to be more effective than traditional intravenous iron therapy in the treatment of iron deficiency anaemia with fewer adverse side effects. Thus, liposomal formulations can be useful in athletes suffering from anaemia.

Furthermore, the vitamin C due to its antioxidant properties are of high significance in sports nutrition. The incidences of upper respiratory tract infections are found to be higher in elite athletes due to the intense physical stress experienced by them. Studies have shown that vitamin C can lower both the incidence and duration of colds as it improves the ability of the body to fight infection. Vitamin C also enhances endurance and exercise performance by improving fatty acid utilization during activity and ameliorating exercise-induced bronchoconstriction.

Liposomal technology is not limited to vitamin C and iron. It is also used to enhance the delivery of other nutrients and phytochemicals that are capable of imparting positive effects on exercise performance. Certain chemicals possess antioxidant and anti-inflammatory properties and can be highly beneficial for sportsmen. Examples of such chemicals include phenolic compounds like resveratrol, quercetin, catechins etc. Liposomal technology can also be used to deliver these compounds as it will increase their absorption and utilization.

An athlete's nutritional requirement is much higher in comparison to normal individuals because of their high physical activity. As a result, meeting that nutritional need is a challenge. Nanocarriers, particularly liposomal technology can address this challenge by improving the bioavailability and stability of various nutrients. The applications of liposomes in enhancing nutrient absorption and exercise performance can open a new avenue in the field of sports nutrition. More research is required to fully understand the usage of liposomal nutrients to enhance athletic performance [39].

Nowadays, certain nutritional nano-compounds like resveratrol-loaded lipid nanoparticles, redox-active nanoparticles with nitroxide radicals, folic acid incorporated into layered hydroxide nanoparticles, and iron in liposomal form etc. are used as an innovative way to enhance endurance and exercise performance. They are more effective in comparison to their non-encapsulated counterparts and placebos. Along with improved performance, it also caused favourable physiological and metabolic changes like heightened antioxidant capacity, lowered oxidative stress, reduced inflammation, maintenance of muscular integrity, improved mitochondrial function and quality, and higher glucose levels during exercise. An increase in both muscle and liver glycogen stores, as well as elevated serum and liver iron content, has also been observed. However, it is noteworthy that these findings are based on animal studies, with little to no direct evidence from human trials. Conventional nutritional supplements have low bioavailability when compared to liposomal nutrients.

as they rapidly get degraded in the alimentary canal, which leads to their diminished efficacy. On the other hand, nano-compounds due to their greater solubility and stability, have greater absorption rates and hence bioavailability. Modification of chemical and physical properties of these nano-compounds can lead to better site-specific and controlled release.

The application of nanotechnology in the field of sports nutrition needs deeper understanding. The use of nutritional nano-compounds as ergogenic aids in exercise and performance needs to be optimized by more human clinical trials.

The efficacy and safety of the nanocarriers in enhancing athletic performance and recovery is still an area of further exploration. The limited studies indicate that nanocarriers are capable of causing potential shifts in dietary supplementation practices for athletes, because of their advantages over traditional supplements [40].

6 Future Perspectives

Nanocarriers have attracted substantial interest across many disciplines, especially in drug and nutrient delivery. Nanoparticle-based therapeutics can overcome biological barriers and effectively deliver hydrophobic drugs to target the disease sites in the human body. Despite that, there are an extremely limited number of nanoparticle-based medicines that have been approved for clinical use.

Till now numerous challenges and obstacles exist in different stages of the development of these nanocarriers. Consequently, these challenges and safety concerns must be addressed properly to ensure their effective and safe application [41].

One of the major challenges in the development of nanocarriers is the need for comprehensive characterisation and standardisation. A few properties of nanocarriers, such as particle size, shape, and surface charge, are very crucial. These characteristics impact the effectiveness and safety profile of nanocarriers. Therefore, the inherent complexity of nanoparticles as multi-component three-dimensional entities needs accurate design and engineering. The implementation of detailed orthogonal analysis methods, along with a reproducible scale-up and manufacturing process is required to achieve a consistent product with the intended physicochemical characteristics, biological behaviours, and pharmacological profiles [42].

The safety of nanocarriers is still a matter of concern as they have some toxic effects. The physicochemical characteristics of nanoparticles, such as their size and surface charge, can influence their interactions with the biological system, which may result in various undesirable health consequences. Certain nanoparticles can induce oxidative stress or inflammatory responses within the biological system [43]. The long-term effects of nanocarriers that are intended for human use need to be understood which requires extensive long-term toxicity studies. Additionally, the effect of nanocarriers on various organs must be monitored closely. Nanocarriers might accumulate in different tissues or organs, causing toxicity. Rigorous toxicological and biocompatibility examinations need to be conducted before using nanocarriers to deliver medicines in disease treatment [44].

Discovering the science behind the mechanism of controlled release of drugs encapsulated in a nanocarrier is a critical challenge.

The release of drugs from the nanocarriers is a complicated process that can be regulated by a variety of factors. It can be the composition of the nanocarriers, environmental conditions, and the presence of bodily fluids. It can influence the flexibility of the timing and dosing of therapeutic administration via nanocarriers. Another very important aspect is the stability of nanocarriers. It can be referred to as one of the most significant characteristics as various environmental factors such as temperature, humidity, and light exposure can adversely affect the stability of nanocarriers. As a result, the encapsulated compounds get degraded. Therefore, one of the key challenges associated with nanocarriers is retaining their structural integrity throughout their shelf life and during storage [45].

So, efforts should be made to optimise the design and formulation of nanocarriers while ensuring their safety and efficacy. It can lay the foundation for their successful application in clinical and nutritional fields.

7 Conclusion

Although the application of nanocarriers in sports nutrition is limited, they can be highly promising as they can enhance the absorption of nutrient manifolds, thus improving exercise performance. They enhance bioavailability by preventing their degradation during digestion, regulating controlled release and ensuring a larger proportion of nutrients enter the systemic circulation. There are different types of nanocarriers and their usage varies depending on the chemical to be delivered and the environment in which it will be delivered.

Nanocarriers can efficiently enhance the solubility of poorly soluble ingredients. Nanoencapsulation is a popular technique that can shield nutrients and bioactive compounds from undesirable interactions with other food components and other inhibiting biomolecules, thus facilitating their delivery.

This novel approach can enhance the absorption of nutrients via several mechanisms. For example, by forming a micelle network, nanocarriers can increase the water solubility of nutrients.

Many studies have been carried out to understand nano-compounds' capability in enhancing endurance and exercise performance. They have been found to improve antioxidant activity, reduce inflammation, and preserve muscle integrity in sportsmen. Moreover, the overall benefits of nanocarriers extend beyond nutrient absorption. As a result, the judicious usage of nano-encapsulated nutrients can help athletes optimise their training and recovery.

A very crucial aspect of lipid-based nanocarriers is their biocompatibility and biodegradability. It is a very important criterion, especially in the food and nutraceutical industries. This ensures the safety of its consumption without adverse effects. This characteristic makes it extremely appealing in sports nutrition formulations.

Nanocarriers have a profound physiological impact due to their efficient nutrient delivery.

References

1. Nile SH, Baskar V, Selvaraj D, Nile A, Xiao J, Kai G (2020) Nanotechnologies in food science: applications, recent trends, and future perspectives. Nano-Micro Lett 12:45
2. Vieira IRS, Tessaro L, Lima AKO, Velloso IPS, Conte-Junior CA (2023) Recent progress in nanotechnology improving the therapeutic potential of polyphenols for cancer. Nutrients 15:3136. https://doi.org/10.3390/nu15143136
3. Abed SN, Deb PK, Surchi HS, Kokaz SF, Jamal SM, Bandopadhyay S, Tekade RK (2019) Nanocarriers in different preclinical and clinical stages. In: Basic fundamentals of drug delivery. Academic Press, pp 685–731. https://doi.org/10.1016/B978-0-12-817909-3.00017-0
4. Din FU, Aman W, Ullah I, Qureshi OS, Mustapha O, Shafique S, Zeb A (2017) Effective use of nanocarriers as drug delivery systems for the treatment of selected tumors. Int J Nanomed 5(12):7291–7309. https://doi.org/10.2147/IJN.S146315. PMID: 29042776; PMCID: PMC5634382
5. Sawant RB, Nikam SP, Roy A, Kumar A, Mohammed OA, Sharma K, Rai AK, Roy A, Gaur A, Verma R (2024) Nanocarriers for nutraceutical delivery: a miniaturized revolution in health. Nano-Struct Nano-Objects. 39:101321. https://doi.org/10.1016/j.nanoso.2024.101321
6. Duncan TV (2011) Applications of nanotechnology in food packaging and food safety: barrier materials, antimicrobials and sensors. Curr Opin Colloid Interface Sci 363:1–24. https://doi.org/10.1016/J.JCIS.2011.07.017
7. Lôbo GCNB, Paiva KLR, Silva ALG, Simões MM, Radicchi MA, Báo SN (2021) Nanocarriers used in drug delivery to enhance immune system in cancer therapy. Pharmaceutics 28, 13(8):1167. https://doi.org/10.3390/pharmaceutics13081167. PMID: 34452128; PMCID: PMC8399799
8. Anajafi T, Mallik S (2015) Polymersome-based drug-delivery strategies for cancer therapeutics. Ther Deliv 6(4):521–534. https://doi.org/10.4155/tde.14.125
9. Rawding PA, Bu J, Wang J, Kim DW, Drelich AJ, Kim Y, Hong S (2022) Dendrimers for cancer immunotherapy: avidity-based drug delivery vehicles for effective anti-tumor immune response. Wiley Interdisciplinary Reviews: Nanomedicine and Nanobiotechnology. 14(2):e1752. https://doi.org/10.1002/wnan.1752
10. Shi Y, Lammers T, Stor G, Hennink WE (2017) Physico-chemical strategies to enhance stability and drug retention of polymeric micelles for tumour-targeted drug delivery. Macromol Biosci 17(1):1600160. https://doi.org/10.1002/mabi.201600160
11. Pada AK, Desai D, Sun K, Goverdhanam NP, Tornquist K, Zhang J, Rosenholam JM (2019) Comparison of polydopamine-coated mesoporous silica nanorods and spheres for the delivery of hydrophilic and hydrophobic anticancer drugs. Int J Mol Sci 20(14):3408. https://doi.org/10.3390/ijms20143408
12. Neamtu I, Rusu AG, Diaconu A, Nita LE, Chiriac AP (2017) Basic concepts and recent advances in nanogels as carriers for medical applications. Drug Delivery 24(1):539–557. https://doi.org/10.1080/10717544.2016.1276232
13. Alyassin Y, Sayed EG, Mehta P et al (2020) Application of mesoporous silica nanoparticles as drug delivery carriers for chemotherapeutic agents. Drug Discovery Today 25(8):1513–1520. https://doi.org/10.1016/j.drudis.2020.06.006
14. Nemati A (2020) Quantum dots in therapeutic, diagnostic and drug delivery applications "a brief review." Iran J Mater Sci Eng 17(2):80–91. https://doi.org/10.22068/ijmse.17.2.1
15. Hooshmand S, Hayat SMG, Ghorbani A, Ghorbani A, Khatami M, Pakravana K, Darroudi M (2021) Preparation and applications of superparamagnetic iron oxide nanoparticles in novel

drug delivery systems: an overview article. Curr Med Chem 28(4):777–799. https://doi.org/10.2174/0929867327666200123152006
16. Yafout M, Ousaid A, Khayati Y, Oatmani IS (2021) Gold nanoparticles as a drug delivery system for standard chemotherapeutics: a new lead for targeted pharmacological cancer treatments. Scientific African 11:e00685. https://doi.org/10.1016/j.sciaf.2020.e00685
17. Balaji V, Manasa B, Aakash N, Chandrakaanth BS, Kiran Kumar KC (2007) Nanocomposites and their applications. Int J Eng Res Mech Civil Eng 2(11):81–85
18. Osorno LL, Brandley AN, Maldonado DE, Yiantsos A, Mosley RJ, Byrne ME (2021) Review of contemporary self-assembled systems for the controlled delivery of therapeutics in medicine. Nanomaterials 11(2):278. https://doi.org/10.3390/nano11020278
19. Yasamineh S, Yasamineh P, Kalajahi HG et al (2022) A state-of-the-art review on the recent advances of niosomes as a targeted drug delivery system. Int J Pharm 624:121878. https://doi.org/10.1016/j.ijpharm.2022.121878
20. Sastri KT, Radha GV, Pidikiti S, Vajjhala P (2020) Solid lipid nanoparticles: preparation techniques, their characterization, and an update on recent studies. J Appl Pharmaceutical Sci 10(06):126–141
21. Choradiya BR, Patil SB (2021) A comprehensive review on nanoemulsion as an ophthalmic drug delivery system. J Mol Liq 339:116751. https://doi.org/10.1016/j.molliq.2021.116751
22. Fan Y, Marioli M, Zhang K (2021) Analytical characterization of liposomes and other lipid nanoparticles for drug delivery. J Pharm Biomed Anal 192:113642. https://doi.org/10.1016/j.jpba.2020.113642
23. Altemimi AB, Farag HAM, Salih TH, Awlqadr FH, Al-Manhel AJA, Vieira IRS, Conte-Junior CA (2024) Application of nanoparticles in human nutrition: a review. Nutrients 25, 16(5):636. https://doi.org/10.3390/nu16050636. PMID: 38474764; PMCID: PMC10934945
24. Sozer N, Kokini JL (2014) Use of quantum nanodot crystals as imaging probes for cereal proteins. Food Res Int 57:142–151. https://doi.org/10.1016/j.foodres.2013.12.031
25. Chen L, Remondetto GE, Subirade M (2006) Food protein-based materials as nutraceutical delivery systems. Trends Food Sci Technol 17:272–283. https://doi.org/10.1016/j.tifs.2005.12.011
26. Boostani S, Jafari SM (2021) A comprehensive review on the controlled release of encapsulated food ingredients; fundamental concepts to design and applications. Trends Food Sci Technol 109:303–321. https://doi.org/10.1016/j.tifs.2021.01.040
27. Liu Q, Huang H, Chen H, Lin J, Wang Q (2019) Food-Grade nanoemulsions: preparation, stability and application in encapsulation of bioactive compounds. Molecules 24:4242. https://doi.org/10.3390/molecules24234242
28. Patel A, Mohanan A, Ghosh S (2019) Effect of protein type, concentration and oil droplet size on the formation of repulsively jammed elastic nanoemulsion gels. Soft Matter 15:9762–9775. https://doi.org/10.1039/C9SM01650C
29. McClements DJ, Li Y (2010) Structured emulsion-based delivery systems: controlling the digestion and release of lipophilic food components. Adv Coll Interface Sci 159:213–228. https://doi.org/10.1016/j.cis.2010.06.010
30. Gonzales GB, Smagghe G, Grootaert C, Zotti M, Raes K, Van Camp J (2015) Flavonoid interactions during digestion, absorption, distribution and metabolism: a sequential structure-activity/property relationship-based approach in the study of bioavailability and bioactivity. Drug Metab Rev 47:175–190
31. Pei R, Liu X, Bolling B (2020) Flavonoids and gut health. Curr Opin Biotechnol 61:153–159
32. McClements DJ (2020) Recent advances in the production and application of nano-enabled bioactive food ingredients. Curr Opin Food Sci 33:85–90. https://doi.org/10.1016/j.cofs.2020.02.004
33. Dudefoi W, Villares A, Peyron S, Moreau C, Ropers MH, Gontard N, Cathala B (2017) Nanoscience and nanotechnologies for biobased materials, packaging and food applications: new opportunities and concerns. Innov Food Sci Emerg Technol 46:107–121. https://doi.org/10.1016/j.ifset.2017.09.007

34. Cushen M, Kerry J, Morris M, Cruz-Romero M, Cummins E (2012) Nanotechnologies in the food industry-recent developments, risks and regulation. Trends Food Sci Technol. 24:30–46. https://doi.org/10.1016/JTIFS.2011.10.006
35. He X, Deng H, Hwang H-M (2018) The current application of nanotechnology in food and agriculture. J Food Drug Anal 2018(27):1–21
36. Robles-García MA, Rodríguez-Félix F, Márquez-Ríos E, Aguilar JA, BarreraRodríguez A, Aguilar J, Ruiz-Cruz S, Del-Toro-Sánchez CL (2016) Applications of nanotechnology in the agriculture, food, and pharmaceuticals. J Nanosci Nanotechnol 16:8188–8207
37. Kaur S, Rana N (2021) Nanochemistry in food science: development of nutrient delivery systems for enhanced bioefficiency. Int J Food Nutrit Sci 10(8):658–670
38. Chaaban H, Ioannou I, Paris C, Charbonnel C, Ghoul M (2017) The photostability of flavanones, flavonols and flavones and evolution of their antioxidant activity. J Photochem Photobiol, A 336:131–139
39. Higgins MF, Da Boit M (2016) Liposomal nanotechnology—a new frontier for sport and exercise nutrition? J Nanomed Res 4(4):00098. https://doi.org/10.15406/jnmr.2016.04.00098
40. Lima MR, Moreira BJ, Bertuzzi R, Lima-Silva AE (2024) Could nanotechnology improve exercise performance? Evidence from animal studies. Brazilian J Med Biol Res 19:57, e13360. https://doi.org/10.1590/1414-431X2024e13360. PMID: 38656076; PMCID: PMC11027182
41. Alshawwa SZ, Kassem AA, Farid RM, Mostafa SK, Labib GS (2022) Nanocarrier drug delivery systems: characterization, limitations, future perspectives and implementation of artificial intelligence. Pharmaceutics 14(4):883. https://doi.org/10.3390/pharmaceutics14040883
42. Desai N (2012) Challenges in development of nanoparticle-based therapeutics. AAPS J 14(2):282. https://doi.org/10.1208/s12248-012-9339-4]
43. Teixeira MI, Lopes CM, Amaral MH, Costa PC (2024) Navigating neurotoxicity and safety assessment of nanocarriers for brain delivery: strategies and insights. Acta Biomater 189:25–56. https://doi.org/10.1016/j.actbio.2024.09.027
44. Lu H, Zangh S, Wang J, Chen Q (2021) A review on polymer and lipid-based nanocarriers and its application to nanopharmaceutical and food-based system. Front Nutr 8:783831. https://doi.org/10.3389/fnut.2021.783831
45. Sunanda K, Praveenanjali T (2024) Nanotechnology in food industry: current trends and future prospects. Int J Sci Res 13(3):544–549

The Science of Nano-Enhanced Hydration in Sports Nutrition

W. Vinu, Rameer Thaslim Khan, N. K. Ashif Ali, Medha Samineni, P. Muneer, and Dilshith A. Kabeer

Abstract Nano-enhanced hydration is a novel method in sports nutrition that integrates nanoparticles to improve hydration. Hydration plays the most significant role in sports nutrition, affecting physical performance, overall health and recovery by helping transport oxygen and nutrients to cells, removing waste from cells and regulating body temperature. Incorporating nanotechnology in hydration can help in faster absorption of nutrients, targeted cellular hydration and electrolyte management. For example, encapsulation of essential nutrients in nano-emulsions or lipid nanoparticles aids in the targeted delivery of nutrients to a specific cell or tissue and increases the bioavailability. Therefore, nanoparticle-enhanced hydration will be much more effective than the traditional hydration methods. This chapter discusses about the science behind nano-enhanced hydration, discussing its role in overcoming dehydration, supporting thermoregulation during high-intensity activities and maintaining optimal osmotic balance. Emerging studies highlight the efficiency of nano-structured water molecules by incorporating nano-emulsions and lipid nanoparticles, which improving intercellular hydration compared to other conventional methods. Nano-enhanced hydration stands as a revolutionary development, which offers athletes to achieve peak physical performance by combining sports nutrition with cutting edge nanoscience. The chapter further explores the evidence-based insights on nano-enhanced hydration for improving endurance, recovery and overall performance.

Keywords Nano-enhanced hydration · Nanoparticles · Nano-emulsion · Lipid nanoparticle · Electrolytes · Bioavailability

---

W. Vinu · N. K. Ashif Ali · D. A. Kabeer (✉)
Department of Physical Education & Sports, Pondicherry University, Kalapet, Puducherry, India
e-mail: dilshith.dpe@pondiuni.ac.in

R. T. Khan
Department of Food Science and Technology, Pondicherry University, Kalapet, Puducherry, India

M. Samineni
University of Houston-Clear Lake, Houston, TX, USA

P. Muneer
Department of Physical Education, University of Calicut, Thenhipalam, India

A. Chatterjee et al. (eds.), *Nanofuel: The Future of Sports Nutrition*, Smart Nanomaterials Technology, https://doi.org/10.1007/978-981-96-5471-0_3

1 Introduction

1.1 Overview of Nano-Enhanced Hydration

Water is a fundamental nutrient for all living things; it plays an important role in stabilizing life and health. Human body weight consists of 50–60% of water also the brain composed of 75% water. Hydration refers to the water consumption through beverage and food [1]. It is further elaborated as maintaining the balance between water consumption and water loss, ensuring a sufficient level of water in the human body to support physiological functions. It is essential to maintain the required water balance in the body for overall health and cerebral functions [2, 3]. It is also essential to transport nutrients across membranes, proper metabolism, cellular homeostasis, blood flow and body temperature regulation. The water we intake is primarily absorbed by the small intestine in the gastrointestinal tract. It has been recorded that, approximately 10 litters of water enter the intestine daily, in which only 100 ml are being eliminated through excretion, indicating high absorptive efficiency [4, 5]. After the water absorption is distributed throughout the human body via interstitial fluid and bloodstream. It is mainly divided into two compartments: extracellular fluid (ECF), which contains plasma (20–25% of ECF) and interstitial fluid, which transports nutrients, hormones, and waste between circular systems and cells [6]. Intercellular fluid (ICF), makes up 60–70% of total body water and is located within the cell to keep cell volume and support metabolic functions [7]. similarly, cellular hydration is very important in our lives. It maintains proper cellular functions and overall physiological functions such as optimal cell function, ATP production and overall metabolic efficiency [8]. This underlines the relative significance placed on human hydration and its effects on overall health and survival. As for the need to intake water and its relationship to sports performance, there is little discussion about the importance and benefit of appropriate hydration necessary to enhance and maintain performance under diverse conditions [9].

The athletes must know the best way to maintain water balance in the body because hydration is an essential part of sports practice or exercises, and they do not underestimate the significance of water intake in terms of quality, quantity and frequency [10]. All sports persons are different, so they should focus on personalized techniques tailored to their specific goals, such as types of physical and physiological conditions. Through these strategies, they improve their performance and avoid hydration-related injuries and problems [11]. At first pass, athletes and experts applied conventional procedures to maintain water balance in the human body like drinking plain water, consuming water-content foods and electrolyte solutions that include salt and sugar [12, 13]. These traditional hydration techniques often fail to work effectively for everyone, especially for individuals with certain health issues. Moreover, another important thing is that these techniques encounter challenges such as poor bioavailability, delayed absorption rate and insufficient retention, limiting their ability to maintain the body's hydration. In contrast, these new and innovative methods, such as nanotechnology, offer more effective solutions for hydration in sports activities among athletes.

1.2 Definition of Nano-Enhanced Hydration

Nano-hydration applies advanced nanotechnologies transformed into hydration methods in sports nutrition. This method balances nanoparticles' unique properties to support hydration's effectiveness in the body. Nano-hydration improves fluid absorption and preservation in the body by using nanotechnology to build ultra-fine water molecules or nano-sized nutrients. This method facilitates nutrient supply and cellular hydration and make sure of overall hydration efficiency, which is vital for maintaining athletic performance and blocking risks of injuries from physical activities or exercises. Traditional hydration methods have more challenges on the basis of absorption and bioavailability; therefore, the application of nanoparticles that supply hydration more effectively and efficiently. The usage of nanoparticles can boost the bioavailability of nutrients, water absorption rate, and fluid and electrolyte retention. This delivers particles that allow the body to stay and maintain fluid balance in the body more effectively [14], such particles can be designed to target specific cells or tissues to help enhance water patterns in the body. Application of this targeted delivery can help achieve the maximum levels of hydration during prolonged physical activities involving bodily exertion [15].

1.3 Advantages and Scope of Nano-Enhanced Hydration Methods in Sports

Based on hydration, nanotechnology gives more advantages and scope in sports nutrition or sports performance in the modern era. For example, nanoparticles can enhance the supply of poorly water-soluble drugs, supplying better absorption and water maintenance at the cellular level [16]. When nanoparticles enter the human body, they can impact different tissues, necessitating detailed biosafety studies to avoid potential health hazards [17]. Another important scope is that when nano-sized molecules are incorporated into hydration, they can keep chemical and physical properties that improve the bioavailability of nutrients. Apart from this, hydration-based liquid nanoparticles are used to enhance the absorption of supplements by increasing the surface area of minerals and nutrients [18]. Hence, these physiological developments increase sports performance and the overall health of athletes.

1.4 History of the Development of Nano-Enhanced Hydration Methods in Sports

Earlier hydration methods were superficial; athletes mainly focused on consuming plain water to maintain water balance during physical activities. These fundamental practices formed the basis for pointing during athletic performance [19]. After the 1960s, the rise of the sports drink industry made a turning point in sports drinks; it started to provide drinks that combined carbohydrates and electrolytes to better help with hydration and energy restoration [20]. As biotechnology advanced in the twenty-first century, researchers built more specialized hydration drinks and supplements for better hydration. These revolutions were driven by an enhanced understanding of nutrient-sensitive cell-signaling mechanisms and their effect on sports performance. Researchers focused on developing formulations to improve the efficacy of hydration solutions in helping maximal physical functions during physical activities [21]. In the modern context, maintaining water balance in the body with the help of nanotechnology has emerged as a promising field in sports nutrition for better hydration, ensuring substantial developments in the delivery and effectiveness of hydration solutions. By manipulating materials at the nanoscale, nanotechnology increases the absorption and bioavailability of hydration compounds by controlling the nanoscale, thus optimizing nutrient and water delivery for better sports performance and recovery [22].

1.5 The Scientific Foundation of Nano-Enhanced Hydration in Sports Nutrition

The concept of nano-hydration is a process of treating matters at the nanoscale to improve the contacts between hydrating agents and biological cells. This is mainly obtained through nano-encapsulation, where maintaining water balance agents in the body are covered in nanocarriers to protect them from degradation and enhance can their delivery efficiency [23]. Like this, nanotechnology expedites the design of smart materials that can act on specific physiological stimuli that deliver hydration agents only when needed. Studies also indicate the central role of nanoporous materials in retaining water for prolonged periods, which ensures maintained hydration. The combination of nano-hydration with wearable technologies and biosensors further emphasizes its potential to monitor hydration in real life [24]. Nano hydration can show a groundbreaking intersection of nanotechnology and hydration science, providing solutions to long-term health, medicine and sports challenges. Its rise reflects the potential of nanoscale novelty to redefine fundamental biological methods.

2 Types of Nano-Enhanced Hydration in Sports Nutrition

Nano hydration is a cutting-edge enhancement in sports nutrition that uses nanoparticles to make water balance solutions more effective and easier to absorb into the body. These systems can help athletes maintain proper hydration in their bodies and support faster recovery along with enhanced athletic performance, better performance, and sustained hydration. Nanotechnology develops a modernized hydration technique that focuses on providing the delivery and absorption of essential nutrients or elements to make the process more effective. The inventors have sensibly developed advanced systems based on nanoparticles that are designed to enhance both the efficiency and the uptake of nutrients in the body. These systems involve innovative carriers like liposomes, nano-emulsions, PLGA nanoporous, and solid lipid nanoparticles, efficiently transferring hydration-promoting elements. In addition, innovative materials such as silica and chitosan nanoparticles provide antimicrobial stabilization and protection hydration solutions, although delivering essentials. Cutting-edge strategies like fullerenes and gold nanoporous offer antioxidant benefits. The dendrimer provides multifunctional abilities for nutrient supply. Such nanoparticles show the ability of nanotechnology to transform traditional hydration techniques, giving concentrated and sustained hydration solutions.

2.1 Liposomes

Liposomes are spherical shaped vesicles made up of phospholipids surrounding an aqueous core, which makes them versatile in encapsulating both fat soluble (lipophilic) and water soluble (hydrophilic) substances, the phospholipids forming the bilayer which is containing a hydrophilic head and a hydrophobic tail, it enables them to create a stable barrier can trap different type molecules [25]. This uniqueness of liposomes to hold both water-based and fat-based molecules makes them an ideal system for delivering drugs, vitamins, electrolytes or other active ingredients [26]. The liposome formation occurs when the phospholipids are exposed to an aqueous environment, where they spontaneously assembling into bilayer structure that enclose a water filled compartment [27].

Liposome can be classified into different types by its surface charge, structure, functionalization, and composition. Based on the surface charge liposomes are classified into three types which are neutral, anionic and cationic; whereas neutral liposomes carry no net charge and it generally used for drug delivery, while anionic and cationic liposomes have negative and positive charges, lower and higher cytotoxicity respectively [28]. Structurally, liposomes categorized into four different types, small unilamellar liposomes (SUVs) which ranges from a diameter of 20–100 nm, whereas large unilamellar liposomes (LUVs) ranges from 100–1000 nm and multilamellar liposomes having multiple concentric lipid bilayers which resembling an onion and larger than the unilamellar forms of liposomes, while liposomes contains multiple internal compartments known as multicompartment liposomes which helps in better drug protection and controlled release [29]. Based on the function the liposomes classified into conventional, stealth (PEGylated), targeted and stimuli responsive liposomes, where the conventional liposomes are the basic liposomes which doesn't have any modifications and generally used for drug delivery [30]. Stealth (PEGylated) liposomes are modified by coating polyethylene glycol (PEG) which allowing them to avoid immune detection and extend circulation time, while targeted liposomes have ligands or antibodies modified on their surface to more efficient drug delivery to specific cells or tissues [31]. Liposomes which are releasing the contents in response to the specific stimuli such as temperature, pH, or enzymes are known as stimuli responsive liposomes [32]. In terms of composition, two types of liposomes are there which are phospholipid-based liposomes which is the most common type made from natural or synthetic phospholipids which can encapsulate both hydrophilic and lipophilic molecules, while non-phospholipid liposomes are made from lipids other than phospholipids and it is mainly used in cosmetics and agriculture [33].

Liposomes contributes significant benefits in hydration by enhancing the delivery and absorption of water-soluble vitamins and electrolytes, which are essential components in maintaining proper hydration levels in human body. Controlled and sustained release of liposome helps in prolonged hydration which is very useful for athlete or sports persons during their extended period of physical activity [34]. In conclusion liposomes shows a multifaced approach in sports nutrition, especially in hydration. They protect the major hydration components such as vitamins, electrolytes, from its degradation and also gives prolonged hydration by the controlled and sustained release.

2.2 *Solid Lipid Nanoparticles (SLNs)*

Solid lipid nanoparticles (SLNs) are composed of solid biocompatible and biodegradable lipids [35]. These are mainly designed for delivering bioactive compounds to improve bioavailability, stability and controlled release of the compounds [36]. They do not leave any harmful residues inside the body due to the biodegradable and metabolizable lipids which are used as the main components of SLNs. Based on the composition and structure SLNs categorized into several types, which are solid lipid nanoparticles (SLNs), nanostructured lipid carriers (NLC), Lipid-coated nanoparticles (LNPs) and Lipid drug conjugated nanoparticles (LDC) [37].

SLNs are made up of biodegradable lipids such as waxes, fatty acids and glycerides which helps them to act as colloidal drug carrier and offers controlled drug release and bioavailability [38]. It has a highly ordered crystalline spherical structure which can encapsulate drugs like hydrophilic and lipophilic [39]. They have a size of 1–1000 nm range [40]. Its high encapsulation efficiency helps in controlled release and protecting drugs from degradation and they can be administered via several routes such as nasal, oral, ophthalmic and pulmonary, it makes them highly versatile [41]. NLCs are modified form of SLNs designed to overcome the limitations, and it contains both solid and liquid lipids, the addition of liquid lipid along with the solid lipid helps in creating more space for drug molecules and improving its loading capacity [37]. It helps in controlled drug release and targeted delivery of drugs. LNPs ranging from a size of 10–1000 nm and contains a core material with lipid coating, this hybrid structure of LNPs making it highly versatile for several medical applications [42]. The core material of LNPs can be customized to enhance specific drug delivery while the lipid coating delivers improved stability, biocompatibility and improved bioavailability [43]. The lipid layer also helps in protection of therapeutic component from the environmental degradation and reducing the premature release of drug which ensures the active component reaches the targeted site [43]. LNPs are mainly used in drug delivery, clinical medicine and cosmetics [37]. This properties of LNPs helps in improve the delivery of nutrition and hydration agents ensuring efficient hydration in athletes [44]. Lipid drug conjugate nanoparticles (LDCs) formed by conjugating lipids such as glycerides, fatty acids or phospholipids with drugs [45]. As like the other types this one also has the controlled release and targeted delivery to specific tissues or organs, such as lymphatic system or tumors, this property of it helps in using mainly in medical field for drug delivery [46].

SLNs helps in hydration of athletes by its improved bioavailability and solubility which can enhance the solubility of poorly soluble compounds which is essential in nutrient delivery in sports nutrition, this helps athletes to fully absorb and utilize the maximum nutrient supplements they take as sports drinks [35]. Their controlled release can maintain steady levels of nutrients and hydration in athletes over extended period, it helps the athletes in hydration during prolonged physical activities [47]. SLNs can prevent the chemical degradation of bioactive compounds, it ensures the effective absorbance of nutrients to the body and it helps in maintaining the effectiveness of minerals and vitamins [47]. SLNs have the solid lipid matrix which enhances the physical stability and preventing leakage of encapsulated bioactive compounds ensuring stability and delivery of nutrients in athletes helps in overcomes various environmental conditions [48]. In conclusion SLNs provides multiple advantage in hydration by their ability to improve bioavailability, protection of bioactive compounds and controlled release.

2.3 Poly (Lactic-Co-Glycolic Acid) (PLGA)

Poly (lactic-co-glycolic acid) (PLGA) is a biodegradable and biocompatible functional polymer made from the polymerization of lactic acid (LA) and glycolic acid (GA) [49]. It can be degraded into components which are naturally metabolize inside the body such as non-toxic byproducts, lactic acid and glycolic acid [49]. PLGA synthesized from lactic acid and glycolic acid through ring-opening polymerization of its monomers, it catalyzed by different agents such as tin-based catalysts and organ catalysts [50]. The properties of PLGA can be affected by the ratio of glycolic acid and lactic acid, degradation rate and hydrophilicity will be increased by the higher content of GA in PLGA meanwhile the higher LA content leads to increase in hydrophobicity and slower in degradation [51].

PLGA can be classified into 3 types, standard PLGA: the PLGA which is used for drug delivery because of its controlled degradation rate and biocompatibility also they can carry both hydrophobic and hydrophilic drugs [52]. Second type is porous PLGA which is developed mainly to overcome the slow degradation of the standard PLGA, it can enhance the drug loading and efficient release by its higher surface area [53]. PLGA with additives is another type which is incorporated with additives like polyethylene glycol (PEG) or poly (propylacrlic acid), it enhances the targeting capability and delivery efficiency of PLGA [54].

Potential benefits of PLGA in hydration of athletes are its sustainable release, customizable degradation rates and biocompatibility [52]. The sustainable release ability of PLGA can provide controlled release of encapsulated nutrients, also vitamins, electrolytes or any other hydration promoting agents can be encapsulated and released slowly can help in hydration in athletes during long time physical activities [55]. PLGA has customizable degradation rate ability, it can be used as an advantage in sports nutrition for long duration and short duration physical activity by tailoring the degradation rate of PLGA by adjusting the GA and LA ratio [52]. Biocompatibility and FDA approval of PLGA making it safer to use on humans [56]. It ensures that it can be used for hydration of athletes too.

2.4 Chitosan Nanoparticles

Chitosan nanoparticles (CNPs) are the polysaccharide which is obtained by partial deacetylation of chitin from crustacean shells [57]. It is a copolymer of D-glucose amine and N-acetyl-D-glucose amine and the structure can be modified for enhancing the solubility and mechanical strength [58]. CNP's smaller in sizes higher in surface area and cationic nature makes them versatile in biodegradable and biological applications [59]. The linear polysaccharide backbone with amine group structure of chitosan allows them for various chemical modifications and enhancing its functionality [60]. They can be synthesized by using several methods such as ionic gelation, reverse micelles and emulsion cross linking [57].

CNPs can be classified into five different types which are simple chitosan nanoparticles, chemically cross-linked chitosan nanoparticles, ionically cross-linked nanoparticles, chitosan derivatives and chitosan-based nanocomposites [61]. Simple chitosan nanoparticles are the simple form which is biodegradable and biocompatible, they generally used for drug delivery [62]. Chemically cross-linked chitosan nanoparticles are formed by creating covalent bonds between the chitosan molecules and enhancing their stability in drug release [61]. This crosslinking of chitosan molecules increases the stability of nanoparticles and makes sure the encapsulated substances are intact until it reaches the targeted site [63]. They can help in extended period of hydration by the controlled release of encapsulated agents due to the covalent bonds in cross-linked nanoparticles [63]. The high drug loading capacity of chemically cross-linked chitosan nanoparticles can help in carrying significant number of hydrating agents for effective hydration [63]. The biocompatibility of chitosan polymer reduces the adverse reactions and makes it safer to use in humans [64]. Ionically cross-linked chitosan nanoparticles are formed through the ironic gelation method by interaction of chitosan with tripolyphosphate (TPP) [65]. The choice of cross-linking agent can influence the size of the nanoparticles formed through ionic cross-linking [66]. It also enhances the stability and mechanical properties of the nanoparticles [67].

The ability to controlled and sustained release of hydration promoting substances can help in effective hydration of athletes during workout and post workout. They can encapsulate electrolytes and release gradually which can enhance prolonged hydration in players [62] CNPs can improve the solubility of hydrophilic compounds which can help in delivering hydration related nutrients which are poorly soluble in water [68].

2.5 Nano-Emulsions

Nano-emulsions are less than 200 nm sized ultrafine nanosized droplets dispersions in continuous phase, they can solubilize both hydrophilic and hydrophobic substances which have low solubility at continuous phase [69]. These ultrafine droplets are composed of mainly two immiscible liquids such as water and oil, and their structure has a dispersed phase and a continuous phase also each droplet surrounded with emulsifiers to prevent coalescence [70]. They are not only used in medical, pharmaceutical and cosmetic but also used in food industry for encapsulating, stabilizing and delivering flavors, vitamins, omega-3 fatty acids nutraceuticals and preservatives [70].

Nano-emulsions can be classified into three types based on their nature of dispersed and continues phase such as oil in water nano-emulsions, water in oil nano-emulsions and bi-continuous nano-emulsions [71]. Oil in water nano-emulsions are the types of nanoparticles which are nanometric solvent droplets are dispersed in water with help of proper surfactant mixtures, it can be formed by different techniques such as both high energy and low energy methods and also, they have high kinetic stability due to their monodispersity and small droplet size which can minimize ostwarld ripening effect and it leads to withstand the density-based separation [72]. They are able to form stable and transparent delivery systems with high oral bio availability which leads them to very much suitable for encapsulation of lipophilic nutraceuticals [73]. Water in oil nano-emulsions is typically less than 500 nm in size and they are colloidal dispersions where water droplets dispersed within oil phase and stabilized by surfactants [74]. It can be used for encapsulating bioactive hydrophilic substances alone or as a structural unit of a double emulsion to protect them from degradation and enhances stability [75]. They are used in food industry for enhancing the nutrition value and stability of hydrophilic bioactive ingredients [76]. And the Bi-continuous nano-emulsions or bijels have a complex structure made up of two immiscible liquids which are tangled and interconnected and these structures are stabilized by colloidal particles attached to the interface between water and oil [77]. They can encapsulate both hydrophilic and lipophilic substances effectively due to their unique structure and it makes them highly versatile various applications such as food processing and drug delivery [78]. Also, their high surface area and small droplet size enhances bioavailability and stability and helps in encapsulating active compounds [78].

They can potentially improve the hydration and nutrition absorption by better interaction with biological cells due to their higher surface area [79]. It helps athletes to get efficient rehydration and nutrient replenishment during and after the exercise. Lipophilic nutrients are vital in recovery of athletes and nano-emulsions helps in enhancing the absorption like stabilizing short fibrils of whey protein isolate which leads to improved lipid digestion and absorption rates, it helps the athletes in prolonged endurance training [80]. Development of functional beverages such as sports drinks by incorporating oil in water nano-emulsions can help in providing enhanced hydration and nutrients and bioactive ingredients such as amino acids, minerals and vitamins to improve the recovery and performance of athletes [81]. The smaller droplet size of nano-emulsions prevents the aggregation and gravitational separation also enhances their stability which leads to consistent hydration in athletes [79].

2.6 Silica Nanoparticles

Silica nanoparticles (SNPs) are 1–100 nm sized nanoparticles which are made up of primarily silicon dioxide (SiO2) [82]. They have high porosity, biocompatibility and high surface area which makes them unique and suitable for various application fields like medicine, environmental and industrial processes [83]. There are several methods used to synthesize silica nanoparticles such as sol–gel method, stober method, microemulsion, flame synthesis, biogenic methods, based on the synthesis method their structure will vary significantly, generally they have a core shell structure, where the core is made of silica and the surface area covered and functionalized with various chemical group for the enhancement of their properties [83].

Based on their structure SNPs can be classified into several types which are Non-porous Silica Nanoparticles, Mesoporous Silica Nanoparticles, Core- Shell Silica Nanoparticles, Hollow Silica Nanoparticles and Dendritic Mesoporous Silica Nanoparticles. Among those the Non-porous Silica Nanoparticles are solid silica nanoparticles without internal pores, due to the absence of the pores they have lower surface area and a solid core [84]. They are thermally and chemically stable because of their structure and it making them safe for ingestion and use in sports nutrition products [84]. Their surface can be modified with various functional groups allows them to attach hydration promoting molecules such as vitamins and electrolytes [85]. Mesoporous silica nanoparticles are inorganic nanoparticles characterized by their mesoporous structure which having a pore ranging from 2 to 50 nm [86]. Their large surface area and pore volume helps them to load significant number of therapeutic agents [87]. The structure of mesoposus silica nanoparticles helps them in controlled and sustained release of drugs and it also can be further enhanced by functionilize the surface area with various molecules [88]. They can also be engineered to respond specific stimuli such as redox conditions, pH, light and enzymes for enabling controlled and targeted release [89]. Core–Shell Silica Nanoparticles are composite nanoparticles with a core material encapsulated within a silica shell, and the core material can be composed of various materials such as metals like gold or silver and metal oxides like Fe3o4 or other nanoparticles such as hydroxyapatite [90]. Hallow silica nanoparticles have a hollow core and a silica shell, and their shell thickness can vary the size ranging from 15 to 35 nm depends on their synthesize method [91]. Their silica shell is often mesoporous, allowing for the influx of gases and small molecules and it helps in drug delivery and catalysis [92]. The hollow and porous enhancing their effectiveness due to the large surface area. Their low toxicity makes them suitable for biomedical applications. The last type is dendritic mesoporous silica nanoparticles, which are a class of silicon based mesoporous materials characterized by their unique structural features such as distinctive dendritic structure which provides them a surface area and tunable pore sizes [93]. They have a three-dimensional dendritic superstructure with center-radial channels this helps in enhancing mass transfer and accessibility to internal surfaces [94]. They are used in biomedical applications such as drug delivery and gene delivery due to their high biocompatibility [95]. Also, their surface can be easily functionalized by various metal nanoparticles or organic groups and it helps in targeted and controlled delivery of therapeutic agents [96]. Template-assisted and room temperature synthesis are the methods used to synthesize dendritic mesoporous silica nanoparticles.

Silica nanoparticles can be used to develop hydration monitoring systems, their low toxicity and stability makes them suitable for using as coating for the sensors helping the athletes to maintain optimal hydration during training by providing real time feedback on hydration levels [97]. The stability of SNPs in physiological conditions helps in maintaining the integrity of hydration and nutritional supplements [98]. They can also use to develop advanced hydration drinks and supplements which ensures sustained release of electrolytes and other essential nutrients to make sure athletes remain hydrated during their activities [99].

2.7 Gold Nanoparticles

Gold nanoparticles are small gold particles with a diameter range of 1–100 with a face centered cubic crystal structure, they also show other structures such as cuboctaheadral, icosaheadral and decahedral [100]. They have a surface coating which can be coated with various molecules or phospholipids and polymers to enhance their stability in biological fluids [101]. They usually synthesized by chemical reduction or green synthesis methods [102]. It can be used as carrier in drug and gene delivery and also targeted delivery of therapeutic agents [103].

They can be categorized based on their shape and structure such as nanospheres, nanorods, nanoshells, nanocages and hybrid nanostructures [104]. Gold nanospheres are common type of gold nanoparticles which having nearly spherical shape and a size range of 1–100 nm in diameter [105]. Their surface can be readily modified by chemical and biological molecules to allowing attachment of biomacromolecules via chemical conjugation [106]. They are generally low toxic and biocompatible which makes them suitable for in vivo applications [107]. The Nanorods have elongated structure which differentiate them from other nanostructures like nanospheres and this shape gives them unique optical properties [108]. Nanoshells are a dielectric core which is coated with a thin gold shell, due to their tunable optical properties they mainly used in drug delivery and photothermal therapy [109]. Meanwhile the nanocages are hollow and porous structure, due to their large surface area and tunable optical properties they are also used in drug delivery [104]. And the hybrid nanostructures are complex structures which combining gold with other materials for advanced applications in diagnostic therapy [110].

Gold nanoparticles help in hydration of athletes by using it in development of calorimetric sensors for real-time hydration monitoring, these are the calorimetric sensors which can detect dehydration and overhydration in athletes by a color-based results from the sweat and provides immediate feedback [111]. Their ability to coat with different molecules helps to enhance their stability and solubility in biological fluids and it makes the suitable for developing sensors like hydration monitoring unit [101]. This kind of monitoring of hydration levels in athletes helps in maintaining the performance levels and also preventing the health issues due to dehydration or improper hydration.

2.8 Silver Nanoparticles

Silver nanoparticles are the nanoparticles of silver which is ranging from a size of 1–100 nm, some of them are composed of large percentage of silver oxide due to their large ratio of surface to bulk silver atoms [112]. The structure of silver nanoparticles may vary widely and their size typically ranges from 1 to 100 nm in size and the particles smaller than 5 nm can be unstable under certain conditions [112]. They exhibit various shapes including cubic, spherical, prismatic, trigonal–pyramidal, icosahedral and decahedral forms, these shapes effect their catalytic properties [113]. Also, they have a crystalline structure which can be can be single crystal, polycrystalline or multiple twinning structure [114].

They are classified based on their synthesize methods and shapes. The silver nanoparticles synthesized by physical, chemical and biological methods, in physical method they synthesized by vacuum thermal evaporation which is a method produces nanoparticles by evaporating silver in vacuum condition and getting various shapes and size of particles [114]. Radiation method is another method used in physical methods which his using UV, gamma or X-ray radiations to induce the formation of silver nanoparticles [115]. In other hand the chemical methods are chemical reduction and seed mediated growth methods using, in this the chemical reduction is reducing silver salts with reducing agents like hydrate which results in different morphological particles such as sphere, cube and rods [116]. Second method in the chemical method which is seed mediated growth has a twostep process where small seed particles are first formed and then further, they form into larger nanoparticles [117]. Biological method is an eco-friendly method, in this the green synthesis is used to form silver nanoparticle by using natural products like plant extracts, bacteria and yeast [118].

Silver nanoparticles contributes in hydration of athletes as enhancing the nutrient absorption by improving bioavailability and increasing solubility, it helps the athletes to absorb proper hydration agents like electrolytes and vitamins [119]. Their anti-inflammatory effect can reduce muscle soreness and inflammation post workout, and help in easy recovery times and enhances the performance [120].

2.9 Fullerens

Fullerens can be described as molecules made up of carbon atoms which can be in the shapes of ellipsoid, tube or a hollow sphere [121]. Fullerenes was first discovered in 1985 and the scientists won Nobel prize of Chemistry for the marvelous discovery [122, 123]. They discovered Buckminsterfullerene (C60) which looks like a football made up of 60 carbon atoms in a sphere form. Till day this is one of the most common fullerenes [124]. Structurally fullerenes are very similar to graphite which means it looks like pile of graphene sheets with hexagonal ring linkages, stacked together [125]. Fullerenes have different structural forms such as spherical fullerenes also known as buckyballs, ellipsoid fullerenes and carbon nanotubes (CNT) [126]. The most common spherical fullerene is the C60 fullerene discovered in 1985. Other types of spherical fullerenes also called buckyballs have been synthesized by scientists. The smallest of the them contain only 20 carbon [124]. Fullerenes can also be in ellipsoid structures forming elongated elliptical structures [125]. The cylindrical structures of Fullerenes are known as carbon nanotubes or buckytubes. These buckytubes are in the shape of rolled graphene sheets forming into tubular structures [123].

Based on the shapes and structural differences we can categorize fullerenes into different types. Fullerenes can be in the form of spherical shape made up of 60 and 70 carbon atoms They form a hybrid sp2 in the form of a conjugated system. This organized structure contributes to high aromaticity and improved electrical properties Each fullerene is composed of a closed networks of hexagons and pentagons in fused form [127]. Carbon nanotubes are a type of cylindrical shaped fullerene made up of carbon atoms forming a hexagonal pattern. CNT can be of two types namely single-walled CNT and multi-walled CNT. This depends on the number of graphite sheets present in the structure. CNT is one the unique innovation of the recent decades which has several great applications in many fields such as nanotechnology [121]. Another type of fullerenes is fullerene derivatives. These derivatives are fullerene structures attached with different types of chemical groups such as forming micellar structures of fullerene rich core and outer shell of polyethylene glycol [128]. Attachment of different functional groups to the fullerene groups can influence the properties of fullerene and exhibit antimicrobial, antioxidant, antineoplasmatic and several other activities [129]. There has been constant advancement in the field of nanotechnology. One such is Core–Shell Nanoparticles where core–shell material made up of nanoparticles such as core–shell Pb@GaS nanostructures. Gallium sulfide, an inorganic material when layered together can form make nanoparticles and nanotubes which are alike to fullerene morphology. These kinds of materials can possess excellent characteristics like optical, mechanicals and electrical properties [130].

In the field of sports nutrition, fullerenes have many applications specially in the benefits of hydration of athletes. Fullerenes are of hydrophobic nature but several advancements have led to innovations of fullerene for newer approaches. Conjugation of fullerene with cyclodextrin and isosteric acid, encapsulation of fullerene with water-soluble structures like polyvinylpyrrolidone nanoparticles and modification with various functional groups and form water-soluble fullerene like of carboxyfullerenes and carboxyfullerenes. These modifications give structural stability to fullerenes which increase its bioavailability and helps to interact with biological environment [131]. Fullerene is biological system can reduce reactive oxygen species which is a major reason for muscle soreness in athletes [132]. Fullerenes in hydrosoluble form as functional beverages can be given to athletes to provide daily requirement of vitamins and minerals and also as a mode of hydration [133].

2.10 Dendrimers

Dendrimers are a set of synthetic macromolecules which are made up of series of branches from the inside core to form new polymeric materials [134, 135]. These branches form a unique three-dimensional architectural structure with properties like uniformity in terms of size, molecular weight, degree of branching, hydrophilic and internal hollow spaces. These properties make them an excellent candidate for several industrial, pharmaceutical and biomedical applications [135, 136]. The structure of the dendrimers is a highly organized unique structure made up of three different sections- the inner shell, the outer shell and a symmetric inside core [137]. The inner shell is the layer of branches that migrate from the inner core. The outer shell comprises of the outer layer of branches. These branches mostly contain various functional groups that helps in involves in imparting different potential benefits. The branches extending from core–shell engages with monomers which have two dormant or any reactive molecules [137, 138]. Due to these structural properties, they are ideal for targeted drug delivery, gene therapy and molecular detectors in biological systems [139].

Based on the structure and composition, dendrimers can be differentiated into various types such as polyamidoamine and polypropyleneimine dendrimers. They are one of the most extensively studied dendrimers. Polyamidoamine dendrimers contains ethylenediamine core and amide bonds which helps the dendrimers to form linkage with drugs and facilitate in drug delivery. Polypropylenimine dendrimers contains core made up of diaminobutane and branches with propyleneimine. Due to its higher surface reactivity property, it is widely used as catalyst [140]. Multiple Antigen Peptide Dendrimers has a core made of lysine amino acid and it has peptide branches which extends outwards. These have several biological applications including development of vaccines [141]. Hybrid dendrimers are complex structures formed by combination of different dendrimers to impart involved characteristics. They can be paired with different drug delivery systems to form novel compounds with various properties such has controlled release of genes. Liquid crystalline dendrimers have a skeleton, also known as mesophage, has rod shaped (calamitic) and disc shaped (discotic) molecules which helps in formation of Nanoparticle interfacial layer [142]. Chiral dendrimers have a chiral core which helps in recognizing chiral molecule and helps in different catalysis process and several drug delivery systems. In electrokinetic chromatography, it helps to isolate optically active molecules for further analysis. Whereas Fréchet-type Dendrimers have a core of poly-benzyl ether with carboxylic group branches which helps extensively in drug delivery [140, 142]. Poly (carbosilane) dendrimers are a type of hydrophobic dendrimers made of silicon and carbon atoms. These compounds are formed due to hydrosilylation and alkenylation reactions. Variation of functional groups with dendrimers introduces more versatile compounds. This helps in several biomedical and pharmaceutical applications. Cationic dendrimers act as antibacterial agents and helps in gene therapy, anionic and glycocarbosilane dendrimers acts as antivirals, metallodendrimers and polyphenoxo dendrimers acts as anti-cancer agents. Heterofunctional dendrimers acts as excellent drug delivery systems for delivering drugs like ibuprofen and penicillin [143].

Dendrimers have several roles in biomedical and sports field. Dendrimers helps to increase the bioavailability of nutrients and therefore increases their absorption in biological systems. Due to hydrophilic properties, it can help in hydrophobic nutrient delivery in case of aqueous conditions [144, 145]. Figure 1 describes The science of nano-enhanced hydration in sports nutrition.

3 Relevance of Nano-Enhanced Hydration in Sports Nutrition

Nanoparticles used with the Nano-hydration methods are emerging in sports nutrition. These methods can improve water absorption, maximize hydration levels in the body and enhance overall athletic performance. Nanotechnology is used in food and nutrition science to augment the bioavailability of bioactive food components, thus enhancing water absorption and optimizing water levels in athletes. Nanotechnology is a more well-organized and effective strategy for supplying nutrients straight to the cellular level and provides a great approach to enhancing water balance in the body delivery at the cellular level. Nanotechnology provides a higher approach to hydration, supporting improved performance and promoting faster recovery during

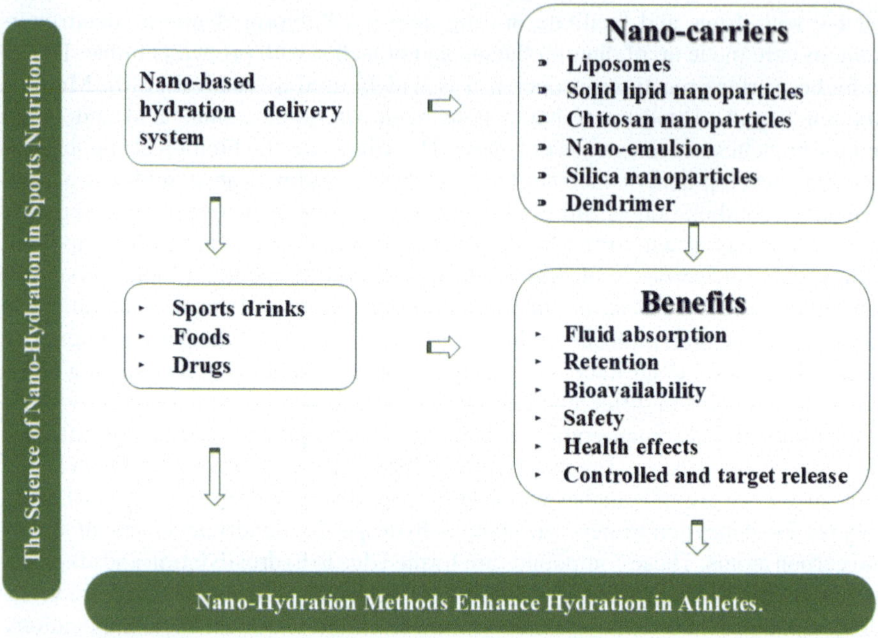

Fig. 1 The science of nano-enhanced hydration in sports nutrition [146]

physical activities or sports performance [147]. Nanoparticles play a crucial role in enhancing the bioavailability of nutrients by protecting them from the negative environment of the stomach and ensuring their maintained delivery in the intestines. This procedure will have more effective absorption and utilization by the body [148]. Also, nanoparticles can be assembled to target the delivery of nutrients to specific body parts, which improves their efficacy in reducing wastage [149]. The encapsulation method within nanoparticles can enhance nutrient stability. It also protects them from deterioration of nutrients and safeguards them from remaining active until arriving at their mark target. Through maintained delivery, nanotechnology provides a steady supply of nutrients for a long duration, which is important for balancing energy levels and promoting recovery in pre-, during, and post-vigorous physical activities or sports performance [150]. In sports nutrition, with the help of nanotechnology, has various key applications. It can enhance electrolyte and fluid balance in the body with the help of the encapsulating electrolytes method, ensure adequate water absorption and prevent dehydration and muscle pulls during long-duration activities [151]. Furthermore, nanotechnology facilitates energy replenishment by enhancing the transfer of energy-boosting nutrients, such as carbohydrates and amino acids, ensuring their availability to maintain sports performance and promote recovery [152]. At last, nanotechnology helps with sufficient recovery by enhancing the target delivery of nutrients that aid in muscle repair and reduce oxidative stress, ultimately reducing recovery time and avoiding the risks of injuries during activities [153].

Table 1 describes Comparative Advantages of traditional hydration methods and nano-enhanced hydration methods in sports nutrition.

3.1 Impact of Nano-Enhanced Hydration on Athletes in Relation to the Practice of Sports Nutrition

Nano-hydration, it is an innovative and sophisticated approach in the context of sports nutrition, uses nanotechnology to enhance methods that aim at improving athletic performance. This method helps engineering the electrolyte molecules, fluid carriers, and water particles in nano-size to increase their bioavailability, retention, and absorption are boots, these things highly affect and influence athletic or sports performance.

The following sections explain the impacts of nano-hydration on sports performance.

3.1.1 Enhancement of Endurance

Athletes who utilize nano-hydration methods often exhibit countless benefits, including endurance in sports activities. Those methods provide significant advantages for athletes in their performance by enhancing fluid retention, improving body thermal regulation, and keeping plasma volume. These benefits collectively deliver belated fatigue onset and sustained performance during prolonged activities, making nano-hydration an important strategy in sports nutrition [163]. Nano-hydration strategies, such as the use of glycerol, make an osmotic drive that improves fluid retention, deferring hydration and maintaining plasma volume during exercise [164]. This can enhance endurance time and overall activities or performance. Studies have released that nutritional nano-sized particles, such as resveratrol-load lipid nanoparticles, it can enhance endurance by improving mitochondrial function and quality, developing muscle and hepatic glycogen levels, and maintaining muscle integrity [165]. Nano-carriers increase sports performance by delivering steady energy transfer, which offers electrolyte balance and also transfers antioxidants effectively [156]. They encapsulate carbohydrates and amino acids, promising a gradual energy supply during long sports activities. They also help balance essential electrolytes for better muscle functions and endurance [166]. And it efficiently transfers antioxidants and delays oxidative stress and inflammations, which can delay performance. These benefits improve energy management and stability in athletes [156].

Table 1 Comparative advantages of traditional hydration methods and nano-enhanced hydration methods in sports nutrition

Aspects	Traditional hydration methods	Nano-Enhanced hydration methods	Impact on sports performance	Reference
Hydration efficiency	Traditional hydration methods rely on consuming plain water and electrolytes frequently, which might not fully meet individual wants and needs during intense sports	Nano-hydration techniques improve bioavailability and fluid absorption, directing to sufficient hydration balance in the body	With the help of nano-hydration methods, enhanced hydration status and performance along with maintained better fluid balance in the body during activity	[18, 111, 154–156]
Fluid balance	Tradition hydration methods many times involve consuming water and electrolyte solutions based on dehydration or scheduled intervals. More risks include hypohydration (dehydration) and hyperhydration (overhydration) both of which can weaken sports performance	Nano-enhanced hydration utilizes nano-sized molecules to improve fluid retention and absorption at cellular level. This way can potentially enhance the efficacy of water balance in the body. through this method, avoid both hydration risks	Sufficient hydration is crucial for balancing sports performance, as dehydration can damage muscular abilities, cognitive functions and overall endurance in athletes. The nano-enhanced hydration may lead to superior water balance maintenance, therefore reaching better sports performance outcomes	[11, 157, 158], (2014)
Nutrient delivery	Traditional sports drinks provide essential nutrients like protein, carbohydrates, and electrolytes, but their efficacy is limited by the rate of gastrointestinal absorption	Nano-hydration methods, like nano-encapsulation methods enhance nutrients delivery and stability, enhance bioactive compounds, improving overall nutrient uptake and sports performance	Improved nutrient compound uptake and utilization. increase performance and rapid recovery	[34, 154, 156, 159]

(continued)

Table 1 (continued)

Aspects	Traditional hydration methods	Nano-Enhanced hydration methods	Impact on sports performance	Reference
Health risks	Traditional hydration methods can cause dehydration or overhydration, that may reach to health problems such as gastrointestinal discomfort and hyponatremia	The nano-hydration strategy poses possible toxicity risks and biosecurity concerns with nano-materials, demanding special monitoring and further research	Nano-hydration is for cellular damage and oxidative stress if not completely managed. Need for new research on long-term effects	[155, 160–162]
Practicality and adoption	Traditional hydration methods are globally acceptable in different sports and easy to introduce with existing hydration products	Nano-hydration methods are offering but requirement new hydration product development and regulation support. This developing technology is obtaining interest and needs further advancement	Promising for improving sports performance and immediate recovery. Potential to set advanced standard in sports nutrition	[18, 155, 156]

3.1.2 Optimizing Body Thermoregulation and Control

Maintaining adequate hydration is very important for regulating body heat and it also plays a crucial role in regulating the body between intense physical activities. Dehydration can undesirably affect cardiovascular and thermoregulatory functions, reducing physical and mental performance [167]. The nano-hydration methods can help maintain the body temperature as they ensure sufficient blood flow to the skin, which is important for temperature regulation of the body, thereby preventing overheating and cardiovascular strain [168]. The body's ability to regulate its temperature directly affects sports performance and safety; overheating can cause fatigue, decrease efficiency, and even serious health risks like heatstroke. Nano-hydration can help address these challenges by enhancing the utilization of water and retention within the body. The nano-sized water particles facilitate rapid water absorption and distribution throughout the cells, ensuring that the body retains sufficient water levels for thermoregulation. By ensuring better hydration, nano-hydration supports maintaining lower core body heat during physical activities, thereby improving performance and reducing the risk of the body overheating [169]. Moreover, nano-hydration decreases the physiological strain associated with heat stress and encourages sustained activity by maximizing energy expenditure. By enhancing the efficiency of thermoregulatory processes, this new hydration method minimizes the energy loss usually required for heat regulation. Consequently, athletes benefit from enhanced physiological balance, increased endurance capacity in prolonged activities and accelerated recovery rates. These benefits underscore the crucial role of nano-hydration as an advancive innovation in sports nutrition and an essential tool for increasing sports performance.

3.1.3 Enhances Cellular Hydration

Cellular hydration plays an important role in a healthy body and is a vital factor in maximizing sports performance. The technology leverages nano-sized water particles that can effectively and efficiently penetrate cell membranes and simplify faster and more effective hydration at the levels of cellular. During physical activities, cellular hydration is critical for balancing physiological functions [157]. Proper water balance in the body is essential for maintaining heart rate, maximum oxygen volume (VO2MAX), and overall performance. Sufficient hydration supports prevent problems like hyponatremia, early exhaustion and fatigue [158]. Nano-hydration could potentially improve these benefits by warranting better water retention and absorption at the cellular level. Nanoparticles achieve a considerable increase in the efficacy of cellular uptake, for example, the presence of a hydrate gel-like pericellular matrix (PCM) on the cell surface can catch and store nanoparticles, increasing their cellular uptake efficacy through receptor-mediated endocytosis [170]. Moreover, nanoparticle surfaces are very famous for improving the cellular uptake of difference vectors, such as nanoparticles, mainly through the process of micropinocytosis. Traditional hydration techniques often face more challenges with retention and absorption, especially under conditions of high physical activity or sports performance. By enhancing cellular uptake, nano-hydration technology may offer a more reliable solution for athletes. Particularly in maintaining hydration status and stopping dehydration-related performance declines.

3.1.4 Improve Electrolyte Balance

Electrolyte balance is very important for athletic performance and recovery due to its role in balancing nerve function, muscle functions and overall cellular activity. Electrolytes like sodium, potassium and calcium are important for nerve signal delivery and muscle contraction. Imbalances can lead to muscle pulls, weakness, and harm performance [171]. During intensive physical exertion, athletes mainly experience significant electrolyte loss through sweating. Electrolyte support regulates water balance within cells, which is vital in providing cellular function and overall homeostasis [172]. Traditional techniques of replenishing electrolytes through drinks or supplements can be applicable but may not always be effective in speed or extent of absorption at the cellular level.

Nanotechnology involves manipulating materials at the molecular or atomic level to make particles that are significantly smaller than those used in traditional [173]. By decreasing the size of water and electrolyte particles to the nanoscale, these particles can very effectively penetrate the cell membrane, which helps to improve absorption and utilization within the body [174]. Nanotechnology-enhanced solubility and hydration molecular dynamics simulations have exposed that confinement at the nanoscale can effectively affect the solubility and hydration of electrolytes, which is critical for their sufficient absorption [175]. In nanoscale environments, hydration shells form especially around ions, stabilizing them and enhancing their transport across cell membranes [174]. The usage of nanotechnology can improve the transportation of electrolyte properties, and it creates them more sufficient in maintaining hydration and electrolytes in the body [176]. Nano molecules can enhance the efficacy of electrolyte solutions by transferring more surface area for interactions, which can lead to rapid and effective absorption [177]. Nano-hydration solutions facilitate the transport of essential minerals and vitamins at the nano-sized molecules level, this process enhancing the absorbing rate of electrolyte molecules. This rapid absorption is important for maintaining electrolyte balance during physical activities or sports performance [178]. Proper hydration and electrolytes balance the body, which are very important for preventing muscular injuries and fatigue during sports activities. Hydration methods, including the use of nano-hydration solutions, can transfer and maintain physiological levels of plasma electrolytes, which are crucial for muscle functions and endurance. Sufficient hydration is connected to optimal sports performance. By ensuring rapid, sufficient transfer of electrolytes, nano-hydration solutions can help sports athletes maintain their performance level for a long period of time [179]. The interaction of electrolytes with water at the particle level can impact hydration properties, affecting athlete's performance stability during physical exertion [180]. Maintenance of electrolyte balance in the body is not only critical during physical activity but also for cognitive and physiological recovery [179]

3.1.5 Accelerates Recovery

Nano-hydration denotes advanced and innovative engineering in sports nutrition that provides sufficient hydration as per the need which in turn helps athletes to speed up the recovery process and enhance sports performance. Utilizing the specific properties of nanoparticles, these methods can notably enhance the hydration of the athletes and their athletic performance. Nano-hydration provides greater benefits for nutrition by increasing water balance in the body's efficacy and rapid recovery. Efficient hydration aids in recovery at the cellular level and mitigates both oxidative stress and inflammation. Effective hydration techniques involving water enriched with suitable nutrients and electrolytes help in this process [181]. Nano-hydration uses fluid with small nano-sized molecules to optimize water and nutrient absorption at the cellular level [182]. This strategy provides hydration efficiency and supplies nutrients to tissue, providing faster recovery. It also helps for water retention by incorporating maintained levels of electrolytes and low carbohydrate concentrations, which can enhance fluid retention and maintain plasma volume-a crucial factor for better recovery [182]. The usages of the nano-technology in hydration and nutrient molecules supply accelerate the transfer of essential properties to muscles and tissues, facilitating immediate recovery [157]. By ensuring proper hydration nutrient supply, nano-hydration enables athletes' peak performance, preventing dehydration-related problems and enhancing overall strength and endurance [17].

3.1.6 Improves Cognitive Function

Dehydration can damage cognitive functions and contain vigilance and attention. Studies have revealed that mild dehydration can lead to confusion, cognitive loss and fatigue in adults, which can be preserved for several hours post-exercise [183]. Athletes maintain who euhydration show better attention and vigilance compared to those who are dehydrated [184]. Dehydration harmfully affects decision-making abilities. Research employing that experiencing dehydration shows impaired decision-making and cognitive functions, mainly under heat stress conditions [185]. Nano-hydration, which contains the use of innovative hydrations methods and technologies can potentially improve cognitive functions and reduce errors during sports performance. Figure 2 depicts the effect of Nano-Enhanced Hydration. The common principles of balancing maximal hydration suggest various benefits:

- *Decreasing Cognitive Errors*: by safeguarding athletes endure well-hydrated nano-hydration could support for decrease cognitive errors and enhance overall sports performance, mainly in high-stress and high-temperature conditions [186].

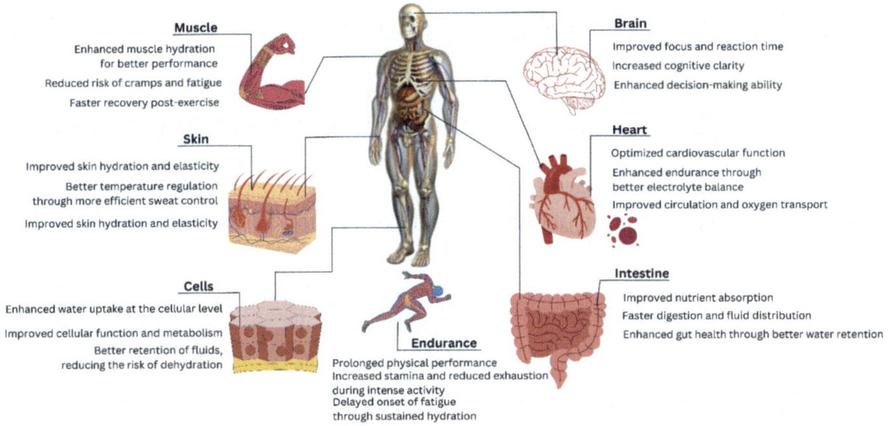

Fig. 2 Effect of nano-enhanced hydration

- *Improved Absorption*: Enhanced water absorption at the cellular level could support maintaining water balance more efficiently, hence helping cognitive functions like, decision-making and focus [187].
- *Minimized Cognitive Impairment*: Innovative hydration methods can help athletes avoid the harmful cognitive effects associated with dehydration, Like impaired reaction time abilities [188].
- *Decrease Oxidative Stress*: Nano-emulsions of centella asiatica (NanoSECA) have been shown to improve memory and mental functions in rats by enhancing acetylcholine and antioxidant levels while decrease oxidative stress [189].
- *Enhanced Mental Clarity*: Hydration favorably influences mood, which can indirectly improve mental function by delaying fatigue and enhancing overall mental clarity [186].

4 Practical Applications of Nano-Enhanced Hydration in Sports Nutrition

Nano-hydration is an advanced technology that uses super tiny molecules to improve how fluids and nutrients are distributed to the body. This method is becoming more popular among professional athletes because it helps them perform better during intense activities and promotes recovery faster. It offers a more efficient and targeted way to keep hydration for a longer duration, thereby solving many of the fluid-related problems that traditional hydrational methods often fail. Here are some products found in sports drinks that made using nanomaterials.

4.1 Buriti Oil Isotonic Beverages

Bruit oil is a natural element that is used for making isotonic drinks that enhance health and make them appealing by replacing artificial dyes with different colors. Its complete antioxidants, it helps to protect the body from injury from exercise or sports activities. The healthy fats in buriti oil also help to make better skin condition, enhance the overall the immune system of the body and provide more energy during prolonged activities. Buriti oil works well with help of elements such as sodium, potassium to make a balanced hydration drink without includes electrolytes. This builds isotonic beverages a better choice for those looking for natural. Healthy options that hydrate and help better heath.

4.2 Nano-Emulsion Functional Drink

Nano-hydration method products utilize small droplets (20–200 nm) to enhance hydration by creating drinks that help the body attract nutrients better. It is very important for sportspersons, as maintaining water balance enhances better performance and faster recovery. Nano-emulsion drinks hold the ability to support a supply of electrolytes, vitamins, and antioxidants functioning better by encapsulating them in tiny droplets. These beverages rapidly replace fluids and electrolytes missing during activities [190]. These drinks are tastier than traditional methods and last longer, making them a consistent choice for athletes during exercise and competitions. Nano-hydration products ensure the growing demand for healthy drinks in sports nutrition for better hydration.

4.3 Hydratable Nanocarrier Powders

Hydratable nano-carrier powders rapidly become stable, non-toxic carriers when mixed with water. These carriers come in different shapes and forms, such as tiny ball shapes or tubes, which support carrying essential nutrients or medicines to the right parts of the body. The powders use different ingredients, such as polymers and carbohydrates. That helps them to stay stronger and work better at supplying these nutrients. When mixing the powder with fluid it also supports the body to absorb water better and stay water balanced in the body. These methods are useful for athletes to stay hydrated during, before and after. They also help faster recovery from sports injuries and better performance during sports [191].

4.4 Nano-Hydroxyapatite in Sports Drinks

Hydroxyapatite (HA) is not used directly to hydrate the body. This indirect effect plays a supporting role in hydration in the body by maintaining the mineral balance and overall physical condition necessary for effective water balance in the body and sports performance. Hydroxyapatite is a naturally occurring mineral in the nano-sized particles, indicating greater solubility and bioavailability. Adding nan-HA to sports drinks enables the release of calcium and phosphate ions, which can enhance the sports drink's efficacy in replenishing essential elements or minerals lost during sports activities from the body. Also, nano-HA can improve hydration by providing a more balanced electrolyte profile, making them a better option for optimal athletic performance. Fruit juices are popular drinks as they contain antioxidants, vitamins, and minerals that are essential for human being and play important role in the prevention of heart diseases, cancer, and diabetes. They contain essential nutrients which support the growth of acid tolerant bacteria, yeasts, and molds. In the present study, we have conducted a microbiological examination of freshly prepared juices (sweet lime, orange, and carrot) by serial dilution agar plate technique. A total of 30 juice samples were examined for their microbiological quality. Twenty-five microbial species including 9 bacterial isolates, 5 yeast isolates, and 11 mold isolates were isolated from juices. Yeasts and molds were the main cause of spoilage of juices [192].

4.5 Silica Nanoparticles Mixed in Fruit Juice

The aims of fruit juice concentration are not only to avoid microbiological spoilage and chemical variations but also to provide health and sport-related considerations like nutritional values and sports performance. Concentrated fruit juices are usually utilized as a source of vitamins, minerals and natural sugar in sports beverages and recovery drinks ration of silica nanoparticles in the polymer membrane utilized for the reverse osmosis (RO) system can have extra health benefits. Silica nanoparticles have been explored for their capacity to improve nutrient absorption and provide cellular health. Using these nano-sized particles in the membrane helps the concentration procedure not only attain the anticipated thickness but also has the potential to increase the bioavailability of the nutrients present in the peach juice. The usage of RO and incorporation of silica nano-sized particles in fruit extract concentration have implications for both health and sport, these ensuring a means to preserve the nutritional value of the juice through supplying to the needs of players and individuals involved in physical activities. Fruit juice can offer better hydration and carbohydrates for athletes, helping their health and athletic performance [193].

5 Potential Concerns and Safety Regulations

The use of nanotechnology in sports nutrition and other applications arrives with different potential concerns and challenges. This method's major problem is toxicology and safety. Nano-sized particles can act unpredictably when they are introduced into the human body [194]. These small particles may interact with human cells and tissues in ways that could face health difficulties and require severe safety studies to confirm their safety for human use; it is also critical to conduct a wide-ranging study to confirm their safety for life. An additional challenge is directing regulatory hurdles. Therefore, nano-hydration products must meet safety standards and obtain approvals from significant authorities [195]. This procedure often takes more time and is complicated, delaying advance and availability. Furthermore, public perception plays a very important role in these methods; many people lack knowledge about nanotechnology and its effect on the body and may have a misunderstanding about its safety and health benefits. This distrust often leads to misinformation, highlighting the need for knowledge initiatives to inform the public about the science behind nanotechnology or nano-hydration methods and their potential benefits [196]. Building trust through clean and clear communication is vital for promoting the acceptance and utilization of nanotechnology products. Enabling these challenges efficiently requires collaboration among exporters, such as regulatory agencies, industry leaders, and public health advocates, to make sure that nanotechnology is both safe and understood by the community [197]. This nanotechnology provides innovative solutions for enhancing hydration and nutrition, improving nutrient transfer, absorption, retention, recovery and enhancing athletic performance [198].

6 Emerging Recommendations for Nano-Enhanced Hydration in Sports Nutrition

6.1 Innovative Research Trends in Nano-Enhanced Hydration

Recent findings in nanotechnology, which is a faster developing field, have shown promising ability in enhancing hydration strategies specially designed for athletes. Nanoparticles, due to their distinctive physical and chemical properties that are unique to their scale, have been exhibited to prominently improve both the bioavailability and overall efficiency of hydration solutions. For example, nano-encapsulated compounds such as resveratrol-loaded lipid nanoparticles and iron-loaded liposomes have been demonstrated to enhance performance in endurance exercises by increasing antioxidant activity, reducing oxidative stress, and thereby helping maintain muscle integrity [156].

6.2 Broader Application and Use

The application of nanotechnology in sports nutrition, which has been an edge-cutting development, extends beyond hydration to also encompass overall performance improvement as well as recovery. Nanoparticles, which are incredibly small and versatile, can be unitized to transport vital minerals and nutrients in a manner that is more efficient, thereby talking challenges such as the lower bioavailability and higher excretion rates that are usually observed with traditional supplements or methods [18]. For example, nano-trace minerals, including zinc, copper, and selenium, have demonstrated significant potential in improving gut health and enhancing the performance of poultry, findings that they can be thoughtfully extrapolated to applications in the human body [18]. Not only that, the integration of nano-materials into sports equipment can serve to improve athletic performance [199].

6.3 Evolving Hydration Science in Sports Nutrition

Innovates in hydration science through nano-used technology focus on maximizing water balance and electrolyte management. Studies have employed that nano-enhanced hydration method solutions can enhance the body's capacity to balancing plasma electrolyte levels and improve water retention and absorption [181]. For example, the use of highly alkaline fluid with nano-minerals has been found to enhance acid–base balance and aerobic exercise performance in athletes [200]. Moreover, the incorporation of the nano-silica in hydration solutions has been demonstrated to improve the hydration rate and enhance the nano-sized hydration products, promoting to faster and better hydration and performance.

7 Conclusion

This chapter has discussed broadly and extensively nanotechnology use with nano-enhanced hydration, which has revealed its outstanding potential to transform the science of hydration in revolutionary and deeply impactful ways. By leveraging the specific and unique properties of nano-size molecules, which allows for enhanced bioavailability and target delivery of essential nutrition, investigators have successfully paved the pathway for solutions that not only enhance water absorption but also indicate the specific physiological needs of athletes and individuals involved in intensive and demanding physical activities. These developments, which represent an important departure from the traditional hydration techniques, strongly emphasize the transformative potential of nano-used technology in overcoming persistent challenges such as inefficient absorption and inadequate management of electrolytes.

Furthermore, the promise of nano-enhanced hydration, which surely represents a paradigm shift in the field of hydration research, is to offer more precise, highly targeted, and increasingly successful techniques to preserve fluid balance and enhance athletic performance. A main area of technological advancement, combining nano-minerals, nano-silica, and other novel materials into hydration solutions, holds great promise for sports nutrition, which is also constantly developing with technological advancements.

Although the development that has been created thus far is definitely impressive. It's also clear that the field is still in the initial stages and requires ongoing effort to fully recognize the capacity of these advanced innovations. Future research studies are very important to focus on enhance these technologies, but also on thoroughly discovering their long-term effects and identifying new and advanced applications that could meaningfully enhance both human performance and recovery in meaningful ways.

In Summary, in this chapter nano-enhanced hydration methods represent much more than a scientific improvement in sports nutrition. They stand as a truly transformative tool that is uniquely controlled to redefine the understanding and the practical application of hydration science in significant ways. It is a developing process that efficiently bridges the gap that exists between advanced innovation and real-world practical use. It also holds the promise of making meaningful and considerable contributions to improving overall health and athletic performance.

References

1. Perales-García A, Estévez-Martínez I, Urrialde R (2016) Hidratación: determinados aspectos básicos para el desarrollo científico-técnico en el campo de la nutrición. Nutr Hosp. https://doi.org/10.20960/nh.338
2. Popkin BM, D'Anci KE, Rosenberg IH (2010) Water, hydration, and health. Nutr Rev 68(8):439–458. https://doi.org/10.1111/j.1753-4887.2010.00304.x
3. Wittbrodt MT, Millard-Stafford M (2018) Dehydration impairs cognitive performance: a meta-analysis. Med Sci Sports Exerc 50(11):2360–2368. https://doi.org/10.1249/MSS.0000000000001682
4. Montrose MH (2004) Small intestine, absorption and secretion. In: Encyclopedia of gastroenterology. Elsevier, pp 477–481. https://doi.org/10.1016/B978-0-12-812460-4.00650-9
5. Schiller LR (2014) Absorption and secretion of fluid and electrolytes. In: Gastrointestinal anatomy and physiology. Wiley, pp 92–107. https://doi.org/10.1002/9781118833001.ch7
6. Chohan AS, Davidow EB (2015) Clinical pharmacology and administration of fluid, electrolyte, and blood component solutions. In: Veterinary anesthesia and analgesia. Wiley, pp 386–413. https://doi.org/10.1002/9781119421375.ch21
7. Schrier RW (2011) The science behind hyponatremia and its clinical manifestations. Pharmacother J Hum Pharmacol Drug Ther 31(5P2). https://doi.org/10.1592/phco.31.5.9S
8. Schäfer C et al. (2007) Osmotic regulation of betaine homocysteine- S -methyltransferase expression in H4IIE rat hepatoma cells. Am J Physiol-Gastrointest Liver Physiol 292(4):G1089–G1098. https://doi.org/10.1152/ajpgi.00088.2006
9. Marino FE (2024) Evolution of the thirst mechanism in Homo: the need and limitations of thirst and hydration. Comp Biochem Physiol A Mol Integr Physiol 298:111745–111745. https://doi.org/10.1016/j.cbpa.2024.111745

10. Marinier FC, Buxeraud J (2024) L'importance de l'hydratation sur la performance sportive. Actual Pharm 63(637):37–41. https://doi.org/10.1016/j.actpha.2024.04.018
11. Belval LN et al (2019) Practical hydration solutions for sports. Nutrients 11(7):1550–1550. https://doi.org/10.3390/nu11071550
12. Priya KV (2022) Analysis of hydration level estimation strategies using deep learning. In: Presented at the 2022 6th international conference on electronics, communication and aerospace technology, IEEE, pp 872–877. https://doi.org/10.1109/ICECA55336.2022.10009296
13. Sharp RL (2007) Role of whole foods in promoting hydration after exercise in humans. J Am Coll Nutr 26(sup5):592S-596S. https://doi.org/10.1080/07315724.2007.10719664
14. Heuberger MP, Zachariah Z, Spencer ND, Espinosa-Marzal RM (2017) Collective dehydration of ions in nano-pores. Phys Chem Chem Phys 19(21):13462–13468. https://doi.org/10.1039/C7CP01439B
15. Saharkhiz S, Zarepour A, Nasri N, Cordani M, Zarrabi A (2023) A comparison study between doxorubicin and curcumin co-administration and co-loading in a smart niosomal formulation for MCF-7 breast cancer therapy. Eur J Pharm Sci 191:106600–106600. https://doi.org/10.1016/j.ejps.2023.106600
16. Ali HS, Hanafy A, Alqurshi A (2019) Engineering of solidified glyburide nanocrystals for tablet formulation via loading of carriers: downstream processing, characterization, and bioavailability. Int J Nanomedicine 14:1893–1906. https://doi.org/10.2147/IJN.S194734
17. Kong F, Ren S (2022) Analysis on the influence of nanotechnology development on sports health industry. Int J Nanotechnol 19(6/7/8/9/10/11):1034–1034. https://doi.org/10.1504/IJNT.2022.128986
18. Hassan S, Hassan F, Rehman MS (2020) Nano-particles of trace minerals in poultry nutrition: potential applications and future prospects. Biol Trace Elem Res 195(2):591–612. https://doi.org/10.1007/s12011-019-01862-9
19. Beltrami FG, Hew-Butler T, Noakes TD (2008) Drinking policies and exercise-associated hyponatraemia: is anyone still promoting overdrinking? Br J Sports Med 42(10):796–801. https://doi.org/10.1136/bjsm.2008.047944
20. Galaz GA (2013) An overview on the history of sports nutrition beverages. In: Nutrition and enhanced sports performance. Elsevier, pp 205–210. https://doi.org/10.1016/B978-0-12-396454-0.00020-5
21. Jonvik KL, King M, Rollo I, Stellingwerff T, Pitsiladis Y (2022) New opportunities to advance the field of sports nutrition. Front Sports Act Living 4. https://doi.org/10.3389/fspor.2022.852230
22. Close GL, Kasper AM, Morton JP (2019) From paper to podium: quantifying the translational potential of performance nutrition research. Sports Med 49(S1):25–37. https://doi.org/10.1007/s40279-018-1005-2
23. Gopi Krishna P, Kameswaran S, Sri Ranjani T, Gunavathi Y (2020) Recent developments in nanoencapsulation of bioactive compounds of microbial sources and their biomedical applications. Recent Dev Appl Microbiol Biochem 2:141–152. https://doi.org/10.1016/B978-0-12-821406-0.00014-X
24. Vo D-K, Trinh KTL (2024) Advances in wearable biosensors for healthcare: current trends, applications, and future perspectives. Biosensors 14(11):560–560. https://doi.org/10.3390/bios14110560
25. Lorin A, Flore C, Thomas A, Brasseur R (2004) Liposomes: description, preparation and applications. Biotechnol Agron Soc Environ 8(3):163–176. Available: https://www.scopus.com/inward/record.uri?eid=2-s2.0-80155167053&partnerID=40&md5=859ae785221bdfc0499f7e6dd219c8f7
26. Kaur S, Singh D (2023) A sojourn on liposomal delivery system: recent advances and future prospects. Assay Drug Dev Technol 21(2):48–64. https://doi.org/10.1089/adt.2022.089
27. Shah SR, Mori DD, Prajapati B, Vyas AJ, Soniwala MM (2023) Liposomal drug delivery systems: concept and recent advances in clinical applications. In: Lipid-based drug delivery systems: principles and applications, pp 47–81. https://doi.org/10.1201/9781003459811-3

28. Dass CR (2008) Drug delivery in cancer using liposomes. Methods Mol Biol 437:177–182. https://doi.org/10.1007/978-1-59745-210-6_9
29. Sharma S, Islam A, Khan MM, Ahmad U (2024) Liposomes: history, properties, and applications. In: Liposomes: advances in research and applications, pp 1–32. [Online]. Available: https://www.scopus.com/inward/record.uri?eid=2-s2.0-85196246228&partnerID=40&md5=778a14adceee1886eaa2451029472c59
30. Nagarani P, Babu MK (2014) Liposomes on drug delivery system. J Chem Pharm Sci 7(3):205–209. Available: https://www.scopus.com/inward/record.uri?eid=2-s2.0-84929159420&partnerID=40&md5=5e65175dd50a2ebf156b0c32fd587d22
31. Bonde S, Nair S (2017) Advances in liposomal drug delivery system: fascinating types and potential applications. Int J Appl Pharm 9(3):1–7. https://doi.org/10.22159/ijap.2017v9i3.17984
32. Nsairat H, Khater D, Sayed U, Odeh F, Bawab AA, Alshaer W (2022) Liposomes: structure, composition, types, and clinical applications. Heliyon 8(5). https://doi.org/10.1016/j.heliyon.2022.e09394
33. Dhillon A, Singh R, Senwar KR (2024) An extensive review on novel liposomes: classification, methodology, characterization, current formulations. Int J Drug Deliv Technol 14(3):1842–1852. https://doi.org/10.25258/ijddt.14.3.83
34. Orrù S et al (2018) Role of functional beverages on sport performance and recovery. Nutrients 10(10):1470–1470. https://doi.org/10.3390/nu10101470
35. Sivakamasundari SK, Leena M, Moses JA, Anandharamakrishnan C (2020) Solid lipid nanoparticles: formulation and applications in food bioactive delivery. In: Innovative food processing technologies: a comprehensive review, pp 580–604. [Online]. Available: https://www.scopus.com/inward/record.uri?eid=2-s2.0-85102474453&partnerID=40&md5=e46d8c105db44718f1bc2b1dcffc3e7e
36. DhimanS, Singh TG, Asthana A, Arora S, Jindal M (2013) Solid lipid nanoparticles: a current approach to new drug-delivery systems in nanotechnology. In: Therapeutic delivery methods: a concise overview of emerging areas, pp 110–124. https://doi.org/10.4155/EBO.13.331
37. Naseri N, Valizadeh H, Zakeri-Milani P (2015) Solid lipid nanoparticles and nanostructured lipid carriers: structure preparation and application. Adv Pharm Bull 5(3):305–313. https://doi.org/10.15171/apb.2015.043
38. Souto EB, Fangueiro JF, Müller RH (2013) Solid lipid nanoparticles (SLN^{TM}). In: Fundamentals of pharmaceutical nanoscience, pp 91–116. https://doi.org/10.1007/978-1-4614-9164-4_5
39. Musielak E, Feliczak-Guzik A, Nowak I (2022) Optimization of the conditions of solid lipid nanoparticles (SLN) synthesis. Molecules 27(7). https://doi.org/10.3390/molecules27072202
40. Samal RP, Sahu PK (2021) Formulation and evaluation of solid lipid nanoparticle of felbamate for improved drug delivery. Res J Pharm Technol 14(1):285–288. https://doi.org/10.5958/0974-360X.2021.00051.2
41. Naga Sujan M, Patil AB, Gowda DV (2020) A review on methods of preparation and characterisation of the solid lipid nanoparticles. Res J Pharm Technol 13(7):3433–3441. https://doi.org/10.5958/0974-360X.2020.00610.1
42. Villafuerte Robles L, García BF, MDL Garzón S, Hernández AL, Vázquez MLR (2008) Solid lipid nanoparticles. Rev Mex Cienc Farm 39(1):38–52. [Online]. Available: https://www.scopus.com/inward/record.uri?eid=2-s2.0-42949165861&partnerID=40&md5=a31ee8d1f92040bc7f79e262fd9d41d7
43. Chauhan S, Nagpal K (2024) Latest delivery advancements of lipid nanoparticles for cancer treatment. Assay Drug Dev Technol 22(6):340–360. https://doi.org/10.1089/adt.2024.019
44. Battaglia L, Gallarate M (2012) Lipid nanoparticles: state of the art, new preparation methods and challenges in drug delivery. Expert Opin Drug Deliv 9(5):497–508. https://doi.org/10.1517/17425247.2012.673278
45. Shrivastava P, Gautam L, Jain A, Vishwakarma N, Vyas S, Vyas SP (2020) Lipid drug conjugates for improved therapeutic benefits. Curr Pharm Des 26(27):3187–3202. https://doi.org/10.2174/1381612826666200311124003

46. Irby D, Du C, Li F (2017) Lipid-drug conjugate for enhancing drug delivery. Mol Pharm 14(5):1325–1338. https://doi.org/10.1021/acs.molpharmaceut.6b01027
47. Vandervoort J, Van Den Bossche M, Ludwig A (2007) Solid lipid nanoparticles: possible applications. Farm Tijdschr Voor Belg 84(2):22–27 [Online]. Available: https://www.scopus.com/inward/record.uri?eid=2-s2.0-34447114628&partnerID=40&md5=a5eddaf2db1830ac057abc36bbdcf176
48. Murali M, Balamurgan SD, Venkatesh DN (2020) A brief review on recent advancement of solid lipid nanoparticle. Int J Pharm Res 12(3):1200–1207. https://doi.org/10.31838/ijpr/2020.12.03.174
49. Lu Y, Cheng D, Niu B, Wang X, Wu X, Wang A (2023) Properties of poly (Lactic-co-Glycolic Acid) and progress of poly (lactic-co-glycolic acid)-based biodegradable materials in biomedical research. Pharmaceuticals 16(3). https://doi.org/10.3390/ph16030454
50. Qian H, Wohl AR, Crow JT, MacOsko CW, Hoye TR (2011) A strategy for control of 'random' copolymerization of lactide and glycolide: application to synthesis of PEG- b -PLGA block polymers having narrow dispersity. Macromolecules 44(18):7132–7140. https://doi.org/10.1021/ma201169z
51. Mei F et al (2015) Synthesis and characterization of biodegradable poly(lactic-co-glycolic acid). J Macromol Sci Part B Phys 54(5):562–570. https://doi.org/10.1080/00222348.2014.1002325
52. Han FY, Thurecht KJ, Whittaker AK, Smith MT (2016) Bioerodable PLGA-based microparticles for producing sustained-release drug formulations and strategies for improving drug loading. Front Pharmacol 7. https://doi.org/10.3389/fphar.2016.00185
53. Kim YH, Byun J, Kim B, Lee K, Lee J-S, Kim TH (2019) One-Pot synthesis of highly monodisperse poly(lactic-co-glycolic acid) particles with controlled porosity as efficient drug delivery vehicles. Bull Korean Chem Soc 40(9):851–856. https://doi.org/10.1002/bkcs.11833
54. Fernando LP, Lewis JS, Evans BC, Duvall CL, Keselowsky BG (2018) Formulation and characterization of poly(propylacrylic acid)/poly(lactic-co-glycolic acid) blend microparticles for pH-dependent membrane disruption and cytosolic delivery. J Biomed Mater Res Part A 106(4):1022–1033. https://doi.org/10.1002/jbm.a.36298
55. McCall RL, Sirianni RW (2013) PLGA nanoparticles formed by single- or double-emulsion with vitamin E-TPGS. J Vis Exp JoVE 82:51015. https://doi.org/10.3791/51015
56. Lanao RPF, Jonker AM, Wolke JGC, Jansen JA, Van Hest JCM, Leeuwenburgh SCG (2013) Physicochemical properties and applications of poly(lactic-co-glycolic acid) for use in bone regeneration. Tissue Eng Part B Rev 19(4):380–390. https://doi.org/10.1089/ten.teb.2012.0443
57. Perera UMSP, Rajapakse N (2014) Chitosan nanoparticles: preparation, characterization, and applications. In: Seafood processing by-products: trends and applications, vol 9781461495901, pp 371–387. https://doi.org/10.1007/978-1-4614-9590-1_18
58. Ghattavi S, Homaei A (2024) Chitosan-based nanomaterials: structure, characterization, and applications. In: Chitosan-based hybrid nanomaterials, pp 47–71. https://doi.org/10.1016/B978-0-443-21891-0.00003-2
59. Purohit G, Rawat DS (2022) Characterization techniques for chitosan and its based nanocomposites. In: Chitosan-Based nanocomposite materials: fabrication, characterization and biomedical applications, pp 79–101. https://doi.org/10.1007/978-981-19-5338-5_3
60. Yhee JY, Koo H, Lee DE, Choi K, Kwon IC, Kim K (2011) Multifunctional chitosan nanoparticles for tumor imaging and therapy. Adv Polym Sci 243(1):139–162. https://doi.org/10.1007/12_2011_119
61. Onishi H (2010) Chitosan microparticles. J Drug Deliv Sci Technol 20(1):15–22. https://doi.org/10.1016/S1773-2247(10)50002-9
62. Singh K, Mishra A (2015) Chitosan nanoparticulate and their applications: a review. Int J Pharma Bio Sci 6(2):P557–P566. [Online]. Available: https://www.scopus.com/inward/record.uri?eid=2-s2.0-85051686274&partnerID=40&md5=a91ab84181855829f7b8c178b42f34d3

63. Jafari AM, Morsali A, Bozorgmehr MR, Beyramabadi SA, Mohseni S (2024) Modeling and characterization of lenalidomide-loaded tripolyphosphate-crosslinked chitosan nanoparticles for anticancer drug delivery. Int J Biol Macromol 260. https://doi.org/10.1016/j.ijbiomac.2024.129360
64. Hamidi M, Azadi A, Rafiei P (2008) Hydrogel nanoparticles in drug delivery. Adv Drug Deliv Rev 60(15):1638–1649. https://doi.org/10.1016/j.addr.2008.08.002
65. Koukaras EN, Papadimitriou SA, Bikiaris DN, Froudakis GE (2012) Insight on the formation of chitosan nanoparticles through ionotropic gelation with tripolyphosphate. Mol Pharm 9(10):2856–2862. https://doi.org/10.1021/mp300162j
66. Gonçalves ACR et al (2023) Improving chitosan properties through ionic and chemical cross-linking and their impact on emulsified systems. Int J Food Sci Technol 58(8):4324–4331. https://doi.org/10.1111/ijfs.16533
67. Mikušová V, Mikuš P (2021) Advances in chitosan-based nanoparticles for drug delivery. Int J Mol Sci 22(17):9652. https://doi.org/10.3390/ijms22179652
68. Popletaeva SB, Arslanova LR (2021) Use of chitosan nanoparticles loaded with biologically active substances for pre-harvest plant protection from pathogens (a review). J Phys Conf Series. https://doi.org/10.1088/1742-6596/1942/1/012077
69. Chuo SC, Mohd Setapar SH (2022) Application of nanoemulsion in cosmetics. In: Nanotechnology for the preparation of cosmetics using plant-based extracts, pp 355–371. https://doi.org/10.1016/B978-0-12-822967-5.00001-1
70. Goindi S, KaurA, Kaur R, Kalra A, Chauhan P (2016) Nanoemulsions: an emerging technology in the food industry. In Emulsions, pp 651–688. https://doi.org/10.1016/B978-0-12-804306-6.00019-2
71. Lu Y, Qi J, Wu W (2012) Absorption, disposition and pharmacokinetics of nanoemulsions. Curr Drug Metab 13(4):396–417. https://doi.org/10.2174/138920012800166544
72. Braccalenti E, Del Gaudio L, Belloni A, Albonico P, Radaelli E, Bartosek M (2017) Enhancing oil recovery with nanoemulsion flooding. In: presented at the offshore mediterranean conference and exhibition 2017, OMC 2017. [Online]. Available: https://www.scopus.com/inward/record.uri?eid=2-s2.0-85051526109&partnerID=40&md5=b69ddb77605b2f05ee335c39637284c6
73. Guttoff M, Saberi AH, Mcclements DJ (2015) Formation of vitamin D nanoemulsion-based delivery systems by spontaneous emulsification: Factors affecting particle size and stability. Food Chem 171:117–122. https://doi.org/10.1016/j.foodchem.2014.08.087
74. Singh Y et al (2017) Nanoemulsion: concepts, development and applications in drug delivery. J Controlled Release 252:28–49. https://doi.org/10.1016/j.jconrel.2017.03.008
75. Tabibiazar M, Hamishehkar H (2015) Formulation of a food grade water-in-oil nanoemulsion: factors affecting on stability. Pharm Sci 21(4):220–224. https://doi.org/10.15171/PS.2015.40
76. Negi A, Vanmathi Mugasundari A, Moses JA (2024) Nanoemulsion: a potential strategy toward edible coatings. In: Food coatings and preservation technologies, pp 339–365. https://doi.org/10.1002/9781394237623.ch10
77. Huang C et al (2017) Bicontinuous structured liquids with sub-micrometre domains using nanoparticle surfactants. Nat Nanotechnol 12(11):1060–1063. https://doi.org/10.1038/nnano.2017.182
78. Bahuguna A, Ramalingam S, Kim M (2020) Formulation, characterization, and potential application of nanoemulsions in food and medicine. In: Nanotechnology in the life sciences, pp 39–61. https://doi.org/10.1007/978-3-030-31938-0_3
79. Salvia-Trujillo L, Soliva-Fortuny R, Rojas-Graü MA, McClements DJ, Martín-Belloso O (2017) Edible nanoemulsions as carriers of active ingredients: a review. Annu Rev Food Sci Technol 8:439–466. https://doi.org/10.1146/annurev-food-030216-025908
80. Leng J, Li L, Zhao W, Bian Y (2025) Enhancing lipid digestion and absorption rate: the role of whey protein isolate short fibrils in emulsions. Int J Biol Macromol 285. https://doi.org/10.1016/j.ijbiomac.2024.138351
81. Gupta A et al. (2023) Trends in functional beverages: functional ingredients, processing technologies, stability, health benefits, and consumer perspective. Food Res Int 170. https://doi.org/10.1016/j.foodres.2023.113046

82. Pieszka M, Bederska-łojewska D, Szczurek P, Pieszka M (2019) The membrane interactions of nano-silica and its potential application in animal nutrition. Animals 9(12). https://doi.org/10.3390/ani9121041
83. Tahmasbi L, Abbasi Z (2024) Silica nanoparticles: synthesis, applications and environmental impact. In: Adv Nanotechnol 30:171–192. [Online]. Available: https://www.scopus.com/inward/record.uri?eid=2-s2.0-85200980466&partnerID=40&md5=cc818520d6f2d76a7f1eef368355597e
84. Petushkov A, Ndiege N, Salem AK, Larsen SC (2010) Toxicity of silica nanomaterials: zeolites, mesoporous silica, and amorphous silica nanoparticles. Adv Mol Toxicol 4(C):223–266. https://doi.org/10.1016/S1872-0854(10)04007-5
85. El Sayed S, Otri I (2023) Silica nanoparticles for sensing applications. In: Fundamentals of sensor technology: principles and novel designs, pp 591–630. https://doi.org/10.1016/B978-0-323-88431-0.00013-2
86. Padiyar N, Nainwal N, Ale Y, Galwan D (2024) Formulation, characterization and stability aspects of mesoporous silica nanoparticles. Int J Drug Deliv Technol 14(2):1217–1224. https://doi.org/10.25258/ijddt.14.2.88
87. Girija AR, Palaninathan V (2020) Multifunctional mesoporous silica nanoparticles for biomedical applications. In: Nano medicine and nano safety: recent trends and clinical evidences, pp 213–235. https://doi.org/10.1007/978-981-15-6255-6_9
88. Sábio RM, Meneguin AB, Ribeiro TC, Silva RR, Chorilli M (2019) New insights towards mesoporous silica nanoparticles as a technological platform for chemotherapeutic drugs delivery. Int J Pharm 564:379–409. https://doi.org/10.1016/j.ijpharm.2019.04.067
89. Bakhshian Nik A et al. (2020) Smart drug delivery: capping strategies for mesoporous silica nanoparticles. Microporous Mesoporous Mater 299. https://doi.org/10.1016/j.micromeso.2020.110115
90. Uamran DA, Ubaid QH, Humud HR (2021) Core-shell SiO_2/Ag composite spheres prepared by electrical exploding wire plasma technique: synthesis and characterization. NeuroQuantology 19(10):82–88. https://doi.org/10.14704/nq.2021.19.10.NQ21161
91. Vishwanath S, Kottantharayil A, Mallick S (2024) Development of anti-soiling coating for PV cover glass using hollow silica nanoparticles. In: Conference record of the IEEE photovoltaic specialists conference, pp 612–614. https://doi.org/10.1109/PVSC57443.2024.10749446
92. Mendez N et al (2017) Assessment of in vivo systemic toxicity and biodistribution of iron-doped silica nanoshells. Nanomedicine Nanotechnol Biol Med 13(3):933–942. https://doi.org/10.1016/j.nano.2016.10.018
93. Bao Y, Wu Y, Gao L (2022) Development on synthesis of dendritic mesoporous silica nanoparticles. Kuei Suan Jen Hsueh Pao J Chin Ceram Soc 50(9):2538–2550. https://doi.org/10.14062/j.issn.0454-5648.20211007
94. Xu C, Lei C, Wang Y, Yu C (2022) Dendritic mesoporous nanoparticles: structure, synthesis and properties. Angew Chem Int 61(12). https://doi.org/10.1002/anie.202112752
95. Malekmohammadi S et al. (2022) Nonordered dendritic mesoporous silica nanoparticles as promising platforms for advanced methods of diagnosis and therapies. Mater Today Chem 26. https://doi.org/10.1016/j.mtchem.2022.101144
96. Wang Y, Huang L, Li S, Liu C, He H (2024) The capture and catalytic conversion of CO_2 by dendritic mesoporous silica-based nanoparticles. Energy Environ Mater 7(2). https://doi.org/10.1002/eem2.12593
97. Paulkumar K, Arunachalam R, Annadurai G (2011) Biomedical applications of organically modified bioconjugated silica nanoparticles. Int J Nanotechnol 8(8–9):653–663. https://doi.org/10.1504/IJNT.2011.041436
98. Bonacchi S et al (2011) Luminescent silica nanoparticles: extending the frontiers of brightness. Angew Chem Int Ed 50(18):4056–4066. https://doi.org/10.1002/anie.201004996
99. Urdampilleta A, Gómez-Zorita S, Soriano JM, Martínez-Sanz JM, Medina S, Gil-Izquierdo A (2015) Hydration and chemical ingredients in sport drinks: food safety in the European context. Nutr Hosp 31(5):1889–1899. https://doi.org/10.3305/nh.2015.31.5.7867

100. Cervellino A, Giannini C, Guagliardi A, Zanchet D (2004) Quantitative analysis of gold nanoparticles from synchrotron data by means of least-squares techniques least-squares analysis of gold nanoparticles. Eur Phys J B 41(4):485–493. https://doi.org/10.1140/epjb/e2004-00342-3
101. Chhour P, Naha PC, Cheheltani R, Benardo B, Mian S, Cormode DP (2016) Gold nanoparticles for biomedical applications: Synthesis and in vitro evaluation. In: Methods in pharmacology and toxicology 39:87–111. https://doi.org/10.1007/978-1-4939-3121-7_5
102. Boruah JS et al (2021) Green synthesis of gold nanoparticles using an antiepileptic plant extract: in vitrobiological and photo-catalytic activities. RSC Adv 11(45):28029–28041. https://doi.org/10.1039/d1ra02669k
103. Mieszawska AJ, Mulder WJM, Fayad ZA, Cormode DP (2013) Multifunctional gold nanoparticles for diagnosis and therapy of disease. Mol Pharm 10(3):831–847. https://doi.org/10.1021/mp3005885
104. Ogarev VA, Rudoi VM, Dement'eva OV (2018) Gold nanoparticles: synthesis, optical properties, and application. Inorg Mater Appl Res 9(1):134–140. https://doi.org/10.1134/S2075113318010197
105. Gadge AN, Wadher SJ, Landge AD (2020) Gold nanoparticle 2
106. Jeong EH, Jung G, Hong CA, Lee H (2014) Gold nanoparticle (AuNP)-based drug delivery and molecular imaging for biomedical applications. Arch Pharm Res 37(1):53–59. https://doi.org/10.1007/s12272-013-0273-5
107. Iravani S, Soufi GJ (2019) Gold nanostructures in medicine and biology. In: Nanoparticles in medicine, pp 175–183. https://doi.org/10.1007/978-981-13-8954-2_6
108. Rizwan Younis M, He G, Gurram B, Lin J, Huang P (2021) Recent advances in gold nanorods-based cancer theranostics. Adv NanoBiomed Res 1(12). https://doi.org/10.1002/anbr.202100029
109. Jeong H-H, Choi E, Ellis E, Lee T-C (2019) Recent advances in gold nanoparticles for biomedical applications: FRom hybrid structures to multi-functionality. J Mater Chem B 7(22):3480–3496. https://doi.org/10.1039/c9tb00557a
110. Mohapatra S, Nguyen TA, Nguyen-Tri P (2019) Noble metal-metal oxide hybrid nanoparticles: fundamentals and applications. In: Noble metal-metal oxide hybrid nanoparticles: fundamentals and applications, p 659. https://doi.org/10.1016/C2017-0-00847-9
111. Zhou Y et al (2016) Real-time colorimetric hydration sensor for sport activities. Mater Des 90:1181–1185. https://doi.org/10.1016/j.matdes.2015.06.078
112. Krishnamoorthy P, Jayalakshmi T (2012) Preparation, characterization and synthesis of silver nanoparticles by using phyllanthusniruri for the antimicrobial activity and cytotoxic effects. J Chem Pharm Res 4(11):4783–4794. [Online]. Available: https://www.scopus.com/inward/record.uri?eid=2-s2.0-84878321183&partnerID=40&md5=064694997 57a6355154356c35456a417
113. Mikhailov OV, Mikhailova EO (2019) Elemental silver nanoparticles: biosynthesis and bio applications. Materials 12(19). https://doi.org/10.3390/ma12193177
114. Grishina YS, Borgardt NI, Volkov RL, Gromov DG, Savitskiy AI (2019) Electron microscopy study of silver nanoparticles obtained by thermal evaporation. Semiconductors 53(15):1986–1991. https://doi.org/10.1134/S1063782619150089
115. Karmakar B (2016) Functionality of reversible glass nanocomposites and their applications. In: Glass nanocomposites: synthesis, properties and applications, pp 265–278. https://doi.org/10.1016/B978-0-323-39309-6.00011-0
116. Mohamed DE, Mohamed AS, El-Dib FI (2022) Role of mixed surfactants system in preparation of silver nanoparticles. Tenside Surfactants Deterg 59(5):400–408. https://doi.org/10.1515/tsd-2022-2421
117. Hu Q, Wu C (2020) Brief introduction and applications on synthesis of silver nanoparticles by chemical reduction. Gongneng Cailiao J Funct Mater 51(4):04046–04053. https://doi.org/10.3969/j.issn.1001-9731.2020.04.008
118. Shydlovska O, Kharchenko Y (2022) Review of green methods of synthesis of silver nanoparticles. In: ICAMS proceedings of the international conference on advanced materials and systems, pp 61–64. https://doi.org/10.24264/icams-2022.I.8

119. Altemimi AB et al (2024) Application of nanoparticles in human nutrition: a review. Nutrients 16(5):636. https://doi.org/10.3390/nu16050636
120. nanoSPACE (2025) Everything you should know about silver nanoparticles. Nanospace.store. Accessed: 04 Jan 2025. [Online]. Available: https://www.nanospace.store/blog/silver-nanoparticles/
121. Bhakta P, Barthunia B (2020) Fullerene and its applications: a review. J Indian Acad Oral Med Radiol 32(2):159–163. https://doi.org/10.4103/jiaomr.jiaomr_191_19
122. Kroto et al (1985) C60: Buckminsterfullerene [Online]. Available: https://www.nature.com/articles/318162a0#citeas
123. Madkour LH (2024) Fullerene-based modifier: biosensing and nanobiotechnological industrial applications. In: Emerging engineering technologies and industrial applications, pp 59–96. [Online]. Available: https://www.scopus.com/inward/record.uri?eid=2-s2.0-85194715052&doi=10.4018%2f979-8-3693-1335-0.ch004&partnerID=40&md5=9e9bc89b5582ba822b3413e4930ac2fb
124. Chae S-R, Hotze EM, Wiesner MR (2014) Possible applications of fullerene nanomaterials in water treatment and reuse. In: Nanotechnology applications for clean water: solutions for improving water quality: second edition, pp 329–338. [Online]. Available: https://www.scopus.com/inward/record.uri?eid=2-s2.0-84941767779&doi=10.1016%2fB978-1-4557-3116-9.00021-4&partnerID=40&md5=97c5d77068f6637d3a121a7faaf6cc49
125. Ellis SB (2014) Fullerenes: chemistry, natural sources and technological applications. In: Fullerenes: chemistry, natural sources and technological applications. [Online]. Available: https://www.scopus.com/inward/record.uri?eid=2-s2.0-84958657157&partnerID=40&md5=a1a1b1eab1aadc54d2174d6ab87cec47
126. Bhaskar et al. (2022) Self-Assembled fullerene nanostructures: synthesis and applications. [Online]. Available: https://onlinelibrary.wiley.com/action/showCitFormats?doi=10.1002%2Fadfm.202106924
127. Sáenz A, López LI, Castañeda A, Farías L (2014) Chemistry of fullerene. In: Fullerenes: chemistry, natural sources and technological applications, pp 189–215. [Online]. Available: https://www.scopus.com/inward/record.uri?eid=2-s2.0-84958609703&partnerID=40&md5=cbf1bd4e70dd8df2d924452ea93d294a
128. Eom T, Barát V, Khan A, Stuparu MC (2021) Aggregation-free and high stability core-shell polymer nanoparticles with high fullerene loading capacity, variable fullerene type, and compatibility towards biological conditions. Chem Sci 12(13):4949–4957. https://doi.org/10.1039/d1sc00602a
129. Babavea E, Bibli SI, Tsantili-Kakoulidou A (2008) Nanomaterials: fullerenes and carbon nanotubes. Structure, physical and chemical properties-biological and therapeutic applications. Pharmakeftiki 21(1):10–21. [Online]. Available: https://www.scopus.com/inward/record.uri?eid=2-s2.0-77953415813&partnerID=40&md5=eb659360c7d06fc7c067ef5edad0a2fb
130. Brontvein O et al. (2017) Synthesis and characterization of Pb@GaS core-shell fullerene-like nanoparticles and nanotubes. Nano 12(3). https://doi.org/10.1142/S1793292017500308
131. Djekic L, Primorac M (2014) Biomedical application of fullerenes. In: Fullerenes: chemistry, natural sources and technological applications, pp 239–261. [Online]. Available: https://www.scopus.com/inward/record.uri?eid=2-s2.0-84958657110&partnerID=40&md5=f6044baf9bd4e501f1b07fed1f13c712
132. Iohara D (2019) Preparation and evaluation of fullerene based nanomedicine. Yakugaku Zasshi 139(12):1539–1546. https://doi.org/10.1248/yakushi.19-00172
133. Svidlo et al (2013) echnology of functional juice-containing non-alcoholic drinks. [Online]. Available: https://keypublishing.org/jhed/wp-content/uploads/2020/07/02.-Full-paper-Karyna-Svidlo.pdf
134. Crespo L, Sanclimens G, Pons M, Giralt E, Royo M, Albericio F (2005) Peptide and amide bond-containing dendrimers. Chem Rev 105(5):1663–1681. https://doi.org/10.1021/cr0304491

135. Doshi M (2011) Dendrimer and its application. Int J Pharm Sci Rev Res 7(2):104–111. [Online]. Available: https://www.scopus.com/inward/record.uri?eid=2-s2.0-79955566864&partnerID=40&md5=ac421217e1eaadec3498a1b0768f3a06
136. Drabu S, Khatri S, Babu S (2010) Dendrimers: nanopharmaceuticals for drug delivery. Res J Pharm Biol Chem Sci 1(2):464–471. [Online]. Available: https://www.scopus.com/inward/record.uri?eid=2-s2.0-79952605230&partnerID=40&md5=cb593ffe07d6973fe571911932a86246
137. Erdem A, Eksin E, Kesici E, Yarali E (2018) Dendrimers integrated biosensors for healthcare applications. In: Nanotechnology and biosensors, pp 307–317. [Online]. Available: https://www.scopus.com/inward/record.uri?eid=2-s2.0-85054893397&doi=10.1016%2fB978-0-12-813855-7.00010-6&partnerID=40&md5=9960f54dc0690a92a4f8a8dafdbd5828
138. Mittal P et al (2021) Dendrimers: a new race of pharmaceutical nanocarriers. BioMed Res Int 2021. https://doi.org/10.1155/2021/8844030
139. Podlogar F, Kristl J (2004) Dendrimers and their future usage in pharmacy. Farm. Vestn 55(2):197–205. [Online]. Available: https://www.scopus.com/inward/record.uri?eid=2-s2.0-3342932332&partnerID=40&md5=fdad5fb0d6d78f83ae9df5a7687ce5bc
140. Wang X et al (2012) Characterizations of polyamidoamine dendrimers with scattering techniques. Polymers 4(1):600–616. https://doi.org/10.3390/polym4010600
141. Niederhafner P, Šebestík J, Ježek J (2005) Peptide dendrimers. J Pept Sci 11(12):757–788. https://doi.org/10.1002/psc.721
142. Filipczak N, Yalamarty SSK, Li X, Parveen F, Torchilin V (2021) Developments in treatment methodologies using dendrimers for infectious diseases. Molecules 26(11). https://doi.org/10.3390/molecules26113304
143. Ortega P, Sánchez-Nieves J, Cano J, Gómez R, De La Mata FJ (2020) CHAPTER 5: Poly(carbosilane) dendrimers and other silicon-containing dendrimers. Monogr Supramol Chem 2020-January(29):114–145. https://doi.org/10.1039/9781788012904-00114
144. Mendes LP, Pan J, Torchilin VP (2017) Dendrimers as nanocarriers for nucleic acid and drug delivery in cancer therapy. Molecules 22(9). https://doi.org/10.3390/molecules22091401
145. Sarode RJ, Mahajan HS (2024) Dendrimers for drug delivery: an overview of its classes, synthesis, and applications. J Drug Deliv Sci Technol 98. https://doi.org/10.1016/j.jddst.2024.105896
146. Teixeira-Lemos E, Oliveira J, Teixeira-Lemos LP, Reis-Lima MJ, Pinheiro JP (201) New trends in food science: the use of nutraceuticals as an antiinflammatory therapeutic tool in exercise. In: Nutraceuticals, Elsevier, pp 669–714. https://doi.org/10.1016/B978-0-12-804305-9.00017-8
147. Srinivas PR et al (2010) Nanotechnology research: applications in nutritional sciences. J Nutr 140(1):119–124. https://doi.org/10.3945/jn.109.115048
148. Turasan H, Kokini JL (2020) Chapter 6—delivery of bioactives using biocompatible nanodelivery technologies. In: Mustansar Hussain C (ed) Handbook of functionalized nanomaterials for industrial applications. In Micro and Nano Technologies, Elsevier, pp 133–166. https://doi.org/10.1016/B978-0-12-816787-8.00006-5
149. Spleis H, Sandmeier M, Claus V, Bernkop-Schnürch A (2023) Surface design of nanocarriers: key to more efficient oral drug delivery systems. Adv Colloid Interface Sci 313:102848–102848. https://doi.org/10.1016/j.cis.2023.102848
150. Nutrient delivery. Elsevier (2017). https://doi.org/10.1016/C2015-0-01048-6
151. Ziegenfuss TN, Landis J, Willoughby D, Greenwood M (2015) Nutritional supplements to enhance recovery. In: Nutritional supplements in sports and exercise.Springer International Publishing, Cham, pp 273–301. https://doi.org/10.1007/978-3-319-18230-8_12
152. Adetunji CO et al (2023) Biological nanocarriers for nutrient delivery. In: Nanobiotechnology for the livestock industry, Elsevier, pp 99–110. https://doi.org/10.1016/B978-0-323-98387-7.00009-4
153. Silveira PCL et al (2016) Effects of phonophoresis with gold nanoparticles on oxidative stress parameters in a traumatic muscle injury model. Drug Deliv 23(3):916–922. https://doi.org/10.3109/10717544.2014.923063

154. Magro DO (2019) Nutrition and hydration. In: The sports medicine physician. Springer International Publishing, Cham, pp 85–98. https://doi.org/10.1007/978-3-030-10433-7_8
155. Garth AK, Burke LM (2013) What do athletes drink during competitive sporting activities? Sports Med 43(7):539–564. https://doi.org/10.1007/s40279-013-0028-y
156. Lima MR, Moreira BJ, Bertuzzi R, Lima-Silva AE (2024) Could nanotechnology improve exercise performance? Evidence from animal studies. Braz J Med Biol Res 57. https://doi.org/10.1590/1414-431x2024e13360
157. Ansari AA (2023) Nanotechnological advancements in sports rehabilitation to elevate athletic performance levels, pp 377–381. https://doi.org/10.1007/978-981-99-4685-3_54
158. Batista MCC, dos Santos MAP (2020) Impact of hydration on exercise performance and physiological responses. Curr Nutr Food Sci 16(9):1346–1352. https://doi.org/10.2174/1573401316666200309113907
159. Kalinina I, Potoroko I, Nenasheva A, Velyamov M, Bagale U (2019) Prospects for the application of taxifolin based nanoemulsions as a part of sport nutrition products. Hum Sport Med 19(1):100–107. https://doi.org/10.14529/hsm190114
160. Dubnov-Raz G, Lahav Y, Constantini NW (2011) Non-nutrients in sports nutrition: fluids, electrolytes, and ergogenic aids. E-SPEN Eur E-J Clin Nutr Metab 6(4):e217–e222. https://doi.org/10.1016/j.eclnm.2011.05.001
161. Kai Y (2013) Study of biosafety of nanomaterials in sports engineering. Appl Mech Mater 340:348–352. https://doi.org/10.4028/www.scientific.net/AMM.340.348
162. Zhu Y (2020) Application and biosafety of nano-copper in sports engineering. Int J Nanotechnol 17(2/3/4/5/6):442–442. https://doi.org/10.1504/IJNT.2020.110728
163. Muñoz CX, Johnson EC (2019) Hydration for athletic performance. In: Nutrition and enhanced sports performance, Elsevier, pp 533–543. https://doi.org/10.1016/B978-0-12-813922-6.00045-X
164. van Rosendal SP, Coombes JS (2.012) Glycerol use in hyperhydration and rehydration: scientific update, pp 104–112. https://doi.org/10.1159/000341959
165. Sun J et al (2019) Resveratrol-loaded solid lipid nanoparticle supplementation ameliorates physical fatigue by improving mitochondrial quality control. Crystals 9(11). https://doi.org/10.3390/cryst9110559
166. Abdelnour SA et al (2021) Nanominerals: fabrication methods, benefits and hazards, and their applications in ruminants with special reference to selenium and zinc nanoparticles. Animals 11(7):1916–1916. https://doi.org/10.3390/ani11071916
167. Desbrow B, Irwin C (2024) Hydration. In: Nutrition for sport, exercise, and performance. Routledge, London, pp 140–148. https://doi.org/10.4324/9781003321286-13
168. Kilduff LP et al (2004) The effects of creatine supplementation on cardiovascular, metabolic, and thermoregulatory responses during exercise in the heat in endurance-trained humans. Int J Sport Nutr Exerc Metab 14(4):443–460. https://doi.org/10.1123/ijsnem.14.4.443
169. Tucker MA, Ganio MS (2024) What effect does hydration status have on body temperature during exercise? In: Quick questions heat-related illness. Routledge, New York pp 199–203. https://doi.org/10.4324/9781003526124-42
170. Zhou R, Zhou H, Xiong B, He Y, Yeung ES (2012) Pericellular matrix enhances retention and cellular uptake of nanoparticles. J Am Chem Soc 134(32):13404–13409. https://doi.org/10.1021/ja304119w
171. Yang Z, Xia S (2020) The effect of exercise fluid supplement on human metabolism after load exercise. J Coast Res 104(sp1). https://doi.org/10.2112/JCR-SI104-088.1
172. Hosey RG, Glazer JL (2004) The ergogenics of fluid and electrolyte balance. Curr Sports Med Rep 3(4):219–223. https://doi.org/10.1249/00149619-200408000-00008
173. Gupta MK, Tandon PK, Shukla N (2019) Nanotechnology: environmentally sustainable solutions for water treatment, pp 225–242. https://doi.org/10.1007/978-3-030-33745-2_8
174. Ohba T, Kanoh H (2013) Energetic contribution to hydration shells in one-dimensional aqueous electrolyte solution by anomalous hydrogen bonds. Phys Chem Chem Phys 15(15):5658–5658. https://doi.org/10.1039/c3cp44671a

175. Malani A, Ayappa KG, Murad S (2006) Effect of confinement on the hydration and solubility of NaCl in water. Chem Phys Lett 431(1–3):88–93. https://doi.org/10.1016/j.cplett.2006.09.071
176. He X, Ye A, Fu X, Yang W, Wang Y (2022) Achieving low-energy-barrier ion hopping in adhesive composite polymer electrolytes by nanoabsorption. Macromolecules 55(16):7117–7126. https://doi.org/10.1021/acs.macromol.2c00928
177. Wang W, Alexandridis P (2016) Composite polymer electrolytes: nanoparticles affect structure and properties. Polymers 8(11):387–387. https://doi.org/10.3390/polym8110387
178. Fernandes LAL, Azadmehr M, Johannessen EA, Hafliger P (2016) An osmotic pressure sensor for monitoring the level of hydration in biological fluids. IEEE Sens J 16(11):4331–4337. https://doi.org/10.1109/JSEN.2016.2548361
179. Paternoster N, Baggio E, Pelosi E (2020) Personalized hydratation status in endurance and ultra-endurance: a review. Mediterr J Nutr Metab 13(3):197–214. https://doi.org/10.3233/MNM-200434
180. Liu L (2015) Aggregation of silica nanoparticles in an aqueous suspension. AIChE J 61(7):2136–2146. https://doi.org/10.1002/aic.14839
181. Pence J, Bloomer RJ (2020) Impact of nuun electrolyte tablets on fluid balance in active men and women. Nutrients 12(10):3030–3030. https://doi.org/10.3390/nu12103030
182. Pérez-Castillo ÍM et al (2023) Compositional aspects of beverages designed to promote hydration before, during, and after exercise: concepts revisited. Nutrients 16(1):17–17. https://doi.org/10.3390/nu16010017
183. Adan A (2012) Cognitive performance and dehydration. J Am Coll Nutr 31(2):71–78. https://doi.org/10.1080/07315724.2012.10720011
184. Rosinger AY, John JD, Murdock KW (2024) Ad libitum dehydration is associated with poorer performance on a sustained attention task but not other measures of cognitive performance among middle-to-older aged community-dwelling adults: a short-term longitudinal study. Am J Hum Biol 36(6). https://doi.org/10.1002/ajhb.24051
185. Srinivasan K et al (2024) Impact of heat stress on thermal balance, hydration and cortical response among outdoor workers in hot environment—an exploratory report from North East India. J Basic Clin Physiol Pharmacol 35(1–2):79–84. https://doi.org/10.1515/jbcpp-2024-0003
186. Dube A, Gouws C, Breukelman G (2022) Effects of hypohydration and fluid balance in athletes' cognitive performance: a systematic review. Afr Health Sci 22(1):367–376. https://doi.org/10.4314/ahs.v22i1.45
187. Zeng J et al (2019) A nanoencapsulation suspension biomimetic of milk structure for enhanced maternal and fetal absorptions of DHA to improve early brain development. Nanomedicine Nanotechnol Biol Med 15(1):119–128. https://doi.org/10.1016/j.nano.2018.09.006
188. Garrett DC et al (2018) Engineering approaches to assessing hydration status. IEEE Rev Biomed Eng 11:233–248. https://doi.org/10.1109/RBME.2017.2776041
189. Jusril NA, Mukhtar SM, Abu Bakar SI, Md Saad WM, Wen NK, Adenan MI (2024) Nanoemulsion of ethanolic extract of centella asiatica (NanoSECA) ameliorates learning and memory performance by enhancing cholinergic activities, increasing antioxidative levels, and attenuating oxidative stress markers in adult rats. Pharm Nanotechnol 12(4):329–339. https://doi.org/10.2174/2211738511666230831122844
190. de Barros Fernandes RV, Marques GR, Borges SV, Botrel DA (2014) Effect of solids content and oil load on the microencapsulation process of rosemary essential oil. Ind Crops Prod 58:173–181. https://doi.org/10.1016/j.indcrop.2014.04.025
191. Bobbala S, Vincent MP, Scott EA (2021) Just add water: hydratable, morphologically diverse nanocarrier powders for targeted delivery. Nanoscale 13(26):11349–11359. https://doi.org/10.1039/D1NR02188E
192. Aneja KR, Dhiman R, Aggarwal NK, Kumar V, Kaur M (2014) Microbes associated with freshly prepared juices of citrus and carrots. Int J Food Sci 2014:1–7. https://doi.org/10.1155/2014/408085
193. Han L, Mohammadi R, Lei H, Jazbizadeh MH, Semiromi D (2023) An investigation of the performance of polyamide membrane for the concentration of pomegranate and effect on the

physical and sports health of patients with physical weakness: a novel study. Food Biosci 56:103336–103336. https://doi.org/10.1016/j.fbio.2023.103336
194. Xuan L, Ju Z, Skonieczna M, Zhou P, Huang R (2023) Nanoparticles-induced potential toxicity on human health: applications, toxicity mechanisms, and evaluation models. MedComm 4(4). https://doi.org/10.1002/mco2.327
195. Simran et al (2023) Regulatory barriers to the marketing authorization of nanodrug delivery system. In: Nanotechnology and drug delivery. Jenny Stanford Publishing, New York, pp 61–100. https://doi.org/10.1201/9781003430407-2
196. Sahu SC, Hayes AW (2017) Toxicity of nanomaterials found in human environment. Toxicol Res Appl 1. https://doi.org/10.1177/2397847317726352
197. Handy RD, Shaw BJ (2007) Toxic effects of nanoparticles and nanomaterials: implications for public health, risk assessment and the public perception of nanotechnology. Health Risk Soc 9(2):125–144. https://doi.org/10.1080/13698570701306807
198. Altemimi AB et al (2024) Application of nanoparticles in human nutrition: a review. Nutrients 16(5). https://doi.org/10.3390/nu16050636
199. Chunyan L, Xingliang L, Sijin M, Yanfen X (2011) Study on application and biosafety of nano-materials in sports engineering. In: Presented at the proceedings of the 2011 international conference on future computer science and education, ICFCSE 2011, pp 131–134. https://doi.org/10.1109/ICFCSE.2011.40
200. Chycki J, Kurylas A, Maszczyk A, Golas A, Zajac A (2018) Alkaline water improves exercise-induced metabolic acidosis and enhances anaerobic exercise performance in combat sport athletes. PLoS ONE 13(11):e0205708–e0205708. https://doi.org/10.1371/journal.pone.0205708

Nanoparticles: Unlocking the Peak Athletic Performance

Dilshith A. Kabeer, W. Vinu, A. K. Safad, Nafih Cherappurath, Borem Siri Reddy, and Kappat V. Sunooj

Abstract Nanoparticles are small particles sized between 1 to 100 nm. They possess diverse properties because of their high surface area compared to their size and quantum effects, which makes them suitable for a variety of applications. Nanoparticles offer promising features and mechanisms for nutrient delivery that makes them capable of improving absorption and customised delivery, enabling controlled and sustained energy release, muscle regeneration, all of these leading to better athletic performance. The mechanics underlying these particles are thoroughly examined in this chapter, including nutritional absorption, cellular uptake, and anti-inflammatory actions. Ethical issues, regulatory requirements, and health implications are also discussed, emphasising the importance of proper safety practices and quality control. The constraints of accessibility, cost and public awareness as well as the prospective development of personalised Nano-nutrition based on genetic profiles are addressed. Collaborative research is required between sports scientists and nanotechnology specialists to guarantee safe and equitable nanoparticle inclusion in sports nutrition. As research develops, nanoparticles are set to become a fundamental element of sports nutrition that provides athletes an effective and ethical tool for long-term performance

D. A. Kabeer (✉) · W. Vinu (✉) · A. K. Safad
Department of Physical Education & Sports, Pondicherry University, Puducherry, India
e-mail: dilshith.dpe@pondiuni.ac.in

W. Vinu
e-mail: vinu@pondiuni.ac.in

A. K. Safad
e-mail: isafadak@gmail.com

N. Cherappurath
Department of Physical Education, Amal College of Advanced Studies, Nilambur, India
e-mail: nafihch@amalcollege.ac.in

B. S. Reddy
Conestoga College Institute of Technology, Kitchener, ON, Canada
e-mail: sborem9797@conestogac.on.ca

K. V. Sunooj
Department of Food Science and Technology, Pondicherry University, Puducherry, India
e-mail: sunooj4u@gmail.com

A. Chatterjee et al. (eds.), *Nanofuel: The Future of Sports Nutrition*, Smart Nanomaterials Technology, https://doi.org/10.1007/978-981-96-5471-0_4

enhancement. This chapter explores what nanoparticles can bring to sports nutrition, focusing more on their potential to unlock the peak athletic performance.

Keywords Sports nutrition · Athletic performance · Nanoparticles · Nutrient delivery

1 Introduction

1.1 Overview of Nanoparticles

Nano particles are nano-scopic particles with a dimension in the range of 1–100 nm [1]. It has reformed various fields of modern science and is gaining wide acceptance in sports nutrition because of its influential role in sports performance. These materials at nano-scale have distinct molecular properties that make them highly effective in nutrient delivery systems. At present, Nanoparticles provide innovative solutions to address the high performance demands of athletes through precision, effectiveness, and targeted delivery, which are paramount to sports nutrition and athletic performance.

The unique features of nanoparticles, such as their minute size similar to biological molecules, help them to penetrate easily in to cellular membranes that enable close interaction with cellular components [2, 3]. The larger surface area, helps them to enhance solubility and bioavailability, ensuring better absorption of nutrients thus making them capable of working well with biological systems [4]. Their structural stability assist in the encapsulation of bioactive compounds is facilitated by their structural stability, which protect nutrients from digestive breakdown and degradation, thereby ensuring their strength and stability under several conditions [5]. Nanoparticles can seamlessly integrate with the body's natural systems because of their biocompatibility, thus ensuring safety as well as broadening their functionality [6].

Various nanoparticles are available for use in sports nutrition, specifically organic, inorganic and hybrid nanoparticles. Organic nanoparticles, which include proteins, lipids, carbohydrates, or other natural-based materials, are recognised for their safety and natural composition. They provide essential nutrients such as proteins, vitamins, which help in muscle recovery, hormonal balance, and overall performance. Lipid-based nanoparticles, like SLNs, play significant role in delivering fat-soluble nutrients and antioxidants that may help improve both physical and mental endurance. Likewise, Inorganic nanoparticles like mineral and metal-based nanoparticles target enhancing uptake of the vital micronutrients, calcium, zinc, and magnesium, critical for building a strong skeleton and muscle functions while protecting against impact from vigorous exercise. Furthermore, metal-based nanoparticles based on gold and silver possess the added benefit of lowering inflammation, increasing endurance, and inhibiting infection. Added benefits to silica and titanium include its utility in reducing friction in sportswear.

The organic–inorganic composite nanoparticle provides further functionality. For example, nano-clays and nano-cellulose fibres improve supplement texture and the nutrient release. Carbon-based nanoparticles such as carbon nanotubes are currently being investigated to increase energy levels and reduce oxidative stress. Since nanoparticles provide both precision and stability, personalized sports supplements that meet particular performance goals with minimal waste and side effects can be achieved. Hence, they are ideal for sustainable and efficient nutrition solutions. Even though, nanoparticles have made ground breaking contributions to the sports field, there are still gaps for extensive research to ensure their long-term safety, efficacy and environmental compatibility. However, nanoparticles reformed the incorporation of these nutrients in sports nutrition, so that athletes can achieve higher ranges of performance ranges while still maintaining their wellness.

Nanoparticles bridge the gap between conventional dietary practices and advanced supplementation, in turn, redefining sports nutrition. Nanoparticles are emerging as the base and foundation for enhancing athleticism and fulfilling the complex physiological needs of athletes. The integration of nanoparticles with sports nutrition is a revolutionary stride that meets the very complex needs of athletes by outperforming all the conventional methods.

1.2 Background and Emergence of Nanotechnology in Sports

Nanotechnology in sports science marked a new dawn of sports because of its incredible contributions to redesigned human performance and recovery boundaries. They bring innovative solutions to improve performance, recovery, and safety for athletes. Its unique properties-like high strength, flexibility, and reactivity have really modernized sports science by changing everything from equipment to apparel and rehabilitation.

Nanotechnology has brought various scientific solutions to different domains of sports performance. Carbon nanotubes (CNTs) are one of the remarkable innovation applied to equipment such as tennis rackets and mountain bikes. It is lightweight, yet has the highest strength and durability for athletes to perform better while their equipment lasts longer [7, 8]. CNTs are also used in sportswear material engineering for enhancing strength and stiffness with abrasion resistance at the same time to achieve better safety with optimal performance [7, 8]. Since graphene combines strength and flexibility, it is highly appreciated when used as an element in sportswear and sporting equipment, where it can improve strength, comfort, and durability. Nanotechnology is not limited to devices and wearables. For example, nanophosphate lithium-ion batteries are used for energy-recovery in sports. Lightweight but efficient, they improve performance in energy-dependent tasks [8]. Green nanomaterials that are added to green sportswear enhance fabric efficiency for more sustainable use and helps in protection of athlete from UV rays [9]. Silver nanoparticles are widely utilised for sports apparels that is known for sweat absorbing ability, protection from ultra-violet rays, and electrical conductivity, thereby facilitating more comfort, protection and enhancing athletic performance [9]. It is also been consumed these nanoparticle for substance delivery, and succeeded in enhancing physiological capabilities such as cardiovascular and lung functions through a drug-free method [10].

From betterment of the equipment to the novel modes of recovery, nanotechnology is shaping the future of sports. It enables athletes to go that extra mile, yet stay safe, comfortable, and eco-conscious. Hence, it is an integral part of modern-day sports science.

1.3 Relevance to Athletic Performance

The contributions of nanoparticles in sports nutrition are beyond exceptional. They are changing the fate of nutritional supplementation as they specifically address the needs of athletes, thus enhancing endurance, recovery, and overall performance. Conventional supplements usually fail in application because they have poor absorption and variable results. Nanoparticles overcome this challenge by delivering nutrients more effectively, hence getting absorbed faster and utilized more efficiently by the body at the targeted site.

Nanoparticles are available in different forms that potentially contribute to different contexts of sports performance. Organic nanoparticles, such as liposomes and polymeric nanoparticles, can encapsulate fat-soluble vitamins, amino acids, and antioxidants, which shield them from degradation and help them to be delivered appropriately to the most needed tissues, throughout the exercise and recovery phases. Continuous release mechanisms from the nanoparticles ensure a steady supply of energy and nutrients over time, critical for sustaining activity or supporting recovery after training.

Nanoemulsions, solid lipid nanoparticles and organic nanoparticles play a significant role in hydration and electrolyte balance. With better bioavailability, nanoemulsions rapidly replenish fluids and electrolytes lost in the course of intense physical activity and prevent fatigue for sustained performance. With the delivery of antioxidants and enzymes, solid lipid nanoparticles ensure that cells are protected against oxidative damage, thus recovery happens fast and cellular health prevails for long. Likewise, metal-based nano-fuels like magnesium and gold nanoparticles aid the recovery capabilities by enhancing durability, decreasing inflammation and improving vascular function by impacting muscle performance and repair mechanism performance. Protein based nanoparticles deliver amino acids and peptide to the targeted muscle fibres, which further facilitates fast recovery from intense training. Synergetic use of these nanoparticle technologies optimises performance and recovery and general well-being.

Nanoparticles will make it possible for athletes to peak in performance, recover adequately and maintain high levels of performance with time, considering issues related to hydration, energy management, muscle repair, and cellular protection. This is one of the critical advances into personalized sports nutrition: a leap to allow the athletes to perform at their optimum and remain healthy.

These all are some of the glimpses what nanoparticle can achieve to enhance the sports performance. Nanotechnology fills the gap between physical demands of athletes and what traditional nutrition can deliver. It constantly pushes athletes to the limit without causing harm. In other words, nanotechnology represents a game-changer revolution where peak performance and recovery rates happen at faster speeds while preserving long-term health and well-being. The relevance of nanoparticles in athletic performance is thoroughly explored in detail in this chapter.

2 Nanoparticles: Types and Mechanisms

Nanotechnology has revolutionized almost every domain and is now emerging as a game changer for the athletic field. It has introduced various forms of nano-scale particles that is modifiable for different purposes, with unique features that offer infinite opportunities to athletic field through optimizing nutrient delivery system which in turn enables athletes to perform beyond their ordinary capacity. Their unique molecular properties such as enhanced bioavailability, sustained release, targeted delivery and safety, facilitate recovery and rehabilitation. Nanoparticles are very versatile and are being utilized widely in sports, from encapsulation of nutrients, to facilitating focused delivery of therapeutic agents, hence reforming recovery and rehabilitation. Being biocompatible and sustainable they offer that extra push beyond the scope of conventional methods. Several categories of nanoparticles are available. For their comprehensive understanding, nanoparticles are presented in this chapter as organic nanoparticles and inorganic nanoparticles. The emergence of nanoparticle in sports nutrition is an ongoing area of interest for researchers so it is not still fully explored. The following section discusses some of the popular nanoparticles with the properties and features to be used in nutrient delivery, their mechanism and how they work in human body, which in turn pointing towards how nanoparticles can unlock the peak athletic performance.

2.1 Organic Nanoparticles

Organic nanoparticles are nano-sized particles that is measured between 1 and 100 nm in diameter and composed of organic compounds like protein, lipids, polymers and carbohydrates [11]. These nanoparticles characterized in a special way, like their tiny structure and high surface area to volume ratio, and they can be manipulated to supplement various therapeutic and performance-enhancement agents through surface modification and functionalization [12, 13].

2.1.1 Liposomes

Liposomes are regarded as one of the first generation nanoparticles that have been considered as a trustable tool for the widespread usage of drug and nutrient delivery systems. They are nanoscopic spherical vesicles made up of organic or synthetic lipids that contain one or more phospholipid bilayers, which resemble human cell membranes and an aqueous core, possessing diverse properties such as biocompatibility, biodegradability, non-toxicity, and sustainability, thereby outperforming traditional delivery systems [14, 15]. They are well known for their versatility in delivering both water-soluble and fat-soluble (hydrophilic and hydrophobic) molecules with better solubility and absorption rates [15–17]. Along with their unique features, their structural similarity to cellular membrane helps them to integrate effectively with biological systems thus facilitating effective supply of nutrients [18, 19]. They are very efficient in targeted delivery, such a way that they are being capable of shielding encapsulated substance, whether drug or nutrient, from external degradation such as light and heat. This capability improves bioavailability of the entrapped molecules and offers an opportunity for controlled release, thus giving sustained energy and recovery support [20, 21].

Liposomes have been developed in different types based on their size, composition and the way they structured, each with unique benefits for athletic performance and nutrition (Figs. 1 and 2). Based on the size, they are categorized in to small unilamellar vesicle, large unilamellar vesicles and multilamellar vesicles. Small unilamellar vesicle (SUVs) is sized 50 nm in diameter, which is best for delivering water soluble molecules due to their small size and interactivity with cells [22]. So, it can be used to deliver water soluble vitamins and other nutrients according to the needs of athletes. Large unilamellar vesicles (LUVs) are another variety, which is comparatively large in size with 50–500 nm, thus providing larger surface area for nutrient delivery [22]. As the name indicates, multilamellar vesicles (MLVs) larger and bigger when compared to other forms, and it is crafted from two or more lipid layers, which are known for their enhanced stability and controlled release as like other nanoparticles [22]. According to the way liposomes are composed and their surface properties, they are further classified into conventional liposomes, sterically stabilized liposomes, immuno-liposomes and polymorphic liposomes. Conventional liposomes are a natural version, which is being crafted from natural phospholipids and holds more potential to interact with their environment in a broader and unique way [15]. Sterically stabilized liposomes, are structurally different from others that is coated with polyethylene glycol (PEG), through which they enhance their structural stability and minimize reactivity, toxicity and side effects, thus facilitating controlled and sustained nutrient supply [15]. Immunoliposomes are possessing unique capabilities such as they can be modified with antibodies or ligands on their surface, as a result, it assist them to target specific cells or tissues for more efficient delivery without wastage [15]. Polymorphic liposomes are known for their high reactivity and they can also be engineered to interact with specific agents. Liposomes are modifiable and designable for specific purposes for example, sterically stabilized liposomes can be manipulated for prolonged circulation, immunoliposomes can be used for targeted delivery to specific tissues [15]. By considering these diverse types and features of liposomes, they open an endless opportunity for supplementation in sports nutrition, based on the specific needs of athletes, thus making the growth and recovery of muscles more faster which in turn optimize their fitness and performance.

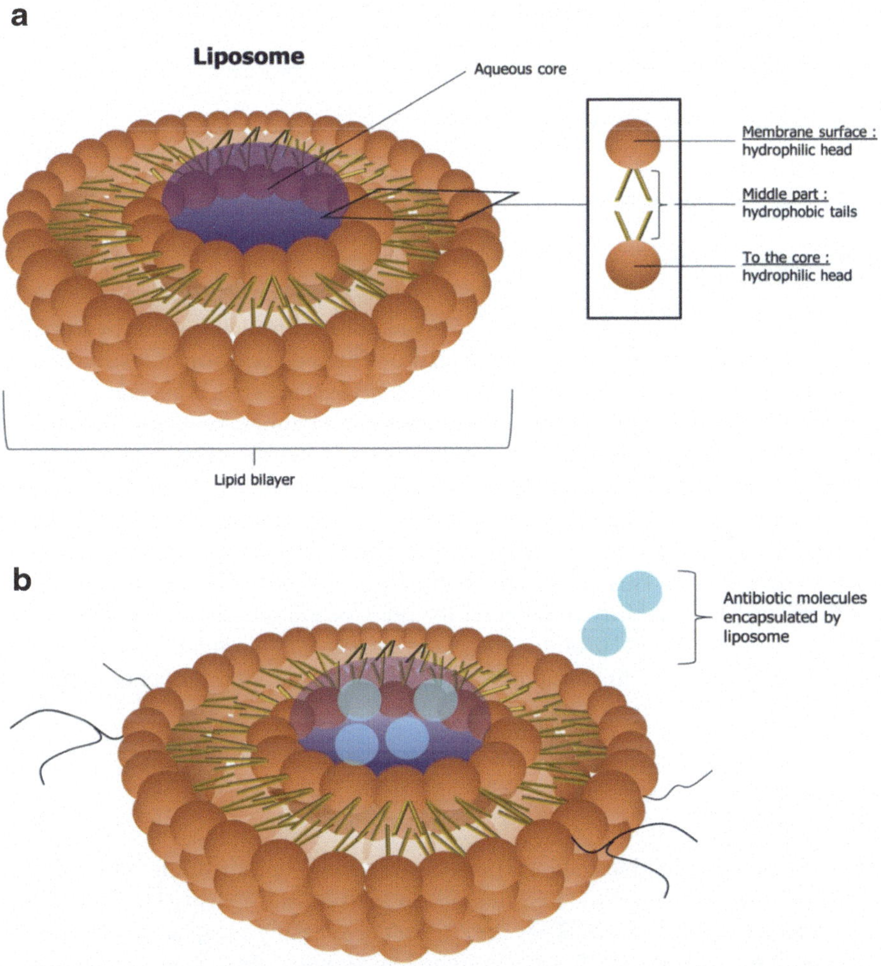

Fig. 1 Schematic representation of a liposome, with a lipid bilayer (**a**) resembling human cell membranes and an aqueous core (**b**) for encapsulating hydrophilic drug molecules [23]

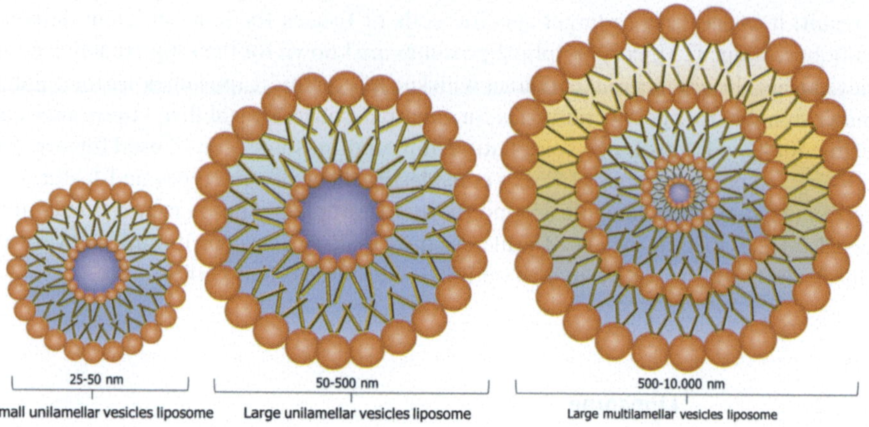

Fig. 2 Types of liposomes based on their molecular size [23]

2.1.2 Mechanism of Action

Encapsulation and Protection

As we discussed earlier, liposomes are capable of encapsulating both hydrophilic and hydrophobic bioactive compounds because of their diverse structures and unique properties, such as their nano-scale size and high surface area. This encapsulation facility helps them to deliver essential nutrients like vitamins, minerals, and antioxidants with high-end protection. The lipid bilayers of liposomes are responsible for the high-end protection that acts as a shield against the threats that destroy molecules, such as heat, pH, enzymic activity and variations in ionic strength. With protective abilities of the lipid bilayer, liposomes sustain the molecules from environmental degradation [24–26]. This protection enables the encapsulated molecules to maintain their structural stability throughout the process and efficacy in the sustainable delivery of molecules [21]. In the context of sports, this mechanism of liposomes will help tremendously to entrap and supply various essential nutrients faster and more effectively than the conventional methods.

Controlled Release

Liposome as self-assembled nanocarriers, have shown solid capabilities in controlled release of bioactive compound like vitamins and minerals, which can be manipulated as per athletes' specific needs. They are efficient in releasing encapsulated molecules in response to specific stimuli, such as light, temperature, pH or acidic background, redox environment and specific enzymes, thus helping them to achieve a predictable release profile that is been proven and widely used for cancer and therapeutic treatments [27]. This mechanism can contribute to the supplementation systems of sports nutrition such as essential nutrients, minerals, antioxidants and other enzymes, to be released steadily and controlled, which is particularly very beneficial in athletic context. Likewise, small uni-lamellar liposomes (SUVs) composed of neutral lipid molecules, can remain in blood stream for an extended period, thus these particle facilitating the availability of nutrients for longer time in a controlled manner [28].

Enhanced Absorption and Bioavailability

The liposome's bilayers ability to attach effectively to cell membranes is a unique feature that facilitates the bioavailability, absorption and direct delivery of encapsulated compounds into cells [29, 30]. By using this mechanism, athletes can ensure sufficient nutrients and other bioactive substances are delivered to specific body parts at the correct time, to achieve optimum growth and recovery. Liposomes are very advanced in their protection capabilities from the degradation of bioactive compounds because of the environmental factors that we discussed earlier such as heat, pH, light, enzyme activities etc., which ensure the availability of a higher proportion of the encapsulated compounds to where we exactly needed [14, 31]. Since liposomes are equipped to face the harsh conditions of the digestive system they can considerably improve the bioavailability and absorption of nutrients [25, 32], thereby offer countless benefits to athletes, who need rapid and sustained nutrient supply for prolonged duration that help directly to their performance and recovery.

Targeted Delivery

Nanoparticles are famous for their targeted delivery capacity along with controlled release and better absorption. Liposome is not different in this context, since they carry the unique ability to deliver the entrapped molecules in the specific region of the body. They can release nutrients to particular muscle fibres in response to specific stimuli with absence of side effects. Liposome can be attached specifically to receptors of targeted cells when it is modified with ligands such as antibodies, peptides, or carbohydrates [33]. This targeted delivery mechanism is better achieved when the nutrient or other drugs are combined with ligand targeting, stimuli-responsive release and strategic surface modification, thus facilitating the sustainability and efficacy of therapeutic and healing outcomes in our body while minimising adversities [31, 34, 35]. This unique mechanism gives a leap to the sports nutrition field for the process of supplementation and therapeutic delivery, which, obviously outperforms the conventional methods and tremendously enhances athletic performance

2.1.3 Solid Lipid Nanoparticles (SLNs)

Solid Lipid Nanoparticle (SLN) is a lipid based nano-carrier system that is mainly used for drug delivery, but also can be utilised in sports nutrition with their diverse properties. SLNs are structurally spherical and are developed from nano-sized solid lipids such as glycerides waxes, and fatty acids which are also stabilized by surfactants [36, 37]. They can also encapsulate water-soluble (hydrophilic) and fat-soluble (lipophilic) compounds like other lipid bases nanoparticles, which shows their versatility for various substance delivery systems [38]. Apart from their encapsulation capabilities, they can also supply those compounds through the mechanism of controlled release that has the potential to ensure long-duration protection against cellular damage. As known for structurally stable, they are also being efficient in bioavailability and targeted delivery of encapsulated molecules.

There are modified types of SLNs are available, slightly different from SLN, such as Nanostructured Lipid Carriers (NLCs) and Lipid Drug Conjugate (LDC). NLCs are composed of a combination of solid and liquid lipids, blending agents and water, which helps them form a partially crystallized system that enhances drug loading capacity and stability. Whereas LDCs are prepared by covalently linking a substance with a lipid moiety such as fatty acids, glycerides or phospholipids which in turn enhancing their ability to dissolve in fat (lipophilicity), better bioavailability and targeted delivery [39, 40]. They are prepared by different methods, such as High pressure homogenization (HPH), microemulsion method, solvent emulsification-diffusion, ultrasonication, solvent injection, hot homogenization, cold homogenization and electrosparaying, all of this including different crafting process and mechanisms. HPH method is one of the main technique used for SLN preparation, which involves the application of high pressure to create nanoparticles [41]. In microemulsion method, forming a microemulsion at an elevated temperature, and then cooled, to form suspensions of solid nanoparticles [42]. Solvent emulsification-diffusion is another main method, which consists of dissolving the lipid in an organic solvent, blending it in water that contains surfactants, and then evaporating solvent under reduced pressure [43]. Through the ultra-sonication method, SLNs are being created using ultrasonic waves to breakdown the lipid particles in nanoparticles [44]. As per hot homogenization and cold homogenization technique, SLNs are prepared by utilizing heat and cold [41]. According to elecrospraying method, SLNs are composed from a lipid solution by using an electric field [45].

Adding SLNs to sports nutrition offer athletes a remarkable opportunity in supplementation and therapeutic drug delivery that promote peak athletic performance and rehabilitation process. Some SLNs have already applied and proven their positives in sports nutrition, for example, Resveratrol-loaded SLNs have already shown their solid abilities such as enhanced bioavailability and efficacy of resveratrol which is a natural polyphenol with antioxidant properties [46, 47]. These nano-scale particles also capable to improve mitochondrial function, reduce muscular fatigue, and mitigate oxidative stress, thus making them a dependable tool to promote endurance, facilitate recovery and overall performance of athletes [46–48]. So, it is evident that SLNs are one of the counterstones in the advancement of sports nutrition towards the journey of peak athletic performance

2.1.4 Mechanism of Action

Encapsulation, Protection and Bioavailability

SLNs are designed from a solid lipid medium, enabling them to encapsulate lipophilic and hydrophilic molecules and labile compounds such as proteins and peptides. This lipid core is also stabilized by surfactants in an aqueous medium, thus helping them to form a protective barrier around the bioactive compounds [49]. SLNs protect the compounds from degradation that happens with environmental factors such as light, oxygen and moisture [50]. The solid lipid matrix of SLN provides a stable environment for the molecules and enhances their physical stability, preventing premature degradation and ensuring effective delivery. They are also efficient in improving the bioavailability of encapsulated nutrients, acting as a protective barrier against the harsh gastrointestinal environment, thereby enhancing metabolism and ensuring a higher proportion of the compounds reach systemic circulation [50].

Targeted Delivery, Controlled Release and Solubility

According to the current evidence, SLNs are modifiable and can be engineered to target specific tissues in the body by manipulating their particle size and surface properties. Usually, smaller particles have higher surface area to volume ratio, which is highly effective in interacting entrapped compounds with targeted tissues and improves bioavailability. The surface of SLN can be changed and modified with ligands or antibodies to target specific receptors on cells, which facilitate targeted delivery [51]. The solid lipid core of SLN allows controlled release of the encapsulated compound overtime because of the capability of the lipid matrix of SLNs to diffuse or release the bioactive compounds like nutrients in a sustained manner [52]. The efficiency of encapsulation and the gastrointestinal stability of the SLN are other factors crucial for controlled release [53]. The way the lipid matrix is composed and the technique used also affects their release profile, which can also be controlled by manipulating them [52]. Along with this, SLN also extends their capacity to improve the solubility of poorly water-soluble drugs by using the ability to disperse them in a lipid matrix, thereby enhancing the solubility and permeability of the drugs and facilitating their absorption [54]. These mechanisms are widespread in almost every nanoparticle that is used in biomedicine, through which they propose the widespread possibility of utilizing these lipid-based nanoparticles in sports nutrition to deliver nutrients such as vitamins and minerals, making them an indispensable tool to achieve optimum performance.

2.1.5 Nanoemulsion

Nanoemulsions are nano-sized droplets that are 20–500 nm in size, with two liquids, such as oil and water, which are normally immiscible and stabilized by surfactants and co-surfactants to form a single phase [55]. Some studies have shown even a narrower size of 20–200 nm for nanoemulsions [55]. Like other nanoparticles, they possess typical features such as larger surface area, better kinetic stability, high optical clarity and improved compound bioavailability [56–58]. In their structure, water acts as the continuous phase in oil-in-water nanoemulsions, while droplets containing active ingredients like oils are in the dispersed phase, and emulsifiers or surfactants stabilize the interface between these phases [59, 60]. They are mainly categorized into three types. The first one is oil-in-water (O/W) nanoemulsions, which are composed of oil droplets dispersed in a continuous water phase, and they are highly compatible with encapsulating hydrophobic substances. The second one, water-in-oil (W/O) nanoemulsions, involves water droplets dispersed in a continuous oil phase that is capable of penetrating water-soluble compounds through oily barriers along with controlled release properties [61, 62]. Bi-continuous nanoemulsions are another type with a structure of both oil and water phases, and they are less common. Nevertheless, they can be used for specialized applications where both hydrophilic and hydrophobic molecules are needed [61, 62].

They can be formed by using different methods such as the high-energy method, which include the high-pressure homogenization method to break down larger particles in a liquid to create a uniform mixture, ultra-sonication technique in which high-frequency sound waves used to break apart or disperse particles in a liquid, and micro-fluidization method to create fine uniform particles with the help of high-pressure fluid collision in small chambers, and low-energy method that includes phase inversion temperature, spontaneous emulsification and solvent displacement, which offer better control over droplet size and are more economical and energy-efficient [63–65].

In the case of sports nutrition and athletic performance, nanoemulsions are ideal for formulating energy drinks and supplements that can be used for quick and sustained energy as well as to supply necessary vitamins and minerals during strenuous sports activities. For instance, it is found that the nanoemulsions stabilized with whey protein isolates WPI promote the lipid digestion and fast absorption, which benefits the performance goals of athletes such as enhanced endurance [66]. Nanoencapsulation of amino acids and antioxidants, such as astaxanthin, can improve the delivery, structural stability, and bioavailability which in turn decline inflammation, oxidative stress and ultimately enhance performance [67, 68].

2.1.6 Mechanism of Action

Bioavailability, Better Stability and Efficient Delivery

The nano-scale-sized droplets of nanoemulsions facilitate enhanced bioavailability because of its larger surface area that can hold more substances, which in turn enhances absorption. These structural advantages can be used to deliver poorly soluble bioactive compounds to specific tissues, ensuring a higher concentration of the components in the blood stream [69, 70]. They can protect sensitive bioactive compounds from oxidative and thermal degradation, enhancing their stability and bioavailability [71]. They provide higher stability against particle aggregation and gravitational separation, which is beneficial for maintaining the homogeneity and efficacy of the encapsulated compounds [72].

Targeted Delivery and Controlled Release

Nanoemulsions can be carefully engineered to respond to various stimuli, such as pH, temperature, light and magnetic fields, all of which can trigger the precise release of the encapsulated compound at the desired target site [73]. The mechanism of targeted delivery utilized by nanoemulsions includes passive targeting, which primarily consists of accumulating the substances to specific cells with the help of permeability and retention effect (EPR) that is usually applied in tumor treatment, and active targeting, which comprised of modifying the surface with ligands or antibodies that can bind specifically to receptors found on targeted cells [71, 74]. Within the context of athletics, these mechanisms can be effectively used to deliver specific nutrients, such as branched-chain amino acids, creatine or antioxidants, etc., directly to the target muscles, which not only fasten the recovery but also enhance overall athletic performance.

2.1.7 Polymeric Nanoparticles

Polymeric nanoparticles are formulated from synthetic or polymer materials like polylactic acid (PLA) and polyglycolic acid (PGA), with sizes ranging from 1 to 1000 nm; these are another great discovery in the realm of nanoparticles, as they are also famous for their biodegradability and compatibility with living things [75, 76]. Different types of PNPs are in the business and are being used widely in various fields, such as biopolymeric nanoparticles, molecularly imprinted polymers (MIPs), and polymer-metal nanocomposites (PMNCs). Bio-polymeric nanoparticles are one of the common types that is crafted from natural polymers, for example, protein, polysaccharides, and such biopolymers, which possess solid potentialities to encapsulate, protect, and release the given compound to the specific area where we need, through a sustained and controlled mechanism. Molecularly imprinted polymers (MIPs) are formed with the help of synthetic materials and engineered to function as artificial receptors with high specificity for target molecules. On the other hand, polymer-metal nano-composites are shaped in a unique way, in which polymers are used as a matrix and metal nanoparticles are taken as fillers [77, 78]. Because of their high surface area, these nanoparticles can incorporate many bioactive compounds such as proteins, vitamins, minerals and other therapeutic agents, thus making it easier for broad encapsulation [79]. This tiny particle can also be modifiable with ligands that enhance their targeting capabilities to specific cells and interactions with biological systems [80, 81].

The applications of these nanoparticles in the domain of sports nutrition are still in the infancy stage. Their diverse capabilities, such as sustained release, bioavailability and targeted delivery, are crucial for effectively supplementing nutrients that are vital for boosting athletic performance. Nevertheless, these nanoparticles have already been proven their success in drug delivery especially for cancer treatment. The unique features like superior absorption, better stability and bioavailability are making them up to the mark in the drug delivery system, thus proposing a great avenue to be used in sports for performance enhancement,

2.1.8 Mechanism of Action

Controlled and Sustained Release of Nano-Compounds

The capability of PNPs to release nano-compounds in a controlled and sustained mechanism makes them more appealing and different from conventional drug delivery systems. PNPs can supply essential nutrients in a controlled and sustained way, thus ensuring a steady supply of compounds for longer, which is beneficial for athletes who need consistent nutrient intake. This mechanism in PNPs is facilitated by using different sophisticated processes like diffusion, degradation, and external triggers for supplying bioactive compounds like nutrients [82]. The unique properties of polymers control this process so that the active ingredient in PNPs are spread in the polymer and diffuses the molecules gradually over time to the external surface, thereby facilitating the controlled delivery [83].

Encapsulation and Protection

Polymeric nanoparticles (PNPs) are highly regarded for their ability to encapsulate and safeguard essential bioactive compounds, making them versatile for applications such as supplementation. They are equipped to encapsulate an adequate amount of nutrient molecules and guard them against degradation. The encapsulation process of PNPs is done through complex procedures, which include trapping the active ingredient, such as drugs, nutrients or even liquids, inside a polymer shell, which acts as a protective shield from harsh conditions like stomach acid or enzymes [84]. These protective shells are formed through different methods, such as nano-precipitation, complex coacervation, and layer-by-layer assembly [85]. This protective shell not only keeps the nutrient stable but also allows for controlled release as per the triggers, such as by changes in pH or specific enzymes.

Targeted Delivery

PNPs are very efficient when it comes to the process of target delivery. This process is facilitated through different mechanisms, such as active targeting and passive targeting methods. Under the active targeting mechanism, the surface of NPs is attached to molecules like antibodies or folic acid, which adds precision, thereby allowing them to bind directly to specific cells [86, 87]. This method is practically beneficial to sports nutrition; essential vitamins, minerals, antioxidants, and other nutrients like branched-chain amino acids (BCCAs) can be supplied directly to the needed cell. However, the passive targeting method is not applicable in sports nutrition, which includes the enhanced permeability and retention (EPR) effect widely used in drug delivery systems but unsuitable for utilizing nutrient delivery and application in sports due to the physiological difference in pathology. As we discussed in the targeted release mechanism of various NPs, PNPs are also modifiable to release nano-compounds in response to multiple stimuli like changes in pH, enzymic activity or external signals such as light or magnet, which facilitate the sustained release and delivery of compounds where it is exactly needed [88]. This mechanism significantly contributes to the efficiency of the delivery system by eliminating the wastage of compounds to unnecessary sites.

Higher Bioavailability, Solubility and Permeability

Enhanced bioavailability, the ability to dissolve easily in targeted sites, and better permeability are other mechanisms that put PNPs ahead of conventional drug delivery systems. PNPs can improve the bioavailability of nutrients by shielding them from environmental degradation, thereby ensuring stability and effectiveness until they reach their target [89]. Along with this protection, they can also contribute to the bioavailability of encapsulated compounds by helping them to pass transportation barriers easily across different systems and fasten absorption. For example, hydrophobic nutrients can be encapsulated in hydrophilic polymers, thus boosting their capacity for solubility and quick absorption [90]. The tiny size and wider surface area of PNPs further help, and their modification with hydrophilic molecules also enhances solubility by aiding water dispersion, which in turn makes them able to speed up the dissolution [91]. The permeability of PNPs depends on a balance between their unique features, such as the composite's structural arrangement and their compatibility with surrounding polymers [92]. With controlled and sustained release, polymer nanoparticles can ensure that nutrients are delivered efficiently, as such facilitating bioavailability and supporting athletes in unlocking their optimum potential.

2.1.9 Chitosan Nanoparticles CSNPs

Chitosan nanoparticles are one of an advanced polymer-based nano-carrier derived from chitosan, a natural biopolymer developed through the process of deacetylating chitin, which is found in the shell of crustaceans like crabs and shrimps [93]. CSNPs hold unique features, such as polycationic nature from amine groups, reactive sites for modification and nano-scale size with a high surface-to-volume ratio

[94, 95]. These well-praised features help them work well with different morphologies like nanoparticles, nano-composites, nano-capsules and fibers [96, 97]. They are proven to be biocompatible, biodegradable, and non-toxic with the mechanism of targeted delivery, thus enabling them to get encapsulated in different fields such as biomedicine and also can be used in sports nutrition [98, 99].

CSNPs are also famous for their flexibility in their distinct forms and applications. Simple chitosan nanoparticles are one of the type of CSNPs that is shaped by using techniques such as emulsion cross-linking, ionotropic gelation and molecular self-assembly, which is applied broadly for controlled release and site-specific delivery of drugs [99]. Chemically cross-linked chitosan nanoparticles are also crafted by a chemical cross-linking process, which involves creating covalent bonds between chitosan molecules [100]. Ionically cross-linked chitosan nanoparticles are another variation in which chitosan is cross-linked with polyanions, thus making them better for molecule delivery with enhanced absorption capacity [95]. Polyelectrolyte complex chitosan nano-particles are formed by mixing chitosan with oppositely charged polyelectrolytes, thus facilitating them to encapsulate bioactive components by improving their stability and bioavailability [100]. Likewise, conjugated chitosan particles are prepared through conjugation with other molecules to enhance specific properties, making them modifiable for specific nutrient delivery systems. In turn, they can be used to improve the efficacy of sports performance.

2.1.10 Mechanism of Action

Contribution to Mitochondrial Biogenesis and Energy Metabolism

Chitosan nanoparticles (CSNPs) are known for improving mitochondrial function by stimulating the activities of mitochondrial enzymes. CSNPs activate the ortholog 1 (Sirt1) and AMP-activated protein kinase (AMPK) pathways, which facilitates mitochondrial biogenesis there by increasing mitochondrial content and functionality in skeletal muscles, thus supporting high energy output [101]. Mitochondrial biogenesis is a complex process that regulates the maintenance of cellular energy mechanisms and adapts to the varying energy demands of organisms [102]. In this process, the coordinated interaction of both nuclear and mitochondrial genomes helps to meet these cellular energy demands. In spite of this, CSNPs can be delivered to the targeted site of mitochondria, leading to reduced oxidative stress and improved nanoparticle stability, enhancing endurance and optimizing energy metabolism and performance. [101, 103].

Anti-Oxidant and Anti-Inflammatory Mechanism

CSNPs have solid anti-inflammatory compounds that reduce oxidative stress in muscle tissues. CSNPs protect muscles from oxidative damage by reducing reactive oxygen species (ROS) and increasing total anti-oxidant capacity (TAC). They have anti-inflammatory effects by targeting inflammatory pathways. CSNPs can surpass cytokine releases such as TNF-α and IL-8 and inhibit activation of NF-κB through which inflammation is eliminated [104]. CSNPs activate anti-oxidative defense mechanisms through the regulation of enzymes like glutathione peroxidase (GPx), catalase (CAT) and superoxide dismutase (SOD), which have the potential for the reduction of oxidative stress [105]. Their ability to reduce lipid peroxidation and reactive oxygen species alongside boosting Nrf2 activity highlights their potential to shield cellular health, optimizing performance [106].

Better Muscle Strength, Tissue Integrity and Cellular function

CSNPs have been shown to enhance the process of muscular regeneration, structural integrity, and reduction in fibrosis of injured muscles, thus contributing to the development of muscular strength and functionality. These mechanisms are achieved by decreasing the fibrosis index and expanding the average myotube diameter, which leads to better muscle fibre formation and growth [107]. Chitosan supports cell adhesion, proliferation and differentiation when used in various forms, such as hydrogels, scaffolds and nanoparticles. This is more effective when it is combined with stem cells, which, in turn, leads to improvement in tissue regeneration and repair [108]. Chitosan constructs loaded with stem cells can integrate better with host tissues there by repairing damaged tissues, which is crucial for maintaining muscle integrity and function.

Optimization of Lipid Profile and Weight Management

Chitosan nanoparticles (CSNPs) can potentially optimize lipid profiles and support weight management, contributing considerably to athletic performance. This is accomplished when the positively charged chitosan interacts with negatively charged lipid molecules, thereby accumulating aggregates that are less accessible to digestive enzymes [109]. Being able to bind dietary fats, CSNPs prevent lipid absorption in the digestive system. They reduce serum cholesterol and triglyceride levels by forming aggregates with lipid molecules, promoting cardiovascular health [110]. They influence the expression of proteins like sterol regulatory element binding protein 1a (SREBP1a), which regulates lipid biosynthesis, thereby enhancing lipid metabolism [111]. This highly efficient metabolic activity contributes to effective energy utilization and minimal fat accumulation. They also manage weight by regulating satiety and reducing overall food intake. Studies on animals have reported a significant reduction in body weight and fat mass alongside improvement in insulin sensitivity and metabolic efficiency [110].

Encapsulation, Controlled Release, Bioavailability and Targeted Delivery

Chitosan nanoparticles are biodegradable, biocompatible, safe and compatible with encapsulation, controlled release, and targeted delivery, just as other NPs [112]. They can encapsulate water-soluble and fat soluble nutrients that can be diffused to specific cells [113]. One of the key benefits of CSNPs is their ability to adhere to the lining of the stomach, thus improving the availability of the nutrient or drug for a longer duration, enhancing absorption and steady release [114]. CSNPs can be designed to target specific tissues or cells using special markers like folic acid and specific ligands, thus reducing side effects, toxicity and wastage. CSNPs also facilitate bioavailability by helping drugs cross cell membranes more efficiently, which is helpful for larger molecules like proteins. They are also equipped to shield capsules from degrading agents such as enzymes and other harsh conditions and ensure effectiveness until they reach the desired site [113, 114]. All these features collectively hold the potential for their broader application to sports nutrition to modernize the nutrient delivery system, which optimizes athletic performance and recovery.

2.1.11 Protein-Based Nanoparticles (PNPs)

Protein-based nanoparticles (PNPs) are another polymer-based nanoparticles that are nanoscopic in structure, made up of proteins, and they are capable of being self-assembled into uniform, stable structures within a diameter of 1–1000 nm [115–117]. They are known for their diverse properties, such as high surface-to-volume ratio, biocompatibility, biodegradability, and ease of functionalization, thus making them ideal for nutrient delivery and other applications [117, 118].

PNPs are built with a protein-rich core such as whey, casein, or albumin; their surface is chemically or genetically modifiable to improve stability and functionality [117, 119]. Whey protein nanoparticles are sourced from milk and characterized as biodegradable, biocompatible and cost-effective, potentially improving the stability of bioactive compounds like astaxanthin [120]. Casein nanoparticles are also formulated from milk, which is an affordable, non-toxic, and highly stable compound suitable for delivering nutrients and therapeutic agents [121]. Albumin nanoparticles are made from human or bovine serum albumin, which also exhibits stability and effectiveness in binding drugs by promoting encapsulation and targeted delivery [122]. Soy protein and corn zein nanoparticles derived from soy and corn also show biodegradability and bioavailability thus being used widely for food supplements to preserve flavor, colour and nutrients [119]. Egg albumin nano particles developed from eggs stand out as an excellent compound with their solid drug binding capacity, hence making them valuable in supplementation [123].

In sports nutrition, these nanoparticles can be utilized for supplementation of amino acids, based on the dietary needs of athletes to improve recovery and rebuilding of muscles and performance, facilitated through their unique mechanism.

2.1.12 Mechanism of Action

Encapsulation and Delivery of Bioactive Compounds

Encapsulation capabilities are one of the key mechanisms of most nanoparticles being used in biomedical applications. PNPs are not different from that, which is an indispensable tool for encapsulating and delivering bioactive compounds because of their unique structural features, such as their nanoscopic size and high surface area to volume ratio. Proteins are an essential macronutrient needed by the human body, and that is highly relevant in the context of athletic performance because of their building and repairing abilities. Protein can be extracted from different sources such as animal, plant or recombinant sources, and these extracted substances can be encapsulated and delivered with the help of PNPs because of their biocompatibility, biodegradability, and unique functional characteristics, such as amphiphilicity, self-assembly and binding potential [124, 125]. PNPs are also known for protecting proteins from degradation and ensuring effective delivery through encapsulation. There are different encapsulation techniques used to create these nanoparticles, such as emulsification, desolvation, coacervation, nanoprecipitation, liquid–liquid dispersion, and electro-hydrodynamic atomization; each of these is second to none, which provide critical mechanisms to facilitate efficient delivery [124, 126]. These methods can enhance the bioactivity and bioavailability of encapsulated nutrients but also influence structural stability and the release mechanism [126]. For athletes, this method can be used to provide nutrients such as vitamins, minerals, and antioxidants by encapsulating them, thus optimizing health and performance.

Improved Bioavailability, Controlled Release and Better Absorption

Protein-based nanoparticles are proven to protect bioactive compounds from degradation, thus ensuring their better stability and absorption in the body. Their nanoscopic structure makes it easier for them to pass through barriers, their wide surface area helps for efficient delivery, and their compatibility with living organisms and biodegradability facilitates safe absorption without side effects, thereby making them ideal for supplementation in sports nutrition [116, 127]. With surface modification by adding special ligands or peptides, PNPs can further improve absorption, as such helping them navigate to cells more easily [128]. Like other nanoparticles, these are modifiable to release nutrients steadily over time, supporting sustained energy during prolonged activity. The density of the polymer coating and packing in PNPs helps to control the nutrient delivery mechanism. The density of the coatings and tightly packed structures, allowing the NPs to release slowly and gradually through surface erosion in the body [129]. The desolvation process and stimulus-responsive design also enable the targeted release of substances triggered by changes in the environment like pH or temperature, thereby ensuring time-bound delivery of nutrients when needed most [130]. PNPs are finely equipped to play a vital role in unlocking peak athletic performance through faster and more efficient delivery of essential nutrients.

Targeted Delivery and Negligible Toxicity

PNPs have been found to be a potential tool for the targeted delivery of nutrients because of their modifying abilities with specific ligands or proteins, which facilitate the diffusion of nutrients to the targeted tissues or cells. Similar to other nanoparticles, the modifiable surface of the PNPs also enables the compounds to attach to the over-expressed targeted cells, thereby improving the absorption of nutrients and reducing wastage and side effects [131]. The way PNPs interact with cells and serum proteins depends upon the modified surface design, thus optimizing effective delivery and targeting. The biocompatible and biodegradable nature of PNPs also adds an extra layer to the exclusion of toxicity, thereby preventing harmful immune responses [132, 133]. PNPs safely break down into harmless byproducts, by ensuring the safety of nutrient delivery and targeting specific tissues, reducing systemic toxicity [134].

2.1.13 Dendrimers

Dendrimers are nano-sized, globular or spherical, multi-functional, monodispersed, synthetic macromolecules characterized by their branched, three-dimensional architecture with a symmetric core, an inner shell and an outer shell [135, 136]. The structure of dendrimers is highly organized and can be broken down into several key components. It consists of a central part known as the core, from which the branches emanate, and this core can be a single atom or a small group of atoms [137]. The branching units that extend from the core are known as Dendron, which comprises multiple layers, each representing a new generation. The outermost layer of dendrimers is called surface groups, which contain functional groups that can be modified for specific applications such as nutrient delivery [138].

2.1.14 Mechanism of Action

Encapsulation of Nutrients with Better Bioavailability, Sustained Release and Targeted Delivery

The three-dimensional structure of dendrimers facilitates efficient encapsulation of a wide range of bio active components while keeping their stability and bioavailability. They can encapsulate drug molecules within their interior through physical encapsulation, electrostatic interactions and hydrophobic effects [139]. Compounds can also covalently attach to the surface functional groups of dendrimers, thereby allowing efficient substance delivery [140]. Their properties like low polydispersity, biocompatibility and better water solubility collectively contribute to ideal bioactive compounds supply [136].

Dendrimers are well known for their ability to get attached to ligands, such as antibodies, peptides, or vitamins, which are intended to target specific tissues or cells, thereby improving the delivery of bioactive compounds. This targeted delivery can happen through active targeting, where ligands bind to receptors on the targeted cells, or through passive targeting with the EPR effect, both helping dendrimers to accumulate in the specific tissues where exactly the substance is needed [141, 142]. They are also potentially modifiable to respond to multiple stimuli that help them to facilitate more precise control over the release of substances; for example, multi-stimuli-responsive dendrimers can release drugs in response to both internal triggers such as pH, redox potential as well as external triggers such as light and temperature [143, 144]. By using dendrimers, athletes can take advantage of enhancing sports performance by ensuring the efficient delivery and absorption of nutrient supplements that support muscle recovery, energy production, and overall performance.

2.1.15 Cyclodextrin Nanoparticles (CDNPs)

Cyclodextrin nanoparticles (CDNPs) are nanoscopic particles obtained from cyclodextrins, which are cyclic oligosaccharides produced as a result of the enzymic breakdown of starch [145, 146]. CDs are crafted from glucose units joined by α-(1,4) glycosidic bonds and are in toroidal shape. They are formed with a hydrophobic interior to trap hydrophobic molecules with inclusion forming capacity and a hydrophilic exterior composed of hydroxyl groups, making them water-soluble [146].

CDs can be classified in terms of their structural difference and specific applications. The basic form of CDs is Native cyclodextrins, which differ in the number of glucose units they contain. For example, α-cyclodextrin has six glucose units, β-cyclodextrin possesses seven glucose units, and γ-cyclodextrin holds eight glucose units [146, 147]. Modified cyclodextrin is another type of CD, which includes both amphiphilic cyclodextrins and cyclodextrin nanosponges (CD-NS). Amphiphilic cyclodextrins are modifiable to carry both hydrophobic and hydrophilic molecules, which in turn enables them to form micelles, vesicles, and other complex structures [146–149]. Cyclodextrin nanosponges (CD-NS) are cross-linked cyclodextrin polymers that create a three-dimensional network [146, 150]. Cyclodextrin aggregates include inclusion complexes which trap guest molecules in the hydrophobic surface of the cyclodextrin and more complexed cyclodextrin-based supramolecular assemblies, such as polypseudorotaxanes and nanotubes [151, 152]. These structural specialities of CDs make them more flexible and adaptable in drug delivery systems, thus offering immense possibilities for their widespread usage in unlocking peak athletic performance.

2.1.16 Mechanism of Action

Encapsulation, Better Solubility, Structural Stability and Bioavailability

CDNPs have been proven successful in drug delivery systems because of their encapsulation abilities, structural stability, and solubility of attached molecules and their improved bioavailability. The structural speciality of CDNPs, including their hydrophobic inner layer and hydrophilic outer surface, allows them to encapsulate water-soluble molecules that are poorly soluble and enhance their solubility and bioavailability [153]. It helps them to form inclusion complexes with different guest molecules like nutrients and drugs [151]. The encapsulation process of CDs involves the guest molecule reaching the cavity, forming a non-covalent complex that enhances solubility, stability and bioavailability [154, 155]. CDs also prove stabilization by inhibiting molecules' aggregation, which is vital for maintaining stability under various conditions such as heat and pH changes [156]. They shield sensitive compounds from degradation caused by light, heat, and oxidation, thus ensuring the long-term efficacy of active ingredients in sports nutrition and dietary supplements [157].

Controlled Release and Targeted Delivery

CDNPs can release the drugs or nutrients in a controlled way through their unique ability to form inclusion complexes, thereby enhancing both the solubility and stability of the molecules. The release rate of encapsulated molecules can be influenced by several factors, such as pH sensitivity, where the release process is faster in acidic conditions (pH 5.3) than in neutral conditions (pH 7.4) [158]. It can also be controlled by external stimuli like temperature variations and ultrasound, where focused ultrasound is mainly effective in enhancing compound release [159]. Furthermore, host–guest interaction between CDNPs and guest molecules can be disrupted by ionic strength, competitive displacement, or protein binding, which can disrupt the host–guest interaction, offering a controlled release mechanism [160]. CDNPs also use several strategies to achieve targeted delivery, such as ligand conjugation, where nanoparticles are modified with specific ligands like folic acid or mannose, which bind to receptors on target cells, thus allowing for receptor-mediated endocytosis [161]. Surface modification, in which nanoparticles are coated with polymers or other functional groups, improves targeting capabilities and minimizes unwanted distribution [162]. These well-known mechanisms of CDNPs propose broad possibilities for utilizing these nanoparticles in sports contexts as per the nutritional needs of athletes, thus facilitating their utmost potential. Figure 3 describes the Mechanism of Controlled Release of Organic Nanoparticles-based nutrient delivery.

2.2 Inorganic Nanoparticles

Inorganic nanoparticles are nanoscale particles of inorganic materials; they are nanoscopic in size, 1–100 nm, and have a high surface area, making them useful in various applications, including sports nutrition. These nanoparticles can be made from metals such as gold and silver, metal oxides, semiconductors like cadmium selenide and many other inorganic compounds. Most inorganic nanoparticles have a core–shell structure, where the core is composed of one material, and the shell is composed of another, which generally enhances stability and functionality.

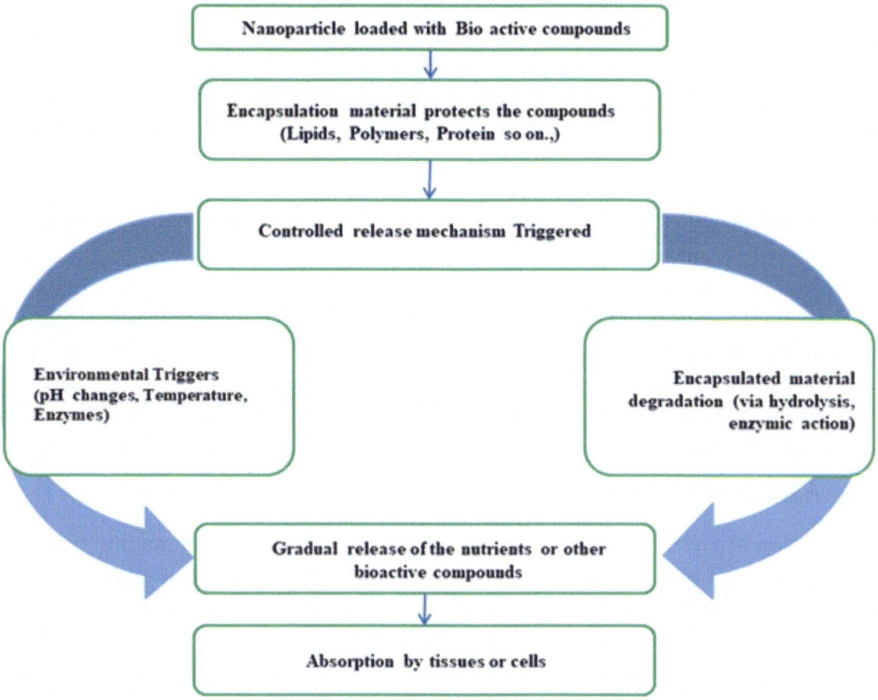

Fig. 3 Mechanism of controlled release of organic nanoparticles-based nutrient delivery

2.2.1 Metal-Based Nanoparticles (MNPs)

Metal-based nanoparticles are tiny-sized particles of metals such as gold, silver, copper, iron, etc. or their compounds like oxides, sulfides, etc., ranging between 1 and 100 nm. They are prepared through various physical, chemical and biological methods and formed in various shapes like spheres, rods, cubes, platelets and branched structures. High surface area to volume ratio with a capacity to react and interact with biological systems and physio-chemical features such as optical, electrical and chemical properties make them versatile in different domains and can be used in sports.

Metallic Nanoparticles

Metallic nanoparticles are tiny clusters or aggregates of metal atoms, such as gold, silver, iron, zinc, and other metals. These nanoparticles have a nanoscopic structure that allows them to interact more effectively with biological systems. Their unique properties enable them to carry various compounds, enhancing their potential application in biomedicine, nutrition and other nanotechnology-induced fields.

Gold Nanoparticles (AuNPs)

Gold nanoparticles are microscopic particles of gold, usually ranging in size between 1 and 100 nm, and possess a high surface area relative to their volume [163]. These AuNPs, formulated with gold atoms, which are inert and work well with biological systems, and they can be coated with molecules, polymers or phospholipids to enhance their solubility and stability in biological fluids that help them improve their overall effectiveness [164]. AuNPs are appreciated for their ability to defend against oxidative stress and inflammation [48] and their interesting optical properties, such as strong and tunable light absorption and fluorescence, along with the capability to convert light to heat, which broadens their capacity in therapeutic and diagnostic applications [165]. These capacities can be utilized for performance enhancement in sports, where AuNPs can assist in delivering nutritional nano-compounds specifically focused in anti-inflammatory, antioxidative and antimicrobial benefits, with better efficacy, that help them boost athletic performance by fastening the recovery.

Silver Nanoparticles (AgNPs)

Silver nanoparticles are ultrafine particles of silver, with dimension ranging between 1 and 100 nm. [166]. These interesting characteristics, such as their small size and unique composition, offer unique properties such as electronic, magnetic and plasmonic properties, which enable them to be stabilized with various agents such as gemini surfactants, which in turn enhance their biological activity and stability, thereby influencing their functionality and application in different fields [167]. They also can fight against a wide range of bacteria, viruses and fungi with their antimicrobial and bactericidal effects, opening a new avenue for nutritional supplementation in sports [168]. They can also be integrated into sportswear to provide these properties, reducing the risk of infections and enhancing overall performance.

2.2.2 Metal Oxide Nanoparticles (MONPs)

Metal Oxide Nanoparticles are nano-structured particles derived from metal oxides such as iron oxides, zinc oxide, magnesium oxide, etc., with dimensions ranging between 1 and 100 nm. They are beneficial for various applications because of their diverse features, such as their small size and high surface area.

Iron Oxide Nanoparticles (IONPs)

Iron oxide nanoparticles (IONPs) are a type of nanomaterial composed of an iron core, which can be in the form of magnetite (Fe_3O_4) or maghemite ($Y–Fe_2O_3$) surrounded by a polymer coating that enhances their stability and functionality [169]. IONPs are superparamagnetic when their size is reduced below 20 nm, making them useful in various applications, and they also possess intrinsic enzyme-like activities, acting as nano enzymes, thereby facilitating biological application [169, 170]. IONPs can increase cell viability at low concentrations, indicating potential benefits from health supplements.

Zinc Oxide Nanoparticles (ZnONPs)

Zinc oxide nanoparticles (ZnONPs) are ultrafine particles of zinc oxide within the size of 1–100 nm, crystalline in structure, and they possess a high surface area to volume ratio with enhanced reactivity and interaction with biological organisms [171]. They are known for strong antifungal and antibacterial properties, which are very helpful in preventing infections [172]. They can generate reactive oxygen species (ROS) under light, which can be used for antimicrobial applications [172]. ZnONPs have the potential to be used as dietary supplements in sports, thereby enhancing zinc bioavailability, which is crucial for various physiological processes such as protein and carbohydrate metabolism, growth and immune function. Their immune-modulatory features also help boost the immune system, promoting wound healing, recovery from injuries and overall health.

Magnesium Oxide Nanoparticles (MgONPs)

Magnesium oxide nanoparticles (MgONPs), composed of magnesium oxide, are nano-sized particles that measure between 5 and 100 nm. They are well-suited to living things and possess diverse features like structural stability and bioavailability [173]. Despite this, they also have solid antimicrobial, antioxidant, anti-cancer, and anti-diabetic capabilities. The anti-oxidant properties of MgONPs help reduce oxidative stress, which is common in athletes due to intense physical activities, thereby aiding faster recovery and growth, and their antimicrobial capability protects from infections [174]. Apart from this, their anti-inflammatory properties can potentially reduce muscle soreness and inflammation after exercise. They are suitable for nutrient delivery in sports nutrition due to their biocompatibility, stability and protection.

2.2.3 Other Mineral Based Nanoparticles

Selenium Nanoparticles (SeNPs)

Selenium Nanoparticles, which are nanoscopic structured elemental selenium within the dimensions from 1 to 100 nm, are well known for their biocompatibility, bioactivity, and low toxicity compared to traditional forms of selenium [175]. They are usually formed in different ways, such as physical, chemical and biological methods, and they can work well with biological systems. Their distinctive mechanism includes combating inflammation and countering oxidative stress arising from their antioxidant, anti-inflammatory and immune-modulating properties [176]. Adequate selenium levels support mitochondrial health, better immune responses and antioxidant defense, and efficient thyroid hormone synthesis and metabolism, which in turn helps in recovery, energy production and metabolic rate that directly impact athletic performance [177–179]. Current investigations are more experimented with animal physical performance, and more human trials are necessary to ensure the efficacy of athletes.

Calcium Phosphate Nanoparticles (CaPNPs)

Calcium Phosphate Nanoparticles (CaPNPs) are tiny particles composed of small mineral crystals that contain calcium cations, phosphate anions, and, in some cases, hydrogens or hydroxide ions [180]. These nanoparticles are known for their chemical similarity to the natural mineral constituents found in bones and teeth, which makes them bio-degradable and compatible with living things [181]. Because of their chemical similarity to human hard tissues, these are effectively working with biological systems, thus making them suitable for various biomedical and nutritional applications. CaPNPs can carry and deliver therapeutic agents such as nucleic acid, drugs, protein and enzymes into cells, thus opening an avenue for nutrients and supplements delivery such as calcium and phosphate, in sports nutrition that will contribute to better athletic performance.

2.2.4 Mechanisms of Action of Inorganic Nanoparticles

Antioxidant and Antimicrobial Properties

Inorganic metal-based nanoparticles can defend antioxidants that improve performance and aid recovery. They can directly scavenge and neutralize reactive oxygen species (ROS), which are harmful byproducts of intense physical activities, by enhancing the activity of antioxidant enzymes like superoxide dismutase (SOD) and catalase [182, 183]. They can also produce ROS such as hydrogen peroxide, hydroxyl radicals and superoxide, which damage microbial cell components, thus leading to cell death. They release ions that can disrupt microbial cell membranes and interfere with cellular processes, thus causing structural damage and inhibiting such cell growth. Being bioavailable, these nanoparticles improve the absorption of antioxidants [184–186].

Anti-Inflammatory Pathways

Mineral-based nanoparticles possess anti-inflammatory properties. They can easily intersect biological barriers because of their tiny nano-scale size and directly interact with cells, making them fight inflammation without needing additional drug payload [187]. Inflammation and injuries are among the most common things for athletes, and efficiently dealing with them is very important to improve their performance. These nanoparticles have proven successful in entrapping and delivering anti-inflammatory agents directly to the site of inflammation, thereby mitigating the pain. So, these tiny materials are very suitable for encapsulating and delivering anti-inflammatory agents like Omega-3 fatty acids and prebiotics. These nanoparticles can stabilise and solubilise them to fasten the recovery process and shield against exercise-induced inflammations through their unique anti-inflammatory pathways.

Better Immune Response

Inorganic mineral-based nanoparticles can stimulate the release of various cytokines, such as pro-inflammatory and anti-inflammatory cytokines, which enhance the immune response, thus making the body more efficient in fighting against injuries and infections and facilitating healing [188]. Being effective in activating the natural immune cells, these nanoparticles promote the secretion of immune molecules. They also enhance the efficacy of vaccines, thus ensuring prolonged immunity, and interact with gut microbiota by influencing its composition and activity, thereby improving recovery, nutrient absorption and even mental health [189]. These features of these nanoparticles can help athletes recover faster from damage.

Encapsulation, Targeted Delivery and Controlled Release

Inorganic mineral-based nanoparticles can be attached to ligands that have a high affinity for specific receptors on targeted cells [190]. Like any nanoparticle that is used in biomedical applications, these nanoparticles' surfaces are also modifiable to specific stimuli such as pH, redox conditions, light, or enzymes, which enable them to release the specific substance to the needed area in a controlled manner [191]. To reduce toxicity, these particles are often encapsulated with biocompatible materials, which ensure safety and save from potential side effects. These features make them potential tools for mineral and other nutrient delivery in sports nutrition. Table 1 Categorises nanoparticles and their synthesis, mechanisms and potential applications in Sports Nutrition.

Table 1 Categorisation of nanoparticles and their synthesis, mechanisms and potential applications in sports nutrition

Nanoparticle type	Structure	Synthesis methods	Mechanisms	Potential applications in sports nutrition	Ref.
Liposomes	Spherical vesicles with a phospholipid bilayer	Thin-film hydration, sonication, and extrusion	Biocompatibility, encapsulation, surface modification, and controlled release	Delivery of hydrophobic nutrients, encapsulation, antioxidants, polyphenols, protein and amino acids	[192, 193]
Solid lipid nanoparticles (SLNs)	Solid lipid core stabilized by surfactants	High-pressure homogenization, microemulsion technique	Encapsulation, High stability, controlled release, protection of active compounds and targeted delivery	Encapsulation of lipophilic, vitamins, antioxidant, anti-inflammatory substance delivery and supplementation	[41, 194–197]
Nano emulsions	Nanoscale droplets dispersed in a liquid	High-shear mixing, ultrasonic emulsification, low energy method	High surface area, enhanced solubility, sustained release, stability and better functional performance	Delivery of fat-soluble vitamins in the form of functional beverages and other supplements	[63, 70]
Polymer nanoparticles	Nano-sized particles made from polymers	Emulsion polymerization, nanoprecipitation	Customizable size, targeted delivery, biodegradability, controlled release, bioavailability	Controlled therapeutic drug delivery and nutrient release	[48, 198]
Chitosan nanoparticles	Derived from chitosan, a natural biopolymer	Ionotropic gelation, self-assembly, and emulsion cross-linking	Biocompatibility, absorption, antioxidant activity, efficient metabolic functions, weight management, mucoadhesion, and muscle health	Improved absorption of minerals, gut health, weight management, supplementation of nutrients, antioxidants and enzymes	[99, 199–201]
Protein-Based nanoparticles	Particles formed from proteins	Desolvation, Nanoprecipitation, coacervation, emulsification, electro-spraying	Encapsulation Biodegradability, low immunogenicity, stability, bioavailability, controlled and targeted release	Encapsulation of amino acids, peptides, proteins, other nutrients, antioxidants and enzyme delivery	[68, 116, 126]

(continued)

Table 1 (continued)

Nanoparticle type	Structure	Synthesis methods	Mechanisms	Potential applications in sports nutrition	Ref.
Cyclodextrin nanoparticles	Cyclic oligosaccharides forming inclusion complexes	Kneading, co-precipitation, freeze-drying spray drying	Biocompatibility, solubility enhancement, controlled release, stability	Delivery of hydrophobic bioactives, flavor masking, nutrient supplementation	[154, 155, 202]
Dendrimers	Highly branched, tree-like structures	Divergent and convergent synthesis	Encapsulation, multiple functional groups, bioavailability, targeted delivery. Controlled release	Multifunctional nutrient delivery, antioxidant transport	[136, 203–205]
Metal-Based nanoparticles	Nano-sized particles of metals	Chemical method, biological method, physical vapor deposition	High reactivity, antimicrobial properties, anti-inflammatory, antioxidant activity and stability Thermal stability, catalytic activity, UV protection	Essential mineral supplementation, antioxidants, anti-microbial agents delivery, enhancing bone health, bio-active sports wear	[206–210]

3 Application of Nanoparticle for Exercise Performance: Current Insights

The physiological and metabolic efficiency of nano-enhanced bioactive compounds is well-known in drug delivery for different disease treatments. This opens the door to a similar application to performance-enhancement substances in the sports nutrition domain for peak athletic performance. NPs are predominantly utilised to supply nutritional bio active compounds to enhance physical performance. Only limited human trials are available in nanoparticle-based nutrition applied to athletic performance enhancement, pointing to the need for more experiments with athletes. Some of the trials are discussed below (Table 2).

Table 2 Application of nanoparticle utilized to enhance exercise performance

Nanoparticle type	Application	Benefits	Concerns	Ref.
Resveratrol-loaded lipid nanoparticles	Endurance exercise	Improved performance, antioxidant activity, reduced inflammation	Evidence from animal studies. human studies needed	[46–48]
Iron-loaded liposomes	Endurance exercise	Increased serum and liver iron content, higher glycogen levels	Researched on animal model. human studies needed	[48, 211]
Resveratrol-loaded solid lipid nanoparticles (RSV-SLNs)	Anti-fatigue supplement	Prolonged exercise duration, improved antioxidant defenses	Study on animal models, human studies needed	[46–48]
Curcumin nanoparticles	Exercise recovery	Reduced muscle soreness, damage, oxidative stress	Human trial, further human studies needed	[212]
Natural-mineral-based nanomaterials	Cardiovascular function	Suppressed heart rate elevation during exercise	Human trial mechanisms need further investigation	[10]
Chitosan-coated L-arginine nanoparticles	Muscle strength and recovery	Improved muscle strength, reduced oxidative damage	Animal model, further studies on diverse measurements needed	[201]

3.1 Nano-based Nutrition for Exercise Performance: Evidence from Human Trials

3.1.1 Effects on Cardiovascular and Pulmonary Function in Athletes

Nanoparticle possess broad capabilities to be used in sports nutrition; however, only limited studies are examining their use in enhancing athletic performance. Cardio-respiratory fitness is one of the core components that unlock the optimum athletic performance. NPs can entrap water, fat-soluble nutrients, and a wide range of minerals and coenzymes.

A natural mineral-based nanomaterial developed by Japanese hot springs was tested for its potential fitness improvement capabilities on 18 students from athletic clubs at Fukushima university in Japan. It has been recorded that these materials significantly suppressed the rise of heart rate during treadmill tests without affecting other respiratory metrics such as maximal oxygen consumption, CO_2 production, etc. [10]. This research proposed that nano-mineral can improve cardio-respiratory function through unique mechanisms of enhancing endothelial nitric oxide synthase activity, which optimises haemoglobin efficiency to carry oxygen and stimulate chemical signals that help to relax and expand the blood vessels, thereby controlling functions of blood vessels and oxygen delivery [10]. This application has proven the potential of nanomaterial to enhance cardiovascular function without drugs or physical intervention in an eco-friendly way.

3.2 Nano-based Nutrition for Exercise Performance: Evidence from Animal Models

Some animal-based experiments have recorded impactful physiological benefits in their exercise performance. In the absence of human trials, few studies that reported potential enhancement in animal physiology that is closely related to humans are explained below.

3.2.1 Performance Enhancement with Nano-nutritional Compounds

Nutritional nano-compounds are one of the great discoveries in sports nutrition with the help of nanotechnology, which have shown considerable promise in enhancing exercise performance by improving glucose levels during physical activity, increasing glycogen stores in both muscles and the liver, and boosting mitochondrial function. Commercially available nano-formulated vitamins, such as Nano-Vitamin E and CoQ10 enzymes, have demonstrated performance-enhancing effects in racehorses that indicate similar benefits may extend to human athletes [213].

Studies that experimented with equines have showcased the immense capability of nano-nutrition in optimizing nutrition and performance. For example, the nano-encapsulation of vitamins A, C and E has been found to enhance absorption, mitigate inflammation, and facilitate recovery from exertion or injury. Nano-forms of vitamin E specifically play a critical role in supporting neurological and muscular functions, which are essential for the performance and recovery of race horses, breeding stallions and lactating mares [213, 214]. Likewise, the CoQ10 enzyme in nano-suspension form has been shown to improve oxidative stress management and enhance energy production [213–215]. Other nanoparticles like casein and nano-minerals have proven their abilities in supplying nutrients to young horses, optimizing nutrient bioavailability and reducing side effects, which all together enhances potential in sports performance [213]. These advancement in equine sports nutrition not only enhance the performance and recovery of equine athletes but also provide valuable insight that could be applied to human athletic performance, recovery and muscle health through innovative nano-based strategies.

3.2.2 Antioxidant and Anti-inflammatory Effects

- *Resveratrol-Loaded Lipid Nanoparticles (RSV-SLNs)*

Resveratrol-loaded lipid nanoparticles offered considerable contributions to the enhancement of physiological performance in animal models, which opened a broad opportunity to be incorporated into sports nutrition and enhancement of athletic performance. Resveratrol is a type of natural compound found in plants, and it is famous for its antioxidant mechanism [46]. The use of resveratrol in biomedicine is limited because of their poor water solubility, and they cannot stay in the bloodstream for an extended duration. However, nanoparticles like SLNs can improve these substances' solubility absorption rate and bioavailability by increasing circulation time [47]. While resveratrol is entrapped with the help of SLNs, they have shown promising progression of endurance in animal models with their bioavailability and antioxidant activity, thereby reducing oxidative stress during exercise [48]. This oxidative stress management promoted overall performance and aided recovery, opening a great avenue to experiment with in athletes. They can also boost the functional capacity of mitochondria, which is crucial for energy production and supply during endurance activities [48, 216]. They are also effective in removing damaged cells and promoting cell repair through the regulation of specific proteins, as well as maintaining an increased serum and liver iron content, higher glucose levels and muscle glycogen stores at the time of exercise, thus providing sustained energy [48, 216].

- *Iron-Loaded Liposomes*

Iron-loaded liposomes are a special type of nutritional nano-compound in which iron is encapsulated within liposomes to enhance its absorption and bioavailability and reduce side effects [217]. Studies with rats showed that iron-loaded liposomes can fix exercise-related anaemia, a condition where iron levels in the blood become low due to intense exercise. They have been proven to boost blood count, serum iron, and liver iron levels after vigorous activity [48, 211]. They are also equipped to eliminate oxidative stress, common after exercise. Iron is one of the essential nutrients for athletes that support oxygen transport, energy production and muscle function, but it is very prone to be deficient due to factors like sweating and gastrointestinal bleeding in enduring sports activities [218]. When iron is encapsulated with liposomes, it offers diverse benefits such as better absorption and reduced toxicity compared to traditional supplementation techniques, optimising muscles' functional capacity and helping unlock peak athletic performance. They also improve mitochondrial function in energy production, thus making them a promising solution for maintaining iron status and preventing fatigue in athletes.

- *Layered Double Hydroxide Nanoparticles for Folic Acid Delivery (LDH-NPs)*

LDH nanoparticles are ultra-small particles designed with special layers that can hold drugs or genes and release them in a controlled way. They can be mixed with folic acid to target specific cells with more folate receptors. These particles get into cells through clathrin-mediated endocytosis, and their nano-size helps them penetrate through cell walls more easily [219]. In animal studies, folic acid-LDH nanoparticles helped to improve exercise performance. For example, a study with mice has shown that, with the help of folic acid-LDH, they swim 32% and 51% longer compared to mice treated with just folic acid or no treatment [220]. This was because the nanoparticles helped to lower harmful substances in the body, such as blood urea, nitrogen, lactic acid, etc., and improved energy stores in the muscles and liver. This endurance enhancement also facilitates antioxidant defence [48, 220].

- *Redox-Active Nanoparticles with Nitroxide Radicals Used in Rodents*

Redox-active nanoparticles (RNPs) with nitroxide radicals are nano-size particles that help to defend against oxidative damage in the body caused by harmful molecules called ROS (reactive oxygen species). These nanoparticles act like superoxide dismutase (SOD) that is a special protein that protects cells by breaking down harmful molecules called superoxide, which is a waste product made when our bodies use oxygen, and boost catalase enzymes in proteins, thereby mitigating oxidative stress [221, 222]. Studies with rats and mice showed that RNPs can enhance exercise capacity by facilitating superior endurance, reducing oxidative stress and maintaining red blood cell (RBC) levels [48, 223]. They also considerably contributed to increasing glucose and glycogen levels, which assists in sustained energy release during prolonged physical activity [48]. These mechanisms not only improve the muscular integrity but also accelerating recovery from exercise induced oxidative stress. Even though they are working well with animal, more trials are needed with humans to assure these effects in athletics performance.

- *Chitosan-Coated L-Arginine Nanoparticles (Nano L-Arginine)*

Chitosan-coated L-arginine nanoparticles (Nano L-arginine) is another novel nano-compound experimented with in rats to explore nanoparticle-based nutrition's capacity to improve exercise performance and muscle health. The research centered on the combined effect of swimming exercise and chitosan-coated L-arginine nanoparticles that examine their impact on physiological activities such as mitochondrial function, oxidative stress, and specific autophagy markers within the soleus muscle [201]. Ageing rats, as expected, have experienced a noticeable decline in muscle strength and structural integrity, and also more oxidative damage and shortage of key proteins like LC3, Nix, and Bnip3, which are closely linked to autophagy processes [201]. However, the improvements were striking and highly relevant when the rats were treated with nano L-arginine in conjunction with exercise. This combination led to enhanced muscle strength, better histological structure and predominantly reduced oxidative stress levels, evident from the lower presence of reactive oxygen species (ROS) in the muscle tissue [201]. Protective proteins like Bcl-xL, crucial for muscle preservation, were better sustained with this dual approach. This trial strengthens the notion that nanoparticle-based nutrient delivery is more powerful in improving exercise performance when effectively paired with exercise and can mitigate the negative effects of aging on muscle tissue, thereby enhancing functional capacity of muscles and physical performance. Figure 4 describes the Enhancement of Exercise performance with Nutritional nano-compounds.

4 Ethical, Safety, and Regulatory Considerations for Nanoparticles in Sports Nutrition

Nanoparticle-based nutrient delivery is an emerging trend in sports nutrition due to its ability to enhance athletic performance, improve recovery, and increase nutrient absorption capacity. Sports people and big companies are looking to harness its power as it has wide useful mechanisms. However, as with any new technology, there are important issues to consider, such as safety concerns, ethics and regulation of its usage. This section delves into the core issues that are related to the use of nanoparticles in sports nutrition in the light of considerations of safety, ethics, and regulations to be followed.

4.1 Safety Concerns

As we discussed throughout the chapter, nanoparticles are well known for their nano-scale size and larger surface area, thus making them able to penetrate easily to the cell membranes and offering a high surface area for encapsulation, which helps them to deliver poorly soluble nutrients with better absorption. Even though

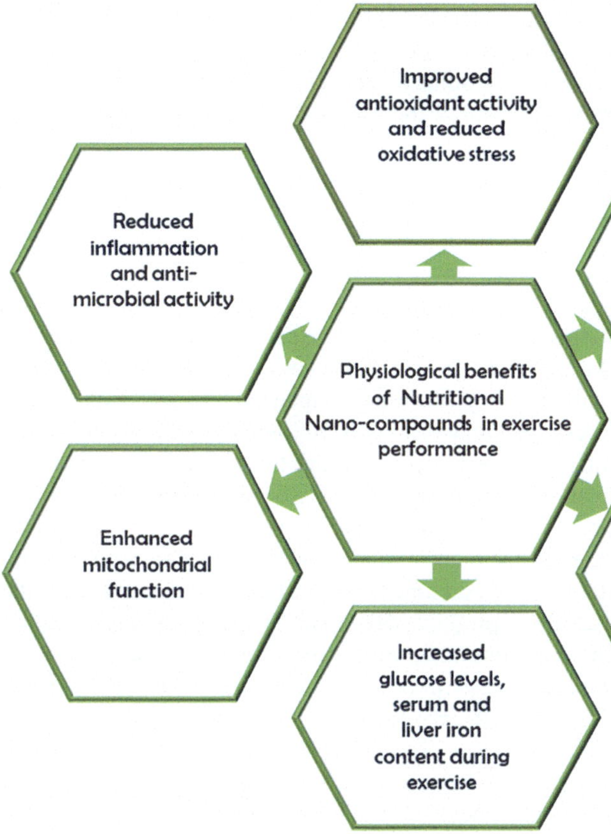

Fig. 4 Enhancement of exercise performance with nutritional nano-compounds

these properties are appreciable, there is some concern about potential health risks, such as side effects or allergies. Nano-based nutrition possesses distinct behaviours compared to its bulk equivalents, including differences in particle agglomeration, bio-kinetics, and biological interactions [224]. These features can lead them towards unique toxicological effects that must be carefully evaluated. Current research on the interactions between nanoparticles and biological systems, especially in the context of athletic performance, is limited; therefore, a careful case-by-case evaluation of formulations and their applications is necessary to address safety concerns regarding nano-nutrients in athletes [224].

The safety assessment of nanoparticles should extend beyond their chemical nature to include their physical form, fate and behaviour in the body [224]. This includes understanding whether they can retain their nano-scale structure and translocate from the gastrointestinal tract (GI) tract to other body parts. It is essential to understand how these vehicles behave in the GI tract. The surfactant and vehicle components used in the formulation also must be assessed for safety [224]. They are primarily composed of natural materials and nano-vehicles derived from liposomes and emulsion and are considered lower risk, but those based on synthetic or more stable materials may require additional safety considerations. So, any product containing nanoparticles must be ensured they are safe for consumption through rigorous standardised testing.

4.2 *Regulations*

Most current regulatory bodies treat nano-based products similarly to traditional food products; however, this approach is not fair enough to ensure the broader safety aspects of nano-nutritional or performance-enhancing compounds. Nano-nutritional compounds are very different from conventional ingredients since they hold unique structures, features and mechanisms, and existing regulatory frameworks are not entirely equipped to address the unique risks associated with nanoparticles. Regarding current regulatory bodies, it is important to recognize their crucial role in ensuring nanoparticle-based nutritional products follow the safety and efficacy standards. In the United States, the Food and Drug Administration deals with the safety, effectiveness and labelling of products developed by using nanoparticles and certifying them as safe [225]. Environmental Protection Agency oversees the environmental hazards and compliance associated with nanoparticles under the Toxic Substances Control Act (TSCA) [226]. By focusing on labelling and food additives, the food safety regulations in the European Union govern the use of such particles in dietary products [227]. The sports anti-doping agencies and sports-specific governing bodies like the International Olympic Committee (IOC) are more connected with sighting the safety and ethical regulations of nano-particle-based nutritional compounds used in sports, specifically formulated for enhancing performance and recovery processes since they are equipped to detect and ban the stuff if it contains any prohibited and toxic substances that give athletes any unfair advantage or health issues [228, 229]. But still, there are significant gaps in the global regulations that specifically address the challenges associated with nano-based performance enhancement aids. A specific and comprehensive regulatory approach formulated for nano-based performance-inducing substances is essential, dealing with its distinct properties to ensure these products are safe and effective.

4.3 *Ethical Considerations*

Athletes deserve the right to make informed choices about what they take into their bodies, which makes autonomy and transparency essential in sports nutrition [230]. Products containing nanoparticles must be clearly labelled so consumers know what they consume. As the use of nanotechnology in supplements is still relatively new, many athletes may not even realize that these ingredients are present in the product they use, its composition and how it will affect their health. Ethical practices in the industry must prioritise this transparency, which is vital to create trust between manufacturers and users, thus enabling athletes to make right choices that align with their preferences, such as personal health, performance goals, values and concerns [231]. This underscores the importance of transparent labelling to support the choice of customers. The products should be rigorously tested, and the claims should back up scientific evidence from studies. The industry should follow ethical marketing practices and refrain from exaggerated and unrealistic claims regarding nanoparticle-based solutions' efficacy [232]. It should be managed to ensure that these technologies do not create any disparities between athletes in access and affordability from diverse socio-economic backgrounds, as this could undermine the principles of fairness and inclusivity in sports [233].

5 Future Perspectives and Conclusion

Nanoparticles are changing the outlook of the athletic world by unlocking new dimensions in sports nutrition. These outstanding technological innovations are not just passing a trend; they represent a promising leap forward in how athletes can enhance their performance and overall well-being. From modernizing sports equipment, apparel and tracking devices to facilitating nutrient supplementation, nanoparticles are reshaping the frontiers of athletic performance by improving nutrient bioavailability, absorption, targeted delivery with sustained and controlled release, and accelerating recovery times. Together, these advancements provide that extra push athletes need to achieve peak athletic performance. Rather than just delivering nutrients, nanoparticle-based supplements offer additional mechanisms that are able to optimize athletic performance, such as anti-inflammatory, anti-microbial and antioxidant properties. This is particularly important in the context of athletic performance since oxidative stress, inflammation, and microbial activity can hinder athletic performance by causing infections and impairing the recovery process. Nanoparticles are equipped to outperform the limitations of conventional supplements by promoting faster recovery, enhancing endurance, reducing fatigue and minimizing injuries through these mechanisms.

The inclusion of nanoparticles in sports nutrition is still in its preliminary phase, especially in enhancing athletic performance. With more trials, Future research could explore broader applications in sports like personalised nano-nutrition. Imagine a world where supplementation is formulated based explicitly on factors such as an athlete's genetic makeup, training intensity and performance goal. As these advancements continue to grow, more collaborations between sports scientists, nutritionists, and nanotechnology experts will take the athletic field to a golden phase. The safety and ethical aspects of using nanoparticles in athletes is a crucial issue to be taken care of. Only a handful of research is available, and more focused, well-conducted research on athletes is needed to ensure these nanoparticles' safety and long-term impact.

Ethical considerations such as transparency, informed consent and fair competition must also be followed to maintain trust and integrity in the sports domain.

In brief, nanoparticles possess promising abilities that can carry any field to tremendous success, including sports nutrition. Their remarkable structure, features, and mechanisms are well equipped for athletic needs like nutrition, therapeutic medicines, sports equipment, apparel, electronic devices, and many more, which optimize their training capacity, sports performance, recovery and rehabilitation. Let's look forward to a future full of opportunities where the athletic realm will progress further towards unprecedented levels of excellence within the wings of nanotechnology.

References

1. Singh D (2020) Cyanobacteria as a source of nanoparticle: application and future projections. In: Advances in cyanobacterial biology, pp 319–331. https://doi.org/10.1016/B978-0-12-819311-2.00021-8
2. Mertz L (2014) Size matters: in labs now and clinics soon, nanotech is a hot ticket. IEEE Pulse 5(2):8–20. https://doi.org/10.1109/MPUL.2013.2296796
3. Toscano F, Torres-Arias M (2023) Nanoparticles cellular uptake, trafficking, activation, toxicity and in vitro evaluation. Curr Res Immunol 4. https://doi.org/10.1016/j.crimmu.2023.100073
4. Krpetić Ž, Anguissola S, Garry D, Kelly PM, Dawson KA (2014) Nanomaterials: impact on cells and cell organelles. Adv Exp Med Biol 811:135–156. https://doi.org/10.1007/978-94-017-8739-0_8
5. Gopi Krishna P, Kameswaran S, Sri Ranjani T, Gunavathi Y (2020) Recent developments in nanoencapsulation of bioactive compounds of microbial sources and their biomedical applications. In: Recent developments in applied microbiology and biochemistry, vol 2, pp 141–152. https://doi.org/10.1016/B978-0-12-821406-0.00014-X
6. Ji A, Hamid A, Andrabi SAH, Haq E, Tombuloglu H (2024) Interaction of nanoparticles with biomolecules. In: Molecular impacts of nanoparticles on plants and algae, pp 143–157. https://doi.org/10.1016/B978-0-323-95721-2.00008-7
7. Chen Z (2023) Application of nanocomposite material in the tennis equipment to avoid the injury. Adv Nano Res 14(3):235–246. https://doi.org/10.12989/anr.2023.14.3.235
8. Da Costa LP (2020) Engineered nanomaterials in the sports industry. In: Handbook of nanomaterials for manufacturing applications, pp 309–320. https://doi.org/10.1016/B978-0-12-821381-0.00014-4
9. Zhao X, Li M, Zhao J, Wang X (2024) Intelligent sportswear design: innovative applications based on conjugated nanomaterials. Sci Eng Comp Mater 31(1). https://doi.org/10.1515/secm-2024-0032
10. Akiyama T, Hatakeyama S, Kawamoto K, Nihei H, Hirata T, Nomura T (2022) An externally-applied, natural-mineral-based novel nanomaterial IFMC improves cardiopulmonary function under aerobic exercise. Nanomaterials 12(6). https://doi.org/10.3390/nano12060980
11. Ealias AM, Saravanakumar MP (2017) A review on the classification, characterisation, synthesis of nanoparticles and their application. In: Presented at the IOP conference series: materials science and engineering. https://doi.org/10.1088/1757-899X/263/3/032019
12. Eker F et al (2024) A comprehensive review of nanoparticles: from classification to application and toxicity. Molecules 29(15). https://doi.org/10.3390/molecules29153482
13. Gupta T, Jaiswal A, Kumar R, Kumar R (2024) Organic nanomaterials. In: Advancements in nanomaterials for energy conversion and storage, pp 184–197. https://doi.org/10.1201/9781003481157-11

14. Saha A, Biswas A, Islam SKM (2023) Transport of drugs using liposomes. In: Green sustainable process for chemical and environmental engineering and science: recent advances in nanocarriers, pp 127–140. https://doi.org/10.1016/B978-0-323-95171-5.00003-0
15. Vanić Ž (2012) Liposomes as drug carriers: structure properties and classification. Farm Glas 68(6):391–400
16. Didandeh M, Hasanpour Souderjani A, Montazeri M, Rahimi F, Amoabediny G (2023) Applications of liposomes in nanomedicine. In: Nanomedicine: technologies and applications, pp 381–395. https://doi.org/10.1016/B978-0-12-818627-5.00013-0
17. Vozza G, Khalid M, Byrne HJ, Ryan S, Frias J (2016) Nutrition—nutrient delivery. In: Nutrient delivery, pp 1–42. https://doi.org/10.1016/B978-0-12-804304-2.00001-9
18. Guimarães D, Cavaco-Paulo A, Nogueira E (2021) Design of liposomes as drug delivery system for therapeutic applications. Int J Pharmaceutics 601. https://doi.org/10.1016/j.ijpharm.2021.120571
19. Siler-Marinkovic S (2016) Liposomes as drug delivery systems in dermal and transdermal drug delivery. In: Percutaneous penetration enhancers chemical methods in penetration enhancement: nanocarriers, pp 15–38. https://doi.org/10.1007/978-3-662-47862-2_2
20. Altemimi AB et al (2024) Application of nanoparticles in human nutrition: a review. Nutrients 16(5). https://doi.org/10.3390/nu16050636
21. Luo M, Oomah BD, Akoetey W, Zhang Y, Daneshfozoun H, Hosseinian F (2024) Liposomes as sustainable delivery systems in food, cosmetic, and pharmaceutical applications. JAOCS J Am Oil Chem Soc. https://doi.org/10.1002/aocs.12907
22. MacLachlan I (2007) Liposomal formulations for nucleic acid delivery. In: Antisense drug technology: principles, strategies, and applications, 2nd edn, pp 237–270. [Online]. Available: https://www.scopus.com/inward/record.uri?eid=2-s2.0-84975143207&partnerID=40&md5=113204cac987fbcc24f86561ad435d68
23. Rizkita LD, Putri RGP, Farid M, Rizkawati M, Wikaningtyas P (2024) Liposome drug delivery in combating the widespread topical antibiotic resistance: a narrative review. Beni-Suef Univ J Basic Appl Sci 13(1):90. https://doi.org/10.1186/s43088-024-00545-2
24. Emami S, Azadmard-Damirchi S, Peighambardoust SH, Valizadeh H, Hesari J (2016) Liposomes as carrier vehicles for functional compounds in food sector. J Exp Nanosci 11(9):737–759. https://doi.org/10.1080/17458080.2016.1148273
25. Taylor TM, Davidson PM, Bruce BD, Weiss J (2005) Liposomal nanocapsules in food science and agriculture. Crit Rev Food Sci Nutr 45(7–8):587–605. https://doi.org/10.1080/10408390591001135
26. Tian H (2022) Research and development of nano-lipid functional food and its impact on sports and health industry. Int J Nanotechnol 19(6–11):885–894. https://doi.org/10.1504/IJNT.2022.128984
27. Koçer A (2007) A remote controlled valve in liposomes for triggered liposomal release. J Liposome Res 219–225. https://doi.org/10.1080/08982100701528203
28. Senior J, Gregoriadis G (2018) Methodology in assessing liposomal stability in the presence of blood, clearance from the circulation of injected animals, and uptake by tissues. In: Liposome technology volume III: targeted drug delivery and biological interaction, pp 263–282. https://doi.org/10.1201/9781351074117
29. Dabral D, Coorssen JR (2017) Phospholipase A2: potential roles in native membrane fusion. Int J Biochem Cell Biol 85:1–5. https://doi.org/10.1016/j.biocel.2017.01.011
30. Rabe M, Schwieger C, Zope HR, Versluis F, Kros A (2014) Membrane interactions of fusogenic coiled-coil peptides: implications for lipopeptide mediated vesicle fusion. Langmuir 30(26):7724–7735. https://doi.org/10.1021/la500987c
31. Dutta S, Moses JA, Anandharamakrishnan C (2020) Biomedical and food applications of biopolymer-based liposome. In: Biopolymer-Based Formul Biomed Food Appl, pp 167–192. https://doi.org/10.1016/B978-0-12-816897-4.00008-4
32. Shade CW (2016) Liposomes as advanced delivery systems for nutraceuticals. Integrative Med (Boulder) 15(1):33–36

33. Du S, Cui S-H, Chen X-Y, Zhang S-B (2023) Targeting modification of liposome drug delivery system. Prog Biochem Biophys 50(6):1272–1285. https://doi.org/10.16476/j.pibb.2022.0182
34. Bibi S, Lattmann E, Mohammed AR, Perrie Y (2012) Trigger release liposome systems: local and remote controlled delivery? J Microencapsul 29(3):262–276. https://doi.org/10.3109/02652048.2011.646330
35. Kumar V, Kewlani P, Singh A, Gautam AK, Mahalingam Rajamanickam V (2023) Multifunctional liposomes to attain targeting, stimuli sensitive drug release and imaging cancer. In: Stud Mechanobiol Tissue Eng Biomater 26:49–87. https://doi.org/10.1007/978-981-99-6564-9_3
36. Sachin J, Vishal Gupta N (2013) Solid lipid nanoparticles—preparation, applications, characterization, uses in various cancer therapies: a review. Res J Pharm Technol 6(8):825–837
37. Souto EB, Fangueiro JF, Müller RH (2013) Solid lipid nanoparticles (SLN^{TM}). In: Fundamentals pharmaceutical nanoscience, pp 91–116. https://doi.org/10.1007/978-1-4614-9164-4_5
38. More SD, Wadhokar AS, Bedjawalge RS (2024) A review on solid lipid nanoparticles as nano drug delivery transporters. Curr Nanosci 20(5):644–670. https://doi.org/10.2174/1573413719666230605120659
39. Wissing SA, Kayser O, Müller RH (2004) Solid lipid nanoparticles for parenteral drug delivery. Adv Drug Deliv Rev 56(9):1257–1272. https://doi.org/10.1016/j.addr.2003.12.002
40. Yoon G, Park JW, Yoon I-S (2013) Solid lipid nanoparticles (SLNs) and nanostructured lipid carriers (NLCs): recent advances in drug delivery. J Pharm Investig 43(5):353–362. https://doi.org/10.1007/s40005-013-0087-y
41. Corrias F, Lai F (2011) New methods for lipid nanoparticles preparation. Recent Pat Drug Delivery Formulation 5(3):201–213. https://doi.org/10.2174/187221111797200597
42. Mojahedian MM, Daneshamouz S, Samani SM, Zargaran A (2013) A novel method to produce solid lipid nanoparticles using n-butanol as an additional co-surfactant according to the o/w microemulsion quenching technique. Chem Phys Lipid 174:32–38. https://doi.org/10.1016/j.chemphyslip.2013.05.001
43. Zhang J, Fan Y, Smith E (2009) Experimental design for the optimization of lipid nanoparticles. J Pharm Sci 98(5):1813–1819. https://doi.org/10.1002/jps.21549
44. Hu L-D, Tang X, Cui F-D (2005) Preparation of solid lipid nanoparticles loaded with all-trans retinoic acid and their evaluation in vitro and in vivo. Yaoxue Xuebao 40(1):71–75
45. Trotta M, Cavalli R, Trotta C, Bussano R, Costa L (2010) Electrospray technique for solid lipid-based particle production. Drug Dev Ind Pharm 36(4):431–438. https://doi.org/10.3109/03639040903241817
46. Qin L et al (2020) In vivo effect of resveratrol-loaded solid lipid nanoparticles to relieve physical fatigue for sports nutrition supplements. Molecules 25(22). https://doi.org/10.3390/MOLECULES25225302
47. Sun J et al (2019) Resveratrol-loaded solid lipid nanoparticle supplementation ameliorates physical fatigue by improving mitochondrial quality control. Crystals 9(11). https://doi.org/10.3390/cryst9110559
48. Lima MR, Moreira BJ, Bertuzzi R, Lima-Silva AE (2024) Could nanotechnology improve exercise performance? Evidence from animal studies. Brazil J Med Biol Res 57. https://doi.org/10.1590/1414-431X2024e13360
49. Yazgan H, Kuley E (2023) Lipid nanoparticles: the future of antifungal agents. In: Nanofungicides: novel applications in plant disease control, pp 271–292. https://doi.org/10.1016/B978-0-323-95305-4.00013-3
50. Harish V et al (2023) Unravelling the role of solid lipid nanoparticles in drug delivery: journey from laboratory to clinical trial. J Drug Deliv Sci Technol 85. https://doi.org/10.1016/j.jddst.2023.104616
51. Islan GA, Cacicedo ML, Rodenak-Kladniew B, Duran N, Castro GR (2017) Development and tailoring of hybrid lipid nanocarriers. Curr Pharmaceutical Des 23(43). https://doi.org/10.2174/1381612823666171115110639

52. Campos JR et al (2020) Solid lipid nanoparticles (SLN): prediction of toxicity, metabolism, fate and physicochemical properties, prediction of toxicity, metabolism, fate and physicochemical properties. In: Nanopharmaceuticals: volume 1: expectations and realities of multifunctional drug delivery systems, pp 1–15. https://doi.org/10.1016/B978-0-12-817778-5.00001-4
53. Ezhilarasi PN, Muthukumar SP, Anandharamakrishnan C (2016) Solid lipid nanoparticle enhances bioavailability of hydroxycitric acid compared to a microparticle delivery system. RSC Adv 6(59):53784–53793. https://doi.org/10.1039/c6ra04312g
54. Rehman M et al (2024) Lipid-Based nanoformulations for drug delivery: an ongoing perspective. Pharmaceutics 16(11). https://doi.org/10.3390/pharmaceutics16111376
55. Das P, Panda JR, Patro CN, Sahu B, Patnaik SS (2023) A comprehensive review of nanoemulsion applications and their recent advancements. Current Nanomaterials 8(3):209–223. https://doi.org/10.2174/2405461508666221103121439
56. Barkat MD, Rizwanullah MD, Pottoo FH, Beg S, Akhter S, Ahmad FJ (2020) Therapeutic nanoemulsion: concept to delivery. Curr Pharm Des 26(11):1145–1166. https://doi.org/10.2174/1381612826666200317140600
57. Gupta A, Eral HB, Hatton TA, Doyle PS (2016) Nanoemulsions: formation, properties and applications. Soft Matter 12(11):2826–2841. https://doi.org/10.1039/c5sm02958a
58. Rubio-Ríos A et al (2019) Biobased nanoemulsions: concept, formulation, and applications. In: Nanotechnology in the life sciences, pp 1–31. https://doi.org/10.1007/978-3-030-17061-5_1
59. Dewangan N, Kothale D, Verma U, Jain D (2021) Nanoemulsion as a nano carrier system in drug delivery: a review. Int J Pharmaceutical Sci Nanotechnol 14(2):5353–5363. https://doi.org/10.37285/ijpsn.2021.14.2.1
60. Nicută D, Rosu A-M, Voicu R-E (2023) Nanoemulsion formulations for food processing and enhancing the nutritional quality and shelf life of food. In: Nanoformulations for sustainable agriculture and environmental risk mitigation, pp 36–62. https://doi.org/10.1079/9781800623095.0003
61. Mangale MR, Pathak SS, Mene HR, More BA (2015) Nanoemulsion: as pharmaceutical overview. Int J Pharmaceutical Sci Rev Res 33(1):244–252
62. Sarker DK (2005) Engineering of nanoemulsions for drug delivery. Curr Drug Deliv 2(4):297–310. https://doi.org/10.2174/156720105774370267
63. Chavda VP et al (2024) Nanoemulsions: summary of a decade of research and recent advances. Nanomedicine 19(6):519–536. https://doi.org/10.2217/nnm-2023-0199
64. Koroleva MY, Yurtov EV (2012) Nanoemulsions: the properties, methods of preparation and promising applications. Russian Chem Rev 81(1):21–43. https://doi.org/10.1070/RC2012v081n01ABEH004219
65. Kumar M, Bishnoi RS, Shukla AK, Jain CP (2019) Techniques for formulation of nanoemulsion drug delivery system: a review. Preventive Nutrit Food Sci 24(3):225–234. https://doi.org/10.3746/pnf.2019.24.3.225
66. Leng J, Li L, Zhao W, Bian Y (2025) Enhancing lipid digestion and absorption rate: the role of whey protein isolate short fibrils in emulsions. Int J Biol Macromolecules 285. https://doi.org/10.1016/j.ijbiomac.2024.138351
67. Bhatti HS, Khalid N, Uemura K, Nakajima M, Kobayashi I (2017) Formulation and characterization of food grade water-in-oil emulsions encapsulating mixture of essential amino acids. Eur J Lipid Sci Technol 119(6). https://doi.org/10.1002/ejlt.201600202
68. Zanoni F, Vakarelova M, Zoccatelli G (2019) Development and characterization of astaxanthin-containing whey protein-based nanoparticles. Marine drugs 17(11). https://doi.org/10.3390/md17110627
69. Donsì F, Velikov KP (2019) Encapsulation of food ingredients by single O/W and W/O nanoemulsions. In: Lipid-based nanostructures for food encapsulation purposes: volume 2 in the nanoencapsulation in the food industry series, pp 37–87. https://doi.org/10.1016/B978-0-12-815673-5.00002-7

70. Negi A, Vanmathi Mugasundari A, Moses JA (2024) Nanoemulsion: a potential strategy toward edible coatings. In: Food coatings and preservation technologies, pp 339–365. https://doi.org/10.1002/9781394237623.ch10
71. Mostafa DM, Abd El-Alim SH, Kassem AA (2017) Nanoemulsions: a new approach for enhancing phytonutrient efficacy. In: Nanotechnology applications in food: flavor, stability, nutrition and safety, pp 107–127. https://doi.org/10.1016/B978-0-12-811942-6.00006-6
72. Bahuguna A, Ramalingam S, Kim M (2020) Formulation, characterization, and potential application of nanoemulsions in food and medicine. In: Nanotechnology in the life sciences, pp 39–61. https://doi.org/10.1007/978-3-030-31938-0_3
73. Boostani S, Jafari SM (2020) Controlled release of nanoencapsulated food ingredients. In: Release and bioavailability of nanoencapsulated food ingredients, pp 27–78. https://doi.org/10.1016/B978-0-12-815665-0.00002-3
74. Chude-Okonkwo UK (2022) Enhancing nanocarrier trigger-sensitivity for targeted drug delivery application using ligand-receptor residence time factor. Int Symb Med Inf Commun Technol ISMICT. https://doi.org/10.1109/ISMICT56646.2022.9828241
75. Pandey P, Patel J, Kumar S (2022) Toward eco-friendly nanotechnology-based polymers for drug delivery applications. In: Sustainable nanotechnology: strategies, products, and applications, pp 89–116. https://doi.org/10.1002/9781119650294.ch6
76. Wu J et al (2014) Development of multinuclear polymeric nanoparticles as robust protein nanocarriers. Angewandte Chemie - International Edition 53(34):8975–8979. https://doi.org/10.1002/anie.201404766
77. Dmitrienko EV, Pyshnaya IA, Martyanov ON, Pyshnyi DV (2016) Molecularly imprinted polymers for biomedical and biotechnological applications. Russ Chem Rev 85(5):513–536. https://doi.org/10.1070/RCR4542
78. Faisal N, Kumar K (2017) Polymer and metal nanocomposites in biomedical applications. Biointerface Res Appl Chem 7(6):2286–2294
79. Vasile C, Nistor MT, Cojocariu A-M (2014) Nano-sized polymeric drug carrier systems. In: Nanotechnology and drug delivery: volume 1: nanoplatforms in drug delivery, pp 81–141. https://doi.org/10.1201/b17271
80. Choudhury H, Pandey M, Gorain B, Nirmal J, Kesharwani P (2022) Surface engineering of nanoparticles for imparting multifunctionality. In: Nanoparticle therapeutics: production technologies, types of nanoparticles, and regulatory aspects, pp 181–210. https://doi.org/10.1016/B978-0-12-820757-4.00001-6
81. Yoo J, Park C, Yi G, Lee D, Koo H (2019) Active targeting strategies using biological ligands for nanoparticle drug delivery systems. Cancers 11(5). https://doi.org/10.3390/cancers11050640
82. Van Herk A, Forcada J, Pastorin G (2015) Controlled release systems: advances in nanobottles and active nanoparticles. In: Controlled release systems: advances in nanobottles and active nanoparticles, p 558. https://doi.org/10.4032/9789814613224
83. Smith KL (2023) Dispersible microparticles. In: Controlled-release delivery systems for pesticides, pp 137–150. https://doi.org/10.1201/9781315140353-7
84. Vimala Bharathi SK, Moses JA, Anandharamakrishnan C (2018) Nano and microencapsulation using food grade polymers. In: Gutiérrez TJ (ed) Polymers for food applications, Springer International Publishing, Cham, pp 357–400. https://doi.org/10.1007/978-3-319-94625-2_14
85. Turasan H, Kokini JL (2020) Chapter 6—delivery of bioactives using biocompatible nanodelivery technologies. In: Mustansar Hussain C (ed) Handbook of functionalized nanomaterials for industrial applications, Micro and Nano Technologies, Elsevier, pp 133–166. https://doi.org/10.1016/B978-0-12-816787-8.00006-5
86. Basu MK (2010) Polymeric nanoparticles as drug carriers. J Surf Sci Technol 26(3–4):213–220
87. Nayab PS, Arif R, Abid M (2017) Nanoparticles as targeted drug delivery agents: synthesis, mechanism and applications. Adv Struct Mater 83:49–63. https://doi.org/10.1007/978-981-10-3842-6_3

88. Cheng R, Meng F, Deng C, Klok H-A, Zhong Z (2013) Dual and multi-stimuli responsive polymeric nanoparticles for programmed site-specific drug delivery. Biomaterials 34(14):3647–3657. https://doi.org/10.1016/j.biomaterials.2013.01.084
89. dos Santos PP, Flôres SH, de Oliveira Rios A, Chisté RC (2016) Biodegradable polymers as wall materials to the synthesis of bioactive compound nanocapsules. Trends Food Sci Technol 53:23–33. https://doi.org/10.1016/j.tifs.2016.05.005
90. Mohanty S, Panda S, Purohit D, Si SC (2019) A comprehensive review on PLGA—based nanoparticles used for rheumatoid arthritis. Res J Pharmacy Technol 12(3):1481–1488. https://doi.org/10.5958/0974-360X.2019.00245.2
91. Kwok PCL, Chan H-K (2014) Nanotechnology versus other techniques in improving drug dissolution. Curr Pharm Des 20(3):474–482. https://doi.org/10.2174/13816128113199990400
92. Matseevich T (2020) Modeling of water permeability through the polymer nanocomposites. J Phys Conf Ser. https://doi.org/10.1088/1742-6596/1425/1/012151
93. Ghattavi S, Homaei A (2024) Chitosan-based nanomaterials: structure, characterization, and applications. In: Chitosan-Based hybrid nanomaterials, pp 47–71. https://doi.org/10.1016/B978-0-443-21891-0.00003-2
94. Beulah P, Jinu U, Ghorbanpour M, Venkatachalam P (2019) Green engineered chitosan nanoparticles and its biomedical applications-an overview. In: Advances in phytonanotechnology: from synthesis to application, pp 329–341. https://doi.org/10.1016/B978-0-12-815322-2.00015-8
95. Jha R, Mayanovic RA (2023) A review of the preparation, characterization, and applications of chitosan nanoparticles in nanomedicine. Nanomaterials 13(8). https://doi.org/10.3390/nano13081302
96. Ding F et al (2020) Recent advances of Chitosan and its derivatives in biomedical applications. Curr Med Chem 27(18):3023–3045. https://doi.org/10.2174/0929867326666190405151538
97. Purohit G, Rawat DS (2022) Characterization techniques for chitosan and its based nanocomposites. In: Chitosan-Based nanocomposite materials: fabrication, characterization and biomedical applications, pp 79–101. https://doi.org/10.1007/978-981-19-5338-5_3
98. Ahmed F, Soliman FM, Adly MA, Soliman HAM, El-Matbouli M, Saleh M (2019) Recent progress in biomedical applications of chitosan and its nanocomposites in aquaculture: a review. Res Vet Sci 126:68–82. https://doi.org/10.1016/j.rvsc.2019.08.005
99. Perera UMSP, Rajapakse N (2014) Chitosan nanoparticles: preparation, characterization, and applications. In: Seafood processing by-products: trends and applications, vol 9781461495901, pp 371–387. https://doi.org/10.1007/978-1-4614-9590-1_18
100. Onishi H (2010) Chitosan microparticles. J Drug Deliv Sci Technol 20(1):15–22. https://doi.org/10.1016/S1773-2247(10)50002-9
101. Jeong HW et al (2012) Chitooligosaccharide induces mitochondrial biogenesis and increases exercise endurance through the activation of Sirt1 and AMPK in rats. PLoS ONE 7(7). https://doi.org/10.1371/journal.pone.0040073
102. Bouchez C, Devin A (2019) Mitochondrial biogenesis and mitochondrial reactive oxygen species (Ros): a complex relationship regulated by the camp/pka signaling pathway. Cells 8(4). https://doi.org/10.3390/cells8040287
103. Robles-Martínez L, Guerra-Sánchez MG, Hernández-Lauzardo AN, Pardo JP, Velázquez-del Valle MG (2014) Effects of chitosan and oligochitosan on development and mitochondrial function of Rhizopus stolonifera. J Basic Microbiol 54(SUPPL.1):S42–S49. https://doi.org/10.1002/jobm.201300790
104. Tu J, Xu Y, Xu J, Ling Y, Cai Y (2016) Chitosan nanoparticles reduce LPS-induced inflammatory reaction via inhibition of NF-κB pathway in Caco-2 cells. Int J Biol Macromol 86:848–856. https://doi.org/10.1016/j.ijbiomac.2016.02.015
105. Wardani G, Nugraha J, Mustafa MR, Kurnijasanti R, Sudjarwo SA (2022) Antioxidative stress and antiapoptosis effect of chitosan nanoparticles to protect cardiac cell damage on streptozotocin-induced diabetic rat. Oxidative Med Cell Longevity 2022. https://doi.org/10.1155/2022/3081397

106. Sarangapani S, Patil A, Ngeow YK, Elsa Mohan R, Asundi A, Lang MJ (2018) Chitosan nanoparticles' functionality as redox active drugs through cytotoxicity, radical scavenging and cellular behaviour. Integrative Biol (United Kingdom) 10(5):313–324. https://doi.org/10.1039/c8ib00038g
107. Mahdy MAA, Akl MA, Madkour FA (2022) Effect of chitosan and curcumin nanoparticles against skeletal muscle fibrosis at early regenerative stage of glycerol-injured rat muscles. BMC Musculoskeletal Disorders 23(1). https://doi.org/10.1186/s12891-022-05633-x
108. Nanduri LSY (2021) Chitosan–Stem cell interactions. In: Adv Polym Sci 287:343–359. https://doi.org/10.1007/12_2021_83
109. Wydro P, Krajewska B, Hąc-Wydro K (2007) Chitosan as a lipid binder: a langmuir monolayer study of chitosan-lipid interactions. Biomacromolecules 8(8):2611–2617. https://doi.org/10.1021/bm700453x
110. Akheratdoost V, Panahi N, Safi S, Mojab F, Akbari G (2024) Protective effects of silymarin-loaded chitosan nanoparticles in the diet-induced hyperlipidemia rat model. Iran J Basic Med Sci 27(6):725–732. https://doi.org/10.22038/IJBMS.2024.74490.16179
111. Rashidpour A, Wu Y, Almajano MP, Fàbregas A, Metón I (2023) Chitosan-Based sustained expression of sterol regulatory element-binding protein 1a stimulates hepatic glucose oxidation and growth in sparus aurata. Marine Drugs 21(11). https://doi.org/10.3390/md21110562
112. Nagpal K, Singh SK, Mishra DN (2010) Chitosan nanoparticles: a promising system in novel drug delivery. Chem Pharm Bull 58(11):1423–1430. https://doi.org/10.1248/cpb.58.1423
113. Herdiana Y, Febrina E, Nurhasanah S, Gozali D, Elamin KM, Wathoni N (2024) Drug loading in chitosan-based nanoparticles. Pharmaceutics 16(8). https://doi.org/10.3390/pharmaceutics16081043
114. Quiñones JP, Peniche H, Peniche C (2018) Chitosan based self-assembled nanoparticles in drug delivery. Polymers 10(3). https://doi.org/10.3390/polym10030235
115. e Frates K et al (2018) Protein polymer-based nanoparticles: Fabrication and medical applications. Int J Mol Sci 19(6). https://doi.org/10.3390/ijms19061717
116. Gao Q, Zhang S, Tang Z, Peng X, Wang N, Xue Y (2023) Research progress on preparation and application of protein nanoparticles in food field. Sci Technol Food Ind 44(11):30–37. https://doi.org/10.13386/j.issn1002-0306.2022110153
117. Lee EJ (2018) Recent advances in protein-based nanoparticles. Korean J Chem Eng 35(9):1765–1778. https://doi.org/10.1007/s11814-018-0102-0
118. Raeeszadeh-Sarmazdeh M, Hartzell E, Price JV, Chen W (2016) Protein nanoparticles as multifunctional biocatalysts and health assessment sensors. Curr Opin Chem Eng 13:109–118. https://doi.org/10.1016/j.coche.2016.08.016
119. Biswas S, Sengupta S (2022) Current use of nanoprotein and application in the development of food products for functional and nutritional benefits. J Microbiol Biotechnol Food Sci 11(5). https://doi.org/10.55251/jmbfs.1737
120. Ramos OL et al (2017) Design of whey protein nanostructures for incorporation and release of nutraceutical compounds in food. Crit Rev Food Sci Nutr 57(7):1377–1393. https://doi.org/10.1080/10408398.2014.993749
121. Haque S, Patra CR (2023) Casein-based nanosystems for therapeutic applications. In: Polymeric nanosystems: theranostic nanosystems, vol 1, pp 621–655. https://doi.org/10.1016/B978-0-323-85656-0.00019-X
122. Srivastava A, Prajapati A (2020) Albumin and functionalized albumin nanoparticles: production strategies, characterization, and target indications. Asian Biomed 14(6):217–242. https://doi.org/10.1515/abm-2020-0032
123. Taheri ES, Jahanshahi M, Mosavian MTH (2012) Preparation, characterization and optimization of egg albumin nanoparticles as low molecular-weight drug delivery vehicle. Part Part Syst Charact 29(3):211–222. https://doi.org/10.1002/ppsc.201100037
124. Okagu OD, Wang B, Acquah C, Udenigwe CC (2018) Protein-based nanodelivery systems for food applications. In: Encyclopedia of food chemistry, pp 719–726. https://doi.org/10.1016/B978-0-08-100596-5.21864-7

125. Okagu OD, Wang B, Udenigwe CC (2021) CHAPTER 4: food proteins as biomaterial for delivery functions. In: Food chemistry, function and analysis, vol 27, pp 97–126. https://doi.org/10.1039/9781839163425-00097
126. Jiménez-Cruz E, Arroyo-Maya IJ, Hernández-Arana A, Cornejo-Mazón M, Hernández-Sánchez H (2015) Protein-based nanoparticles. In: Food engineering series, pp 69–79. https://doi.org/10.1007/978-3-319-13596-0_5
127. Joye IJ, Davidov-Pardo G, McClements DJ (2014) Nanotechnology for increased micronutrient bioavailability. Trends Food Sci Technol 40(2):168–182. https://doi.org/10.1016/j.tifs.2014.08.006
128. Wang L et al (2022) Enhanced oral absorption and liver distribution of polymeric nanoparticles through traveling the enterohepatic circulation pathways of bile Acid. ACS Appl Mater Interfaces 14(37):41712–41725. https://doi.org/10.1021/acsami.2c10322
129. Ahsan SM, Rao CM (2017) The role of surface charge in the desolvation process of gelatin: implications in nanoparticle synthesis and modulation of drug release. Int J Nanomed 12:795–808. https://doi.org/10.2147/IJN.S124938
130. Wen J et al (2011) Controlled protein delivery based on enzyme-responsive nanocapsules. Adv Mater 23(39):4549–4553. https://doi.org/10.1002/adma.201101771
131. Jain A, Singh SK, Arya SK, Kundu SC, Kapoor S (2018) Protein nanoparticles: promising platforms for drug delivery applications. ACS Biomater Sci Eng 4(12):3939–3961. https://doi.org/10.1021/acsbiomaterials.8b01098
132. Rahimnejad M, Mokhtarian N, Ghasemi M (2009) Production of protein nanoparticles for food and drug delivery system. Afr J Biotech 8(19):4738–4743
133. Rai A, Jenifer J, Upputuri RTP (2017) Nanoparticles in therapeutic applications and role of albumin and casein nanoparticles in cancer therapy. Asian Biomed 11(1):3–20. https://doi.org/10.5372/1905-7415.1101.534
134. Bamburowicz-Klimkowska M, Poplawska M, Grudzinski IP (2019) Nanocomposites as biomolecules delivery agents in nanomedicine. J Nanobiotechnol 17(1). https://doi.org/10.1186/s12951-019-0479-x
135. Abbasi E et al (2014) Dendrimers: synthesis, applications, and properties. Nanoscale Res Lett 9(1):1–10. https://doi.org/10.1186/1556-276X-9-247
136. Mittal P, Kapoor R, Mishra B (2021) Dendrimers: role in novel drug delivery. In: Nanopharmaceutical advanced delivery systems, pp 81–97. https://doi.org/10.1002/9781119711698.ch4
137. Wu L-P, Ficker M, Christensen JB, Trohopoulos PN, Moghimi SM (2015) Dendrimers in medicine: therapeutic concepts and pharmaceutical challenges. Bioconjug Chem 26(7):1198–1211. https://doi.org/10.1021/acs.bioconjchem.5b00031
138. Kareem OO et al (2021) Solution size variation of linear and dendritic bis-MPA analogs using DOSY-1H NMR. Polym Chem 12(10):1507–1517. https://doi.org/10.1039/d0py01070g
139. Markowicz-Piasecka M, Mikiciuk-Olasik E (2016) Dendrimers in drug delivery. In: Nanobiomaterials in drug delivery: applications of nanobiomaterials, pp 39–74. https://doi.org/10.1016/B978-0-323-42866-8.00002-2
140. Caminade A-M, Turrin C-O (2014) Dendrimers for drug delivery. J Mater Chem B 2(26):4055–4066. https://doi.org/10.1039/c4tb00171k
141. Dubey SK et al (2020) Understanding the pharmaceutical aspects of dendrimers for the delivery of anticancer drugs. Curr Drug Targets 21(6):528–540. https://doi.org/10.2174/1389450120666191031092259
142. Ghaffari M et al (2018) Surface functionalized dendrimers as controlled-release delivery nanosystems for tumor targeting. Eur J Pharm Sci 122:311–330. https://doi.org/10.1016/j.ejps.2018.07.020
143. Pathak A, Jain K (2023) Dendrimer–drug conjugates. In: Polymer-Drug conjugates: linker chemistry, protocols and applications, pp 315–345. https://doi.org/10.1016/B978-0-323-91663-9.00005-9
144. Wang H, Huang Q, Chang H, Xiao J, Cheng Y (2016) Stimuli-responsive dendrimers in drug delivery. Biomater Sci 4(3):375–390. https://doi.org/10.1039/c5bm00532a

145. Narayanan G,. Shen J, Matai I, Sachdev A, Boy R, Tonelli AE (2022) Cyclodextrin-based nanostructures. Prog Mater Sci 124. https://doi.org/10.1016/j.pmatsci.2021.100869
146. Varan G, Varan C, Erdoğar N, Hıncal AA, Bilensoy E (2017) Amphiphilic cyclodextrin nanoparticles. Int J Pharm 531(2):457–469. https://doi.org/10.1016/j.ijpharm.2017.06.010
147. Zandarek J, Gumułka P, Starek M, Dąbrowska M (2022) Characteristics and applications of cyclodextrin complexes. Farm Pol 78(6):308–316. https://doi.org/10.32383/farmpol/152762
148. Haimhoffer A et al (2019) Cyclodextrins in drug delivery systems and their effects on biological barriers. Scientia Pharmaceutica 87(4). https://doi.org/10.3390/scipharm87040033
149. Muderawan IW (2022) Modified cyclodextrins and their application in drug delivery. In: Presented at the AIP conference proceedings. https://doi.org/10.1063/5.0122038
150. Gidwani B, Vyas A (2014) Synthesis, characterization and application of Epichlorohydrin-β-cyclodextrin polymer. Colloids Surf, B 114:130–137. https://doi.org/10.1016/j.colsurfb.2013.09.035
151. Braga SS (2022) Cyclodextrin superstructures for drug delivery. J Drug Deliv Sci Technol 75. https://doi.org/10.1016/j.jddst.2022.103650
152. Horský J, Jindřich J (2013) Cyclodextrins in the world of polymers. Chem Listy 107(10):769–776
153. Egawa Y, Seki T (2013) Supramolecular nanomachines for sugar responsive insulin release systems. Yakugaku Zasshi 133(12):1287–1295. https://doi.org/10.1248/yakushi.13-00222-4
154. Arora D, Saneja A, Jaglan S (2019) Cyclodextrin-based delivery systems for dietary pharmaceuticals. Environ Chem Lett 17(3):1263–1270. https://doi.org/10.1007/s10311-019-00878-w
155. Marques HMC (2010) A review on cyclodextrin encapsulation of essential oils and volatiles. Flavour Fragr J 25(5):313–326. https://doi.org/10.1002/ffj.2019
156. Samra HS et al (2010) The effects of substituted cyclodextrins on the colloidal and conformational stability of selected proteins. J Pharm Sci 99(6):2800–2818. https://doi.org/10.1002/jps.22053
157. Pereira AB, Braga SS (2015) Cyclodextrin inclusion of nutraceuticals, from the bench to your table. In: Cyclodextrins: synthesis, chemical applications and role in drug delivery, pp 195–224. [Online]. Available: https://www.scopus.com/inward/record.uri?eid=2-s2.0-84957616517&partnerID=40&md5=a09ae7bbd6bce8dc95fb575d71d68b30
158. Shu C, Li R, Guo J, Ding L, Zhong W (2013) New generation of β-cyclodextrin-chitosan nanoparticles encapsulated quantum dots loaded with anticancer drug for tumor-target drug delivery and imaging of cancer cells. J Nanopart Res 15(12). https://doi.org/10.1007/s11051-013-1927-4
159. Gourevich D et al (2013) Ultrasound-mediated targeted drug delivery with a novel cyclodextrin-based drug carrier by mechanical and thermal mechanisms. J Control Release 170(3):316–324. https://doi.org/10.1016/j.jconrel.2013.05.038
160. Shimpi S, Chauhan B, Shimpi P (2005) Cyclodextrins: application in different routes of drug administration. Acta Pharm 55(2):139–156
161. Han B, Liao X, Yang B (2014) Targeted drug delivery systems based on cyclodextrins. Prog Chem 26(6):1039–1049. https://doi.org/10.7536/PC131127
162. Khattabi AM, Alqdeimat DA (2018) The effect of cyclodextrin on both the agglomeration and the in vitro characteristics of drug loaded and targeted silica nanoparticles. In: Presented at the IOP conference series: materials science and engineering. https://doi.org/10.1088/1757-899X/305/1/012008
163. Chhour P, Naha PC, Cheheltani R, Benardo B, Mian S, Cormode DP (2016) Gold nanoparticles for biomedical applications: synthesis and in vitro evaluation. In: Methods in pharmacology and toxicology, vol39, pp 87–111. https://doi.org/10.1007/978-1-4939-3121-7_5
164. Mieszawska AJ, Mulder WJM, Fayad ZA, Cormode DP (2013) Multifunctional gold nanoparticles for diagnosis and therapy of disease. Mol Pharm 10(3):831–847. https://doi.org/10.1021/mp3005885
165. Glomm WR (2005) Functionalized gold nanoparticles for applications in bionanotechnology. J Dispersion Sci Technol 26(3):389–414. https://doi.org/10.1081/DIS-200052457

166. Krishnamoorthy P, Jayalakshmi T (2012) Preparation, characterization and synthesis of silver nanoparticles by using phyllanthusniruri for the antimicrobial activity and cytotoxic effects. J Chem Pharm Res 4(11):4783–4794
167. Pisarcik M, Zatekova M, Lukac M (2024) Cationic gemini surfactants as efficient capping agents of silver and gold nanoparticles. In: Presented at the world congress on recent advances in nanotechnology. https://doi.org/10.11159/nddte24.005
168. Sousa A, Carvalho F, Fernandes E, Freitas M (2023) Applications of silver nanoparticles in daily life. In: Appl Silv Nanopart, pp 151–169. [Online]. Available: https://www.scopus.com/inward/record.uri?eid=2-s2.0-85161976970&partnerID=40&md5=b69016ab84c83935734541183f5fc7cc
169. Sun S-N, Wei C, Zhu Z-Z, Hou Y-L, Venkatraman SS, Xu Z-C (2014) Magnetic iron oxide nanoparticles: synthesis and surface coating techniques for biomedical applications. Chinese Phys B 23(3). https://doi.org/10.1088/1674-1056/23/3/037503
170. Gao L, Fan K, Yan X (2017) Iron oxide nanozyme: a multifunctional enzyme mimetic for biomedical applications. Theranostics 7(13):3207–3227. https://doi.org/10.7150/thno.19738
171. Xiao X, Hu G, Liang Y, Yan P, Jiang A (2024) A review on the properties of zinc oxide nanoparticles in various industries and biomedical fields: enhancing chemical and physical characteristics. Iran J Chem Chem Eng 43(1):46–65. https://doi.org/10.30492/ijcce.2023.1990407.5889
172. Ringu T, Ghosh S, Das A, Pramanik N (2022) Zinc oxide nanoparticles: an excellent biomaterial for bioengineering applications. Emergent Materials 5(6):1629–1648. https://doi.org/10.1007/s42247-022-00402-x
173. Gatou M-A et al (2024) Magnesium oxide (MgO) nanoparticles: synthetic strategies and biomedical applications. Crystals 14(3). https://doi.org/10.3390/cryst14030215
174. Nagchowdhury P, Krishna M, Patra CR (2024) Biomedical applications of magnesium oxide nanoparticles. In: Nanomaterials for biomedical and bioengineering applications, pp 97–132. https://doi.org/10.1007/978-981-97-0221-3_5
175. Sakr TM, Korany M, Katti KV (2018) Selenium nanomaterials in biomedicine—an overview of new opportunities in nanomedicine of selenium. J Drug Deliv Sci Technol 46:223–233. https://doi.org/10.1016/j.jddst.2018.05.023
176. Zambonino MC, Quizhpe EM, Mouheb L, Rahman A, Agathos SN, Dahoumane SA (2023) Biogenic selenium nanoparticles in biomedical sciences: properties, current trends, novel opportunities and emerging challenges in theranostic nanomedicine, Nanomaterials 13(3). https://doi.org/10.3390/nano13030424
177. Bodnár D et al (2016) Dietary selenium augments sarcoplasmic calcium release and mechanical performance in mice. Nutr Metab 13(1):1–13. https://doi.org/10.1186/s12986-016-0134-6
178. Fernández-Lázaro D, Fernandez-Lazaro CI, Mielgo-Ayuso J, Navascués LJ, Martínez AC, Seco-Calvo J (2020) The role of selenium mineral trace element in exercise: antioxidant defense system, muscle performance, hormone response, and athletic performance. a systematic review. Nutrients 12(6):1–16. https://doi.org/10.3390/nu12061790
179. White SH, Wohlgemuth S, Li C, Warren LK (2017) Rapid communication: dietary selenium improves skeletal muscle mitochondrial biogenesis in young equine athletes. J Anim Sci 95(9):4078–4084. https://doi.org/10.2527/jas.2017.1919
180. Wu K, Zhou H, Yang L (2023) Application and translation of nano calcium phosphates in biomedicine. In: Nanomedicine: technologies and applications, pp 19–57. https://doi.org/10.1016/B978-0-12-818627-5.00004-X
181. Khalifehzadeh R, AramiH (2020) Biodegradable calcium phosphate nanoparticles for cancer therapy. Adv Colloid Interf Sci 279. https://doi.org/10.1016/j.cis.2020.102157
182. Ge X, Cao Z, Chu L (2022) The antioxidant effect of the metal and metal-oxide nanoparticles. Antioxidants 11(4). https://doi.org/10.3390/antiox11040791
183. Singh N (2021) Antioxidant metal oxide nanozymes: role in cellular redox homeostasis and therapeutics. Pure Appl Chem 93(2):187–205. https://doi.org/10.1515/pac-2020-0802

184. Girma A, Abera B, Mekuye B, Mebratie G (2024) Antibacterial activity and mechanisms of action of inorganic nanoparticles against foodborne bacterial pathogens: a systematic review. IET Nanobiotechnol 1:2024. https://doi.org/10.1049/2024/5417924
185. Islam S et al (2021) Nanomaterials and nanocoatings for alternative antimicrobial therapy. In: Handbook of nanomaterials and nanocomposites for energy and environmental applications, volumes 1–4, vol 3, pp 2603–2620. https://doi.org/10.1007/978-3-030-36268-3_3
186. Rosli NA, Teow YH, Mahmoudi E (2021) Current approaches for the exploration of antimicrobial activities of nanoparticles. Sci Technol Adv Mater 22(1):885–907. https://doi.org/10.1080/14686996.2021.1978801
187. Yadav S, Maurya PK (2021) Biomedical applications of metal oxide nanoparticles in aging and age-associated diseases. 3 Biotech 11(7). https://doi.org/10.1007/s13205-021-02892-8
188. Dwivedi PD, Misra A, Shanker R, Das M (2009) Are nanomaterials a threat to the immune system? Nanotoxicology 3(1):19–26. https://doi.org/10.1080/17435390802604276
189. Lamas B, Martins Breyner N, Houdeau E (2020) Impacts of foodborne inorganic nanoparticles on the gut microbiota-immune axis: potential consequences for host health. Particle and Fibre Toxicology 17(1). https://doi.org/10.1186/s12989-020-00349-z
190. Sabu C, Ameena Shirin VK, Sankar R, Pramod K (2022) Inorganic nanoparticles for drug-delivery applications. In: Nanomaterials and nanotechnology in medicine, pp 367–399. https://doi.org/10.1002/9781119558026.ch14
191. Zhang L, Li Y, Yu JC (2014) Chemical modification of inorganic nanostructures for targeted and controlled drug delivery in cancer treatment. J Mater Chem B 2(5):452–470. https://doi.org/10.1039/c3tb21196g
192. Kim E et al (2024) Lipid loss and compositional change during preparation of simple two-component liposomes. Biophys Rep 4(3). https://doi.org/10.1016/j.bpr.2024.100174
193. Trif M, Craciunescu O (2015) Liposome as efficient system for intracellular delivery of bioactive molecules. In: Nanotechnology and functional foods: effective delivery of bioactive ingredients, pp 191–213. https://doi.org/10.1002/9781118462157.ch12
194. Asha Spandana KM,. Natarajan J, Thirumaleshwar S, Hemanth Kumar S (2020) A review of the preparation, characterization and application of nanostructured lipid carriers. Int J Res Pharmaceutical Sci 11(1):1130–1135. https://doi.org/10.26452/ijrps.v11i1.1946
195. Chaturvedi SP, Kumar V (2012) Production techniques of lipid nanoparticles: a review. Res J Pharm, Biol Chem Sci 3(3):525–541
196. De Jesus MB, Radaic A, Zuhorn IS, De Paula E (2013) Microemulsion extrusion technique: a new method to produce lipid nanoparticles. J Nanopart Res 15(10). https://doi.org/10.1007/s11051-013-1960-3
197. Madkhali OA (2022) Perspectives and prospective on solid lipid nanoparticles as drug delivery systems. Molecules 27(5). https://doi.org/10.3390/molecules27051543
198. Kulkarni AA, Rao PS (2013) Synthesis of polymeric nanomaterials for biomedical applications. In: Nanomaterials in tissue engineering: fabrication and applications, pp 27–63. https://doi.org/10.1533/9780857097231.1.27
199. Naskar S, Koutsu K, Sharma S (2019) Chitosan-based nanoparticles as drug delivery systems: a review on two decades of research. J Drug Target 27(4):379–393. https://doi.org/10.1080/1061186X.2018.1512112
200. Singh A, Mittal A, Benjakul S (2021) Chitosan nanoparticles: preparation, food applications and health benefits. ScienceAsia 47(1):1–10. https://doi.org/10.2306/scienceasia1513-1874.2021.020
201. Zargani M, Rahimi A, Mazaheri Tirani Z, Arabzadeh E, Feizolahi F (2023) Swimming exercise and nano-L-arginine supplementation improve oxidative capacity and some autophagy-related genes in the soleus muscle of aging rats. Gene 850. https://doi.org/10.1016/j.gene.2022.146955
202. Bilensoy E, Hincal AA (2009) Recent advances and future directions in amphiphilic cyclodextrin nanoparticles. Expert Opin Drug Deliv 6(11):1161–1173. https://doi.org/10.1517/17425240903222218

203. Christensen JB (2013) Dendrimers in biology and nanomedicine. In: Organic synthesis and molecular engineering, pp 361–392. https://doi.org/10.1002/9781118736449.ch13
204. Singh I, Rehni AK, Kalra R, Joshi G, Kumar M (2008) Dendrimers and their pharmaceutical applications—a review. Pharmazie 63(7):491–496. https://doi.org/10.1691/ph.2008.8052
205. Yousefi M, Orojzadeh P, Jafari SM (2019) Nanoencapsulation of food ingredients by dendrimers. In: Biopolymer nanostructures for food encapsulation purposes, pp 607–625. https://doi.org/10.1016/B978-0-12-815663-6.00020-3
206. Gutiérrez-Wing C, Velázquez-Salazar JJ, José-Yacamán M (2012) Procedures for the synthesis and capping of metal nanoparticles. Methods Mol Biol 906:3–19. https://doi.org/10.1007/978-1-61779-953-2_1
207. Usman O et al (2024) Green synthesis of metal nanoparticles and study their anti-pathogenic properties against pathogens effect on plants and animals. Scientific Reports 14(1). https://doi.org/10.1038/s41598-024-61920-8
208. Poulose S, Panda T, Nair PP, Théodore T (2014) Biosynthesis of silver nanoparticles. J Nanosci Nanotechnol 14(2):2038–2049. https://doi.org/10.1166/jnn.2014.9019
209. Schrand AM, Rahman MF, Hussain SM, Schlager JJ, Smith DA, Syed AF (2010) Metal-based nanoparticles and their toxicity assessment. Wiley Interdisciplinary Reviews: Nanomedicine and Nanobiotechnology 2(5):544–568. https://doi.org/10.1002/wnan.103
210. Wang Z, Tang M (2021) Research progress on toxicity, function, and mechanism of metal oxide nanoparticles on vascular endothelial cells. J Appl Toxicol 41(5):683–700. https://doi.org/10.1002/jat.4121
211. Xu Z, Liu S, Wang H, Gao G, Yu P, Chang Y (2014) Encapsulation of Iron in liposomes significantly improved the efficiency of iron supplementation in strenuously exercised rats. Biol Trace Elem Res 162(1–3):181–188. https://doi.org/10.1007/s12011-014-0143-0
212. Liu Q, Wang C, GuoX, Du Q, Keshavarzi M (2024) Curcumin and its nano-formulations combined with exercise: from molecular mechanisms to clinic. Cell Biochemistry and Function 42(4). https://doi.org/10.1002/cbf.4061
213. Reddy PRK, Yasaswini D, Reddy PPR, Zeineldin M, Adegbeye MJ, Hyder I (2020) Applications, challenges, and strategies in the use of nanoparticles as feed additives in equine nutrition. Veterinary World 13(8):1685–1696. https://doi.org/10.14202/vetworld.2020.1685-1696
214. Nemec Svete A, Vovk T, Bohar Topolovec M, Kruljc P (2021) Effects of vitamin e and coenzyme q10 supplementation on oxidative stress parameters in untrained leisure horses subjected to acute moderate exercise. Antioxidants 10(6). https://doi.org/10.3390/antiox10060908
215. Sinatra ST, Jankowitz SN, Chopra RK, Bhagavan HN (2014) Plasma coenzyme Q10 and tocopherols in thoroughbred race horses: effect of coenzyme Q10 supplementation and exercise. J Equine Vet 34(2):265–269. https://doi.org/10.1016/j.jevs.2013.06.001
216. Su L-Y, Huang W-C, Kan N-W, Tung T-H, Huynh LBP, Huang S-Y (2023) Effects of resveratrol on muscle inflammation, energy utilisation, and exercise performance in an eccentric contraction exercise mouse model. Nutrients 15(1). https://doi.org/10.3390/nu15010249
217. Battistelli M, Salucci S, Falcieri E (2018) Morphological evaluation of liposomal iron carriers. Microsc Res Tech 81(11):1295–1300. https://doi.org/10.1002/jemt.23137
218. Nolte S, Krüger K, Hollander K, Carlsohn A (2024) Approaches to prevent iron deficiency in athletes. Dtsch Z Sportmed 75(5):195–202. https://doi.org/10.5960/dzsm.2024.607
219. Oh J-M, Choi S-J, Kim S-T, Choy J-H (2006) Cellular uptake mechanism of an inorganic nanovehicle and its drug conjugates: enhanced efficacy due to clathrin-mediated endocytosis. Bioconjug Chem 17(6):1411–1417. https://doi.org/10.1021/bc0601323
220. Qin L, Wang W, You S, Dong J, Zhou Y, Wang J (2014) In vitro antioxidant activity and in vivo antifatigue effect of layered double hydroxide nanoparticles as delivery vehicles for folic acid. Int J Nanomed 9:5701–5710. https://doi.org/10.2147/IJN.S74306
221. Prescott C, Bottle SE (2017) Biological relevance of free radicals and nitroxides. Cell Biochem Biophys 75(2):227–240. https://doi.org/10.1007/s12013-016-0759-0

222. Sadowska-Bartosz I, Bartosz G (2024) The cellular and organismal effects of nitroxides and nitroxide-containing nanoparticles. Int J Mol Sci 25(3). https://doi.org/10.3390/ijms25031446
223. Toriumi T et al (2021) An antioxidant nanoparticle enhances exercise performance in rat high-intensity running models. Adv Healthc Mater 10(10). https://doi.org/10.1002/adhm.202100067
224. Chaudhry Q, Castle L (2015) Safety assessment of nano- and microscale delivery vehicles for bioactive ingredients. In: Nanotechnology and functional foods: effective delivery of bioactive ingredients, pp 348–357. https://doi.org/10.1002/9781118462157.ch21
225. Bergeson LL, Hutton CN (2015) FDA's final and draft nanotechnology guidance documents: no big surprises. Nanotechnol Law Bus 12(1):54–58
226. Bashaw J (2009) Regulation of nanoparticles: trying to keep pace with a scientific revolution. Nanotechnol Law Bus 6(4):475–482
227. Baran A (2016) Nanotechnology: legal and ethical issues. Eng Manage Prod Serv 8(1):47–54. https://doi.org/10.1515/emj-2016-0005
228. Collins R, Kalman D (2008) Effect of government regulation on the evolution of sports nutrition. In: Nutritional supplements in sports and exercise, pp 3–31. https://doi.org/10.1007/978-1-59745-231-1_1
229. Collins R, Kalman DS (2015) Effect of government regulation on the evolution of sports nutrition. In: Nutritional supplements in sports and exercise: second edition, pp 3–21. https://doi.org/10.1007/978-3-319-18230-8_1
230. Bunch WH, Dvonch VM (2004) Informed consent in sports medicine. Clin Sports Med 23(2):183–193. https://doi.org/10.1016/j.csm.2004.01.004
231. Reuter T (2021) Nanotechnology in food: ethics, industry practices, and regulatory frameworks. In: Handbook of nanoethics, pp 159–176. https://doi.org/10.1515/9783110669282-009
232. Mekel M (2006) Nanotechnologies: small science, big potential and bigger issues. Development 49(4):47–53. https://doi.org/10.1057/palgrave.development.1100296
233. Souto EB, Dias-Ferreira J, Shegokar R, Durazzo A, Santini A (2020) Ethical issues in research and development of nanoparticles. In: Drug delivery aspects: volume 4: expectations and realities of multifunctional drug delivery systems, pp 157–168. https://doi.org/10.1016/B978-0-12-821222-6.00007-5

Optimizing Protein Absorption with Nano-Encapsulation

Sammra Maqsood, Panchali Dasgupta, and Waseem Khalid

Abstract Proteins are significant to human health because they enhance cellular activity, immunological response, and repair of muscles. Due to poor bioavailability and severe degradation in the intestines, absorption is more difficult. This chapter reviews nano-encapsulation as a potentially valuable technique that improves protein absorption through tailored and protected delivery methods. The use of nanocarriers, including liposomes, polymeric nanoparticles, and lipid-based nanoparticles, provides regulated release, enhances cellular absorption, and protects proteins from digestive degradation. Nano encapulation is applied in a host of products such as pharmaceuticals, nutritional supplements, and cosmetics. Nano-encapsulation allows for the targeted therapeutic protein delivery, improved nutrient delivery, and absorption in skincare products. The functions of most lipid-based, polymeric, and biopolymer carriers for enhancing the absorption of nanomaterials are analyzed. Nano-encapsulation has a good number of advantages such as improved permeability across cell membranes, controlled release, and stability of proteins. However, it is beset by concerns such as the expense in production, legal standards and stability in storage. Among the mentioned topics, the authors also presented contemporary development in nano-encapsulation, case studies, as well as novel approaches that enhance bioavailability and targeting. Looking at the directions for the future, this chapter explores areas like customized protein delivery, environmentally friendly nanomaterials, and nano-encapsulation integrated with other technologies. Such advances imply that encapsulation technology has enormous potential to totally change the face of protein-based products in food, medicine, and cosmetics in ways of nutritional improvement.

S. Maqsood (✉)
National Institute of Food Science and Technology, University of Agriculture, Faisalabad, Pakistan
e-mail: sumramaqsood190@gmail.com

P. Dasgupta
Child Fund India, Kolkata, India

W. Khalid
Department of Organic Chemistry, Faculty of Chemical Sciences and Technologies, University of Castilla La Mancha, Ciudad Real, Spain

A. Chatterjee et al. (eds.), *Nanofuel: The Future of Sports Nutrition*, Smart Nanomaterials Technology, https://doi.org/10.1007/978-981-96-5471-0_5

Keywords Protein absorption · Nano-encapsulation · Bioavailability · Nanocarriers · Controlled release · Lipid nanoparticles · Polymeric nanoparticles · Liposomal delivery · Protein stability · Nutrient delivery systems

1 Introduction

Proteins are the most fundamental macromolecules of the human body needed for all the physiological processes that the body mediates and carries out, such as hormone production [1], immunological assistance, muscle repair, and many more physiological functions that are done through enzyme activity [2, 3]. Therefore, dietary proteins are necessary to enable the body to fulfill all these roles and indeed to provide cellular housekeeping, since amino acids obtained from them satisfy tissue healing, protein synthesis, or even cellular maintenance needs [4, 5]. In fact, proper absorption of dietary proteins is essential because amino acids are utilized in numerous physiological pathways [1]. This chapter would then summarize the methods and processes by which proteins are absorbed, digested, and used by the body before discussing how, in principle, nano-encapsulation techniques may maximise proteins' bioavailability and efficacy [6]. Proteins are essential to homeostasis because they are structural as well as functional molecules. The proteins myosin and actin are crucial in muscle contraction and repair of muscle cells in order to recover after exercise and trauma [1]. Moreover, proteins are involved in the synthesis of enzymes that act as biological catalysts to catalyze metabolic processes. They control pathways for cell signaling, digestion, metabolism, as well as DNA replication [7]. Dietary proteins also support a full and intact, strong immune response because antibodies, or immunoglobulins, are specialized proteins. These help the immune system detect and neutralize pathogens; therefore, they are very essential [7]. Some proteins act as signaling molecules that control physiological processes ranging from metabolism to cellular development and reproduction. Examples of such proteins are growth factors and hormones [1, 8].

1.1 Steps in Protein Digestion and Absorption

Protein digestion and absorption is a multistep, relatively complex procedure. It involves several organs and several enzymes [9]. The final step in this pathway is the absorption of amino acids by body cells. This begins with the stomach and goes on to be continued throughout the small intestine.

1. **Protein Digestion in the Stomach**

The first step in the digestion of protein is its breakdown by digestive enzymes and gastric acids in the stomach into larger molecules of protein. The HCl in the stomach raises the pH of the surrounding solution to the optimum level, where the enzyme pepsin is activated from its precursor, pepsinogen [10]. Pepsin selectively acts on the peptide bonds of proteins and breaks them down into smaller chains known as peptides. This first step in the stomach primes protein molecules for further enzymatic action in the colon and therefore is critical in protein digestion [10, 11].

2. **Further Digestion in the Small Intestine**

After entering the small intestine, the partly digested peptides get further broken down by enzymes that have been released from the pancreas as well as from the intestinal lining. Pancreatic enzymes such as trypsin, chymotrypsin, and carboxypeptidase are released into the small intestine to hydrolyze peptide bonds and consequently break the peptides into smaller chains of amino acids as well as free amino acids [1, 10]. These pancreatic enzymes are activated by the enzyme enteropeptidase that is released from the small intestine for proper breakdown of proteins. The small intestine provides an alkaline environment to ensure maximum enzymic action and neutralize chyme that enters from the stomach [11].

3. **Absorption of Amino Acids and Peptides**

These amino acids and short peptides are subsequently absorbed into enterocytes, absorptive cells along the inner lining of the small intestine. Certain transporters on the surface of enterocytes actively transport amino acids and dipeptides across the cell membrane [10, 11]. Once inside the enterocytes, dipeptides and tripeptides are further hydrolyzed to release free amino acids, which are transported out of the cell to the bloodstream through the cell's basolateral membrane [12]. The circulatory system is also involved in transport to various tissues, thereby providing cells with amino acids for energy production, synthesis of proteins, and other metabolic processes [11].

4. **Cellular Uptake and Utilization**

The amino acids are transported throughout the body, into all tissue and cells, where they are absorbed by the cells. Certain transport proteins can aid in the movement of amino acids through cell membranes. Once the amino acids have entered the cell, they may either be temporarily deposited or become immediately available for the synthesis of new proteins [12]. The rate and efficiency of absorption of amino acids are affected by the cell type, and this depends on the amount of food consumed and the cell need. For instance, in muscle cells, amino acids are required for the formation of muscle protein, especially in the recovery phase after exercise or during some other forms of recovery [1, 10]. Amino acids can also be deaminated inside cells to produce energy, as they are converted into different intermediates of the citric acid cycle or another metabolic process [8, 11].

1.2 Importance of Efficient Protein Absorption and the Role of Nano-Encapsulation

Health and physiological events-immune response, maintaining muscle integrity, and synthesis of enzymes-all depend on the availability of the bioavailability of proteins [13]. Some drawbacks of traditional pathways of protein uptake are transport and absorption losses and enzymatic breakdown of protein digesta in the gastrointestinal tract [14]. Nano-encapsulation technology, despite potentially conferring resistance of proteins to degradative processes and promoting cellular uptake, offers an opportunity for optimizing protein delivery and absorption [15]. The encapsulating proteins may enhance stability together with inducing regulated release and protect proteins from harsh gastrointestinal conditions [16]. Encapsulated proteins would also increase the bioavailability and controlled release of protein supplements and functional foods in the food and nutraceutical industries [5]. The process of multiplicity steps involving protein digestion and absorption explains the need to maximize protein bioavailability for health and well-being [17]. Future researches on nano-encapsulation technique may reveal new functional meals and supplements according to specific needs in health and therefore improve human nutrition and well-being [18].

1.3 Challenges in Protein Absorption

1.3.1 Digestive Degradation and Limited Bioavailability

Degradation of dietary proteins by digestive processes reduces the bioavailability of certain amino acids, which makes the absorption of proteins challenging [1]. Enzymes and acid environments degrade proteins into shorter peptides and amino acids [10]. However, certain proteins may be degraded to an extent that they could not pass into the bloodstream [7]. Other structural characteristics of some proteins that make them susceptible to degradation include intricate molecular structure or fiber binding [13]. Bioactive peptides cannot achieve their sites of action due to degradation by digestion [19]. Because only some of the amino acids in these lower bioavailability proteins can be used by the body, these proteins may not confer their entire range of health benefits [5, 7].

1.4 Factors Influencing Protein Absorption

All of these below parameters influence the digestion, absorption, and utilization of proteins within the body. These factors are the type of protein, individual physiology, and the composition of the diet.

1. **Protein Type**

   Protein sources differ significantly in terms of their digestibility and bioavailability. For example, animal proteins like whey, casein, and egg protein have superior bioavailability when compared to most plant-based proteins and are easier to digest than many of them as well [7]. Whey protein, as it digests and gets absorbed pretty quickly, is preferred by athletes for direct protein delivery to the muscles to initiate the repair process rapidly. On the other hand, the antinutritional factors such as phytates and tannins found in the plant-based proteins that may inhibit the digestive enzymes and thereby hinder absorption may occur [8]. As a result, the plant proteins, such as soy and pea proteins, are in general less available compared to the proteins from animals [13].

2. **Individual Physiology**

   All these factors-age, health, and genetics-may play a tremendous role in determining how well the body absorbs proteins. For instance, elderly people may have fewer digestive enzymes, leading to inability to adequately digest proteins in the body [19]. Another example is of patients suffering from diseases such as Crohn's disease or irritable bowel syndrome, that may lead to protein absorption failures due to inflammation or destruction of the lining of the gut [15]. This metabolism is also under hereditary influences, as some people have genetic variations that weaken the activity of amino acid transporters or the production of enzymes, which may impair the efficiency of overall protein absorption [10].

3. **Dietary Composition and Interaction with Other Nutrients**

   The amount of fat and fibre in a meal would affect the absorption of proteins. Fats may inhibit stomach emptying and thereby delay the output of amino acids into the circulation [1]. This would also delay the digestion and absorption of proteins [19]. Fibre can bind to proteins or digestive enzymes and inhibit protein breakdown, especially if it is derived from plant source. Some studies indicate that once absorbed, the release of insulin can stimulate the uptake of amino acids into cells; this process may be more enhanced by the concurrent presence of other macronutrients, such as carbs, incorporation [11].

All these factors contribute to interindividual variation in protein absorption and reflect the need for tailored approaches to enhance protein delivery.

1.5 Need for Enhanced Delivery Systems

To enhance bioavailability of protein and ensure reliable absorption, new delivery methods need to be formulated with an awareness of the increasing complexities of gastric breakdown and protein absorption [11, 12]. A method is nano-encapsulation whereby proteins are encapsulated in nanoparticles made of biocompatible materials such as lipids and polymers [19]. These nano-carriers protect the proteins from stomach acid degradation and permit the controlled release of the proteins in the small intestine, where the absorption is most efficient [20]. They also regulate when and how fast the proteins are released; this again leads to a more sustained and prolonged supply of amino acids into the bloodstream [1]. Scientists have extensively researched this process in food and pharmaceutical industries, especially with respect to bioactive compounds like anthocyanins and folic acid. Lipid-based nanoparticles such as liposomes and solid lipid nanoparticles will create a hydrophobic barrier that will protect proteins from severe gastric conditions and ensure their safe passage to the intestines [1, 3–5, 11, 12, 17]. These methods, which provide a regulated, step-by-step release of amino acids that better meet the body's needs for metabolism, could revolutionize protein supplementation as a whole [18]. They improve the efficiency with which proteins are fed to many individuals and conditions of health by mitigating digestive problems but also physiological and dietary variability [2, 19].

2 Basics of Nano-Encapsulation Technology

The technique known as nano-encapsulation allows for an active agent such as vitamins, drugs, or bioactive food component to be encapsulated within nanoscale carriers in order to increase stability, bioavailability, and controlled release [21, 22]. At a very basic level, it is the production of particles between 1 and 1000 nm that enclose the core material within shells in order to protect it from external excitants, including light, oxygen, and digestive enzymes [5, 12]. The functioning of nano-encapsulation depends on the nanocarrier characteristics and the encapsulation technique. These characteristics can be modulated to control the release profile, targeted distribution, and protection of the active ingredient. For instance, through manipulation of the way poorly soluble medicines interact with biological membranes, nano-encapsulation in drug delivery may improve their solubility and absorption and facilitate better passage across cellular barriers [3, 13]. Fortified foods may be able to improve their functional and nutritional quality in the food industry by protecting fragile nutrients during processing, storage, and digestion through nano-encapsulation [5, 14]. Nanocarriers, which constitute the heart of nanoscale components of nano-encapsulation systems, are made up of biocompatible and biodegradable elements such as proteins, lipids, and polymers. Because of these, particular delivery systems with controlled release, targeted delivery, and also stability under particular conditions become feasible [19]. Incorporating technologies that consist of problems like low solubility, poor stability, and faster degradation of nutrients and drugs into nano-encapsulation systems may enhance the bioavailability of encapsulated bioactive compounds [12].

2.1 *Types of Nanocarriers*

The type of nanocarrier chosen also plays a big role in determining whether the nano-encapsulation will be a success because each of them has different features, which might influence the release, absorption, and stability of the molecule encapsulated. There are three major forms of nanocarriers: liposomes, polymeric nanoparticles, and lipid-based nanoparticles, with each having its advantages and disadvantages.

2.1.1 Liposomes

Liposomes are vesicular, spherical structures with a phospholipid bilayer. They may represent cellular membranes and can encapsulate hydrophilic (water-soluble) compounds and lipophilic (fat-soluble) compounds [17, 23]. Liposomes are extremely adaptable and can deliver a wide variety of bioactive molecules. The unique bilayered structure of liposomes that contributes toward biocompatibility and biodegradability can be utilized in the pharmaceutical for site-specific drug delivery [12, 24]. Due to this structure, drugs or nutrients can be added to the lipid bilayer or the aqueous core of liposomes, which also increases bioavailability and shields the contents from environmental deterioration [3, 19].

2.1.2 Polymeric Nanoparticles

Polymeric nanoparticles—a colloid with a solid core that is made from synthetic or naturally-based polymers such as alginate, chitosan, and poly(lactic-co-glycolic acid), PLGA [2, 25]. They can allow for controlled and targeted release and provide a very strong, rigid framework for protecting bioactive agents from hostile environmental exposures like acidic stomach juices [2, 17]. Such nanoparticles have great value in pharmaceutical applications because of their excellent stability and capability of having controlled formulations tailored for specific profiles of drug release. For example, the nanocarrier breakdown rate can be controlled to attain a controlled, sustained release of the encapsulated medication or nutrient by adjusting the polymer composition [26].

2.1.3 Lipid-Based Nanoparticles

The majority composition of lipid-based nanoparticles comprises emulsifier-stabilized lipids, comprising SLNs and NLCs. Such particles may effectively improve the solubility as well as the bioavailability of fat-soluble vitamins and nutraceuticals since they are specifically suitable for the entrapment of lipophilic substances [3, 24]. The lipid composition of these nanocarriers enhances the permeability of bioactive substances via cell membranes, thus enhancing easier absorption by the gastrointestinal tract [18]. The rigid lipid core of SLNs makes them protect drugs from degradation and provide a controlled release mechanism. SLNs have been widely used in drug delivery to improve the stability and bioavailability of active components, mainly in topical and oral formulations [16].

2.1.4 Nanoemulsions

Nanoemulsions are fine water-in-oil or oil-in-water dispersions stabilized by surfactants with droplet sizes ranging between 20 and 200 nm. The small droplet size provides a large surface area, thereby increasing the bioavailability and rate of absorption of the encapsulated substance [2, 12]. Due to smaller droplet size, these systems are highly beneficial for delivering hydrophobic (fat-soluble) chemicals, wherein interaction with biological membranes is particularly important for enhancing transport across cell barriers. Due to enhancement in the stability of volatile substances such as tastes and essential oils and improvement in absorption of bioactive substances in the gastrointestinal tract, nanoemulsions are used in food, cosmetics, and pharmaceutical industries [12].

2.2 Mechanism of Action for Improved Absorption

Nano-encapsulation technology enhances the absorption of proteins since it shields proteins from getting degraded in the digestive tract [3]. Bioavailability of proteins is affected by the highly acidic environment in the stomach and digesting enzymes of the intestines [12]. Proteins are preserved because they are encapsulated in polymeric and lipid-based nanoparticles, which have resistance towards the acidic environments [5]. This protective mechanism is of great importance for therapeutic proteins and peptides as it extends their activity and increases their effective concentration within the target area. Nano-encapsulation makes targeted and effective delivery possible and allows the controlled release of proteins at specific locations inside the gastrointestinal system [27]. The mechanism is likely to allow more concentrated active protein to pass through the small intestine intact [15]. The small size of nano-encapsulated particles increases the chances of contacting cellular membranes, which consequently increases intestinal epithelial absorption [5]. This can increase the effectiveness of oral protein therapies and nutraceuticals since it prevents the degradation of proteins while enhancing the absorption process by optimizing both release and cellular uptake [28]. Table 1 shows about the case studies and their treatment for Nano Encapsulation.

Table 1 Case studies and techniques in nano-encapsulation

Method	Treatment	Effect	Dose	References
Nano-lipid carriers for protein stabilization	Protein delivery systems	Enhanced bioavailability and stability of therapeutic proteins	Variable, depending on study	[29]
Polysaccharide nanoparticles for peptide delivery	Oral delivery of bioactive peptides	Improved gastrointestinal absorption	Not specified	[30]
Hybrid nanocarriers combining polysaccharides/lipids	Targeted anticancer therapy	Increased cytotoxicity toward cancer cells	Not specified	[31]
Liposomal encapsulation of Spirulina protein	Nutritional supplementation	Enhanced controlled release in gastrointestinal conditions	10 mg/kg	[32]
Surface-modified nanoparticles	Anthocyanin delivery	Improved bioavailability and antioxidant activity	5–10 mg/day	[33]
Pectin-based nanocarriers	Bioactive compound delivery	Enhanced stability and slow-release profiles	Not specified	[34]
Nano-encapsulation of phenolic compounds	Cancer prevention	Reduced tumor progression	Not specified	[35]
Nano-emulsions in agricultural biostimulants	Nutritional enhancement of crops	Improved stability and delivery of active agents	Not specified	[36]
Nano-lipid carriers in food	Nutrient fortification in yogurt	Functional benefits including immune modulation	Not specified	[37]
Biopolymer nanoparticles for antioxidants	Drug and nutrient delivery	Enhanced antioxidative properties	Variable	[38]
Polysaccharide/lipid hybrid systems	Targeted therapy	Improved pharmacokinetic profiles	Variable	[39]
Nano-carbohydrate delivery systems	Nutraceutical and food applications	Increased nutrient absorption efficiency	Not specified	[40]
Co-encapsulation in nano-lipid carriers	Functional foods	Prolonged nutrient stability and controlled release	10–50 mg	[34]
Micro/nano-fluidics	Food bioactives	Precise control of encapsulation and release	Variable	[41]
Cyanidin-3-glucoside nanocarriers	Functional food and dietary supplements	Improved bioactivity and health benefits	Not specified	[42]

(continued)

Table 1 (continued)

Method	Treatment	Effect	Dose	References
Nano-fortification in yogurts	Natural bioactive compounds	Multi-functional health benefits	Not specified	[43]
Nano-encapsulation of essential oils	Food preservation	Enhanced flavor stability and controlled release	5 mg/g	[44]
Nanotechnology in tuberculosis vaccines	Polymer-based nano-vaccines	Enhanced immunogenicity and stability	Variable	[45]
Liposome-based encapsulation	Protein drug delivery	Improved structural integrity and targeted release	Not specified	[46]
Co-encapsulation strategies	Anti-cancer drugs	Increased bioavailability and reduced side effects	Variable	[47]
Nutraceutical encapsulation in polymers	Functional beverages	Improved solubility and stability of encapsulated agents	2–5 g/L	[48]
Carotenoid encapsulation	Dietary supplements	Enhanced bioavailability and antioxidant properties	Not specified	[49]
Nano-protein carriers	Immune modulation	Enhanced immunomodulatory effects	Variable	[50]
Encapsulation using bio-polyester carriers	Drug delivery	Improved drug release profiles	Variable	[51]
Advanced co-encapsulation methods	Food bioactive compounds	Stability improvements in processing and storage	Not specified	[48]

2.3 Applications of Nano-Encapsulation for Protein Absorption

State-of-the-art technique with profound assistances for enlightening protein uptake in numerous fields, for example, nutrition, medicines, and cosmetics, the nano-encapsulation comes. This chapter reviews numerous applications of this technique such as cosmeceutics, therapeutic protein delivery and as dietary supplements. The Major applications of Nano Encapsulation shows in Fig. 1.

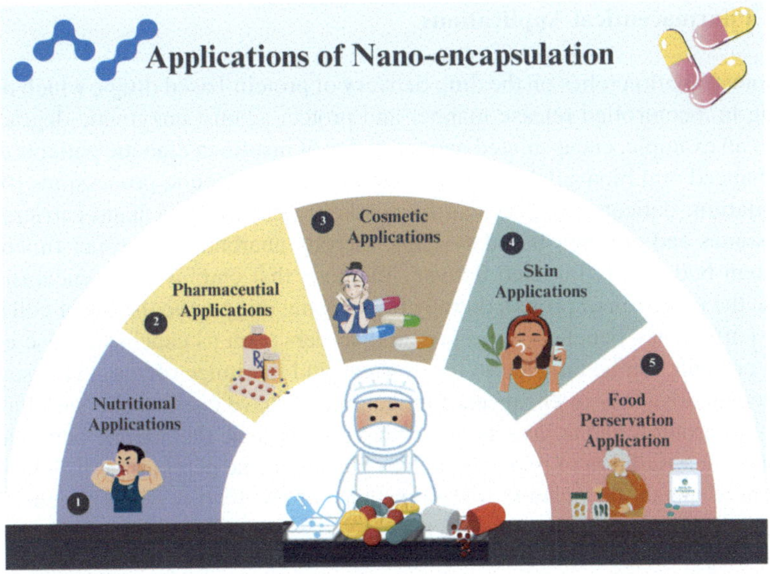

Fig. 1 Applications of nano encapsulation of protein absorption

2.3.1 Nutritional Applications

Nano-encapsulation has completely transformed dietary supplements by cumulative the stability and bioavailability of protein-based nutrients [52, 53]. Proteins can be precisely released and absorbed in the gastrointestinal tract by being encapsulated in nanocarriers, which break them from humiliating during digestion [54]. This invention certifies that essential elements such as vitamins and amino acids are efficiently supplied to customers, making fortified foods extra effective [22, 55, 56]. Useful practices that can be assumed toward solving this worldwide problem comprise nano-encapsulation of whey protein in functional beverages, accumulative the solubility and enhancing its bioactivity [57, 58]. Nano-encapsulation in sports nutrition exploits protein distribution for resolution and muscle recovery. For instance, proteins such as creatine and branched-chain amino acids can be encapsulated to ensure continuous release to the system, thereby ensuring a sustained flow of energy for more extensive exercise [59, 60]. This also enhances the effectiveness of clinical nutrition by administering tailored therapeutic proteins to malnourished or critically ill patients [61, 62]. It has shown that nano-encapsulation could improve hydrophobic proteins' bioavailability and potential use for clinical applications [63, 64].

2.3.2 Pharmaceutical Applications

Nano-encapsulation relies on the drug delivery of protein-based drugs, which deliver the drug in a controlled release manner and protect against enzymatic degradation [65]. As an example, encapsulated nanoparticles of insulin in diabetic patients exhibited enhanced oral bioavailability with long-lasting therapeutic possessions [66]. In accumulation, delicate proteins can be securely delivered via nano-carriers such as liposomes and polymer-based systems, thereby guaranteeing proper function in the human body [67]. Targeted delivery methods that employ nano-encapsulation increase the precision of protein therapy by focusing on specific tissues or cells [68]. This is particularly beneficial for chronic disorders such as cancer because encapsulated proteins can be targeted to tumor areas and minimize off-target possessions [69]. Recent advances in encapsulation technology have enabled the simultaneous delivery of proteins and adjuvants, which have synergistic therapeutic benefits [70]. The nano-encapsulation of peptides such as glucagon-like peptide-1 (GLP-1) for the treatment of metabolic sicknesses is an instance of noteworthy development [71].

2.3.3 Cosmetic and Skin Health Applications

Nano-encapsulation made it possible to distribute the bioactive proteins and peptides into deeper layers of the skin, which has suggestively enhanced the cosmeceutical field [72, 73]. In anti-aging foodstuffs, this technique certifies the effectiveness of delicate mechanisms like collagen and elastin by shielding them from oxidation [74]. Additionally, encapsulated proteins have profiles of prolonged relief, thus providing long-term compensations for hydration and suppleness of the skin [75]. For instance, protein-based nanoparticles have been shown to advance photoprotection and enhance barrier function in the skin when rummage-sale in sunscreens [76]. In line with the trend towards usual and ecological cosmetics, the increased popularity of nano-encapsulated plant-derived proteins and peptides in skincare products also received much attention. These products now become safer and more attractive with the advancement of biodegradable nano-carriers developed for this purpose [77]. Nano-encapsulation has emerged as a playing technique with a wide spread of applications in nutrition and medicine to improve protein absorption besides cosmetics. Enhanced stability and bioavailability with targeted delivery do address significant issues within the domains. Nano-encapsulation stands as a very promising technique that can improve health and well-being, and a future of more research and development may unlock new roads for travel. Table 2 shows about the applications of protein absorption.

Table 2 Applications of nano-encapsulation for protein absorption

Application domain	Specific focus	References
Nutritional applications		
Dietary supplements	Fortified foods with enhanced protein bioavailability	[76–79]
Sports nutrition	Improved nutrient delivery systems for performance enhancement	[80, 79, 30]
Clinical nutrition	Specialized formulations for critical care nutrition	[79, 31, 50]
Pharmaceutical applications		
Protein-based therapies	Advanced delivery systems for therapeutic proteins	[81, 82, 38]
Insulin delivery	Targeted and sustained insulin release mechanisms	[83–85]
Peptide drugs	Enhanced stability and bioavailability of peptide-based drugs	[88, 86, 87, 39]
Cosmetic and skin health		
Cosmeceuticals	Protein-based formulations for anti-aging and skincare	[89, 90, 29]
Dermal delivery	Encapsulation of bioactive proteins for improved skin absorption	[91–93]
Antioxidant delivery	Protein encapsulation systems for oxidative stress prevention	[94, 95, 38]

3 Types of Nanomaterials Used in Protein Encapsulation

3.1 Polymeric Nanoparticles

Because of their tunable physicochemical properties, polymeric nanoparticles are among the furthermost versatile platforms for the encapsulation of proteins. These nanoparticles, which could be derived from either biodegradable or synthetic polymers, provide encapsulated proteins with greater stability and a controlled drug release. Polymers such as poly(lactic-co-glycolic acid) (PLGA) and polycaprolactone (PCL) have been proven in the long term to deliver continuous and tailored protein release [29]. Polymeric nanoparticles shield proteins from environmental stressors such pH and temperature fluctuations as well as enzymatic degradation [96]. When polymer exteriors are functionalized with ligands, the cellular approval and selectivity are better [97]. FDA-approved polymers used in medicinal applications, like PEG and PLGA, greatly minimize the toxicity [31]. These degrade to physiological, non-toxic endogenous metabolites such as lactic and glycolic acids [30]. There are two significant barriers though: residual organic solvents in the preparation of the formulations and the manufacturing process costs are very high [98]. Vaccine delivery, enzyme replacement therapies, and cancer treatment use polymeric nanoparticles. For instance, PLGA nanoparticles containing insulin have shown promising results in the management of diabetes [51].

3.2 *Lipid-Based Nanocarriers*

Due to their biomimetic qualities, lipid-based systems such as liposomes, solid lipid nanoparticles (SLNs), and nanostructured lipid carriers (NLCs) are getting more and more attractive. Such delivery systems are particularly suitable for the delivery of hydrophilic proteins and peptides. In liposomes, proteins are encapsulated within the phospholipid bilayers of the aqueous core [45]. Hydrophilic as well as hydrophobic drugs can be distributed by these systems simultaneously due to *amphiphilia*; thus, bioavailability can be augmented [39]. Notable examples are liposomal provisions of vaccination formulations, which comprise the mRNA COVID-19 vaccines [99].

Lipid Nanoparticles in Solid Form the solid lipids that establish SLNs are structurally steady because they are solid at body temperature [38]. They deliver protection in contradiction of enzymatic deprivation and controlled protein announcement [40]. Low encapsulation competence and drug leakage are the difficulties accompanying with SLNs [33]. Lipid Carriers with Nanostructures Compared to SLNs, NLCs are a blend of liquid and solid lipids that offer better stability and encapsulation effectiveness [34]. In the treatment of brain tumors, therapeutic proteins have been transported into the brain with the help of NLCs, which can permeate finished the blood-brain barrier [100]. Usually, lipid-based systems are tremendously biocompatible and absorbance is improved due to communication with cell membranes. High manufacture costs and poor storing stability remain challenges although [36].

3.3 *Protein-Based and Polysaccharide-Based Nanoparticles*

The most common protein-based precursors for the formation of nanoparticles are casein, albumin, and gelatin [46]. These resources offer the subsequent benefits: they can deliver controlled protein release, stability in physiological circumstances, high biocompatibility and biodegradability, and natural ligand-binding competences that permit selective targeting [32]. Disadvantages of these materials contain some of the protein carriers may induce immunogenicity, and there is batch-to-batch unpredictability in natural proteins that may disturb uniformity [35]. Paclitaxel (Abraxane) was brought effectively as an anticancer drug consuming albumin nanoparticles [42].

Being mucoadhesive and biodegradable, polysaccharides such as chitosan, alginate, and dextran are ideal encapsulation vehicles for proteins. Chitosan the positively charged chitosan nanoparticles, which are derivative from chitin, improve bioavailability and boost adherence to negatively charged cell membranes [47]. They shield proteins from enzymatic collapse and have antibacterial properties [50]. Alginate anionic polysaccharide produces hydrogels when there are divalent cations. Such hydrogels are excellent in encapsulating fragile proteins [101]. Its drawbacks include deprived mechanical strength and rapid squalor in acidic environments [44]. Polysaccharide-based nanoparticles are used for oral delivery of insulin and other

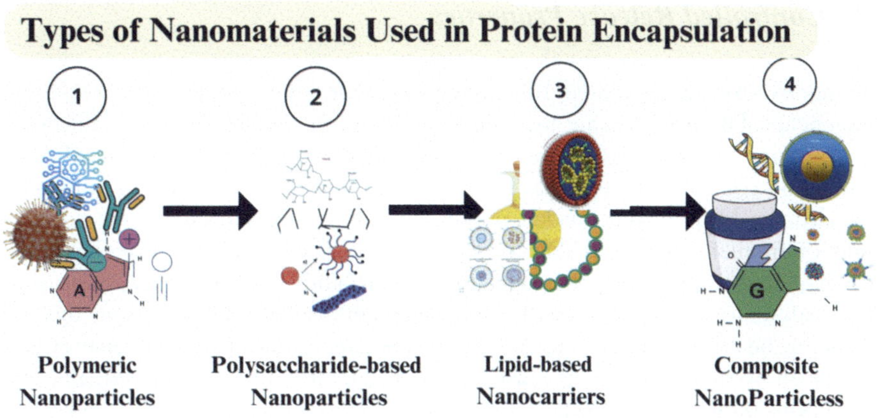

Fig. 2 Types of nano encapsulation

therapeutic proteins, thus avoiding the incomplete bioavailability caused by enzymatic squalor in the gastrointestinal tract [102]. Each type of nanomaterial has its own benefits and hitches regarding protein encapsulation. Protein polysaccharide-based nanoparticles syndicate functionality with inherent biocompatibility, lipid-based systems mimic biological membranes to improve absorption, and polymeric nanoparticles are outstanding in controlled proclamation and durability. The intended use, physiognomies of the protein, and delivery method all influence nanomaterial selection. Figure 2 illustrates about the types of Nano Encapsulation.

4 Advantages of Nano-Encapsulation for Protein Delivery

4.1 Enhanced Stability of Protein Molecules

Nano-encapsulation greatly provides protection to protein molecules against harmful environmental factors such as pH, temperature, and enzymatic squalor. Hydrolysis and denaturation take place in proteins due to extreme pH or high temperatures. The protecting barrier of nano-encapsulation maintains the functionality of proteins for storing and passageway through the gastrointestinal tract [103]. With oxidation and aggregation reduced, the encapsulated proteins have increased shelf-life that has a great importance in both food and pharmaceutical solicitations [80]. The encapsulation of enzymes such as lactase into nanocarriers improves the functionality of dietary supplements by maintaining their activity under acidic stomach conditions [37].

4.2 Controlled Release Properties

Continuous over a stipulated time releases of proteins are possible owing to nano-encapsulation increasing their nutritional and medicinal efficacies. Encapsulated proteins by nanocarriers are released gradually using diffusion-controlled or degradation-controlled methods as in the cases of lipid-based systems or polymeric nanoparticles [41]. Insulin-loaded polymeric nanoparticles decrease the requirements of frequent injections due to a gradual medication release, bringing long-term glycemic control [48]. Nanoencapsulated probiotics, including those encapsulated in alginate nanogels, gradually announcement live cells into the colon, thereby enhancing health properties [104]. This characteristic diminishes adverse effects brought on by rapid release, increases the bioavailability of active ingredients, and diminutions protein waste [49].

4.3 Increased Cellular Uptake and Bioavailability

Nano-encapsulation dramatically increases the bioavailability and cellular absorption of proteins by resolving issues of low permeability and degradation. Because lipids and biological membranes interact, the encapsulation in lipid-based nanoparticles upsurges penetrability across the intestinal lining, easing protein absorption [105]. Nanoparticles functionalized with targeting ligands or surfactants take use of endocytosis mechanisms to guarantee efficient delivery to particular cells or tissues [106]. This permits the evasion of enzymatic barriers by liposome carriers, increasing the bioavailability of therapeutic proteins such as interferons [67]. Since nano-encapsulation offers constancy, controlled release, and enhanced cellular uptake, it offers a revolutionary way for protein distribution. Therefore, this technique is a requirement for contemporary treatments and functional food systems since it solves imperative issues in protein delivery. Table 3 illustrates about the advantages of Protein delivery in Nano Encapsulaition.

5 Challenges and Limitations of Nano-Encapsulation

5.1 Manufacturing and Scalability Issues

High production costs are instigated by the obligation for high-end machinery and acquaintance in nano-encapsulation technologies. Techniques such as nano-spray drying, microfluidization, and supercritical fluid extraction add difficulty, necessitating precise control over parameters [49]. It is also still stimulating to optimize particle size, surface charge, and encapsulation competence at scale-up [105]. Their substantial energy and material inputs further limit the cost-effectiveness of these

Table 3 Advantages of nano-encapsulation for protein delivery

Aspect	Description	Citations
Enhanced stability of protein molecules		
Protection against pH degradation	Nano-encapsulation protects proteins from acidic or basic pH conditions, maintaining functionality	[96, 38, 107]
Protection against temperature extremes	Shields proteins from heat or cold-induced denaturation during processing and storage	[39, 33, 108]
Resistance to enzymatic degradation	Encapsulation prevents premature breakdown by digestive enzymes, increasing protein survival	[99, 47, 109]
Controlled release properties		
Sustained release over time	Enables gradual protein release, maintaining therapeutic levels or prolonged nutritional benefits	[100, 105, 110]
Targeted delivery	Nano-systems allow for site-specific protein release in the body, reducing side effects	[47, 111, 112]
Applications in nutrition and therapeutics	Widely used for vitamin C stabilization, enzyme therapies, and tailored drug release systems	[34, 102, 113]
Increased cellular uptake and bioavailability		
Improved intestinal permeability	Enhances the transport of proteins through tight epithelial junctions in the gut	[98, 50, 114]
Mechanisms enhancing membrane transport	Facilitates protein delivery via receptor-mediated or lipid-based transport pathways	[32, 44, 115]

processes, particularly for commercial food and pharmaceutical solicitations [106]. One of the greatest challenges in mass production is achieving uniformity in particle size and encapsulation effectiveness. Even slight variations can disturb the firmness and bioavailability of the encapsulated proteins [67]. Quality control measures have to be very stringent to guarantee reproducibility and compliance with regulatory standards for nanoparticle size and arrangement, which upsurges the overall cost [116].

5.2 Safety and Regulatory Concerns

The potential toxicity of some nanomaterials does not help in the safe usage of nanoencapsulated products. For instance, nanoparticles could pose a threat to human health through their accumulation in tissues or disruption of biological activities [117]. Other concerns in the area of safety might also result from changes on the surface, like stabilizers or coatings [109]. Because nano-encapsulation regulatory regimes vary from one industry and geographical area to another, manufacturers are cautious. Approvals for products and entering a market are hindered in the absence of common nanomaterial standards for foods and pharmaceuticals [118]. Before commercialization is allowed, more rigorous assessments must be conducted, which would include toxicity testing and assessment of environmental impact [113].

5.3 Stability and Storage Challenges

Environmental factors such as light, humidity, and temperature can impact nanoencapsulated proteins. The functionality of the encapsulated protein may be compromised if the nanoparticles aggregate or lose their active properties due to improper storage conditions [119]. Preservatives are often added to shelf life, which could render preparations more challenging [43]. Proteins encapsulated in nanocarriers may interrelate with other mechanisms in multi-component systems foremost to instability; encapsulated enzymes in efficient food products for example may undergo a response with fats and carbohydrates that may lead to its premature release or degradation [120]. Formulation enterprise also becomes difficult due to interaction of nanoparticles with other particles especially those foods with diverse population profiles [121]. Although protein distribution using nano-encapsulation has revolutionary potential, its wider use is reliant on on resolving issues of stability, safety, and manufacture [112, 122]. Overcoming these footraces and unlocking the full potential of nano-encapsulation technology will require cooperative investigation and supervisory initiatives.

6 Future Directions in Nano-Encapsulation for Protein Delivery

6.1 Personalized Nutrition and Targeted Protein Delivery

This development will surely bring a revolution to the concept of personalized nutrition since nano-encapsulation can enhance the precision of the nutritional intervention by designing it to deliver proteins as needed and in accordance with individual-specific health profiles and diet compositions [123]. Through specific genetic or metabolic profiles, targeted nutritional intake or even filling definite deficits can be made achievable by designing customized protein nano-carriers [124]. These systems hold potential for applications in therapeutic nutrition, where bioactive proteins can be administered individually to patients suffering from a specific disorder or chronic condition [111].

6.2 Development of Biodegradable and Sustainable Nanomaterials

Biodegradable and eco-friendly nano-encapsulation resources are gaining popularity with the development of environmental trepidations. Advances in biopolymer-based carriers, such as cellulose derivatives, alginate, and chitosan, offer an environmentally friendly means of carrying proteins [125]. These materials natural degradation decreases their environmental impact while preserving encapsulation competence [115]. In addition, biomaterials derived from plants and marine life is gaining popularity as sustainable sources for nano-encapsulation [126]. The functionalization of natural nanomaterials to enhance stability, compatibility, and protein-loading capacity has freshly been the focus of interest [108]. Protein or starch-based nanoparticles have established promising in the delivery of fragile bioactives such as peptides and enzymes [127].

6.3 Integration with Other Delivery Technologies

Synergy between nano-encapsulation and other delivery expertise will be accomplished of doing much better. Long-term controlled release, along with increased preservation of proteins, can be offered by layered delivery approaches combining micro- and nano-encapsulation [128]. For instance, while nano-encapsulation potentials targeted absorption at the cellular level, micro-encapsulation may protect proteins from degradation during digestion [129]. Enzymatic degradation that prevails within the gut lumen is overcome by integration of oral peptide delivery technologies along with nano-encapsulation [125]. These hybrid systems are making it feasible to deliver therapeutic peptides at increased bioavailability and with enhanced efficiency [111].

6.4 Potential for Broader Applications Beyond Absorption

Besides improving the protein absorption, nano-encapsulation also offers a prospect for delivering proteins that may provide sophisticated functions, for instance, immunomodulatory encapsulated proteins which could be targeted specifically at a particular organ or tissues in therapy for autoimmune conditions or improving the efficacy of vaccination [115]. The process of nano-encapsulation provides stabilization and protection to the enzymes and therapeutic peptides for their potential use in the processes involved in metabolic regulation and in the treatment of chronic syndromes [123]. For example, nano-carriers can improve the stability of heat-sensitive enzymes, offering them potential applications in both industrial and medicinal aspects [128]. Future progresses on protein delivery by nano-encapsulation will emphasis on multifunctionality, accuracy, and sustainability. The revolutionary potential of such technology in nutrition and treatment is seen with the expansion of biodegradable materials, integration into other technologies, and cumulative applications.

7 Conclusion

Nano-encapsulation is a game-changing technique that advances protein absorption by addressing issues such as physiological variability, restricted bioavailability, and intestinal breakdown. Nano-encapsulation upsurges the absorption efficiency of proteins, protects them, permits for controlled release, and increases cellular uptake. This chapter deliberates how technologies in lipid-based carriers, polymeric nanoparticles, and liposomes enhance the stability and bioavailability of proteins. These advance arenas in areas like nutritional supplements, delivery of therapeutic proteins, and cosmeceuticals. Their solicitations transcend boundaries from nutrition to pharmaceuticals, as well as cosmetics, but still there are encounters and problems that need more innovations and regulatory progress for those areas such as complexity of production, regulatory barriers, and storage steadiness. Eco-friendly nanomaterials, tailored nutrition plans, and incorporation into other delivery systems are some of the bright futures in the use of therapeutic proteins and enzymes beyond absorption. This would be a revolution in the protein-based goods industry through continued research on nano-encapsulation, making it more effective and accessible to different industries.

References

1. Bhutia YD, Ganapathy V (2018) Protein digestion and absorption. In: Physiology of the gastrointestinal tract. Academic Press, pp 1063–1086
2. Ahmed GHG, Fernández-González A, García MED (2020) Nano-encapsulation of grape and apple pomace phenolic extract in chitosan and soy protein via nanoemulsification. Food Hydrocolloids 108:105806
3. Amoabediny G, Haghiralsadat F, Naderinezhad S, Helder MN, Akhoundi Kharanaghi E, Mohammadnejad Arough J, Zandieh-Doulabi B (2018) Overview of preparation methods of

polymeric and lipid-based (niosome, solid lipid, liposome) nanoparticles: a comprehensive review. Int J Polym Mater Polym Biomater 67(6):383–400

4. Assadpour E, Maghsoudlou Y, Jafari SM, Ghorbani M, Aalami M (2016) Optimization of folic acid nano-emulsification and encapsulation by maltodextrin-whey protein double emulsions. Int J Biol Macromol 86:197–207
5. Bratovcic A, Suljagic J (2019) Micro-and nano-encapsulation in food industry. Croatian J Food Sci Technol 11(1):113–121
6. Qin L, Cui Z, Wu Y, Wang H, Zhang X, Guan J, Mao S (2023) Challenges and strategies to enhance the systemic absorption of inhaled peptides and proteins. Pharm Res 40(5):1037–1055
7. Leidy HJ, Clifton PM, Astrup A, Wycherley TP, Westerterp-Plantenga MS, Luscombe-Marsh ND, Mattes RD et al (2015) The role of protein in weight loss and maintenance. Am J Clin Nutr 101(6):1320S-1329S
8. Wu G (2016) Dietary protein intake and human health. Food Funct 7(3):1251–1265
9. Schoenfeld BJ, Aragon AA (2018) How much protein can the body use in a single meal for muscle-building? Implications for daily protein distribution. J Int Soc Sports Nutr 15:1–6
10. Loveday SM (2023) Protein digestion and absorption: the influence of food processing. Nutr Res Rev 36(2):544–559
11. Dupont D, Tomé D (2020) Milk proteins: digestion and absorption in the gastrointestinal tract. In: Milk proteins. Academic Press, pp 701–714
12. Ferreira CD, Nunes IL (2019) Oil nanoencapsulation: development, application, and incorporation into the food market. Nanoscale Res Lett 14:1–13
13. Noorbatcha IA, Jaswir I, Ahmad H (2017) Nano-encapsulation of proteins and peptides. Current Nanomaterials 2(2):76–83
14. Katouzian I, Jafari SM (2016) Nano-encapsulation as a promising approach for targeted delivery and controlled release of vitamins. Trends Food Sci Technol 53:34–48
15. Vinarov Z, Abrahamsson B, Artursson P, Batchelor H, Berben P, Bernkop-Schnürch A, Augustijns P et al (2021) Current challenges and future perspectives in oral absorption research: An opinion of the UNGAP network. Adv Drug Deliv Rev 171:289–331
16. Mahalakshmi L, Leena MM, Moses JA, Anandharamakrishnan C (2020) Micro-and nano-encapsulation of β-carotene in zein protein: Size-dependent release and absorption behavior. Food Funct 11(2):1647–1660
17. Dristant U, Mukherjee K, Saha S, Maity D (2023) An overview of polymeric nanoparticles-based drug delivery system in cancer treatment. Technol Cancer Res Treat 22:15330338231152084
18. Rashwan AK, Karim N, Xu Y, Xie J, Cui H, Mozafari MR, Chen W (2023) Potential micro-/nano-encapsulation systems for improving stability and bioavailability of anthocyanins: an updated review. Crit Rev Food Sci Nutr 63(19):3362–3385
19. Jafari SM (2017) An overview of nanoencapsulation techniques and their classification. Nanoencapsulation technologies for the food and nutraceutical industries, pp 1–34
20. Lu H, Zhang S, Wang J, Chen Q (2021) A review on polymer and lipid-based nanocarriers and its application to nano-pharmaceutical and food-based systems. Front Nutr 8:783831
21. Wang C, Chen X, Nakamura Y, Yu C, Qi H (2020) Fucoxanthin activities motivate its nano/micro-encapsulation for food or nutraceutical application: a review. Food Funct 11(11):9338–9358
22. Darwish AM, Soliman TN, Elhendy HA, El-Kholy WM (2021) Nano-encapsulated iron and folic acid-fortified functional yogurt enhance anemia in albino rats. Front Nutr 8:654624
23. Hosseini SF, Ramezanzade L, McClements DJ (2021) Recent advances in nanoencapsulation of hydrophobic marine bioactives: bioavailability, safety, and sensory attributes of nano-fortified functional foods. Trends Food Sci Technol 109:322–339
24. Mishra DK, Shandilya R, Mishra PK (2018) Lipid based nanocarriers: a translational perspective. Nanomed: Nanotechnol Biol Med 14(7):2023–2050
25. Rostamabadi H, Falsafi SR, Boostani S, Katouzian I, Rezaei A, Assadpour E, Jafari SM (2021) Design and formulation of nano/micro-encapsulated natural bioactive compounds for

food applications. In: Application of nano/microencapsulated ingredients in food products. Academic Press, pp 1–41
26. Livney YD (2017) Nanoencapsulation technologies. Engineering foods for bioactives stability and delivery, pp 143–169
27. Shishir MRI, Xie L, Sun C, Zheng X, Chen W (2018) Advances in micro and nano-encapsulation of bioactive compounds using biopolymer and lipid-based transporters. Trends Food Sci Technol 78:34–60
28. Yewale C, Patil S, Kolate A, Kore G, Misra A (2015) Oral absorption promoters: opportunities, issues, and challenges. Critical reviews™ in therapeutic drug carrier systems 32(5)
29. Araujo ARTS, Rodrigues M, Mascarenhas-Melo F, Peixoto D, Guerra C, Cabral C, Paiva-Santos AC et al (2022). New-generation nanotechnology for development of cosmetics using plant extracts. In Nanotechnology for the preparation of cosmetics using plant-based extracts. Elsevier, pp 301–325
30. Islam MM, Raikwar S (2024) Enhancement of oral bioavailability of protein and peptide by polysaccharide-based nanoparticles. Protein Pept Lett 31(3):209–228
31. Basu A, Halder M, Das A, Das R, Bhattacharjee C (2024) Nanomaterials in drug delivery: application of polysaccharides and protein-based nanomaterials in modern drug delivery. Biogenic nanomaterials for environmental sustainability: principles, practices, and opportunities. Springer International Publishing, Cham, pp 293–317
32. Forutan M, Hasani M, Hasani S, Salehi N, Sabbagh F (2022) Liposome system for encapsulation of Spirulina platensis protein hydrolysates: controlled-release in simulated gastrointestinal conditions, structural and functional properties. Materials 15(23):8581
33. Shen Y, Zhang N, Tian J, Xin G, Liu L, Sun X, Li B (2022) Advanced approaches for improving bioavailability and controlled release of anthocyanins. J Control Release 341:285–299
34. Rehman A, Ahmad T, Aadil RM, Spotti MJ, Bakry AM, Khan IM, Tong Q et al (2019) Pectin polymers as wall materials for the nano-encapsulation of bioactive compounds. Trends Food Sci Technol 90:35–46
35. Rahaiee S, Assadpour E, Esfanjani AF, Silva AS, Jafari SM (2020) Application of nano/microencapsulated phenolic compounds against cancer. Adv Coll Interface Sci 279:102153
36. Chadha S (2021) Recent advances in nano-encapsulation technologies for controlled release of biostimulants and antimicrobial agents. Adv Nano-Fertilizers Nano-Pesticides Agricul 29–55
37. Pooja Pandey M, Kumar T, Goswami H, Kumari R, Kumari S, Chawla PA et al (2024) Nano-nutraceuticals to combat oxidative stress: unlocking newer paradigms in adjuvant therapy. Curr Top Med Chem 24(17):1490–1503
38. Stevanović M, Filipović N (2024) A review of recent developments in biopolymer nano-based drug delivery systems with antioxidative properties: insights into the last five years. Pharmaceutics 16(5):670
39. Khan T, Tiwari D, Sanap N. Design and structure of polysaccharide-based nanoparticles: state of the art. Polysaccharides 81–93
40. Kakade P, Chatterjee A, Pandya A, Disouza J, Patravale V (2022) Carbohydrate anchored lipid nanoparticles. Int J Pharm 618:121681
41. Mehta S, Sangwan S, Dhawan V (2024) Challenges in developing delivery systems of nutraceuticals and nanonutraceuticals. Handbook of nutraceuticals: science, technology and engineering. Springer International Publishing, Cham, pp 1–25
42. Yang M, Ahmad N, Hussain M, Lu X, Xu J, Zhong H, Guan R (2023) A review of recent advances on cyanidin-3-glucoside: the biotransformation, absorption, bioactivity and applications of nano-encapsulation. Food Funct 14(14):6320–6345
43. Shishir MRI, Saifullah M, Hashim SB, Aalim H, Bilal M, Khan S, Zou X (2024) Micro and nano-encapsulated natural products in yogurt: an emerging trend to achieve multifunctional benefits in product quality and human health. Food Hydrocolloids 110124
44. Shaaban HA, Farouk A (2022) Encapsulation of essential oils and their use in food applications. Essential Oils—Advances in Extractions and Biological Applications, 1–15
45. Tafaghodi M, Khademi F, Firouzi Z (2020) Polymer-based nanoparticles as delivery systems for treatment and vaccination of tuberculosis. In: Nanotechnology based approaches for tuberculosis treatment. Academic Press, pp 123–142

46. Subramani T, Ganapathyswamy H (2020) An overview of liposomal nano-encapsulation techniques and its applications in food and nutraceutical. J Food Sci Technol 57(10):3545–3555
47. Tang L, Li J, Zhao Q, Pan T, Zhong H, Wang W (2021) Advanced and innovative nano-systems for anticancer targeted drug delivery. Pharmaceutics 13(8):1151
48. Manocha S, Dhiman S, Grewal AS, Guarve K (2022) Nanotechnology: an approach to overcome bioavailability challenges of nutraceuticals. J Drug Deliv Sci Technol 72:103418
49. Saini RK, Prasad P, Lokesh V, Shang X, Shin J, Keum YS, Lee JH (2022) Carotenoids: dietary sources, extraction, encapsulation, bioavailability, and health benefits—a review of recent advancements. Antioxidants 11(4):795
50. Jampilek J, Kralova K (2020) Potential of nanonutraceuticals in increasing immunity. Nanomaterials 10(11):2224
51. Samrot AV, Sean TC, Kudaiyappan T, Bisyarah U, Mirarmandi A, Abubakar A, Kumar SS et al (2020) Production, characterization and application of nanocarriers made of polysaccharides, proteins, bio-polyesters and other biopolymers: a review. Int J Biol Macromol 165:3088–3105
52. Tahir A, Shabir Ahmad R, Imran M, Ahmad MH, Kamran Khan M, Muhammad N, Javed M et al (2021) Recent approaches for utilization of food components as nano-encapsulation: a review. Int J Food Prop 24(1):1074–1096
53. Chakraborty A, Dhar P (2017) A review on potential of proteins as an excipient for developing a nano-carrier delivery system. Crit Rev™ Therapeutic Drug Carrier Syst 34(5)
54. Paredes AJ, Asensio CM, Llabot JM, Allemandi DA, Palma SD (2016) Nanoencapsulation in the food industry: manufacture, applications and characterization
55. Taha HA, Agamy NF, Soliman TN, Younes NM, El-Enshasy HA, Darwish AM (2024) Preparation and characterization of vitamin E/calcium/soy protein isolate nanoparticles for soybean milk beverage fortification. PeerJ 12:e17007
56. Martins JT, Bourbon AI, Pinheiro AC, Fasolin LH, Vicente AA (2018) Protein-based structures for food applications: from macro to nanoscale. Front Sustain Food Syst 2:77
57. Soni M, Maurya A, Das S, Prasad J, Yadav A, Singh VK, Dwivedy AK et al (2022) Nanoencapsulation strategies for improving nutritional functionality, safety and delivery of plant-based foods: recent updates and future opportunities. Plant Nano Biology 1:100004
58. Jhan F, Gani A, Noor N, Malla BA, Ashwar BA (2022) Nano reduction coupled with encapsulation as a novel technique for utilising millet proteins as future foods. Ultrason Sonochem 86:106006
59. Gharibzahedi SMT, Moghadam M, Amft J, Tolun A, Hasabnis G, Altintas Z (2023) Recent advances in dietary sources, health benefits, emerging encapsulation methods, food fortification, and new sensor-based monitoring of vitamin B12: a critical review. Molecules 28(22):7469
60. Poonia A (2022) Nanoencapsulation methods in iron fortification of dairy food matrices. In: Nanotechnology applications in agricultural and bioprocess engineering. Apple Academic Press, pp 97–115
61. Khatoon S, Kalam N, Shaikh MF, Hasnain MS, Hafiz AK, Ansari MT (2022) Nanoencapsulation of polyphenols as drugs and supplements for enhancing therapeutic profile-a review. Curr Mol Pharmacol 15(1):77–107
62. Peñalva R (2019) Protein nanoparticles for oral delivery of bioactives
63. Sharifan P, Bagherniya M, Bajgiran MM, Safarian M, Vatanparast H, Eslami S, Ghayour-Mobarhan M et al (2021) The efficacy of dairy products fortified with nano-encapsulated vitamin D3 on physical and mental aspects of the health in obese subjects; the protocol of the SUVINA trial. Transl Metab Syndr Res 4:1–9
64. Vieira MV, Pastrana LM, Fuciños P (2020) Microalgae encapsulation systems for food, pharmaceutical and cosmetics applications. Mar Drugs 18(12):644
65. Delshadi R, Bahrami A, Tafti AG, Barba FJ, Williams LL (2020) Micro and nano-encapsulation of vegetable and essential oils to develop functional food products with improved nutritional profiles. Trends Food Sci Technol 104:72–83
66. Trindade LR, da Silva DVT, Baião DDS, Paschoalin VMF (2021) Increasing the power of polyphenols through nanoencapsulation for adjuvant therapy against cardiovascular diseases. Molecules 26(15):4621

67. Al-Khayri JM, Asghar W, Akhtar A, Ayub H, Aslam I, Khalid N, Attimarad M et al (2022) Anthocyanin delivery systems: a critical review of recent research findings. Appl Sci 12(23):12347
68. Karuppaiah A, Annamalai G, Dhasaiyan M, Manikandan S, Jose PA, Rahman H (2024) An updated review of nano techniques for enhancing the bioavailability and therapeutic efficacy of poly phenolic bioactive compounds. Curr Nanomed
69. Ali Redha A, Kodikara C, Cozzolino D (2024) Does encapsulation improve the bioavailability of polyphenols in humans? A concise review based on in vivo human studies. Nutrients 16(21):3625
70. Ozogul F, Elabed N, Ceylan Z, Ocak E, Ozogul Y (2021) Nano-technological approaches for plant and marine-based polysaccharides for nano-encapsulations and their applications in food industry. In: Advances in food and nutrition research, vol 97. Academic Press, pp 187–236
71. Aguilar-Pérez KM, Ruiz-Pulido G, Medina DI, Parra-Saldivar R, Iqbal HM (2023) Insight of nanotechnological processing for nano-fortified functional foods and nutraceutical—opportunities, challenges, and future scope in food for better health. Crit Rev Food Sci Nutr 63(20):4618–4635
72. Hameed A, Fatima GR, Malik K, Muqadas A, Fazal-ur-Rehman M (2019) Scope of nanotechnology in cosmetics: dermatology and skin care products. J Med Chem Sci 2(9)
73. Casanova F, Santos L (2016) Encapsulation of cosmetic active ingredients for topical application—a review. J Microencapsul 33(1):1–17
74. Siddiqui SA, Bahmid NA, Taha A, Khalifa I, Khan S, Rostamabadi H, Jafari SM (2023) Recent advances in food applications of phenolic-loaded micro/nanodelivery systems. Crit Rev Food Sci Nutr 63(27):8939–8959
75. De S, Gopikrishna A, Keerthana V, Girigoswami A, Girigoswami K (2021) An overview of nanoformulated nutraceuticals and their therapeutic approaches. Curr Nutr Food Sci 17(4):392–407
76. Arratia-Quijada J, Nuño K, Ruíz-Santoyo V, Andrade-Espinoza BA (2024) Nano-encapsulation of probiotics: need and critical considerations to design new non-dairy probiotic products. J Funct Foods 116:106192
77. Meng Y, Qiu C, Li X, McClements DJ, Sang S, Jiao A, Jin Z (2024) Polysaccharide-based nano-delivery systems for encapsulation, delivery, and pH-responsive release of bioactive ingredients. Crit Rev Food Sci Nutr 64(1):187–201
78. Ngwuluka NC, Abu-Thabit NY, Uwaezuoke OJ, Erebor JO, Ilomuanya MO, Mohamed RR, Ebrahim NA et al (2021) Natural polymers in micro-and nanoencapsulation for therapeutic and diagnostic applications: part I: lipids and fabrication techniques. Nano Microencapsul Tech Appl 3–54
79. Nallamuthu I, Khanum F, Fathima SJ, Patil MM, Anand T (2017) Enhanced nutrient delivery through nanoencapsulation techniques: the current trend in food industry. In: Nutrient delivery. Academic Press, pp 619–651
80. Dehnad D, Ghorani B, Emadzadeh B, Emadzadeh M, Assadpour E, Rajabzadeh G, Jafari SM (2023) Recent advances in iron encapsulation and its application in food fortification. Crit Rev Food Sci Nutr 1–17
81. Ferraro C, Dattilo M, Patitucci F, Prete S, Scopelliti G, Parisi OI, Puoci F (2024) Exploring protein-based carriers in drug delivery: a review. Pharmaceutics 16(9):1172
82. Yan C, Kim SR (2024) Microencapsulation for pharmaceutical applications: a review. ACS Appl Bio Mater 7(2):692–710
83. Zhang Y (2018) Studies of insulin transport across intestinal epithelium using nano-encapsulation for oral delivery. Doctoral dissertation
84. Wong CY, Luna G, Martinez J, Al-Salami H, Dass CR (2019) Bio-nanotechnological advancement of orally administered insulin nanoparticles: comprehensive review of experimental design for physicochemical characterization. Int J Pharm 572:118720
85. Moncalvo F, Martinez Espinoza MI, Cellesi F (2020) Nanosized delivery systems for therapeutic proteins: clinically validated technologies and advanced development strategies. Front Bioeng Biotechnol 8:89

86. Ngwuluka NC, Abu-Thabit NY, Uwaezuoke OJ, Erebor JO, Ilomuanya MO, Mohamed RR, Ebrahim NA et al (2021) Natural polymers in micro-and nanoencapsulation for therapeutic and diagnostic applications: part II-polysaccharides and proteins. Nano Microencapsul Tech Appl 8(9)
87. Wagle SR, Kovacevic B, Walker D, Ionescu CM, Shah U, Stojanovic G, Al-Salami H et al (2020) Alginate-based drug oral targeting using bio-micro/nano encapsulation technologies. Expert Opin Drug Deliv 17(10):1361–1376
88. Çağlar MY, Demirci M, Bayrambas K, Çakır B, Gülseren İ (2017) Nanoencapsulation of enzymes, bioactive peptides, and biological molecules. Nanoencapsulation of food bioactive ingredients: principles and applications
89. Castro ML, Ferreira JP, Pintado M, Ramos OL, Borges S, Baptista-Silva S (2023) Grape by-products in sustainable cosmetics: nanoencapsulation and market trends. Appl Sci 13(16):9168
90. Javed S, Mangla B, Salawi A, Sultan MH, Almoshari Y, Ahsan W (2024) Essential oils as dermocosmetic agents, their mechanism of action and nanolipidic formulations for maximized skincare. Cosmetics 11(6):210
91. Achagar R, Ait-Touchente Z, El Ati R, Boujdi K, Thoume A, Abdou A, Touzani R (2024) A comprehensive review of essential oil-nanotechnology synergy for advanced dermocosmetic delivery. Cosmetics 11(2):48
92. Sharma A, Agarwal P, Sebghatollahi Z, Mahato N (2023) Functional nanostructured materials in the cosmetics industry: a review. ChemEngineering 7(4):66
93. Chu CC, Chew SC, Nyam KL (2021) Recent advances in encapsulation technologies of kenaf (*Hibiscus cannabinus*) leaves and seeds for cosmeceutical application. Food Bioprod Process 127:99–113
94. Bilal M, Iqbal HM (2020) New insights on unique features and role of nanostructured materials in cosmetics. Cosmetics 7(2):24
95. Gorup LF, Neto FNS, Kubo AM, Souza JAS, Fernandes RA, Fernandes GL, Camargo ER et al (2017) Nanostructured functional materials: silver nanoparticles in polymer for the generation of antimicrobial characteristics. Recent advances in complex functional materials: from design to application, pp 271–292
96. Jamroży M, Kudłacik-Kramarczyk S, Drabczyk A, Krzan M (2024) Advanced drug carriers: a review of selected protein, polysaccharide, and lipid drug delivery platforms. Int J Mol Sci 25(2):786
97. Carrion CC, Nasrollahzadeh M, Sajjadi M, Jaleh B, Soufi GJ, Iravani S (2021) Lignin, lipid, protein, hyaluronic acid, starch, cellulose, gum, pectin, alginate and chitosan-based nanomaterials for cancer nanotherapy: challenges and opportunities. Int J Biol Macromol 178:193–228
98. Tantray J, Patel A, Prajapati BG, Kosey S, Bhattacharya S (2024) The use of lipid-based nanocarriers to improve ovarian cancer treatment: an overview of recent developments. Curr Pharm Biotechnol 25(17):2200–2217
99. Yin L, Yuvienco C, Montclare JK (2017) Protein based therapeutic delivery agents: contemporary developments and challenges. Biomaterials 134:91–116
100. Suja Rani S, Latha C (2024) Nanoencapsulation for controlled release of active components. Handbook of nutraceuticals: science, technology and engineering. Springer International Publishing, Cham, pp 1–41
101. Biswas S, Biswas S, PadmaPriya G, Kaur J, Pramanik A, Islam F, Shah MA (2024) Role of nutraceuticals in viral infections as immunomodulators: a comprehensive review. eFood 5(5), e70000
102. Pachuau L, Laldinchhana, Roy PK, Zothantluanga JH, Ray S, Das S (2021) Encapsulation of bioactive compound and its therapeutic potential. Bioactive Nat Prod Pharmaceutical Appl 687–714
103. Bazana MT, Codevilla CF, de Menezes CR (2019) Nanoencapsulation of bioactive compounds: challenges and perspectives. Curr Opin Food Sci 26:47–56

104. Dahiya D, Terpou A, Dasenaki M, Nigam PS (2023) Current status and future prospects of bioactive molecules delivered through sustainable encapsulation techniques for food fortification. Sustain Food Technol 1(4):500–510
105. Siddiqui SA, Farooqi MQU, Bhowmik S, Zahra Z, Mahmud MC, Assadpour E, Jafari SM et al (2023) Application of micro/nano-fluidics for encapsulation of food bioactive compounds-principles, applications, and challenges. Trends Food Sci Technol 136:64–75
106. Pateiro M, Gómez B, Munekata PE, Barba FJ, Putnik P, Kovačević DB, Lorenzo JM (2021) Nanoencapsulation of promising bioactive compounds to improve their absorption, stability, functionality and the appearance of the final food products. Molecules 26(6):1547
107. Jafari SM, Katouzian I, Akhavan S (2017) Safety and regulatory issues of nanocapsules. In: Nanoencapsulation technologies for the food and nutraceutical industries. Academic Press, pp 545–590
108. Wang X, Li H, Bao Y, Wang Y, Chen C (2024) Overcoming the application limitations of pH-driven encapsulation of bioactive compounds: strategies and perspectives. Current Opinion in Food Science 101212
109. Comunian T, Babazadeh A, Rehman A, Shaddel R, Akbari-Alavijeh S, Boostani S, Jafari SM (2022) Protection and controlled release of vitamin C by different micro/nanocarriers. Crit Rev Food Sci Nutr 62(12):3301–3322
110. Otchere E, McKay BM, English MM, Aryee AN (2023) Current trends in nano-delivery systems for functional foods: a systematic review. PeerJ 11:e14980
111. Marikar SN, El-Osta A, Johnston A, Such G, Al-Hasani K (2022) Microencapsulation-based cell therapies. Cell Mol Life Sci 79(7):351
112. Weisany W, Yousefi S, Tahir NAR, Golestanehzadeh N, McClements DJ, Adhikari B, Ghasemlou M (2022) Targeted delivery and controlled released of essential oils using nanoencapsulation: a review. Adv Coll Interface Sci 303:102655
113. Gali L, Pirozzi A, Donsì F (2023) Biopolymer-and lipid-based carriers for the delivery of plant-based ingredients. Pharmaceutics 15(3):927
114. Rao TJM, Kesharwani RK, Keservani RK, Sharma AK (eds) (2024) Formulations, regulations, and challenges of nutraceuticals. CRC Press
115. Harun-Or-Rashid M, Aktar MN, Hossain MS, Sarkar N, Islam MR, Arafat ME, Yusa SI et al (2023) Recent advances in micro-and nano-drug delivery systems based on natural and synthetic biomaterials. Polymers 15(23):4563
116. Hosseini H, Tajiani Z, Jafari SM (2019). Improving the shelf-life of food products by nano/micro-encapsulated ingredients. In: Food quality and shelf life. Academic Press, pp 159–200
117. Ayyaril SS, Shanableh A, Bhattacharjee S, Rawas-Qalaji M, Cagliani R, Shabib AG (2023) Recent progress in micro and nano-encapsulation techniques for environmental applications: a review. Results Eng 18:101094
118. Esfanjani AF, Jafari SM (2016) Biopolymer nano-particles and natural nano-carriers for nano-encapsulation of phenolic compounds. Colloids Surf, B 146:532–543
119. Saifullah M, Shishir MRI, Ferdowsi R, Rahman MRT, Van Vuong Q (2019) Micro and nano encapsulation, retention and controlled release of flavor and aroma compounds: a critical review. Trends Food Sci Technol 86:230–251
120. Rashwan AK, Younes HA, Karim N, Taha EM, Mozafari MR, Tawfeuk HZ, Chen W (2022) Health properties of bioactive food compounds-loaded micro and nano-encapsulation systems: a review. SVU-Int J Agricult Sci 4(2):178–203
121. Nadaroglu H, Yeşilyurt MK (2022) Micro-/nano-encapsulation and encapsulation applications. In: Science and applications of nanoparticles. Jenny Stanford Publishing, pp 55–102
122. Maurya VK, Shakya A, McClements DJ, Srinivasan R, Bashir K, Ramesh T, Sathiyamoorthi E et al (2023) Vitamin C fortification: need and recent trends in encapsulation technologies. Front Nutr 10:1229243
123. Chudasama M, Goyary J (2024) Nanostructured materials in food science: current progress and future prospects. Next Materials 100206

124. de Alcantara Lemos J, Oliveira AEM, Araujo RS, Townsend DM, Ferreira LAM, de Barros ALB (2021) Recent progress in micro and nano-encapsulation of bioactive derivatives of the Brazilian genus Pterodon. Biomed Pharmacother 143:112137
125. Balde A, Kim SK, Benjakul S, Nazeer RA (2022) Pulmonary drug delivery applications of natural polysaccharide polymer derived nano/micro-carrier systems: a review. Int J Biol Macromol 220:1464–1479
126. Rathee S, Ojha A, Upadhyay A, Xiao J, Bajpai VK, Ali S, Shukla S (2023) Biogenic engineered nanomaterials for enhancing bioavailability via developing nano-iron-fortified smart foods: advances, insight, and prospects of nanobionics in fortification of food. Food Funct 14(20):9083–9099
127. Liu F, Li M, Wang Q, Yan J, Han S, Ma C, McClements DJ et al (2023) Future foods: alternative proteins, food architecture, sustainable packaging, and precision nutrition. Crit Rev Food Sci Nutr 63(23):6423–6444
128. Oprea I, Fărcaş AC, Leopold LF, Diaconeasa Z, Coman C, Socaci SA (2022) Nano-encapsulation of citrus essential oils: methods and applications of interest for the food sector. Polymers 14(21):4505
129. Ahmadzadeh S, Lenie MDR, Mirmahdi RS, Ubeyitogullari A (2023) Designing future foods: harnessing 3D food printing technology to encapsulate bioactive compounds. Critical Rev Food Sci Nutrit 1–17

Vitamins and Minerals: Nanotech Solutions for Enhanced Bioavailability

Nizwa Itrat, Beenish Israr, Rabiya Saroosh, Dua Munawwar, Faiza Ameen, and Suad Ajab

Abstract This chapter delves into the largest breakthroughs in techniques aimed at boosting the bioavailability of vitamins and minerals with a special focus on nanotechnology. Bioavailability is crucial for the effective absorption and utilization of nutrients and cutting strategies have been developed to tackle challenges, like low solubility, stability issues, and poor nutrient bioavailability. Innovative methods such as nanoencapsulation, liposomal carriers, solid lipid nanoparticles, supercritical fluid processes, and no Embulsions have been thoroughly explored. These approaches not only enhance nutrient solubility but also protect them from physiological degradation and environmental ensuring pre-size and controlled delivery within this body. Nanoencapsulation and lysosomal Systems use Nanoscale careers to safeguard nutrients like vitamins A, and, E from oxidative, and enzymatic breakdown. Meanwhile, solid–liquid Nanoparticles improve stability and enable the sustained release of delicate nutrients supercritical fluid Technology offers an eco-friendly way to produce fine particles with superior absorption characteristics. Nano emulsions with their Ultrafine droplets significantly boost the solubility and uptake of lipophilic vitamins. These nanotechnological innovations greatly benefit the development of functional food and nutraceuticals promoting better health outcomes and effectively addressing nutrient efficiency. This chapter highlights the transformative potential of Nanotechnology in enhancing nutrient bioavailability paving the way for innovative efficient and sustainable nutrition solutions.

N. Itrat (✉) · B. Israr
Institute of Home Science, Faculty of Food Nutrition and Home Sciences, University of Agriculture, Faisalabad, Pakistan
e-mail: nizwa.itrat@uaf.edu.pk

R. Saroosh · D. Munawwar
Department of Nutrition and Dietetics, The University of Faisalabad, Faisalabad, Pakistan

F. Ameen
British School of Leadership and Management, Glasgow Caledonian University (London Campus), London, United Kingdom

S. Ajab
Institute of Public Health, College of Medicine and Health Sciences, United Arab Emirates University, Al Ain, Abu Dhabi, United Arab Emirates

A. Chatterjee et al. (eds.), *Nanofuel: The Future of Sports Nutrition*, Smart Nanomaterials Technology, https://doi.org/10.1007/978-981-96-5471-0_6

Keywords Nanoencapsulation · Supercritical fluid · Nanoemulsions · Nutraceutical nanotechnology · Liposomes · Bioavailability

1 Introduction

Micronutrients which include vitamins and minerals are vital for maintaining health promoting growth and supporting overall well-being. Despite being needed in small amounts, these components are crucial for various physiological and biochemical processes. Vitamins, as organic compounds different from minerals which are inorganic elements but both are essential for a balanced diet. Together they play key roles in maintaining structural integrity facilitating enzymatic activity and regulating bodily functions [1].

Vitamins are divided into two categories water-soluble and fat-soluble categories based on their solubility and storage within the body. Water-soluble vitamin C and B-complex groups are not stored in significant amounts and must be regularly obtained through diet. On the other hand, Fat-soluble A, D, E, and K vitamins are stored in adipose tissues the and liver allowing for their gradual utilization over time. These vitamins are essential for blood clotting, Vision and bone health, and antioxidant protection.

These water-soluble vitamins are vital for energy metabolism, nerve functionality, DNA synthesis, red blood cell production, and tissue repair. Dietary sources of these vitamins span a variety of foods, including leafy green vegetables, citrus fruits, dairy products, whole grains, nuts, seeds, and animal-derived foods. Deficiencies of vitamins result in health issues such as night blindness, scurvy, rickets, and neurological disorders (Table 1) [2].

Minerals are classified into macro minerals and micro minerals according to the quantities required by the human body. Macro minerals such as magnesium, sodium, calcium, potassium, chloride, Phosphorus, and sulfur are essential for the formation of robust bones, the transmission of nerve impulses, and the maintenance of fluid homeostasis micro-minerals including iodine, zinc, copper, Cobalt, fluoride chromium, Selenium, iron, manganese, molybdenum, and Boron are required in trace amount are equally crucial for various physiological functions. Iron is crucial for hemoglobin synthesis and oxygen transport, zinc strengthens immune responses, and iodine is fundamental for thyroid hormone production.

A variety of dietary sources supply these minerals, including fruits, whole grains, vegetables, dairy products, meat, and seafood, ensuring a balanced intake to meet physiological needs. Mineral deficiencies can result in significant health issues, such as anemia, goiter, and compromised immunity (Table 2) [3].

The functions of vitamins and minerals go beyond their roles, as they often work synergistically within the body. For instance, vitamin D enhances calcium absorption, which is critical for maintaining bone health. Trace minerals frequently serve as cofactors for enzymes, while vitamins function as coenzymes in various metabolic processes. Together, these nutrients regulate essential physiological functions,

Table 1 Detailed overview of vitamins

Sr. No	Type	Vitamin	Functions	Deficiency effects	Sources
1	Fat-soluble	Vitamin A	Vision, skin health, immune function	Weakened immunity, night blindness, dry skin	Spinach, liver, carrots, sweet potatoes
2	Fat-soluble	Vitamin D	Bone health, immune support, calcium absorption	Rickets, weakened bones, osteomalacia	Fatty fish, fortified milk, sunlight
3	Fat-soluble	Vitamin E	Antioxidant defense, protection against oxidative damage, skin health	Muscle weakness, neurological issues	Seeds, vegetable oils, spinach, nuts
4	Fat-soluble	Vitamin K	Bone metabolism blood clotting	Bone weakness, Bleeding disorders,	Brussels sprouts, leafy greens, broccoli,
5	Water-soluble	Vitamin C	Antioxidant defense, immune support, collagen synthesis,	Slow wound healing, Scurvy, weak immunity,	Bell peppers, broccoli, citrus fruits, strawberries,
6	Water-soluble	Vitamin B1 (Thiamine)	Nerve function, Energy metabolism,	Nerve damage, beriberi, fatigue,	Fish, legumes, whole grains,
7	Water-soluble	Vitamin B2 (Riboflavin)	Energy production, skin health	Cracks at mouth corners, sore throat	Dairy, eggs, lean meat, green vegetables
8	Water-soluble	Vitamin B3 (Niacin)	Energy metabolism, skin health, DNA repair	Pellagra (diarrhea, dermatitis, dementia)	Meat, fish, whole grains, legumes
9	Water-soluble	Vitamin B5 (Pantothenic Acid)	Hormone synthesis energy metabolism,	Numbness, irritability fatigue,	Meat, broccoli, whole grains, avocados
10	Water-soluble	Vitamin B6 (Pyridoxine)	Neurotransmitter synthesis, amino acid metabolism	Depression, confusion, anemia	Bananas, potatoes, fish, poultry
11	Water-soluble	Vitamin B7 (Biotin)	Fat and carbohydrate metabolism, skin and hair health	Hair loss, brittle nails, skin rash	Eggs, nuts, seeds, fish

(continued)

Table 1 (continued)

Sr. No	Type	Vitamin	Functions	Deficiency effects	Sources
12	Water-soluble	Vitamin B9 (Folate)	DNA synthesis, red blood cell production	Neural tube defects in pregnancy, anemia	Leafy greens, legumes, citrus fruits
13	Water-soluble	Vitamin B12	Red blood cell production, DNA synthesis	Anemia, neurological issues	Meat, fish, dairy

including immune response, brain activity, blood pressure, glucose metabolism, and antioxidant defenses, ensuring overall health [4].

Micronutrient deficiencies remain a significant global issue, particularly in developing nations where deficiencies in iron, iodine, zinc, selenium, and vitamin A pose substantial health challenges. Such deficiencies can impair cognitive development, weaken immune systems, and elevate mortality rates, especially among children and pregnant women. To overcome the difficulties associated with integrating bioactive substances in food products, nanoparticle (NP)-based delivery systems offer promising solutions (Table 3) [5].

Nanotechnology is revolutionizing the food supply chain including its processing, production, storage, and consumption phases. Scientific research demonstrates that nanotechnology improves water solubility, thermal stability, and oral bioavailability of nutrients [6]. Bioavailability, defined as the efficiency with which the body absorbs and utilizes nutrients, is significantly improved through nanotechnology by creating smaller, more absorbable particles. Nanoparticles (NPs) used in edible delivery systems are microscopic entities, typically ranging from 1 to 100 nm, and are synthesized through multiple techniques. These nanoparticles exhibit unique chemical and physical properties, including enhanced color, stability, solubility, durability, and a high surface-to-volume ratio [7].

These distinctive properties make NPs applicable in numerous fields, including tissue engineering, cell therapy, drug delivery, diagnostic tools, biomaterials development, and cellular signaling [8, 9]. Recently, a diverse nanoparticle-based delivery system has been engineered to encapsulate nutrients and nutraceuticals extracted from foods [10]. Nanoparticles are primarily categorized into four groups, (i) polymeric nanomaterials, such as polymersomes, polymeric micelles, nano gels dendrimers, and nanospheres (ii) non-polymeric nanomaterials, including silica NPs, carbon nanotubes, quantum dots, nanodiamonds, and metallic nanoparticles; (iii) nanocrystalline materials, including pyramidal and cubic nanocrystals (iv) lipid-based nanomaterials, such as liposomes, solid lipid NPs, exosomes, and nanoemulsions, (Fig. 1) [11–13].

Table 2 Detailed overview of minerals

Mineral	Type	Sources	Functions	Deficiency effects
Calcium (Ca)	Major	Dairy products, leafy greens, fish (sardines, salmon), fortified foods	Muscle contraction, nerve signaling, bone and teeth formation, blood clotting	Osteoporosis, rickets, dental problems, muscle spasms
Magnesium (Mg)	Major	Nuts, seeds, whole grains, green leafy vegetables, legume	Enzyme activation, energy production, muscle and nerve function, bone structure	Weakness, fatigue, muscle cramps, arrhythmias
Potassium (K)	Major	Bananas, oranges, potatoes, beans, spinach, fish	Electrolyte balance, muscle contraction, nerve function	Hypokalemia, muscle weakness, irregular heartbeat
Sodium (Na)	Major	Table salt, processed foods, seafood	Fluid balance, nerve signaling, muscle contraction	Hyponatremia, confusion, muscle cramps, fatigue
Chloride (Cl)	Major	Table salt, seaweed, tomatoes, olives	Fluid balance, stomach acid production	Alkalosis, muscle spasms, respiratory issues
Phosphorus (P)	Major	Meat, dairy, nuts, whole grains, legumes	Bone formation, energy storage, DNA and RNA synthesis	Weak bones, fatigue, muscle pain
Sulfur (S)	Major	Meat, eggs, nuts, legumes, fish	Protein synthesis, detoxification, joint health	Rare; joint pain, weak nails and hair
Iron (Fe)	Trace	Red meat, poultry, beans, fortified cereals, spinach	Oxygen transport (hemoglobin), energy production, immune function	Anemia, fatigue, impaired cognitive function Anemia, fatigue, impaired cognitive function
Zinc (Zn)	Trace	Meat, shellfish, legumes, seeds, nuts	Immune response, wound healing, DNA synthesis	Growth retardation, hair loss, delayed wound healing, increased infections
Selenium (Se)	Trace	Brazil nuts, seafood, meat, eggs	Antioxidant defense, thyroid function	Fatigue, muscle weakness, weakened immunity
Iodine (I)	Trace	Iodized salt, seafood, dairy products, eggs	Thyroid hormone synthesis	Goiter, hypothyroidism, intellectual disability
Copper (Cu)	Trace	Shellfish, nuts, seeds, whole grains	Enzyme activation, iron metabolism, immune function	Anemia, weakened immunity, neurological issues

(continued)

Table 2 (continued)

Mineral	Type	Sources	Functions	Deficiency effects
Manganese (Mn)	Trace	Whole grains, nuts, leafy vegetables, tea	Bone formation, enzyme function, antioxidant activity	Bone deformities, impaired growth, reproductive issues
Fluoride (F)	Trace	Fluoridated water, tea, fish	Tooth enamel strength, prevention of dental cavities	Dental caries
Chromium (Cr)	Trace	Whole grains, broccoli, potatoes, garlic	Regulation of blood sugar levels, insulin sensitivity	Impaired glucose metabolism
Cobalt (Co)	Trace	Meat, fish, dairy	Vitamin B12 synthesis (essential component), red blood cell formation	Anemia, neurological disorders
Molybdenum (Mo)	Trace	Legumes, grains, nuts	Enzyme activation, detoxification of sulfites	Rare; metabolic disturbances

2 Advancement of Nanotechnology for Nutritional Science

Nanotechnology has transformed numerous fields, including pharmacy, agriculture, nutrition, and food processing [14]. Over the past few decades, advancements in nutrition-related nanoscience have evolved rapidly, fostering significant interest in the targeted delivery of micronutrients [15]. Nano-capsules are increasingly utilized to enhance the delivery of micronutrients and drugs, including vitamins and minerals, within the body [16]. Various methods, such as nano-emulsification, nano-composites, nano-structuration, and encapsulating materials into smaller forms, enable more efficient delivery of bioactive compounds [17].

Commercial supplements, such as Nutrition-be-nanotech and nanoceuticals, demonstrate the practical application of this technology. Nanoparticles dispersed in vitamin sprays are believed to enhance nutrient absorption for compounds like curcumin, folic acid, and iron. Similarly, nano-sized powders are employed as nano-cochleate to further improve nutrient uptake. These approaches have proven effective in distributing nutrients to cells without altering the taste or color of food products. Production of food and nutraceutical Supplements primarily relies on encapsulation techniques, utilizing zinc and iron nano-structured capsules to deliver probiotics and other beneficial compounds into the human body. Nanoparticles in food supplements demonstrate in-hand activity compared to traditional supplements interacting more effectively in human cells going to their distinctive size and properties [18].

Nanotechnology contributes significantly to enhancing water solubility, the bioavailability, and antioxidant properties of herbs and spices, while also promoting uniform dissolution of active ingredients [19]. Nano-materials have been recognized for their ability to improve the bioavailability of critical phytochemicals, such as

Table 3 Nutritional deficiency trends and nanotech solutions

Deficiency	Global prevalence (%)	Key causes	Nanotech solution	Example product
Iron deficiency anemia	60%	Poor diet, blood loss, pregnancy	Nano-iron oxide particles for enhanced absorption and reduced gastrointestinal irritation	Iron nano-supplements
Vitamin D deficiency	50%	Limited sunlight, poor dietary intake	Nanoemulsions for improved stability and better intestinal uptake	Vitamin D nano-capsules
Zinc deficiency	30%	Inadequate intake, malabsorption	Zinc oxide nanoparticles for controlled release and higher absorption efficiency	Zinc-fortified food and supplements
Calcium deficiency	40%	Poor diet, vitamin D deficiency	Nano-hydroxyapatite for bone-targeted calcium delivery	Calcium nano-tablets
Iodine deficiency	25%	Low iodine levels in soil and water	Iodine-loaded nanoparticles for sustained release	Iodine-enriched salt
Vitamin A deficiency	30%	Poor intake of animal sources, malnutrition	Nano-encapsulation with lipids for enhanced stability and bioavailability	Vitamin A-fortified food oils
Magnesium deficiency	20%	Low dietary intake, chronic illnesses	Magnesium nanoparticles for rapid intestinal absorption and bioavailability	Magnesium nano-drinks or powders
Vitamin C deficiency	15%	Low intake of fruits and vegetables	Liposomal vitamin C nanoparticles for oxidative stability and sustained release	Effervescent vitamin C nano-formulations
Selenium deficiency	10%	Low selenium content in soil	Selenium nanoparticles for antioxidant protection and bioavailability	Selenium-enriched nano-powders
Vitamin B12 deficiency	15%	Vegan diet, absorption issues	Nano-B12 micelles for sublingual or mucosal delivery	Vitamin B12 nano-sprays
Folic acid deficiency	20%	Poor diet, pregnancy	Nano-folic acid liposomes for protection and efficient delivery	Folic acid nano-fortified supplements
Potassium deficiency	18%	High-sodium diets, diuretics use	Potassium nanoparticles for sustained release and controlled uptake	Potassium nano-drink formulations
Phosphorus deficiency	10%	Starvation, certain medications	Phosphorus-loaded nanoparticles for efficient absorption and bioavailability	Phosphorus nano-capsules

(continued)

Table 3 (continued)

Deficiency	Global prevalence (%)	Key causes	Nanotech solution	Example product
Copper deficiency	5%	Malnutrition, malabsorption disorders	Copper nanoparticles for enzyme-linked bioavailability enhancement	Copper-fortified nano-supplements
Chromium deficiency	12%	Low intake, aging	Chromium nanoparticles for targeted insulin signaling improvement	Chromium nano-fortified foods

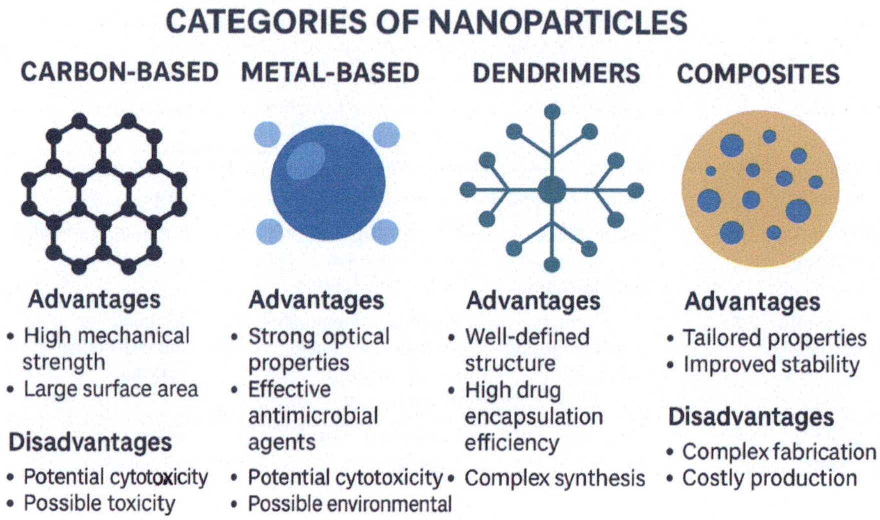

Fig. 1 Four categories of nanoparticles (NPs)

genistein and curcumin [20]. Furthermore, nano-nutraceuticals are now widely available in the form of dietary supplements, herbal products, and bioactive compounds formulated using nanotechnology [21].

3 Techniques in Enhancing Bioavailability of Vitamins and Minerals

Advanced nanotechnology techniques have been developed to enhance the bioavailability of vitamins and minerals, increasing their effectiveness within the body (Table 4). These methods aim to resolve challenges like malabsorption and digestive disorders, thereby improving the delivery and utilization of nutrients. Below is an overview of the primary techniques employed:

Table 4 Examples of functional/nutraceutical food featuring bioactive components in capsulated by nanotechnology [22]

Sr. No	Manufacturer's country of origin	Product name	Bioactive compounds	Health claims
1	Austria	Creatine	Creatine-nano monohydrate	Muscle mass enhancer
2	Netherland	VEGGIE HOT MEALS	Vitamins minerals	Various health benefits
3	India	Nano-Deca	Zinc, Co-enzyme Q10, Ginkgo biloba (antioxidant), L-carnitine, L-arginine	Enhances erectile function
4	New Jersey	NUTRASORB™ PROCES Ingredient	Polyphenols, Gingerols, Alkaloids	Health promotion
5	USA	NanOmega®	Folic acid, Vitamins D3, Vitamins, E	Important for eye, brain, heart health
6	USA	Nano-Mineral by Nano-Greens	Each packet contains 12 minerals	Helps weight loss, improves muscle performance metabolic functionregulates blood glucose
7	USA	Nutri magnesium supplement nano magnesium	Minerals	Muscles, improve optimal functioning of kidneys, and heart
8	Canada	IÖGO®	Source of calcium and vitamin D	Bone growth
9	Hindustan unilever	Red label tea	6-Gingerol	Increased vitality, increased natural killer cell activity

3.1 Nanoencapsulation

Nanoencapsulation entails enclosing vitamins and minerals within nanometer-scale particles, typically composed of biocompatible polymers, proteins, or lipids. This method enhances the surface area of nutrients, promoting improved digestive tract absorption. By reducing particle size to the nanoscale, nanoencapsulation significantly boosts the solubility of vitamins and minerals that are otherwise poorly soluble. Additionally, it protects against degradation by gastric acid and digestive enzymes. Fat-soluble vitamins are prone to oxidation and greatly benefit from this encapsulation approach. Techniques commonly used in nanoencapsulation include solvent evaporation, and solvent diffusion, effectively encapsulating and safeguarding nutrients, and enabling controlled and sustained release within the body [23].

3.2 Liposomal Technology

Liposomal Technology employees' liposomes are small spherical vehicles, composed of lipids by layer to encapsulate vitamins and minerals. these liposomes mimic the structure of the body's cell membrane allowing them to easily merge with intestinal walls and deliver nutrients directly into the bloodstream. This technique significantly enhances the bioavailability of vitamins and minerals by protecting them from stomach acidity and digestive enzymes. Encased within liposomes the nutrients are shielded by the vehicle which acts as a barrier to degradation ensuring the active ingredient reaches the target site. The preparation of liposomes involves dissolving the liquid in an organic solvent and then mixing it with water to form nutrient-encapsulating vesicles. This method is particularly effective for improving the absorption of fat-soluble vitamins and minerals such as calcium and magnesium [24].

3.3 Supercritical Fluid Technology (SFT)

Supercritical fluid Technology employs supercritical Fluids like carbon dioxide to improve the solubility and bioavailability of vitamins and minerals. In this method, CO_2 is used in its supercritical state, where it exhibits liquid and gas properties, allowing it to dissolve minerals and vitamins effectively. Once the nutrients are dissolved the supercritical carbon dioxide rapidly expands creating extremely fine nutrient particles. This process known as micronization increases the nutrients' surface area which enhances the absorption in the digestive system super critical carbon dioxide is an environment-friendly solvent making this method safe and more suitable than traditional solvent waste techniques this technology is especially effective in improving the solubility of fat-soluble nutrient life Vitamin D and mineral such as zinc can Selenium [25].

3.4 Solid Lipid Nanoparticles (SLNs)

Solid lipid Nanoparticles offer novel methods to enhance the bioavailability of vitamins and minerals. Made from solid lipids these nanoparticles in case nutrients protect them from the environment and factors like oxidation and degradation. Solid–liquid nanoparticles are created by dissolving liquid in an organic solvent, then emulsifying them in water and finally incorporating vitamins are minerals. This process produces a nanoparticle formation that can deliver nutrients to the body with controlled release solid liquid Nanoparticles are particularly beneficial for delivering sensitive vitamins such as A and E and minerals like calcium, which are prawn to depletion when exposed to light or oxygen. In addition to improving nutrient stability these nanoparticles in hand absorption by alloying nutrients to pass easily through cell membranes [26].

3.5 Technology of Nanoemulsion

Nano-emulsion Technology combines oil and water phases with an emulsifier, to create tiny emulsions, with droplet sizes between 20 and 200 nm. This technique improves the bioavailability of fat-soluble vitamins are usually poorly absorbed due to their low solubility. By reducing droplet size Nano emulsions increase the surface area of absorption for enhancing nutrient intake in the digestive system.

This process often achieved through energy methods like high-pressure homogenization, allows nutrients to bypass digestive barriers and ensure efficient delivery. Nano emulsions protect against oxidative degradation extending the stability and effectiveness of nutrients. they also enable targeted delivery by allowing nutrients to cross cell membranes more efficiently. Additionally, this technology offers controlled and sustained nutrient release boosting nutrient efficacy. Nanoemulsions, further shield sensitive nutrients from stomach acid and digestive enzymes, leading to good health.

Ultimately, Nano evolution technology enhances the solubility stability and bioavailability of fat-soluble nutrients and provides significant health benefits [27].

4 Micronutrient Bioavailability

Availability means a fraction of taken bioactive compounds that are absorbed and subsequently utilized for physiological processes within the body. The bioavailability of bioactive substances such as vitamins E, and k carotenoids, coenzyme Q10, curcumin, and conjugated linoleic acid, frequently diminishes post-oral administration. This decline is attributed to various physicochemical factors, such as absorption efficiency, transformation processes, and overall bioavailability (Fig. 2) [28].

To address the limited availability of hydrophobic bioactive molecules used in disease management nanotechnology-based delivery systems have been developed to enhance targeted nutrient delivery. These nanoparticles are frequently derived from food-grade macro components such as protein-polysaccharide phospholipids and lipids ensuring they have a non-toxic nature. For incense nano-emulsions use food-grade components to Improve the bioavailability stability, and solubility of these bioactive molecules [29]. The encapsulation properties of Food grade nanoparticles enhance the efficacy stability bioavailability of micronutrients [30].

A range of advanced techniques including Nanofibers, Nanosheets, Nano whiskers, and Nanotubes, are utilized to enhance the effective delivery of nutraceuticals. These Nanostructures are dispersed using carriers such as liposomes solid

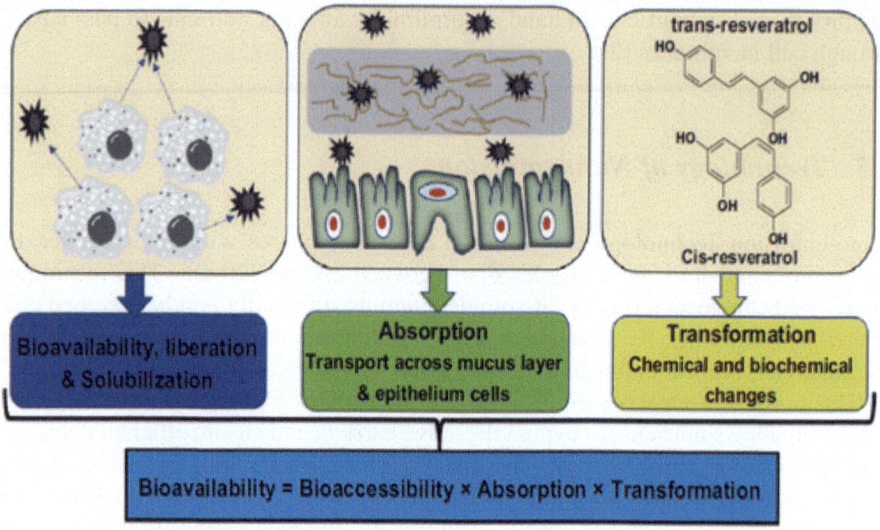

Fig. 2 Mechanism of micronutrients by availability in the human body

liquid nanoparticles, (SLNs), ribosomes solid lipids, nanoparticles, microemulsions, Nanosensors, and fibers [31].

Precisely engineered excipient food is formulated to improve the bioavailability, absorption, and transformation of phenolic compounds within the gastrointestinal track various nanoparticle-based delivery systems have been utilized in hands the ability of both hydrophilic and hydrophobic micronutrients through appropriate encapsulation techniques as depicted in Fig. 3 [32]. For a colloidal delivery system to be effective, it must be meticulously designed for specific applications, taking into account the nature of the functional particle with intended end product [5].

5 Micronutrient Stability in the Products

The stability of micronutrients in food products is largely influenced by their physiochemical and molecular properties along with storage conditions and food composition. The micronutrients are prone to physical, chemical, enzymatic, and stability that can diminish their nutritional value. Chemical alteration such as modification in molecular structure can significantly impact the physiochemical and nutritional attributes of these components of chemical degradation including oxidation, and hydrolysis, by enzymes present in food products [33].

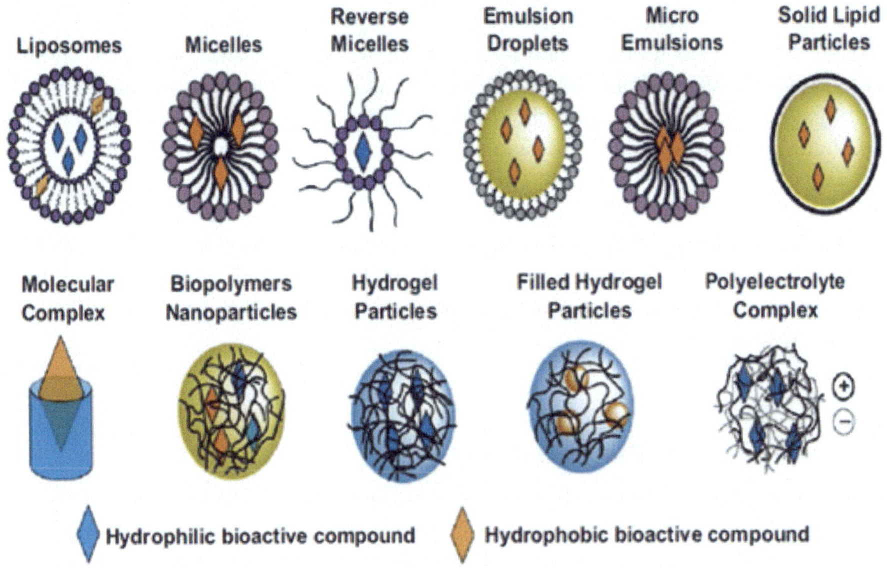

Fig. 3 Enhanced bioavailability of encapsulated micronutrients nano-based delivery systems

Physical instability, on the other hand, refers to changes in location of micronutrients or the physical state. Examples include phase changes like polymorphic transitions, crystallization, melting, gravitational separation, or aggregation. Additionally, key factors such as pH, temperature, and water activity play a critical role in determining the stability and degradation of specific micronutrients [34].

Nanoparticles can be coated with materials that prevent the diffusion of reactant into particles thereby enhancing their stability, for example, lipid oxidation in oil-in-water suspensions can be minimized by embedding lipid molecules within microgels surrounded by protein molecules. A specific instance of this is the reduced oxidation rate of fish oil droplets when encapsulated in casein-pectin microgels compared to the encapsulated form. The microgels limit the oxidation through transition metals and restriction of reactant diffusion [35].

6 Factors Enhancing and Inhibiting Nutrient Bioavailability

Bioactive compounds can interact with each other or modify their bioavailability based on the medium in which they are present. Nutrient likeVitamin D for calcium and vitamin C for iron, can enhance nutrient absorption and mitigate the effects of inhibitors. For instance, vitamin C can increase iron absorption by 2 to 3 times [36]. Conversely, inhibits nutrient absorption by binding nutrients to other substances or rendering them insoluble. A notable example is phytic acid, a mineral inhibitor found in many plant-based foods, which tightly binds to calcium, zinc, and iron, significantly lowering their bioavailability. Nanotechnology offers a promising approach to mitigating the effects of such chemical inhibitors on bioactive compounds [37].

Curcumin encapsulated in colloidal nanoparticles exhibits a much higher absorption capacity compared to natural curcumin powder. Research has shown that beta-carotene dissolved in nano-emulsions has a slower rate of decolonization when exposed to chemical inhibitors, due to the chelation of transition elements like Fe^{2+}, which typically promote oxidation [38]. Additionally, the oral bioavailability of curcumin can increase up to ninefold when combined with an absorption enhancer such as piperine [39].

7 Post-Ingestion Bioavailability

Bioavailability is essential for ensuring the bioefficacy of orally administered drugs and food ingredients. It is affected by the chemical degradation of bioactive compounds during processing, manufacturing, storage, and transport, as well as their exposure within the Gastrointestinal tract. The key processes involved in availability include metabolism, absorption, liberation, distribution, and elimination [37].

Enhancing the bioavailability of nutraceutical compounds can be achieved by designing targeted delivery systems. This objective is accomplished by modifying the composition and structure of nano-based delivery systems. Studies have demonstrated the effectiveness of triglycerides encapsulated in microgels, such as calcium alginate beads, in preventing breakdown within the upper GIT. These beads are digested enzymatically upon reaching the large intestine, thereby releasing bioactive fatty acids [40].

Biopolymer nanoparticles (NPs) and microgels derived from dietary fibers have shown the potential to deliver multiple lipophilic nutraceuticals into the intestinal lumen while maintaining structural integrity within the GIT. However, it is essential to design biopolymer NPs that effectively safeguard micronutrients in food products. When developing bio Polymers nanoparticles for a controlled release system it is essential to conduct comprehensive investigations using in vitro gastrointestinal models, human trials, and in Vivo animal studies [26].

7.1 Vitamin A

Nano-materials designed for the fat-soluble vitamin A significantly enhance its bioavailability by addressing the biological processes that regulate its absorption. Encapsulation of vitamin A using nano-materials has been shown to provide higher bioavailability compared to its free form. This hypothesis was validated through studies where researchers observed improved bioavailability of beta-carotenoids delivered via nano-materials compared to their natural forms [41].

The bioavailability can be further enhanced by Nanomaterials that facilitate paracellular transport modifying the integrity of tight junctions in nanoparticles. Additionally, oral bioavailability can be improved for lipophilic lipophilic compounds absorbed paracellularly without undergoing metabolic transformation by intracellular enzymes. Nanotechnology has proven effective in enhancing the oral bioavailability of Vitamin A by protecting it against physiochemical changes such as PH, temperature, and moisture variations [42].

7.2 *Vitamin B_{12}*

Protein, composite nanoparticles with a three-layer structure consisting of a phospholipid layer a protein layer, surrounding an inner hydrophilic core, have recently been developed as an efficient delivery system for hydrophilic bioactive compounds. These advanced (NPs) have demonstrated over 20-fold the uptake efficacy of the B12 vitamin [43].

This vitamin is absorbed in the last part of small intestine, ileum with the aid of intrinsic factors produced by the parietal cells of stomach. Once absorbed, vitamin B12 acts as a cofactor for enzymes involved in synthesizing fatty acid, DNA, and myelin [44]. In Vivo studies have demonstrated that these modified nanoparticles are significantly more effective at elevating vitamin B12 deficiency in a red model than conventional vitamin supplements however further extensive research is required to develop practical strategies for enhancing the bioavailability of vitamin B12 through nanotechnology [45].

7.3 *Folic Acid*

Nanoparticles hold great promise for the targeted oral delivery of nutrients. when taken orally this nanoparticle stays within the Gastrointestinal tract and protects the mucus layer of the jejunum. Using nanoparticles to encapsulate Folic acid has been shown to double the relevant reality oral bioavailability of Vitamin B9 compared to its Aqueous suspension. This improvement is likely due to the nanoparticle's ability to interact with the intestinal because and form a mucoadhesive Bond. However, despite these encouraging results, there is still a lack of extensive critical trial data on the impact of Nanotechnology on the bioavailability of Folic acid [46].

7.4 Iron

The integration of iron into solid–liquid nanoparticles has been found to increase its bioavailability by over four times compared to traditional iron tablets. The creation of ferrous Fe^{2+} loaded LG net nanoparticles as a targeted oral iron delivery system has shown considerable promise. However, studies and clinical trials are necessary to confirm the effectiveness of Nano-based delivery systems in enhancing the availability of micronutrients, and iron [47].

8 Nanoparticles Toxicity

Although nanotechnology has many beneficial applications, some nanoparticles can be toxic. their unique characteristics, including large surface area, high reactivity, and small size, can occasionally cause cellular damage. Extended exposure to nanoparticles in the Gastrointestinal tract mainly to negative effects [48]. A common toxicity issue with nanoparticles is their propensity to aggregate protein concentrations, by particle size shape, and surface properties. This aggregation can result in biological problems such as protein fibrillation, cross-linking and unfolding, and loss of enzymatic activity, [49].

Furthermore, although nanoparticles have high bioavailability, their larger surface area can lead to different toxicities. Another potential issue is the release of toxic ions influenced by the thermodynamic properties of these materials in bio-biological systems [50]. Although therapeutic nanostructures are still in their infancy, their long-term impact on human health is uncertain and demands research. Despite efforts to reduce the toxicity of synthesized Nanostructures, significant concerns about the potential risk of nanoparticles remain. the production, handling, and storage processes of nanostructures can affect their efficacy and elevate toxicity risks, highlighting the need for extensive investigations [51].

9 Limitations and Future Perspective and Trends

Nanoparticles offer significant promise in human nutrition, especially for antimicrobial applications and active food packaging nutrient Nano encapsulation. Additionally, nanomaterials have been widely studied for biomedical uses, including Diagnostic, disease treatment, and drug delivery, However, there are considerable challenges related to and potential risks of Nanomaterials and the safety of human health which also impact consumer acceptance of Nanotechnology in food [52].

A major concern with incorporating nanoparticles in food is potential toxicity. Due to their tiny size, nanoparticles can enter the human body through inhalation, ingestion, or absorption, potentially causing adverse effects like DNA damage, inflammation, and oxidative stress.

Moreover, the increased bioavailability of nanoparticles might lead to their accumulation in tissue and organs, resulting in long-term harm. NP can interact with biological barriers, such as blood–brain barriers, disrupting their functions, and allowing harmful substances to reach sensitive areas like the brain or a developing fetus [53].

When nanoparticles come into contact with food, they may cause immunogenicity, where the immune system recognizes them as a foreign body and initiates responses like chronic inflammation or allergic reactions. Prolonged exposure to nanoparticles heightens the risk of immunological issues [54].

Additionally, certain nanoparticles can remain in the human body for a prolonged period raising, and potential adverse effects and concerns about long-term exposure. Their tendency to accumulate in specific organs increases the risk of localized toxicity. The unique properties of nanoparticles may interfere with normal biological processes, metabolic pathways, cellular communication, and gene expression, leading to unpredictable health outcomes [55]

The regulation of products containing nanoparticles is still evolving and varies widely across different regions. the lack of standardization in nanoparticle synthesis and properties complicates the assessment of their safety and efficacy. International bodies such as the Food and Drug Administration FDA and the European Food Safety Authority, EFSA are actively working to regulate Nanomaterials in food. The FDA uses a risk-based evaluation approach, bioavailability, relying on company-provided data on toxicity, and potentially adverse effects. while EFSA evaluates scientific data on the toxicity, Exposure, and behavior of Nanomaterials, offering safety guidelines and Recommendations based on these assessments [56, 57].

The risk associated with nanoparticles in food products can vary based on whether they are organic or inorganic. Inorganic nanomaterials, including carbon nanotubes, metals, metal oxides, graphene oxide (GO), reduced graphene oxide, and Nanodiamonds, have demonstrated higher toxicity in both in Vivo studies and in vitro researches. On the other hand, organic Nanomaterials made from carbohydrates, lipids, or proteins such as nanoemulsions and nanocapsules, tend to exhibit lower toxicity. However careful consideration of their ingredients and preparation method is still necessary [58].

Recent research has investigated the cytotoxic effects of multiwall carbon nanotubes on human epithelial adenocarcinoma cells A-549 for potential using composites and food coatings. Coating for biomedical and food packaging applications was found to beat psycho toxic to L 929 fibroblast cells at a concentration of about 60 PPM and with prolonged exposure. The increased cytotoxicity at their higher concentration is attributed to reactive oxygen species ROS generated by GO. These findings highlight the necessity for the research to understand the interaction between Nanomaterials and biological systems ensuring their safe use in food applications [59].

Edible Nano emulsions made from vegetable oil and surface surfactants showed toxicity against $CACO_2$ human intestinal cells with effect influenced by particle size concentration and exposure duration [60]. Evaluating the cytotoxicity of these materials is crucial before integrating them into food matrices, providing companies and consumers with accurate information about their benefits and risks [61].

Nanoencapsulation is a rapidly evolving dynamic and rapidly advancing field, with numerous challenges still to be tackled. It is crucial to understand the properties of various encapsulation materials and their interactions with bioactive substances to enhance delivery, stability, efficiency, stability, and bioavailability. Further research is necessary to investigate controlled release mechanisms and the impact of factors such as pH, temperature, and environmental conditions on capsulated substances in the field forward [62].

Another key area of focus is exploring Nanoencapsulation applications across various sectors including medicine, nutrition, cosmetics, and agriculture. Each field of our unique opportunities and challenges for nanotechnology. Additionally, studies are crucial to ensure that Nanoencapsulation systems are safe for both humans and the environment. Epidemiological and toxicological studies using Vivo approaches are essential to assess exposure risks. Evaluating encapsulation materials and their potential adverse effects will help us understand their behavior in biological and environmental contexts, paving the way for innovative applications and significant advancement in agriculture, health, and other fields [62].

10 Conclusion

The research underscores the potential benefit of utilizing the nanoparticles as a carrier for active food ingredients including anti-microbial and bioactive compounds. Due to their Nano scale dimensions, NPs exhibit a remarkable capacity to interact with microbial cells, thereby enhancing anti-microbial efficacy. Furthermore, NP-encapsulating bioactive substances demonstrate improved digestibility and absorption thereby increasing their bioavailability. However, the optimistic perspective derived from recent studies is based on a limited body of conclusive research, necessitating further investigation and validation. Future research should in compass a broader spectrum of bioactive substances and their ingredients within NPs to provide a comprehensive understanding of the behavior and functionality of various liquid based NPs.

Currently, most studies on the activity of lipid NP utilize simplified models it is therefore imperative to investigate how lipid nanoparticles interact with other macromolecules in food formulations such as protein, Carbohydrate, and fiber. This knowledge is crucial for providing accurate information to both companies and consumers regarding the benefits of nanoparticles in food products. A key step in evaluating the incorporation of nanoparticles in food systems is to assess their functionality within a real food matrix such as says are designed to determine the biological fate and efficacy of nanoparticles under realistic condition.

While lipid Nano particle generally consists of an edible component and exhibits digestibility patterns similar to those of conventionally emulsified lipids, they are toxicological safety cannot be fully assured. Therefore, it is essential to investigate the gastrointestinal faith of lipid Nanoparticles upon digestion injection including their biological effect on tissue distribution and potential for bioaccumulation achieving these objectives will require additional Vivo research and the development of in vitro methods to validate and expand upon existing data.

References

1. Halver JE (2003) The vitamins. InFish nutrition 2003 Jan 1. Academic Press, pp 61–141
2. Ball GF (2008) Vitamins: their role in the human body. John Wiley & Sons
3. Lall SP (2022) The minerals. InFish nutrition 2022 Jan 1. Academic Press, pp 469–554
4. National Research Council, Life Studies, Board on Earth Sciences, Committee on Earth Resources, Committee on Critical Mineral Impacts of the US Economy. Minerals, critical minerals, and the US economy. National Academies Press; 2008 Mar 11
5. McClements DJ (2020) Recent advances in the production and application of nano-enabled bioactive food ingredients. Curr Opin Food Sci 1(33):85–90
6. Ndlovu N, Mayaya T, Muitire C, Munyengwa N (2020) Nanotechnology applications in crop production and food systems. Int. J. Plant Breed. 7(1):624–634
7. Priyanka S (2018) Nanotechnology in food preservation. Food Sci Res J 9(2):441–447. https://doi.org/10.15740/HAS/FSRJ/9.2/441-447
8. Jain KK (2020) Role of nanobiotechnology in drug delivery. In: Jain K (eds) Drug delivery systems, methods in molecular biology 2059. Humana, pp 55–73. https://doi.org/10.1007/978-1-4939-9798-5-2
9. Sajid A, Manzoor Q, Sajid A, Imran M, Khalid S, Arshad Z, Saleem A, Aslam F (2021) Characterization and antibacterial properties of Eriobotrya japonica extract loaded silver-nanoparticles. Curr Bioact Compd 17(7):58–63
10. Vieira IR, Tessaro L, Lima AK, Velloso IP, Conte-Junior CA (2023) Recent progress in nanotechnology improving the therapeutic potential of polyphenols for cancer. Nutrients 15(14):3136
11. Mirbagheri R, Elhamifar D, Shaker M (2021) Yolk–shell structured magnetic mesoporous silica: a novel and highly efficient adsorbent for removal of methylene blue. Sci Rep 11(1):23259
12. Yetisgin AA, Cetinel S, Zuvin M, Kosar A, Kutlu O (2020) Therapeutic nanoparticles and their targeted delivery applications. Molecules 25(9):2193
13. Abolmaali SS, Tamaddon AM, Salmanpour M, Mohammadi S, Dinarvand R (2017) Block ionomer micellar nanoparticles from double hydrophilic copolymers, classifications and promises for delivery of cancer chemotherapeutics. Eur J Pharm Sci 15(104):393–405
14. Nile SH, Baskar V, Selvaraj D, Nile A, Xiao J, Kai G (2020) Nanotechnologies in food science: applications, recent trends, and future perspectives. Nano-Micro Letters 12:1–34
15. Dudefoi W, Villares A, Peyron S, Moreau C, Ropers MH, Gontard N, Cathala B (2018) Nanoscience and nanotechnologies for biobased materials, packaging and food applications: new opportunities and concerns. Innov Food Sci Emerg Technol 1(46):107–121
16. Koo OM, Rubinstein I, Onyuksel H (2005) Role of nanotechnology in targeted drug delivery and imaging: a concise review. Nanomed Nanotechnol Biol Med 1(3):193–212
17. Sing T, Shukla S, Kumar P, Wahla V, Bajpai VK, Rather IA (2017) Corrigendum: application of nanotechnology in food science: perception and overview. Front Microbiol 12(8):2517
18. McClements DJ, Li F, Xiao H (2015) The nutraceutical bioavailability classification scheme: classifying nutraceuticals according to factors limiting their oral bioavailability. Annu Rev Food Sci Technol 6(1):299–327

19. Samah NA, Mahmood MR, Muhamad S (2017) The role of nanotechnology application in antioxidant from herbs and spices for improving health and nutrition: a review. Selangor Sci Technol Rev (SeSTeR) 1(1):13–17
20. Yen FL, Wu TH, Tzeng CW, Lin LT, Lin CC (2010) Curcumin nanoparticles improve the physicochemical properties of curcumin and effectively enhance its antioxidant and antihepatoma activities. J Agric Food Chem 58(12):7376–7382
21. He X, Deng H, Hwang HM (2019) The current application of nanotechnology in food and agriculture. J Food Drug Anal 27(1):1–21
22. Rao PJ, Naidu MM (2016) Nanoencapsulation of bioactive compounds for nutraceutical food. Nanosci Food Agricult 2:129–156
23. Mallakpour S, Hussain CM, Kumar Sahu C, Dhanashri Sanjay S, Kadeppagari RK (2021) Application of nanotechnology for encapsulation of micronutrients. Handbook Consumer Nanoproducts 1–8
24. Apte A, Lubree H, Kapoor M, Juvekar S, Banerjee R, Bavdekar A (2021) Development and implementation of liposomal encapsulated micronutrient fortified body oil intervention for infant massage: an innovative concept to prevent micronutrient deficiencies in children. Front Public Health 25(8):567689
25. Paraman I, Wagner ME, Rizvi SS (2012) Micronutrient and protein-fortified whole grain puffed rice made by supercritical fluid extrusion. J Agric Food Chem 60(44):11188–11194
26. Sharma S, Singh N, Rastogi M (2022) Nanotechnology to address micronutrient and macronutrient deficiency. In: Nanotechnology. CRC Press, pp 35–59
27. Khalid A, Arshad MU, Imran A, Haroon Khalid S, Shah MA (2023) Development, stabilization, and characterization of nanoemulsion of vitamin D3-enriched canola oil. Front Nutr 24(10):1205200
28. Zhang R, McClements DJ (2016) Enhancing nutraceutical bioavailability by controlling the composition and structure of gastrointestinal contents: emulsion-based delivery and excipient systems. Food Struct 1(10):21–36
29. Kumar MN (2000) Nano and microparticles as controlled drug delivery devices. J Pharm Pharm Sci 3(2):234–258
30. Quintanilla-Carvajal MX, Camacho-Díaz BH, Meraz-Torres LS, Chanona-Pérez JJ, Alamilla-Beltrán L, Jimenéz-Aparicio A, Gutiérrez-López GF (2010) Nanoencapsulation: a new trend in food engineering processing. Food Eng Rev 2:39–50
31. Cushen M, Kerry J, Morris M, Cruz-Romero M, Cummins E (2012) Nanotechnologies in the food industry—recent developments, risks and regulation. Trends Food Sci Technol 24(1):30–46
32. McClements DJ, Xiao H (2014) Excipient foods: designing food matrices that improve the oral bioavailability of pharmaceuticals and nutraceuticals. Food Funct 5(7):1320–1333
33. McClements DJ (2015) Nanoscale nutrient delivery systems for food applications: improving bioactive dispersibility, stability, and bioavailability. J Food Sci 80(7):N1602–N1611
34. Joye IJ, Davidov-Pardo G, McClements DJ (2014) Nanotechnology for increased micronutrient bioavailability. Trends Food Sci Technol 40(2):168–182
35. Zhang XP, Le Y, Wang JX, Zhao H, Chen JF (2013) Resveratrol nanodispersion with high stability and dissolution rate. LWT-Food Sci Technol 50(2):622–628
36. Teucher, Olivares, Cori (2004) Enhancers of iron absorption: ascorbic acid and other organic acids. Int J Vitamin Nutrit Res 74(6):403–419
37. Zhou H, Yue Y, Liu G, Li Y, Zhang J, Gong Q, Yan Z, Duan M (2010) Preparation and characterization of a lecithin nanoemulsion as a topical delivery system. Nanoscale Res Lett 5:224–230
38. Mao L, Wang D, Liu F, Gao Y (2018) Emulsion design for the delivery of β-carotene in complex food systems. Crit Rev Food Sci Nutr 58(5):770–784
39. Shaikh J, Ankola DD, Beniwal V, Singh D, Kumar MR (2009) Nanoparticle encapsulation improves oral bioavailability of curcumin by at least 9-fold when compared to curcumin administered with piperine as absorption enhancer. Eur J Pharm Sci 37(3–4):223–230

40. Li Y, Hu M, Du Y, Xiao H, McClements DJ (2011) Control of lipase digestibility of emulsified lipids by encapsulation within calcium alginate beads. Food Hydrocolloids 25(1):122–130
41. Faulks RM, Southon S (2005) Challenges to understanding and measuring carotenoid bioavailability. Biochimica et Biophysica Acta (BBA)-Molecular Basis of Disease 1740(2):95–100
42. Maurya VK, Aggarwal M (2017) Enhancing bio-availability of vitamin D by nano-engineered based delivery systems—an overview. Int J Curr Microbiol App Sci 6(7):340–353
43. Yao M, McClements DJ, Xiao H (2015) Improving oral bioavailability of nutraceuticals by engineered nanoparticle-based delivery systems. Curr Opin Food Sci 1(2):14–19
44. Hoque M, Emon K, Malo PC, Hossain MH, Tannu SI, Roshed MM (2023) Comprehensive guide to vitamin and mineral sources with their requirements. Indiana J Agricult Life Sci 3(6):23–31
45. Liu G, Yang J, Wang Y, Liu X, Chen L (2019) Protein-lipid composite nanoparticles for the oral delivery of vitamin B12: impact of protein succinylation on nanoparticle physicochemical and biological properties. Food Hydrocolloids 1(92):189–197
46. Peñalva R, Esparza I, González-Navarro CJ, Quincoces G, Peñuelas I, Irache JM (2015) Zein nanoparticles for oral folic acid delivery. J Drug Deliv Sci Technol 1(30):450–457
47. Katuwavila NP, Perera AD, Dahanayake D, Karunaratne V, Amaratunga GA, Karunaratne DN (2016) Alginate nanoparticles protect ferrous from oxidation: potential iron delivery system. Int J Pharm 513(1–2):404–409
48. Ibrahim KS (2013) Carbon nanotubes? properties and applications: a review. Carbon letters. 14(3):131–144
49. Bahadar H, Maqbool F, Niaz K, Abdollahi M (2016) Toxicity of nanoparticles and an overview of current experimental models. Iran Biomed J 20(1):1
50. Xia T, Kovochich M, Liong M, Madler L, Gilbert B, Shi H, Yeh JI, Zink JI, Nel AE (2008) Comparison of the mechanism of toxicity of zinc oxide and cerium oxide nanoparticles based on dissolution and oxidative stress properties. ACS Nano 2(10):2121–2134
51. Agrawal P (2016) Potential prospects of future medicine: nano medicine. J Pharmacovigilance 4(1):1000–1149
52. Henchion M, McCarthy M, Dillon EJ, Greehy G, McCarthy SN (2019) Big issues for a small technology: consumer trade-offs in acceptance of nanotechnology in food. Innov Food Sci Emerg Technol 1(58):102210
53. Meng H, Leong W, Leong KW, Chen C, Zhao Y (2018) Walking the line: the fate of nanomaterials at biological barriers. Biomaterials 1(174):41–53
54. Ilinskaya AN, Dobrovolskaia MA (2016) Understanding the immunogenicity and antigenicity of nanomaterials: past, present and future. Toxicol Appl Pharmacol 15(299):70–77
55. Kehrer JP, Klotz LO (2015) Free radicals and related reactive species as mediators of tissue injury and disease: implications for health. Crit Rev Toxicol 45(9):765–798
56. Vieira IR, de Carvalho AP, Conte-Junior CA (2022) Recent advances in biobased and biodegradable polymer nanocomposites, nanoparticles, and natural antioxidants for antibacterial and antioxidant food packaging applications. Comprehensive Rev Food Sci Food Safety 21(4):3673–3716
57. Schoonjans R, Castenmiller J, Chaudhry Q, Cubadda F, Daskaleros T, Franz R, Gott D, Mast J, Mortensen A, Oomen AG, Rauscher H (2023) Regulatory safety assessment of nanoparticles for the food chain in Europe. Trends Food Sci Technol 1(134):98–111
58. Fröhlich EE, Fröhlich E (2016) Cytotoxicity of nanoparticles contained in food on intestinal cells and the gut microbiota. Int J Mol Sci 17(4):509
59. Mohammadi S, Babaei A (2022) Poly (vinyl alcohol)/chitosan/polyethylene glycol-assembled graphene oxide bio-nanocomposites as a prosperous candidate for biomedical applications and drug/food packaging industry. Int J Biol Macromol 15(201):528–538
60. Wooster TJ, Moore SC, Chen W, Andrews H, Addepalli R, Seymour RB, Osborne SA (2017) Biological fate of food nanoemulsions and the nutrients they carry–internalisation, transport and cytotoxicity of edible nanoemulsions in Caco-2 intestinal cells. RSC Adv 7(64):40053–40066
61. Kunjiappan S, Sankaranarayanan M, Kumar BK, Pavadai P, Babkiewicz E, Maszczyk P, Glodkowska-Mrowka E, Arunachalam S, Pandian SR, Ravishankar V, Baskararaj S (2020)

Capsaicin-loaded solid lipid nanoparticles: design, biodistribution, in silico modeling and in vitro cytotoxicity evaluation. Nanotechnology 32(9):095101
62. Contado C (2015) Nanomaterials in consumer products: a challenging analytical problem. Front Chem 3:48
63. Matalanis A, Jones OG, McClements DJ (2011) Structured biopolymer-based delivery systems for encapsulation, protection, and release of lipophilic compounds. Food Hydrocolloids 25(8):1865–1880
64. Rubio L, Barguilla I, Domenech J, Marcos R, Hernández A (2020) Biological effects, including oxidative stress and genotoxic damage, of polystyrene nanoparticles in different human hematopoietic cell lines. J Hazard Mater 5(398):122900
65. Kyriakidou K, Brasinika D, Trompeta AF, Bergamaschi E, Karoussis IK, Charitidis CA (2020) In vitro cytotoxicity assessment of pristine and carboxyl-functionalized MWCNTs. Food Chem Toxicol 1(141):111374

The Role of Nanofibers in Muscle Recovery

Bhaskar Pal, Kushal Roychoudhuri, Saikat Santra, Rajdip Goswami, Mahenderan Appukutty, and Ayan Chatterjee

Abstract The versatility of nanofibers as an emerging exploitable approach has gain attention in muscle recovery process due to their distinct feature of physical absorbent property. These tiny fibers are made from the materials, including polymers, ceramics, and composites; they have a high surface area to volume ratio, strong mechanical properties, and are biocompatible. Nanofibers also support cellular attachment, growth, and differentiation in imitating their role in the extracellular matrix to provide support during remanufacturing of damaged tissue with engineered scaffolds for tissue engineering. The problem is that they can also transport bioactive compounds and growth factors to the damaged muscles directly, cutting days and even weeks off the healing process. In addition, nanofibers could be used in methods like electrospinning to provide muscles with the most advantageous environment for their repair and regeneration. Recent studies show they can reduce inflammation and oxidative stress, helping to speed recovery after injury or surgery. In this review article, we briefly summarize the latest technology for nanofiber, including the mechanism of muscle recovery using nanofiber, and discuss the future research, clinical applications of this technology. Advancing therapeutic approaches for muscle injuries to more effective recovery outcomes is possible leveraging the unique property of nanofibers.

B. Pal · K. Roychoudhuri
Department of Pharmaceutical Technology, Charaktala College of Pharmacy, Charaktala, Mothabari, Malda, West Bengal, India

S. Santra
Department of Pharmaceutics, JRSET College of Pharmacy, Panchpota, Chakdah, Nadia, West Bengal, India

R. Goswami
Department of Pharmaceutical Technology, School of Health and Medical Sciences, Adamas University, Barasat, Kolkata, West Bengal, India

M. Appukutty
School of Sports Science and Recreation, Universiti Teknologi MARA, Shah Alam, Malaysia

A. Chatterjee (✉)
School of Health Science and Technology, Medhavi Skills University, Topakhani, Sikkim, India
e-mail: ayan4189@yahoo.com

A. Chatterjee et al. (eds.), *Nanofuel: The Future of Sports Nutrition*, Smart Nanomaterials Technology, https://doi.org/10.1007/978-981-96-5471-0_7

Keywords Nanofibers · Biocompatible · Extracellular matrix · Inflammation · Oxidative stress

1 Introduction

The volumetric muscle loss (VML) after traumatic injury or other injury caused by external factors serves potential therapeutic challenge often related to reduced healing, notable fibrosis, as well as chronic pain and functional impairments. AMIs (acute muscle injuries) serves a prominent capacity for muscle reformation, potentially sized VML deficiency represent an imbalanced immune microenvironment that saturate innate repair processes leading to pro-fibrotic signaling and chronic inflammatory stimulations. The pathogenesis of muscle injury is complex and involves numbers of potential risk factors that modulate the location and severity of muscle injuries. The standard care in muscle recovery serves the autologous transportation of tissue matrix but this therapy shows limited muscle restoration and also serves complications in both the injury site and donor. Furthermore, the soft tissue injury and accessory damage in collateral blood vessels causing inadequate vascularization to the grafted tissue [1]. Currently, this unsuccessful muscle reformation causing the traumatic injury associated to immune dysfunction. The immunogenic response is induced by pro-inflammatory stimulus that produces phase of inflammation to a point where muscle reformation is unable to take place. Instead, the chronic inflammatory stimulation induced the fibrosis and loss of function. The nanofibers containing nano structural components fabricated with different homo- or hetero-polymers as fiber like assembly which are significantly explored as the predominant treatments for regenerative therapy, especially in context of muscle regeneration. The inherent biomimesis of nanofibers, intimately similar to the intricate structure of native ECM, exhibits promise for adopting cellular mechanisms central in muscle reformation and repair. Most of the studies are reported that, the nanofibers healing the lesioned region owing to their biological natures of high interfacial site to volume ratio, and significantly porous skeleton, mimetic the extracellular matrix (ECM). Furthermore, the properties of nanofibers like fibre diameter, porosity, mechanical properties, and conductivity, helps in manipulation of the cellular microenvironment presented in injured muscles. The physical indication serves by the nanofibers potentially modulate the cellular functions involved the cell adhesion and migration followed by conditional growth of muscles producing comparatively healthy muscle [2]. The aligned structures of nanofibers regulate the parallel inclination of muscle cells, imitating the native tissue growth and improving functional tissue reformation. However, the controlled drug liberation also influenced by different diameter and porosity, and helps in the delivery of bioactive substances, growth factors, or pharmacotherapeutic agents at effective concentrations, promoting nanoscale conditions conducive to myogenic diversity, angiogenesis, and entire tissue reformation in vivo. Nanofibers incorporated with conducting polymers shows electrical conductivity and help in cell to cell interconnection and mimetic the electrical phenomena of muscles, and induced the functional tissue reformation [3].

2 Properties of Nanofibers

The unique properties of nanofibers make them attractive for applications in medicine such as muscle recovery, therefore, they have been a point of focus in biomedical research. Due to their high surface area to volume ratio, porosity, flexibility and their exceptional ability to replicate the natural extracellular matrix (Fig. 1), these fibers present great promise. These types of characteristics are important for helping support the processes of tissue regeneration and healing. The role of nanofibers in muscle recovery is examined in this chapter into the fundamental attributes. It also describes the materials used to fabricate them, the requirements for their production, and the physical and mechanical properties that make them successful. With the aid of these aspects, it points out the potential of nanofibers in the realms of therapeutic strategies for muscle repair and recovery [4].

Figure 1 illustrates key nanofiber properties—porosity, tensile strength, elasticity, and stiffness—enabling nutrient transport, mechanical support, and cellular guidance for effective muscle tissue repair and regeneration

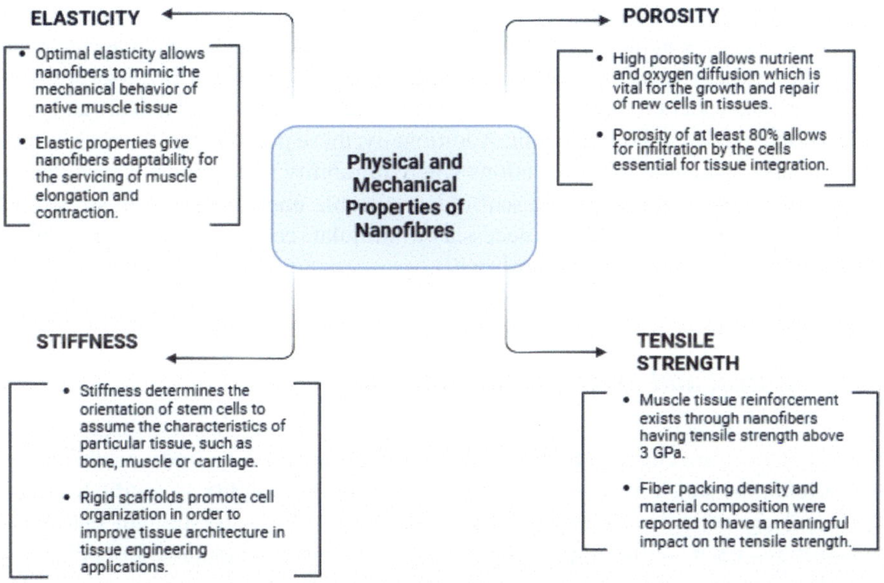

Fig. 1 Physical and mechanical properties of nanofibers for muscle recovery

2.1 *Materials Requirements for Nanofiber Production*

Effective production of nanofibers for muscle recovery applications is largely reliant on choosing suitable materials. Natural and synthetic polymers are used very often with the choice based on properties required and the requirements for a particular application [5].

2.1.1 Natural Polymers

The most commonly used materials, such as collagen, gelatin, and chitosan, exhibit excellent biocompatibility and biodegradability. Such attributes make them especially good candidates for tissue engineering and muscle recovery applications. They are essential for supporting important biological processes including cell adhesion, proliferation, and tissue regeneration, all of which are critical to facilitate healing and recovery. When these polymers mimic the natural extracellular matrix, they create the perfect environment for cellular activities to repair and regenerate damaged muscle tissues. This ensures their suitability for advancement of therapeutic approaches in muscle recovery by their ability to integrate seamlessly into biological systems [6].

2.1.2 Synthetic Polymers

Polymers like polylactic acid (PLA), polyglycolic acid (PGA), polycaprolactone (PCL) are report to be excellent mechanical strength and versatility. Given these properties, they are perfect for constructing scaffolds that can resist mechanical stress and allow cell activities and tissue repair. Additionally, these polymers provide structural integrity and resilience for applications where durability and support within regenerative medicine is required. The scaffolds are stable under mechanical stress and provide the mechanical properties necessary to stimulate cell growth and regeneration to facilitate muscle repair and recovery [7].

2.2 *Physical and Mechanical Properties*

The physical and mechanical properties of nanofibers are fundamental to their success as scaffolds in muscle recovery, directly impacting their ability to facilitate tissue repair and regeneration. These properties include porosity, tensile strength, elasticity, and stiffness, each contributing to the scaffold's overall functionality and effectiveness. Porosity is one of the most essential characteristics, as it ensures efficient transport of nutrients and oxygen to cells while enabling cellular infiltration into the scaffold structure. This property is particularly critical for thick scaffolds, where nutrient diffusion can be challenging. The intended usage of the cells and tissues for which regeneration is sought requires a minimum porosity level of at least 80%. The high porosity not only supports a favourable environment for cells, but also facilitates waste product removal, to ensure a healthy microenvironment for tissue development. This also includes tensile strength, which represents nanofibers' ability to resist mechanical stress placed on them under conditions while moving and repairing muscle. Depending on the materials used and the packing density of the materials, nanofiber mats may achieve tensile strengths of 3 GPa or greater, which is sufficient for retaining the scaffold integrity under physiological conditions. Applications such as dynamic muscle tissues require mechanical stability, and this property is vital. It is also important to note that elasticity plays an important role in mimicking the mechanical properties of native muscle tissue. Natural functioning and recovery is enhanced because proper elasticity of the nanofibers allows them to elongate and contract in response to muscle movement. Stiffness also has a major role in directing cellular behavior during regenerating. Cell alignment, proliferation, and differentiation are all sensitive to optimal stiffness, all of which are important for the proper muscle repair. Therefore these properties make nanofibers excellent scaffolds for muscle recovery. They can support cellular activities, maintain structural stability, and resemble native tissue environment, all of which improves healing outcome and promotes tissue integration, and advance the field of regenerative medicine [8].

3 Nanofibers Role in Muscle Recovery

The muscle reformation focused on developing intricate muscle skeleton in vitro, with ultimately implanting and restoring absent muscles function. ECM is an essential element that verifies tissue stability and regulates the contractile force, thus contribute a principal role in these endeavours. The ECM-generated 3D scaffold produces micro environmentally appropriate platforms to assist distinct cell types and reform the functional tissue matrix in-vitro. Nanofibers exhibits promising role in muscle reformation and restoring the functional capacity of muscles after injuries [9, 10].

3.1 Mechanisms of Action

The Nanofibers are reforming the muscles after injury by mimetic the ECM and serving desired indications, which support cell division, migration, and differentiation. The versatility of electrospinning strategy helps to integrate the protein growth factors (PGF) into polymeric nanofibers, which further regulate the activation and release of growth factors. The proteins are incorporated into these nanofiber centers through coaxial electrospinning that alter the electrospinning strategy by utilizing two concentric needles. This process permits for electrospinning of protein solvent when defending it from the organic solution utilized to solvate the exterior polymeric layer.

A study also indicating that the incorporation of growth factors (GF) into nanofibers by coaxial electrospinning, that particularly focused on model proteins like albumin and lysozyme for muscle recovery. The GF are generally utilized as transient intercellular signaling components, which are routinely used in tissue engineering and nanomedicine to modulate the cells growth and induces repairing through stimulating cell survival, division, and differentiation. However, the efficient delivery of pharmacotherapeutic agent is difficult for their larger dimension compare to traditional drug components and physical instability [11–13].

3.2 Delivery of Bioactive Compounds

The polymeric nanofibers are consists of a wire or string like structure with nano sized diameter, are (1D) one-dimensional nanomaterials. The fiber configurations are abundant in nature as found in human body and in the complex structure of the muscle tissues. To mimetic the nanofiber structure, polymeric nanofibers are fabricated by using electrospinning techniques. PLGA, chitosan-conjugated polycaprolactone (PCL), silk, cellulose nano-whiskers (CnW), and alginate/PCL nanofibers with various nanoscale diameters are generally used to enhance the cell division and differentiation of in vitro C2C12 myoblasts assays. For in vitro *implementation* of nanofibers, gelatin nanofibrous sheets (GNS) are fabricated with sacrificial coaxial electrospinning (SCE) of PCL/gelatin solvent [14–16]. After C2C12 implanting in the nanofiber sheet, a stepwise stretching is employed to induce myotube production in the nanofiber scaffold. The nanofibers or differentiated C2C12 assemble implanted in VML model with deficit muscle shows enhanced muscle reformation and functional improvements [17–20].

4 Nanofibers Fabrication Techniques

Nanofibers, owing to their high surface area to volume ratio, tunable mechanical properties and flexibility, are highly suitable in many applications, including muscle recovery and rehabilitation. Various techniques are used for the fabrication of nanofibers, each having its advantage and disadvantages. The structural and material properties of the resulting nanofibers depend on these fabrication methods (Table 1), and these are critical in determining the performance of the nanofibers in biomedical applications. Here we give an overview of the main techniques in nanofiber fabrication with emphasis on electrospinning and other methods [21].

Figure 2 illustrates the electrospinning process, a common technique for producing nanofibers with diameters spanning from nanometers to micrometers. It works by ejecting a charged polymer solution through an electric field to create fine fibers

Table 1 Fabrication strategies of Nanofibers

Fabrication technique	Description	Advantages	Limitations	References
Electrospinning	It ejects a thin jet of polymer solution into an electric field, which stretches and deposits the polymer solution as nanofibers	It has high precision, allows multiple materials, and is versatile	Energy intensive and may need post processing	[29]
Self-assembly	It relies on the non covalent interactions between molecules or nanoparticles to induce autonomous formation of nanofibers	No environmental footprint, low cost and control over structure	Applicable only to some materials and properties	[30]
Template-based synthesis	The nanofiber formation uses porous templates as molds for the resulting geometry, and structural control	Mass production with high design flexibility and scalability	Template removal complexity and material restrictions	[27]
Phase separation	Involve techniques (TIPS and NIPS) for controlled separation of phases in a material and creating nanofibrous structure	Morphology can be customized and mild processing conditions may be possible	Available only to those materials capable of undergoing phase separation	[31]
Melt blowing	It consists in extruding molten polymer through nozzles, stretching fibers at very high speed using high speed air, and fabricating nonwoven materials	It is efficient and scalable for high through put applications	Mainly applicable for thermoplastic polymers with restricted control	[7]

4.1 Electrospinning

It is one of the broadest used techniques for producing nanofibers because it is capable to generate fibers of diameter from a few nanometers to several micrometers. The process involved ejecting a charged polymer solution with a nozzle, overcoming an electric field and stretching the polymer into fine fibers as it travels to a collector. This technique is highly versatile for that it is possible the fabrication nanofibers from a wide variety of materials (Fig. 2), including polymers, ceramics, and composites. The major advantage in electrospinning is that it is able to spin out fibers with complex structures, i.e. hollow or porous fibers through changing the related parameters, such as solution viscosity and electric field strength. Such flexibility allows nanofibers to perform better in applications such as drug delivery, tissue engineering and muscle

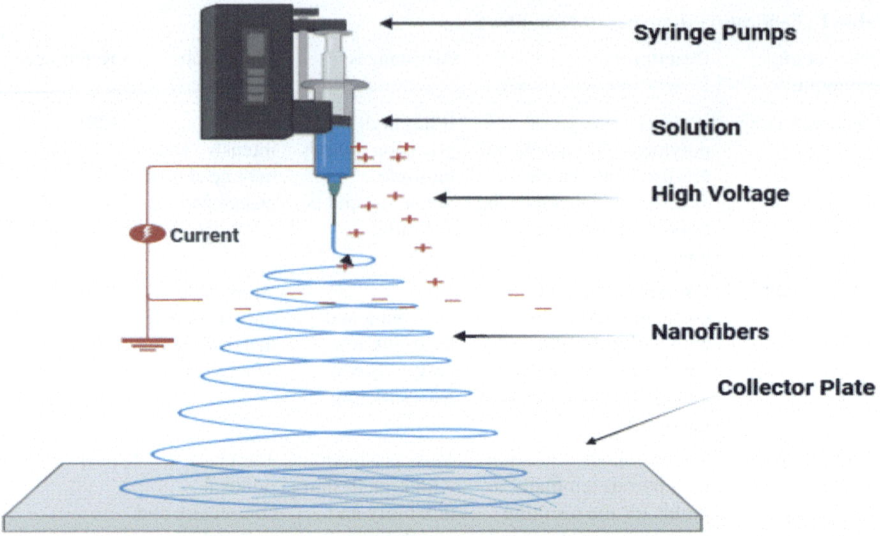

Fig. 2 Electrospinning: a versatile technique for nanofiber production

recovery, where we can tailor the fiber's characteristics to these needs. In addition, electrospinning enables the incorporation of additives, such as nanoparticles and biomolecules, for further functionalities of the nanofibers. Despite its many benefits, challenges with scalability and achieving uniform fiber diameters in larger batches restrict its applicability in high-volume production [22–25].

4.2 Other Techniques to Produce Nanofibers

Some other methods for nanofiber fabrication (Table 1), besides electrospinning, also exist, differing in the benefits and problems they have. Below are some notable techniques:

4.2.1 Self-assembly

Natural tendency of certain molecules or nanoparticles to self-assemble into Nano fibrous structures by non covalent interactions makes the method self-assembly. Examples include peptide based self-assembly and DNA nanofiber formation. They yield highly ordered nanofiber structures with specified functionality. Its simplicity and efficiency are two of self-assembly's primary strengths, which exploit the intrinsic properties of some species of molecule or nanoparticle to assemble themselves into nanofibrous structures autonomously. This self-organization is innate, dispensing with the need for complex and energy intensive processing, reducing the footprint of nanofiber production overall. Additionally, the self-assembly method provides a high degree of control over the produced nanofiber structures [26].

4.2.2 Template-Based Synthesis

Methods based on template include the use of porous templates as molds to create nanofibers. A lot of materials are deposited in the template, and then the template removed, leaving the nanofibers behind. Researchers can control the size, shape and alignment of the resulting nanofibers by using templates with given geometries and dimensions, allowing fabrication of intricate structures. This level of control is vital for applications that demand very high dimensional accuracy, for example photonic devices, and filtration membranes. Moreover, template based synthesis enables production of nanofibers from a variety of materials, such as metals, polymers and ceramics. The method proves to be scalable and applicable to mass production, and is efficient and reproducible which an important factor for industrial applications is. In addition, template-based synthesis can be more environmentally friendlier than some of the most energy intensive methods where operation is under mild conditions with little waste. However, the method is limited by some material types and the problem of nanofibers separation from the template [27].

4.2.3 Phase Separation

Controlled segregation of different material phases through phase separation techniques is important for producing nanofibrous structures. In this category, there are two well-known methods: thermally induced phase separation (TIPS) and nonsolvent induced phase separation (NIPS). Using phase separation as a versatile means, it accommodates a variety of materials, including polymers, ceramics and composites into nanofibers with various properties and functionality. While offering attractions, phase separation is possible only in certain materials that can be formed into distinct phases and, typically, requires tuning for particular purposes. The ability to control nanofiber morphology and porosity with this technique provides a major benefit. Several parameters, such as solution composition, temperature and solvent choice, can be used for customizing the resultant structures. On top of that, phase separation techniques typically require operation at a temperature and pressure that are low enough to minimize the risk of harming sensitive or bioactive components. The simplicity and low cost of these methods are attractive because they go around complex hardware and expensive, energy intense procedures. The extreme practicality of these techniques has led to the widespread use of phase separation methods across many disciplines, and as such, they demonstrate promise in enhancing materials science and engineering [28].

4.2.4 Melt Blowing

One of the widely used techniques to generate nanofiber reinforced nonwoven materials is melt blowing. It is an extrudable procedure in which molten polymer is driven out of very fine nozzles with the aid of high speed air or gases as the fibers are stretched. This efficiency and scalability has led to the recognition of melt blowing for large scale nanofiber production for multiple industrial applications. It is a continuous and high throughput process for rapid nanofiber manufacturing. Being tunable for various polymers makes the operation straightforward, and it can produce nanofibers with varied chemical and mechanical properties. Key advantages of the technique are in its simplicity and cost efficiency, even though the technique is limited to thermoplastic polymers and does not provide complete control of properties, including fiber diameter uniformity. Additionally, melt blowing is fully compatible with other manufacturing processes, enabling incorporation of nanofibers into many different products applied directly to substrates. Melt blowing is relevant to current material science and engineering, and the integration of this process for large scale nanofiber production reinforces the versatility and applicability of nanofibers to many fields (Table 1) [5].

5 Muscle Healing Effects of Nanofibers

Muscle injuries including strains, tears and degenerative forms are very challenging in clinical medicine. Present day treatments do not fully regain muscle structural integrity and function because the process of scaring is not integrated into the muscle assembly during the healing process. New approach in the field of nanotechnology has provided opportunity for fabrication of nanofiber scaffolds that show potential for muscle tissue engineering and regeneration. Due to their small size and high surface area nanofibers mirror the muscle tissue ECM and encourage cell attachment, growth and differentiation. It is biocompatible because they are developed on the nanoscale and from polymers including polycaprolactone (PCL), polylactic acid (PLA) and collagen. Such materials provide predictable degradation so that scaffold is gradually replaced with the native tissue and avoids the late complications. In addition, nanofiber's tensile strength is adequate for mechanical stress experienced during muscle motion. Nanofibers facilitate muscle repair in various ways as will be discussed below. All of these increase cell attachment and growth since are in the form of natural base structure for the cells known as the ECM. They further suggest that subsequently aligned nanofibers help in aligning the muscle cells and encourage growth in an orderly manner like the real muscular fibers. Nanofibers that are functionalized encourage the release of bio-molecules such as the vascular endothelial growth factor (VEGF) and the insulin-like growth factor (IGF-1), which are essential in angiogenesis and muscle formation. Conductive nanofibers also promote electrical connection between the electrodes that is essential for muscle contraction and signal restoration. Applicable fields of nanofiber scaffolds include the promotion of the wound healing process in cases of acute muscle rupture by increasing cell migration and proliferation. They also help in the treatment of other chronic muscular disorders for instance muscular dystrophy through the supply long term mechanical support as well as slow release drug delivery. Furthermore, nanofibers are incorporated in muscle tissue engineered for grafting in the worst-case scenarios. Despite their potential, nanofiber-based therapies face challenges such as ensuring minimal immune response, improving scalability and cost-effectiveness, and optimizing prolonged mechanical and biological stability. Preclinical and clinical studies are needed to establish safety and efficacy. Future research should focus on developing hybrid nanofiber scaffolds that combine synthetic and natural materials and incorporating smart biomaterials that respond to physiological stimuli [32].

5.1 Reduction of Inflammation

Unfortunately chronic inflammation damages tissues and can contribute to diseases ranging from arthritis to cardiovascular and neurodegenerative disorders. Managing these conditions and improving human health may possibly depend on the role that inflammation has to take in these illnesses. However, perhaps one of the best-known anti-inflammatory agents is dietary. A prospective diet is one low in antioxidant and Omega-3 fatty acids, berries, dark and leafy greens, and tomatoes containing anti-inflammatory compounds, and fatty fish, flaxseeds and walnuts, respectively. Processed foods, Transfats, and high sugar intake should be kept as a minimum since they trigger inflammation [33]. This inflammation is also avoided through exercise, which is very vital in anyone's lifestyle. Walking, cycling and other moderate forms of exercise can reduce inflammation that might be experienced throughout the body, though yoga and stretch exercises may lower inflammation caused by stress. Stress is another important variable as stress increases the levels of inflammatory indicators in the body. Stress, which is a major factor in inflammation, can be combatted through the use of such practices as meditation, deep breathing and mindfulness. Also, the lack of proper sleep, usually 7–9 h per night, hampers the body's ability to bring down inflammation and repair. For some forms of inflammation there are medications which are made to treat it such as NSAIDs or corticosteroids for example, these however are best taken under the doctor's prescription so as to avoid the side effects that may be as a result of taking them. Natural anti-inflammatory medications are available in foods such as turmeric, ginger, green tea, as well as Omega-3 supplements. Paying attention to different aspects of lifestyle, including smoking or alcohol consumption, which also cause inflammation and that can lead to chronic diseases. The last one is weight control; obesity is especially deleterious since increased adipose tissue, particularly visceral fat, increases cytokine secretion in the body. By incorporating these strategies into daily life, inflammation can be effectively reduced, leading to improved overall health and well-being [34].

5.2 Reducing Oxidative Stress

Oxidative stress may be defined as a state of excessive production of free radicals, which overwhelms the natural antioxidant activity to cause damage to cells, proteins and DNA. These injuries can lead to a lot of diseases, for instance, aging, cancer, cardiovascular diseases and neurodegenerative disorders. Oxidative stress has to be minimized in order to promote good health and to prevent pathophysiological processes that lead to the development of chronic diseases [35]. They also noted that one of the possible therapies to decrease oxidative stress is by using food products. Some antioxidant foods include fruits (Berries, Oranges, Apples), vegetables (spinach, Kale, Broccoli) and nuts (Walnuts, Almonds). More so, vitamin C and vitamin E in citrus fruits, nuts, and seeds are great antioxidants as they help boost the body immunity system. Long chain unsaturated omega-3 PUFA in fatty fish like salmon, flaxseed and walnut have not only anti-inflammatory activity but also decrease oxidative stress by stabilizing the cell membranes and thereby not allowing damage. Oxidative stress is also prevented by regular physical activity [19]. Moderate exercise also enhance the body's antioxidant enzymes and effectively reducing oxidation injury. But at the same time it is well known that physical load full of hot and severe control raises the oxidative stress; therefore healthy dose is needed. In addition, mechanisms like meditation, mindfulness, and yoga could make a difference in the volume of free radicals produced due to stress [36]. Sufficient sleep ensures proper body healing and helps the body to fight free radicals. Lack of sleep can lead to increased levels of oxidative stress, thus restoring 7–9 h of sound sleep is critical to the body's healing systems. Further, a relativity healthy lifestyle including non-smoking, alcohol and toxic substances, are equally important since they are known to produce free radicals. Supplements like vitamin C, vitamin E, selenium, and coenzyme Q10 can also help reduce oxidative stress when dietary intake may not be sufficient. However, it's important to consult with a healthcare professional before starting supplementation, as excessive intake can sometimes have adverse effects. By incorporating these strategies into daily life, oxidative stress can be significantly reduced, promoting better health and a lower risk of chronic disease [16, 37, 38].

5.3 Accelerated Healing Processes

Optimizing the rate of tissue repair is important in treatment of injuries, operations as well as management of various diseases. It is important to comprehend one basic truth: the body itself heals and for this the best thing is to learn how to help this process along. Diet, exercise and certain life style modifications have a very important role in the healing process. This paper will also looking at how nutrition plays a significant role in the promotion of the recovery process [39]. Of the molecules, proteins are vital ingredients for tissue repair and regeneration processes. The other foods you need to take in are lean meats, fish, eggs, beans, and legumes so as to restock the body with collagen and regenerate new cells [40]. There are essential nutrients that are involved in immune support, collagen formation and wound healing which includes vitamin C, zinc and vitamin A. Citrus fruits, strawberries, as well as bell peppers contain Vitamin C which is vital in collagen formation, sine it is a protein that is used by the body to make tissues. The nutrients that help strengthen the immune system and increase the rate of tissue repair include Zinc found in nuts seeds and whole grains. Another area which is very important in helping the muscle heal is adequate water intake. The cells ought to be in their best condition and adequate water assist in circulation and conveying of nutrient to the required site of body. Loss of water can slow the potential of the body to recover thereby affecting the natural healing process. Exercise when done correctly also has another advantage of recovering from injuries faster. While the patient should rest, simple walks or stretches will go a long way in improving circulation as well as reducing inflammation and stiffness, which are essential for winning the war against the ailment [41–44]. What's unwise though is engaging in strenuous activity or exercise soon after starting with a health or fitness regime as this could lead to straining the muscles and joints too early. This is important for the healing processes because most repair in the body occurs when one is asleep. At night, while sleeping body produces growth hormones, which are essential in repairing damaged tissues. These processes are supported by getting 7–9 h of quality sleep per night, and promote recovery. Additionally, minimizing stress is crucial for the healing process. Chronic stress can impair immune function and delay healing by increasing inflammation and the production of cortisol, a hormone that suppresses the repair process. Practices such as meditation, deep breathing, and mindfulness can help reduce stress and promote healing. In some cases, supplements such as collagen, turmeric, or arnica may be beneficial for supporting the body's healing. However, it's important to consult with a healthcare professional before taking any new supplements. By combining proper nutrition, hydration, physical activity, sleep, and stress management, healing processes can be significantly accelerated, leading to a faster recovery [6, 19, 45].

6 Future Directions in Nanofiber Research

The potential to develop future research on nanofibers for muscle recovery and rehabilitation is huge, and several key developments are of interest. The biocompatibility and bioactivity of nanofibers are a critical area of study. Specifically, they are changing the surface chemistry so they interact better with muscle cells and tissues [46]. Harnessing the wide range of biochemical properties present in nanofibers allows researchers to tailor them to the biological systems in which they will be integrated, thereby facilitating more efficient repair and regeneration of tissue, an important biophysical consideration for patients recovering from injuries or surgery [47]. The other promising direction is to develop 'smart' nanofibers that can sense environmental stimuli. Once designed, these nanofibers can be programmed to release therapeutic agents such as growth factors or anti inflammatory drugs when triggered for specific conditions such as when stressed or injured at the muscle [48]. This ability to dynamically respond allows for targeted treatment that ensures that the right agents are delivered right where they are needed, yielding higher recovery efficiency and lower side effects. This also allows for the release of therapeutic agents over a period of time, enhancing sustained healing and shortening the healing process. In addition, 3D printing is a key area of innovation for nanofiber scaffolds creation [49]. This technology allows for the fabrication of complex structures with high feature fidelity mimicking the natural extracellular matrix (ECM) of tissues. In addition, these scaffolds offer a perfect environment to grow and regenerate muscle cells. Future work should sharpen these scaffolds to have better mechanical properties to withstand stretching forces involved in muscle movement and provide supporting tissue growth. In addition, they will need to be able to adapt to varied muscle injury and rehabilitation requiring levels. Improving performance of nanofibers may also be achieved by incorporating nanofillers (nanoparticles or nanotubes) into them. Introducing nanofillers into nanofibers improves the mechanical strength and functionality, thus enhancing nanofibers' suitability for supporting muscle recovery [50]. The concentration, distribution and compatibility of these fillers could be investigated further, at optimal concentrations, to maximize their benefits. In this field, a defining step is finally moving from research to clinical application. Safety, effectiveness and long term impact of nanofiber based treatment of muscle recovery requires extensive in vivo studies. However, as the area of research evolves and potentially groundbreaking clinical applications emerge, it is absolutely crucial that researchers and clinicians work in tandem to fill this gap between theoretical study and real world applications of these emerging technologies to ensure that these advanced technologies are tested in as many diverse patients as possible. The future research of nanofibers has the ability to increase muscle recovery and improve patient rehabilitation outcomes by focusing on these key areas [51].

7 Conclusion

Nanofibers are therefore a revolutionary enhancement in muscle recovery therapies given their structure and functions. Being able to imitate the ECM their use in tissue engineering is suited because they enable cell attachment, proliferation and differentiation. Being biocompatible, high surface area to volume ratio and mechanical strength, nanofibers favors muscle regeneration and repair including volumetric muscle loss and chronic inflammation. The association of nanofibers with bioactive compounds and growth factors improves the site specific delivery of drugs and controls the healing process in addition to the functional results. Methods such as electro spinning and self-assembling allow desired fiber alignment to be achieved and designed to fulfill certain need in therapy. The discussed fabrication techniques provide information on how nanofiber fabrication lends itself to both clinic and industrial uses. However, further refinement issues are still there such as scale up, best immune response, and long term durability. These problems call for additional studies on hybrid scaffolds, smart biomaterials, and 3D-printed nanofiber structures that have active responses to signals supplied by body fluids. Thus, it is recommended that future investigations focus on in vivo assays with the view of ascertaining safety, effectiveness and practical usability. It is therefore clear that nanofibers are not only a promising tool in regenerative medicine, especially in muscle tissue repair. The potential for further implementation of fabrication solutions, better materials incorporation, and bioengineering approaches may be recognized to improve patient results, making a transition from laboratory discoveries to replacement treatments and cures. 'This novel technology opens up possibilities at the same time giving efficacious means of rehabilitating skeletal muscle injuries.

References

1. Electroactive electrospun nanofibers for tissue engineering—ScienceDirect [Internet]. [cited 2024 Dec 25]. Available from: https://www.sciencedirect.com/science/article/abs/pii/S1748013221001213
2. Damaraju SM, Shen Y, Elele E, Khusid B, Eshghinejad A, Li J et al (2017) Three-dimensional piezoelectric fibrous scaffolds selectively promote mesenchymal stem cell differentiation. Biomaterials 149:51–62
3. Khare D, Basu B, Dubey AK (2020) Electrical stimulation and piezoelectric biomaterials for bone tissue engineering applications. Biomaterials 258
4. Guo HF, Li ZS, Dong SW, Chen WJ, Deng L, Wang YF et al (2012) Piezoelectric PU/PVDF electrospun scaffolds for wound healing applications. Coll Surf B Biointerfaces 1(96):29–36
5. Mei L, Zhang Z, Zhao L, Huang L, Yang XL, Tang J et al (2013) Pharmaceutical nanotechnology for oral delivery of anticancer drugs. Adv Drug Deliv Rev 65(6):880–890
6. Fujioka K, Takada Y, Sato S, Miyata T (1995) Novel delivery system for proteins using collagen as a carrier material: the minipellet. J Control Release 33(2):307–315
7. Ansari A, Hussain A, Wadekar R, Altamimi MA, Malik A, Mujtaba MA et al (2023) Nanovesicles based drug targeting to control tumor growth and metastasis. Adv Cancer Biol—Metastasis 1(7):100083
8. Sahoo SK, Labhasetwar V (2003) Nanotech approaches to drug delivery and imaging. Drug Discov Today 8(24):1112–1120
9. Rodrigues MS, de Paula GC, Duarte MB, de Rezende VL, Possato JC, Farias HR et al (2021) Nanotechnology as a therapeutic strategy to prevent neuropsychomotor alterations associated with hypercholesterolemia. Colloids Surfaces B Biointerfaces 201
10. Wilczewska AZ, Niemirowicz K, Markiewicz KH, Car H (2012) Nanoparticles as drug delivery systems. Pharmacol Reports 64(5):1020–1037
11. Hu W, Zhai C, Sun H, Gong XY, Cui L, Cai L et al (2023) The global burden of disease attributable to metabolic risks in 204 countries and territories from 1990 to 2019. Diabetes Res Clin Pract 196
12. Sarmento B, Ribeiro A, Veiga F, Ferreira D (2006) Development and characterization of new insulin containing polysaccharide nanoparticles. Coll Surf B Biointerfaces 53(2):193–202
13. Alkaladi A, Abdelazim AM, Afifi M (2014) Antidiabetic activity of zinc oxide and silver nanoparticles on streptozotocin-induced diabetic rats. Int J Mol Sci 15(2):2015–2023
14. Khalil LM, Abdallah OY, Elnaggar YSR, El-Refaie WM (2022) Novel dermal nanobilosomes with promising browning effect of adipose tissue for management of obesity. J Drug Deliv Sci Technol 74
15. Hu H, Guo Q, Wang C, Ma X, He H, Oh Y et al (2015) Titanium dioxide nanoparticles increase plasma glucose via reactive oxygen species-induced insulin resistance in mice. J Appl Toxicol 35(10):1122–1132

16. Bahadori E, Farjami Z, Rezayi M, Lngari H, Darroudi M, Avan A et al (2019) Recent advances in nanotechnology for the treatment of metabolic syndrome. Diabetes Metab Syndr Clin Res Rev 13(2):1561–1568
17. Timko M, Koneracká M, Tomašovičová N, Kopčanský P, Závišová V (2006) Magnetite polymer nanospheres loaded by Indomethacin for anti-inflammatory therapy. J Magn Magn Mater 300(1)
18. Dos Santos Giuberti C, De Oliveira Reis EC, Ribeiro Rocha TG, Leite EA, Lacerda RG, Ramaldes GA et al (2011) Study of the pilot production process of long-circulating and pH-sensitive liposomes containing cisplatin. J Liposome Res 21(1):60–69
19. Kuzuya T, Nakagawa S, Satoh J, Kanazawa Y, Iwamoto Y, Kobayashi M et al (2002) Report of the committee on the classification and diagnostic criteria of diabetes mellitus. Diabetes Res Clin Pract 55(1):65–85
20. Pan Y, Neuss S, Leifert A, Fischler M, Wen F, Simon U et al (2007) Size-dependent cytotoxicity of gold nanoparticles. Small 3(11):1941–1949
21. Alam T, Khan S, Gaba B, Haider MF, Baboota S, Ali J (2017) Nanocarriers as treatment modalities for hypertension. Drug Deliv 24(1):358–369
22. Kim SY, Shin IG, Lee YM (1998) Preparation and characterization of biodegradable nanospheres composed of methoxy poly(ethylene glycol) and DL-lactide block copolymer as novel drug carriers. J Control Release 56(1–3):197–208
23. Xue Y, Xu X, Zhang XQ, Farokhzad OC, Langer R (2016) Preventing diet-induced obesity in mice by adipose tissue transformation and angiogenesis using targeted nanoparticles. Proc Natl Acad Sci USA 113(20):5552–5557
24. Elkasabgy NA, Elsayed I, Elshafeey AH (2014) Design of lipotomes as a novel dual functioning nanocarrier for bioavailability enhancement of lacidipine: In-vitro and in-vivo characterization. Int J Pharm 472(1–2):369–379
25. Ambroselli D, Masciulli F, Romano E, Catanzaro G, Besharat ZM, Massari MC et al (2023) New advances in metabolic syndrome, from prevention to treatment: the role of diet and food. Nutrients 15(3)
26. Hiwrale A, Bharati S, Pingale P, Rajput A (2023) Heliyon nanofibers: a current era in drug delivery system. Heliyon [Internet] 9(9):e18917. https://doi.org/10.1016/j.heliyon.2023.e18917
27. Brigger I, Dubernet C, Couvreur P (2012) Nanoparticles in cancer therapy and diagnosis. Adv Drug Deliv Rev 64(Suppl):24–36
28. Chhabra G, Chuttani K, Mishra AK, Pathak K (2011) Design and development of nanoemulsion drug delivery system of amlodipine besilate for improvement of oral bioavailability. Drug Dev Ind Pharm 37(8):907–916
29. Alberti KGMM, Zimmet P, Shaw J (2005) The metabolic syndrome—a new worldwide definition. Lancet 366(9491):1059–1062
30. Stevenson R, Hueber AJ, Hutton A, McInnes IB, Graham D (2011) Nanoparticles and inflammation. Scient World J 9(11):1300–1312
31. Ayala-Fuentes JC, Gallegos-Granados MZ, Villarreal-Gómez LJ, Antunes-Ricardo M, Grande D, Chavez-Santoscoy RA (2022) Optimization of the synthesis of natural polymeric nanoparticles of inulin loaded with quercetin: characterization and cytotoxicity effect. Pharmaceutics 14(5)
32. Cedervall T, Hansson LA, Lard M, Frohm B, Linse S (2012) Food chain transport of nanoparticles affects behaviour and fat metabolism in fish. PLoS ONE 7(2):1–6
33. Jafari-Giv Z, Avan A, Hamidi F, Tayefi M, Ghazizadeh H, Ghasemi F et al (2019) Association of body mass index with serum calcium and phosphate levels. Diabetes Metab Syndr Clin Res Rev 13(2):975–980
34. Assi RA, Abdulbaqi IM, Yee CS (2021) The evaluation of drug delivery nanocarrier development and pharmacological briefing for metabolic-associated fatty liver disease (Mafld): an update. Pharmaceuticals 14(3)
35. Li L, Wang K, Ma X, Liu Z, Wang S, Du J et al (2019) Radiomic analysis of multiparametric magnetic resonance imaging for differentiating skull base chordoma and chondrosarcoma. Eur J Radiol 1(118):81–87

36. Li R, Liu K, Huang X, Li D, Ding J, Liu B et al (2022) Bioactive materials promote wound healing through modulation of cell behaviors. Adv Sci 9(10)
37. Amin SA, Adhikari N, Kotagiri S, Jha T, Ghosh B (2019) Histone deacetylase 3 inhibitors in learning and memory processes with special emphasis on benzamides. Eur J Med Chem 15(166):369–380
38. Kheyrandish M, Bafande F, Sheikh Arabi M (2022) Features and methods of making nanofibers by electrospinning, phase separation and self-assembly. Jorjani Biomed J 10(1):13–25
39. Chen G, Xiao X, Zhao X, Tat T, Bick M, Chen J (2022) Electronic textiles for wearable point-of-care systems. Chem Rev 122(3):3259–3291
40. Karthikeyan E, Sivaneswari S (2024) Advancements in transdermal drug delivery systems: enhancing medicine with pain-free and controlled drug release. Intell Pharm [Internet]. 2024 Oct 11 [cited 2024 Dec 25]; Available from: https://linkinghub.elsevier.com/retrieve/pii/S2949866X24001072
41. Timin AS, Muslimov AR, Zyuzin MV, Peltek OO, Karpov TE, Sergeev IS et al (2018) Multifunctional scaffolds with improved antimicrobial properties and osteogenicity based on piezoelectric electrospun fibers decorated with bioactive composite microcapsules. ACS Appl Mater Interfaces 10(41):34849–34868
42. Xi Y, Pan W, Xi D, Liu X, Yu J, Xue M et al (2020) Optimization, characterization and evaluation of ZnO/polyvinylidene fluoride nanocomposites for orthopedic applications: improved antibacterial ability and promoted osteoblast growth. Drug Deliv 27(1):1378–1385
43. Lysaght MJ, Hazlehurst AL (2004) Tissue engineering: the end of the beginning. Tissue Eng 10(1–2):309–320
44. Yan H, Wang Y, Li L, Zhou X, Shi X, Wei Y et al (2020) A micropatterned conductive electrospun nanofiber mesh combined with electrical stimulation for synergistically enhancing differentiation of rat neural stem cells. J Mater Chem B 8(13):2673–2688
45. Langwald SV, Ehrmann A, Sabantina L (2023) Measuring physical properties of electrospun nanofiber mats for different biomedical applications. Membranes (Basel) 13(5):20–24
46. Xue J, Wu T, Dai Y, Xia Y (2019) Electrospinning and electrospun nanofibers: methods, materials, and applications. Chem Rev 119(8):5298–5415
47. Zhao G, Qing H, Huang G, Genin GM, Lu TJ, Luo Z et al (2018) Reduced graphene oxide functionalized nanofibrous silk fibroin matrices for engineering excitable tissues. NPG Asia Mater 10(10):982–994
48. He J, Liang Y, Shi M, Guo B (2020) Anti-oxidant electroactive and antibacterial nanofibrous wound dressings based on poly(ε-caprolactone)/quaternized chitosan-graft-polyaniline for full-thickness skin wound healing. Chem Eng J 1:385
49. Dvir T, Timko BP, Kohane DS, Langer R (2011) Nanotechnological strategies for engineering complex tissues. Nat Nanotechnol 6(1):13–22
50. Zonari A, Martins TMM, Paula ACC, Boeloni JN, Novikoff S, Marques AP et al (2015) Polyhydroxybutyrate-co-hydroxyvalerate structures loaded with adipose stem cells promote skin healing with reduced scarring. Acta Biomater 15(17):170–181
51. Lutolf MP, Hubbell JA (2005) Synthetic biomaterials as instructive extracellular microenvironments for morphogenesis in tissue engineering. Nat Biotechnol 23(1):47–55

Nanotechnology in Energy Supplements: Sustained Endurance

Elsa Cherian, Rosamma, K. A. Anju, and Bijina Naushad

Abstract Nanotechnology application in energy supplements shows promising potential for improved sustained endurance. Nanotechnology contributes to the creation of novel formulations that increase bioavailability and enable controlled release of energy-providing substances. Nano-encapsulated nutrients and lipid-based nanoparticles allow for targeted and extended delivery of energy resources, helping to maintain stable energy levels over long durations. Nanotechnology principles are examined as they apply to energy supplementation, explaining how nano-sized particles function within biological systems to enhance nutrient delivery. Certain nano-delivery systems significantly boost the efficacy of ingredients like caffeine, amino acids, and B vitamins. Case studies highlight the considerable improvements in endurance and recovery observed in athletes using nano-formulated supplements compared to traditional products. Potential challenges related to regulatory concerns, consumer safety, and long-term health effects of nanomaterials are also addressed. Looking toward future trends, nanotechnology is anticipated to redefine energy supplements, offering enhanced endurance for a broad spectrum of users.

Keywords Nanotechnology · Energy supplements · Sustained endurance · Bioavailability · Controlled release · Nano-encapsulation · Lipid-based nanoparticles · Targeted delivery · Nutrient absorption

1 Introduction

The present world demands for more sophisticated healthy performance among sports persons long with high pressure working environment. This leads to a opportunities of more research in the area of nutritional supplements. This doesn't mean that

---

E. Cherian (✉) · Rosamma · K. A. Anju
Department of Food Technology, Saintgits College of Engineering, Kottayam, Kerala, India
e-mail: elsa.c@saintgits.org

B. Naushad
Applied Food Safety and Quality Management, University of Greenwich, London, England

A. Chatterjee et al. (eds.), *Nanofuel: The Future of Sports Nutrition*, Smart Nanomaterials Technology, https://doi.org/10.1007/978-981-96-5471-0_8

energy supplements are novel. Even in the traditional life also energy supplements played a major role. But the major disadvantages include slow absorption rates, poor bioavailability and varying rate of ingredients delivery. This situation force to have more intensive research and development for the increment in the efficacy of supplements which paved the way for nanotechnology as a revolutionary solution in this area.

Nanotechnology is considered as an interdisciplinary science that connects several disciplines such as biology, mechanical, chemical etc. at the micro level including atomic, molecular or supramolecular levels [1]. This technology includes the study of any material with a minimum one length dimension of 1–100 nm size. This reduction in size results in considerable variation in physicochemical properties which cause more advanced features of macroscale materials [2].

By the utilization of nanoscale carriers, such as liposomes and polymeric nanoparticles, experts can immobilize bioactive compounds and then deliver it directly to target sites within the body thereby enhanced precision. This type of delivery mechanism confirms better absorption and bioavailability and also supporting the reduction in required dosage, thereby lessening potential side effects.

In the framework of energy supplements, nanotechnology keeps its own position by enabling the sustained and precise release of active ingredients like caffeine, creatine etc. The sustained release of supplements provides a smooth pace of delivery avoiding spikes in the energy levels. Even the quality and activity of the supplements can also be retained with the utilization of nano particles. The protection and release of supplements utilizing nanocarriers can increase the efficiency of sensitive products from environmental conditions like heat, light etc. thereby preserving it till it reaches the correct site of action.

This chapter mainly explores the interrelation and dependence between nanotechnology and energy supplements, opening up path to the mechanisms of nanocarrier systems, its role in enhancing bioavailability, and their potential in sustaining endurance. Further recent advancements, practical applications, and the challenges are also discussed.

2 Mechanisms of Nanotechnology in Energy Supplements

Nanotechnology employs various mechanisms, including focused, controlled, and sustained delivery of nutrients, to enhance their bioavailability.

(1) **Improved Bioavailability of Nutrients**

The bioavailability of compounds in traditional energy supplements is often very low due to decreased dispersibility and solubility. These issues can be reduced by using nutrients in nanoscale thereby incorporating nanotechnology in energy supplements. Nanocarriers help in targeted delivery thereby increasing bioavailability and absorption [3]. One of the research studies have shown that curcumin-encapsulated nanoparticles have proven to have significantly better bioavailability and absorption when compared to conventional formulations.

(2) **Targeted Delivery of Nutrients**

Nanotechnology in energy supplements enables the targeted delivery of nutrients to specific organs, cells, or tissues, ensuring maximum nutrient supply with minimal wastage. This is achieved by designing nanoparticles that bind specifically to receptors on target cells. Additionally, nanoparticles release nutrients in a controlled manner to specific tissues, such as muscle tissue during contraction and physical activity [4].

(3) **Sustained Delivery of Nutrients**

Another mechanism by which nanotechnology supports nutrient delivery is by raising the sustained delivery mechanism. It supports sustained and consistent nutrient release in targeted tissues. Nanoparticles in polymeric or nanogel forms can enhance a gradual release of nutrients, thus maintaining consistent release ofenergy levels over time. Thus current technology can efficiently be useful for athletes and sports professionals, as it largely avoid energy depletion and thus causing prolonged physical activity [5].

(4) **Enhanced Solubility and Stability of Nutrients**

Considering the chemical nature of many active nutrients in supplements, it was found to be hydrophobic. This nature can reduce their solubility and absorption in target cells. Thus it is proved that the conversion or reduction of nutrients into nanoscale forms can enhance their solubility and stability, thereby increasing absorption. In one of the study it was proved that, incorporating omega-3 fatty acids, an essential nutrient, into nanoemulsions, improved stability and solubility, making it a valuable addition to energy supplements [6].

3 Key Ingredients and Nanotechnological Approaches

3.1 Types of Lipids Used in Nanoparticle Fabrication

Natural Lipids: Natural lipids, like glycerophospholipids, are the building blocks of the bilayer structure of cell membranes. These comprise of glycerol, two fatty acids, and a phosphate group, which is often attached to choline. The mail sources include egg yolk and soybeans. There are mainly six types based on their polar head groups, which include phosphatidylcholine and phosphatidylserine. Because of the presence of unsaturated fatty acids, natural phospholipids are not as stable as compared with synthetic ones [7, 8].

Synthetic Lipids: Synthetic lipids are produced by the chemical modifications of the naturally existing phospholipids and can be shaped with a customized structure. Examples of such lipids are PEGylated phospholipids and cationic lipids, which ease drug delivery and cellular interaction. The modifications cause rise in stability, bioavailability, and cell adhesion, thus making them as perfect candidates for biomedical applications [9].

Steroids: Steroids, such as cholesterol, are hydrophobic lipids that augment membrane stability and rigidity in nanoparticles. Classically comprising less than 30% of lipid content, steroids lessen membrane fluidity, change zeta potential, and maintain phase transitions, enhancing liposome performance.

3.2 Natural Bioactives in Lipid Nanocarriers

Curcumin, extracted from *Curcuma longa*, prvide numerous health benefits, such as anti-inflammatory, antioxidant, and anti-cancer properties. However, its poor water solubility and bioavailability cause reduction in the usage. To advance these, lipid-based formulations have been developed [10]. Nanocurcumin, made with phospholipids and cholesterol, proved to have high encapsulation efficiency and better stability and release compared to curcumin extract [11]. Other formulations, such as lipid nanoparticles and SLNs, increase bioavailability (up to 155 times) and enhance transdermal penetration. These advances provide promising solutions to improve curcumin's therapeutic effectiveness [12].

Quercetin, a polyphenol present in foods like apples and onions, has the properties such as antioxidant, anti-inflammatory, and cardio-protective nature [13]. To improve its bioavailability and efficacy, polymeric nanocarriers have been created [14]. Quercetin-loaded solid lipid nanoparticles (SLNs), both coated and uncoated, showed enhanced penetration and therapeutic effects for bladder carcinoma treatment [15]. Cholate-modified polymer-lipid hybrid nanoparticles enhanced quercetin bioavailability by 3.7 times [16]. Mesoporous silica nanoparticles and PLGA nanoparticles with hyaluronic acid upgraded its dissolution rate, bioavailability, and colloidal stability [17, 18]. These formulations give promising solutions to optimize quercetin's medicinal potential.

Lycopene, a powerful antioxidant which are found in watermelon, grapefruit, and tomato, has anti-inflammatory and anti-tumor properties but has poor solubility and stability [19, 20]. Studies have proved that lycopene-loaded NLCs with Plantacare 1200 enhanced stability ($t_{1/2} = 192.5$ h) and antioxidant activity, while rice bran oil improved stability further [21].

β-Carotene, a tetraterpenoid essential for photosynthesis, provides antioxidant, anticancer, and immune-boosting benefits but faces problems due to poor solubility and bioavailability [22]. Strategies to improve its stability include integrating vitamin E into NLCs, which protects β-carotene for up to 19 days [23], using high-melting surfactants like lecithin in LNPs to enhance stability [24], and utilizing sodium caseinate-based SLNs to improve stability and cellular uptake [25]. Additionally, Tween 40-maintained nanoemulsions (NEs) proved to have antioxidative effects, while whey protein and soy lecithin enhance NE stability and show antiproliferative effects against breast cancer cells [26–28].

4 Recent Advances

Nanotechnology has found its way into the field of energy supplements improving its efficiency and functionality. Scientists have been manipulating materials at the nanoscale to enhance the bioavailability, absorption and controlled release of energy-releasing compounds. It improves the precise delivery of nutrients like caffeine, amino acids and nutrients. This intersection of advanced technology and nutrition is reshaping the way energy supplements are designed, consumed, and experienced, paving the way for more efficient and personalized solutions in endurance and performance enhancement.

4.1 Nanoemulsions

Emulsions can be defined as a mixture of two immiscible fluids. Emulsions with droplet size in the nanoscale are called nanoemulsions. They are produced by applying pressure or energy externally [29, 30]. They are kinetically stable systems of multiphase colloidal dispersion. The food industry uses nanoemulsions to encapsulate bioactive molecules in capsules having a diameter in the nanoscale. Bioactive molecules in such capsules have enhanced spreading, uniform distribution in targeted locations and improved penetration [31].

Encapsulating hydrophobic ingredients in nanoemulsions enhances their solubility and dispersion. Energy supplements contain a variety of active ingredients like caffeine, coenzyme Q10, L-carnitine, taurine and electrolytes. Delivering these active ingredients through nanoemulsions enhances the stability, efficiency and bioavailability of these ingredients. The improved physical stability of nanoemulsions reduces the chances of degradation of active ingredients during storage or use [32].

4.2 Nanostructured Lipid Carriers (NLC)

NLCs are highly stable matrices created by blending solid and liquid lipids which is used for encapsulating active ingredients. Lipid carriers used traditionally have limited loading capacity and stability. NLCs having unique structures help researchers and scientists to overcome these challenges [33]. Hydrophilic and lipophilic active ingredients can be encapsulated into a stable matrix thus protecting them from environmental degradation due to heat, light or oxidation. They also ensure the consistent delivery of energy over a long period of time, facilitated by the targeted and controlled release of these active ingredients. Thus, NLCs containing active ingredients in energy supplements help to maintain endurance for athletes and prevent sudden energy crashes.

NLCs can also be made by blending natural lipids which are safer for human consumption. Scientific advances have enabled the creation of uniform particles with high encapsulation efficiency using high-pressure homogenisation. These advancements have increased their application in the formulation of energy supplements [34].

4.3 *Nanofibres*

Nanofibres are ultrathin fibres of the nanometer scale, typically less than 1000 nm. They are produced by electrospinning, a simple, cost-effective technique for employing electrostatic forces to produce polymer fibres. These fibres have a high surface-area-to-volume ratio and enhanced porosity.

Bioactive molecules when encapsulated in nanofibers are protected from extreme environmental conditions like heat and moisture and also mask undesirable flavours. They can be designed so that fast-dissolving strips of nutrients for oral delivery dissolve rapidly in saliva and subsequently into the bloodstream [35]. They can also be engineered to release nutrients gradually, ensuring a consistent energy supply throughout the day. They also allow the production of a synergistic mix of energy-boosting compounds which can be incorporated as a single supplement.

Challenges and Limitations

- Nanoparticles can penetrate biological barriers and accumulate in the body. This can lead to gene mutations and have carcinogenic effects.
- Scaling up the production strategies used for nanoparticles from a laboratory scale without compromising their quality and efficiency is challenging.
- Processing equipment and raw materials used for nanotechnology is expensive which increases the production cost.
- The regulatory framework for nanotechnology in food products is uncertain complicating the production process for manufacturers.
- The chemical processes involved in the production process of nanoparticles generate waste and by-products that create environmental concerns [30].

5 Conclusion

Thus, nanotechnology is proved to be a transformative technology in the field of energy supplements, supporting unprecedented precision and efficiency in nutrient delivery. This novel technology highly helps to overcome the limitations of traditional supplement formulations thus enabling the creation of supplements that are not only more effective but also safer and more sustainable. Even the usage of nanocarriers support to have significant enhancements in the sustained release of active compounds. This release strategy supports availability of energy even for prolonged hours making the availability of energy even in crucial hours of performance and recovery. Even though the technology has got many advantages, it too has concerns on safety and long-term effects of nanomaterials, regulatory approvals, and scalability. All these concerns have to be must be addressed through laborious research and proper communication with consumers.

References

1. de Francisco EV, García-Estepa RM (2018) Nanotechnology in the agrofood industry. J Food Eng 238:1–11
2. Chellaram C, Murugaboopathi G, John AA, Sivakumar R, Ganesan S, Krithika S, Priya G (2014) Significance of nanotechnology in food industry. APCBEE Proc 8:109–113
3. Basu B, Garala KK, Patel R, Dutta A, Ash D, Prajapati B, Singh S, Jha SK (2024) Advanced targeted drug delivery of bioactive nanomaterials in the management of cancer. Curr Med Chem 2024. [Online ahead of print]. https://doi.org/10.2174/0109298673283362231220115050
4. Zhang R, Zhang Z, McClements DJ (2020) Nanoemulsions: an emerging platform for increasing the efficacy of nutraceuticals in foods. Colloids Surf, B 194:111202. https://doi.org/10.1016/j.colsurfb.2020.111202
5. Patra JK, Das G, Fraceto LF et al (2018) Nano-based drug delivery systems: recent developments and future prospects. J Nanobiotechnol 16, Article 71. https://doi.org/10.1186/s12951-018-0392-8
6. Arshad R, Gulshad L, Haq I-U et al (2021) Nanotechnology: a novel tool to enhance the bioavailability of micronutrients. Food Sci Nutr 9(6):3354–3361. https://doi.org/10.1002/fsn3.2311
7. Abdalla AM, Hossain S, Azad AT et al (2018) Nanomaterials for solid oxide fuel cells: a review. Renew Sustain Energy Rev 82:353–368
8. van der Veen JN, Kennelly JP, Wan S et al (2017) The critical role of phosphatidylcholine and phosphatidylethanolamine metabolism in health and disease, Biochimica et Biophysica Acta (BBA)-Biomembranes 1859:1558–1572
9. Carmona-Ribeiro AM (2003) Bilayer-forming synthetic lipids: drugs or carriers? Curr Med Chem 10:2425–2446
10. Flora G, Gupta D, Tiwari A (2013) Nanocurcumin: a promising therapeutic advancement over native curcumin. Crit Rev Ther Drug Carrier Syst 30:331–368
11. Chittasupho C, Srisawad K et al (2023) Targeting spike glycoprotein S1 mediated by NLRP3 inflammasome machinery. Pharmaceuticals 16:862
12. Vandita K, Shashi B, Santosh KG, Pal KI (2012) Enhanced apoptotic effect of curcumin-loaded solid lipid nanoparticles. Mol Pharm 9:3411–3421
13. Eichenseher J (2018) Peptic ulcer disease, integrative medicine, 4th edn, Chap 43. Elsevier, Philadelphia, PA, USA
14. Allgrove JE, Davison G (2018) Chocolate/cocoa polyphenols and oxidative stress, polyphenols: mechanisms of action in human health and disease. Elsevier, Hoboken, NJ, USA, pp 207–219
15. Shawky S, Makled S et al (2022) Quercetin-loaded cationic solid lipid nanoparticles. Pharmaceutics 14:2527
16. Yin J, Hou Y, Song X et al (2019) Cholate-modified polymer-lipid hybrid nanoparticles for oral delivery of quercetin. Int J Nanomed 14:4045–4057
17. Elmowafy M, Alruwaili NK et al (2022) Quercetin-loaded mesoporous silica nanoparticle-based lyophilized tablets. AAPS PharmSciTech 24:6
18. Chittasupho C, Junmahasathien T et al (2021) Suppression of intracellular reactive oxygen species. Pharmaceuticals 14:679

19. Stahl W, Heinrich U et al (2006) Lycopene-rich products and dietary photoprotection. Photochem Photobiol Sci 5:238–242
20. Shi J, Maguer M, Bryan M, Kakuda Y (2003) Kinetics of lycopene degradation in tomato puree. J Food Process Eng 25:485–498
21. Riangjanapatee P, Okonogi S (2012) Effect of surfactant on lycopene-loaded nanostructured lipid carriers. Drug Discov Therapeutics 6:163–168
22. Bendich A, Olson JA (1989) Biological actions of carotenoids. FASEB J 3:1927–1932
23. Hentschel A, Gramdorf S, Müller R, Kurz T (2008) β-Carotene-loaded nanostructured lipid carriers. J Food Sci 73:N1–N6
24. Helgason T, Awad TS, Kristbergsson K et al (2009) Impact of surfactant properties on oxidative stability of β-carotene encapsulated within solid lipid nanoparticles. J Agric Food Chem 57:8033–8040
25. Yi J, Lam TI, Yokoyama W, Cheng LW, Zhong F (2014) Cellular uptake of β-carotene from protein stabilized solid lipid nanoparticles. J Agric Food Chem 62:1096–1104
26. Jimenez-Escobar M, Pascual-Mathey L et al (2020) In vitro and in vivo antioxidant properties of paprika carotenoids nanoemulsions. LWT 118:108694
27. Flores-Andrade E, Allende-Baltazar Z et al (2021) Carotenoid nanoemulsions stabilized by natural emulsifiers: whey protein, gum Arabic, and soy lecithin. J Food Eng 290:110208
28. Hsu H-Y, Chen B-H (2022) A comparative study on inhibition of breast cancer cells by carotenoid extract and nanoemulsion. Pharmaceutics 14:980
29. Sneha K, Kumar A (2022) Nanoemulsions: techniques for the preparation and recent advances in food applications. Innov Food Sci Emerg Technol 76:102914
30. Kumar P, Mahajan P, Kaur R, Gautam S (2020) Nanotechnology and its challenges in the food sector: a review. Mater Today Chem 17:100332
31. Priyakshree Borthakur, Boruah PK, Sharma B, Das MR (2016) Nanoemulsion: preparation and its application in the food industry, pp 153–191
32. McClements DJ, Xiao H (2012) Potential biological fate of ingested nanoemulsions: Influence of particle characteristics. Food Function 3(3):202–220
33. Pardeike J, Hommoss A, Müller RH (2009) Lipid nanoparticles (SLN, NLC) in cosmetic and pharmaceutical dermal products. Int J Pharm 366(1–2):170–184
34. Jafarifar Z et al (2022) Preparation and characterization of nanostructured lipid carrier (NLC) and nanoemulsion containing vitamin D3. Appl Biochem Biotechnol 194(2):914–929
35. Maria Leena M, Vimala Bharathi SK, Moses JA, Anandharamakrishnan C (2019) Potential applications of nanofibers in the beverage industry. Academic Press, pp 333–368

Targeted Delivery: Precision Nutrition for Athletes

Nizwa Itrat, Beenish Israr, Javeria Shabbir, Faiza Ameen, and Suad Ajab

Abstract Precision nutrition for athletes represents a cutting approach to optimizing performance and health through individualized dietary strategies. This concept leverages genomics and technology advancement to tailor nutritional intervention based on unique biological and environmental factors. Integrating data from various sources including genetic profiles, metabolic response, and real-time physiological monitoring and nutrition aims to enhance training adaptation recovery and overall athletic performance. This targeted delivery system not only addresses the specific nutritional needs of athletes but also mitigates the risk of nutrient deficiency and over-training. Future research should focus on refining these personalized strategies through well-power clinical trials and the research should focus on refining their customized strategies. This target delivery system not only addresses the specific nutritional needs of athletes but also mitigates the risk of nutrient deficiency and over-training. Further research should focus on refining these personalized strategies through well-powered clinical trials and validating biomarkers to establish robust dietary guidelines for athletes across different sports disciplines.

Keywords Precision nutrition · Athletes · Personalized nutrition · Sports dietary guidelines

N. Itrat (✉) · B. Israr
Faculty of Food Nutrition and Home Sciences, Institute of Home Science, University of Agriculture, Faisalabad, Pakistan
e-mail: Nizwa.itrat@uaf.edu.pk

J. Shabbir
Department of Nutrition and Dietetics, The University of Faisalabad, Faisalabad, Pakistan

F. Ameen
British School of Leadership and Management, Glasgow Caledonian University, London Campus, London, England

S. Ajab
Institute of Public Health, College of Medicine and Health Sciences, United Arab Emirates University, Al Ain, Abu Dhabi, United Arab Emirates

A. Chatterjee et al. (eds.), *Nanofuel: The Future of Sports Nutrition*, Smart Nanomaterials Technology, https://doi.org/10.1007/978-981-96-5471-0_9

1 Introduction

Precision nutrition for athletes is centered on customizing dietary components to maximize performance, recuperation, and general health according to personal requirements [1]. The method uses environmental, ecological, and genetic elements to develop a customized diet plan. Recent developments in the discipline include the use of wearable technology, genomics, proteomics, metabolomics, and other omic technologies to track and modify dietary consumption in real time [2]. To provide more accurate dietary recommendations, the technologies assist in understanding how different athletes react to different nutrients. For instance, endurance athletes can benefit from Persian nutrition by optimizing their intake of carbohydrates to calculate their training and recovery needs, and athletes can also benefit from focus on protein timing and quality to improve muscle growth and repair [3].

When it comes to delivering nutrients, high-intensity exercise can present a number of difficulties. Among the primary problems are gastrointestinal upset, dehydration, and the sequence of nutritional consumption. Diarrhea, cramping, and nausea are symptoms of digestive disorders that can impair the absorption of nutrients [4]. Athletes ought to experiment with different diets and meal time to control them. Reduced performance and impaired nutrition absorption can result from dehydration. Before, during, and after workout, it's important to be properly hydrated by consuming water and electrolyte-rich drinks [5].

Athletes are susceptible to shortages in essential nutrients such as calcium, iron, and vitamin D, making nutrition timing another difficulty. Deficiencies can be avoided with a balanced diet that includes foods high in nutrients and regular monitoring of levels of nutrients [6]. Another problem is individual variability, as every athlete's body reacts to nutrition differently because of variations in metabolism and genetics. Often created in collaboration with sports nutritionists, customized diet regimens can meet specific demands and enhance performance [7].

2 Understanding Athlete Nutritional Needs

2.1 Caloric Need

Athletes need to consume enough calories daily to maintain weight and body composition goals. They consume food, fluids, and supplements for energy expenditure. Adolescent athletes' calorie requirements depend on gender, age, and physical activity level. On average, adolescents spend 10–25% more energy than adults in certain sports. Calorie needs are based on estimated energy needs (EER) by age category [8].

2.2 Carbohydrates

Due to their ability to be processed both aerobically and anaerobically, carbohydrates are essential for athletic performance. They are necessary for blood glucose and muscle glycogen, two important energy sources for working muscles. Consuming a lot of carbohydrates promotes recuperation and maximizes glycogen storage for later training sessions. The amount of carbohydrates needed depends on the volume and intensity of training, and meals with a glycemic index between low and moderate are advised. However, it is OK to include concentrated, nutrient-dense food sources of CHO during demanding and strenuous training sessions [9].

Foods containing carbohydrates are categorized by the glycemic index according to how they affect blood glucose levels. Low pre-exercise glycogen levels can lead to early exhaustion, decreased training intensity, glycogenolysis, impaired muscular contraction, depleted muscle glycogen, and protein breakdown. Maintaining necessary glycogen reserves is required for endurance exercises of moderate to high intensity. Depending on the length and intensity of their workout, athletes should ingest enough carbs to optimize their internal glycogen stores [10].

Athletes often struggle to meet recommended energy and carbohydrate levels, making it crucial to focus on restoring glycogen reserves for optimal performance. Exercise duration, intensity, and glycogen loss all affect the synthesis of glycogen throughout the exercise. While CHO supplementation at 1.0 g/kg/h, administered at regular intervals, is more successful in maintaining raised blood glucose levels, recovery meals within two hours of exercise can improve recovery [11].

Studies on the rates of glycogen resynthesis following varied carbohydrate intake after exercise, however, have shown contradictory findings. According to available data, maximum glycogen resynthesis can be attained by consuming 1.0 to 1.5 g of carbs per kilogram per hour. 3–12 g of carbs per kg of body weight is the recommended daily consumption; the exact quantity depends on the length and intensity of physical exercise. To replicate a standard training regimen and examine the long-term effects of post-exercise nutrition, more study is required [12].

2.3 Proteins

For improved muscularity, recuperation, and athletic performance, skeletal muscle mass is essential. Because skeletal muscle is constantly regulated, the equilibrium between muscle protein production and breakdown is particularly crucial for older persons. The main factors influencing changes in muscle mass over time are shifts in the rates at which muscle proteins are synthesized. Protein consumption, including protein and essential amino acids, is essential for muscle remodeling after training. A protein source's anabolic efficacy depends on a number of variables, including the amounts of amino acids, digestibility, rate of digestion, and absorption kinetics. Proteins derived from plants are incomplete because they lack some necessary amino acids, but animal protein is said to be complete since it contains all the necessary amino acids [9].

Plant sources typically have biological values between 56 and 74, while animal sources have a spectrum of 77–104 on theoretical scales. Protein Digestibility Corrected Amino Acid scores are a common benchmark for assessing protein quality, with animal sources scoring higher than plant sources. Plant protein sources typically display values below 100, with soy protein having a score of 100. The DIAAS approach reveals that animal sources often exceed 100, while most plant sources are below this threshold. Plant-based protein sources often lack adequate amounts of specific amino acids and exhibit reduced levels of essential amino acids. Digestibility rates from plant sources are generally lower than those from animal proteins, affecting the postprandial protein synthesis response [4].

Before, during, and after exercise, protein consumption can support muscle remodeling, repair, and post-exercise strength and growth. Improved anabolic hormone status, less muscle injury, increased muscle cross-sectional area, and a longer time to fatigue are just a few advantages of consuming carbs together with protein or necessary amino acids throughout endurance and resistance exercise. Rehydration and simple carbohydrate consumption are essential during activity since protein is restricted for energy. Each meal should contain 0.31 g/kg of high-quality, readily digested protein, such as whey protein, to promote myofibrillar protein synthesis and reduce amino acid breakdown. For muscle glycogen production, a mixed carbohydrate-protein ratio of 4 to 1 is advised [13].

2.4 Fats

To perform at their best, athletes must ingest enough energy both during activity and during recuperation. Carbohydrate catabolism may happen with or without oxygen, whereas fat oxidation requires it. It is advised to ingest 20–35% of calories from fat in the diet, with less than 10% coming from saturated fat. Dietary supplements are frequently used by athletes to increase metabolic capacity, postpone tiredness, promote muscular growth, and speed up recovery times. An important part of the body's inflammatory response, omega-3 polyunsaturated fatty acids (PUFAs) have been associated with reducing oxidative stress from exercise, enhancing anaerobic endurance capacity, enhancing oxygen efficiency during aerobic exercise, promoting skeletal muscle health, and delaying muscle soreness. In addition to improving fat utilization, a five-day high-fat, low-carb, and low-protein diet enabled athletes to engage in high-intensity, high-volume exercise. Nevertheless, later research failed to demonstrate gains in exercise performance [9].

The impact of high-fat diets on athletic performance has been investigated, however the results vary depending on the individual and the dietary circumstances. Johnson's study could not find any conclusive evidence to support the idea that performance is negatively impacted by lowering intramuscular triglycerides. By improving fatty acid transport and beta-oxidation and increasing the enzymes involved in fatty acid oxidation, a diet high in fats encourages fat oxidation. Exercise performance gains, however, were not always seen, and in several situations, the opposite was seen. Due to their detrimental effects on performance, decreased utilization of carbohydrates, and gastrointestinal discomfort, high-fat diets are generally not advised [14].

2.5 *Micronutrients*

Vitamins, minerals, and other micronutrients are necessary for maintaining life as well as promoting development, well-being, and reproduction. They are divided into water-soluble vitamins (B and C) and fat-soluble vitamins (A, D, E, and K) according to their solubility. Non-organic substances called minerals support physiological processes; macro-minerals need around 100 mg daily, while trace elements need about 20 mg. The validity of these recommendations for athletes is up for discussion, although a nutritionally balanced diet often supplies critical micronutrients in prescribed quantities for routine body functioning [10].

Micronutrient supplements are often used by athletes; over half of them use them in their routine. Athletes' micronutrient requirements may also be influenced by variables such as exercise duration, intensity, and personality, according to research. There is a chance that micronutrient requirements may rise in sports with high energy demands, although accurate measurement is still difficult. When athletes find it difficult to achieve their micronutrient demands through food alone, supplements may be helpful. Nevertheless, micronutrients do not have ergogenic effects, and micronutrient shortages can be caused by low-energy diets, vegetarianism, sickness, and injury [15].

2.6 *Hydration*

For athletes, fluid loss before, during, and after exercise is critical since it can impact cognitive function and performance. Hypohydration, which can be fatal, can occur when the body loses 2% of its fluids. Hyponatremia from excessive fluid intake during endurance exercise might cause dilutional hyponatremia. Dry skin, black urine, fast weight loss, weakness, exhaustion, headache, sunken eyes, cramping in the muscles, a racing pulse, higher injury rates, and trouble recovering are all signs of dehydration. Weight gain, low blood sugar, weakness, nausea, vomiting, hand and foot edema, orthopnea, exhaustion, headaches, seizures, and increased urination are all signs of overhydration. A pre-training routine should include drinking 200–300 mL of water or a sports drink to ensure proper hydration. Ten to twenty minutes prior to working out [16].

Sports drinks improve performance and endurance by rehydrating and replenishing vital electrolytes. The beverage hydration index model compares a drink's capacity to hydrate to that of resting plain water. It makes the assumption that a drink that increases diuresis causes the body to retain less fluid, as shown by a hydration index that is less than 1.0. The model is strong in repeatability and has been repeated by other research groups. Although adding electrolytes to water improves fluid retention, different studies have different minimum salt requirements. While liquids with higher sodium content, such as Pedialyte, often have a higher hydration index, sports drinks with 20 mmol sodium may not have a noticeably higher hydration index than water [17].

3 Targeted Delivery Systems

3.1 *Nanotechnology: Improving Bioavailability and Absorption of Nutrients*

Using nanoparticles to enhance nutrient delivery and absorption is known as nanotechnology. By encasing vitamins, minerals, and various other bioactive substances, these microscopic particles can prevent their deterioration and increase their bioavailability. For instance, omega-3 fatty acids, antioxidants, and other vital nutrients may be delivered more effectively and absorbed by the body more efficiently with the use of nanoencapsulation [18].

3.2 *Time-Release Formulations: Guaranteeing Long-Term Nutrient Provision*

The purpose of time-release formulations, sometimes referred to as sustained-release or extended-release supplements, is to release nutrients gradually over a longer period of time. By keeping circulatory nutrient levels constant, this method lessens the need for regular doses and increases the nutrients' overall efficacy. For example, sustained-release caffeine can give you a consistent energy boost all day long without the crashes and spikes that come with immediate-release types [19].

3.3 *Customized Nutrition Plans: Adapting Diets to Meet Specific Requirements and Objectives*

A person's lifestyle, genetic profile, and particular health objectives are taken into consideration when creating a personalized nutrition plan. This method, called nutrigenomics, takes into account how different individuals react to different nutrients and dietary regimens. Athletes may maximize their performance, recuperation, and general health by customizing their meals. For instance, genetic testing can show how certain nutrients are metabolized by athletes, enabling personalized nutritional advice to improve their distinct metabolic profile [20].

4 Athletes' Precision Nutrition Approaches

4.1 Pre-workout Nutrition: Enhancing Performance and Energy Levels

- Carbohydrates: To receive rapid energy before working out, eat readily digested carbohydrates like fruits, whole grains, or starchy greens two to three hours beforehand.
- Proteins: To aid in maintaining muscle mass, consume a reasonable quantity of protein. Plant-based proteins, dairy products, and lean meats are all excellent sources.
- Hydration: Make sure you're properly hydrated by consuming a lot of water both before and throughout your workout [21].

4.2 Intra-workout Nutrition: Sustaining Hydration and Energy While Working Out

- Carbohydrates: Simple carbohydrates can assist sustain energy levels throughout extended activity. Gels or sports drinks are practical choices.
- Electrolytes: necessary to keep fluid balance and avoid cramping. Seek for beverages with magnesium, potassium, and sodium.
- Hydration: To stay well-hydrated, especially during vigorous or prolonged exercise, keep drinking water or electrolyte drinks [22].

4.3 Post-workout Recovery: Increasing Muscle Replenishment and Repair

- **Proteins**: Essential for both muscle development and repair. Within 30 min after working out, try to consume high-quality protein sources such as chicken, lentils, or whey protein.
- **Carbohydrates**: Aid in restocking glycogen reserves. For the best recovery, mix protein and carbohydrates, such as a fruit and yogurt smoothie.
- **Hydration**: After a really strenuous workout, rehydrate with water and think about adding electrolytes [23].

4.4 *Daily Nutritional Assistance: Fulfilling Continuous Nutritional Requirements for Exercise and Recuperation*

- **Balanced Diet**: To promote general health and performance, make sure your diet includes a variety of fats, proteins, and carbs. Consume a range of whole grains, lean meats, fruits, and veggies.
- **Micronutrients**: Vitamins and minerals are necessary for healthy bones, energy production, and immune system performance. To satisfy these demands, concentrate on nutrient-dense meals.
- **Consistent Hydration**: Drink water all day long, not just when doing out. This promotes the general performance and functioning of the body [24].

5 Case Studies and Applications

Sports nutrition is essential for improving performance, recuperation, and general health of athletes. Depending on their sport, level of training, and physiology, athletes have specific dietary requirements. Performance and recuperation can be greatly enhanced by customized nutrition plans that are based on scientific studies and technical developments (Fig. 1). Athletes and their support networks may attain peak performance and good health by comprehending and putting these strategies into practice [25]. Case studies and implementations of specific dietary approaches for athletes are examined below:

5.1 *Case Study: Endurance Athletes' Carbohydrate Loading*

- **Background**: Before long-distance cycling, triathlons, and marathons, endurance athletes frequently stock up on carbohydrates to increase their glycogen stores.
- **Case**: The effects of a carbohydrate-loading regimen were examined in research involving a group of marathon runners. For three days leading up to the marathon, the runners had meals heavy in carbohydrates (70–75% of total calories).
- **Results**: Compared to runners who did not follow the carbohydrate-loading regimen, those who did exhibited increased endurance performance and a considerable increase in glycogen reserves. The longer-lasting energy from the higher glycogen stores postponed the onset of weariness.
- **Application**: The significance of customized carbohydrate consumption plans for endurance athletes is shown by this case study. Similar procedures can be used by nutritionists and coaches to maximize performance in endurance competitions [12].

Fig. 1 Strategical applications to overcome athletic nutritional deficiencies

5.2 Case Study: Protein Consumption for Strength Athletes' Muscle Recovery

- **Background**: For strength athletes who participate in resistance training, an adequate protein intake is essential for both muscle development and repair.
- **Case**: In research, a group of bodybuilders were split into two groups. A modest amount of protein (1.2 g/kg of body weight) was consumed by one group, whereas 2.2 g/kg of body weight was consumed by the other.
- **Results**: The high-protein group recovered faster and increased greater muscle mass than the moderate-protein group. The study emphasized how protein helps to reduce muscle pain and promote muscular growth.
- **Application**: In order to promote muscle growth and recuperation, strength athletes can gain from consuming more protein. Based on the athlete's training intensity and objectives, customized nutrition regimens should take into account their unique protein needs [26]. Figure 1 describes the strategical applications to overcome athletic nutritional deficiencies

5.3 *Case Study: Team Sports Hydration Strategies*

- **Background**: Because physical effort and environmental variables can cause considerable fluid loss, players engaging in team sports must maintain proper hydration.
- **Case**: The effects of various hydration techniques were assessed in soccer player research. Two groups of players were formed; one group drank a specific electrolyte solution throughout practice and competition, while the other group adhered to a usual hydration regimen.
- **Results**: Compared to the standard hydration group, players who drank the electrolyte solution had improved hydration status, felt less tired, and had fewer muscular cramps. The electrolyte solution assisted in preserving fluid balance and replenishing lost minerals.
- **Application**: Especially in hot and muggy weather, team sport participants might profit from customized hydration plans that incorporate electrolyte solutions. Coaches should keep an eye on players' levels of hydration and modify their fluid intake as necessary [16, 17].

5.4 *Case Study: Female Athletes' Use of Micronutrient Supplements*

- **Background**: Due to dietary constraints and monthly blood loss, female athletes are more likely to experience vitamin deficiencies.
- **Case**: The effects of calcium and iron supplements were evaluated in research including female runners. One set of participants got daily calcium and iron supplements, while the other group got a placebo.
- **Results**: When compared to the placebo group, the supplemented group's iron levels and bone density significantly improved. While calcium supplementation promoted bone health and decreased the incidence of stress fractures, elevated iron levels helped avoid anemia.
- **Application**: To treat deficiencies, female athletes should think about routinely checking their micronutrient levels and using the right supplements. Individualized supplements strategies should be created by nutritionists to guarantee optimum performance and health [27].

5.5 *Case Study: Competitive Cyclists' Timing of Nutrient Intake*

- **Background**: Long races can be difficult for competitive cyclists to sustain their energy levels.

- **Case**: To investigate the effects of nutrition timing, a group of competitive cyclists took part in the study. Both during lengthy rides and right after training sessions, they ate a mix of proteins and carbs.
- **Results**: Cyclists who adhered to the nutrition timing schedule reported increased performance and sustained energy levels. Additionally, they recovered from training sessions more quickly.
- **Application**: The significance of nutrition scheduling for endurance athletes is illustrated by this case study. For cyclists along with other endurance athletes, coaches and dietitians may create precise timing schedules that maximize energy levels and recuperation [28].

5.6 *Case Study: CrossFit Athletes' Use of Antioxidants*

- **Background**: CrossFit and other high-intensity workouts can raise oxidative stress, which can cause muscle damage and slow recovery.
- **Case**: Two groups of CrossFit athletes were split up; one group was given antioxidant supplements (such as vitamins C and E), and the other group was given a placebo.
- **Results**: Compared to the placebo group, the antioxidant-supplemented group had faster recovery times, less muscular discomfort, and lower levels of oxidative stress indicators.
- **Application**: Antioxidant supplements can improve recovery and aid athletes who engage in high-intensity exercise to control oxidative stress [29].

5.7 *Case Study: Powerlifters' Use of Creatine Supplements*

- **Background**: To achieve their maximum lifts, powerlifters need rapid energy bursts, which can be improved by using creatine supplements.
- **Case**: For eight weeks, powerlifters received creatine supplements, and their performance was tracked.
- **Results**: When compared to the placebo group, the creatine-treated group demonstrated much higher maximum strength and power output.
- **Application**: For athletes who need explosive strength, such as powerlifters, creatine supplements can be a useful tactic [30].

5.8 *Case Study: Indoor Athletes' Use of Vitamin D Supplements*

- **Background**: Due to their frequent lack of exposure to sunshine, indoor sportsmen may be deficient in vitamin D.

- **Case**: Basketball players with low vitamin D levels were the subject of research. For six months, they received vitamin D pills.
- **Results**: Players who took vitamin D supplements performed better overall, had higher bone density, and saw fewer injuries.
- **Application**: To preserve bone health and avoid injuries, indoor athletes must take vitamin D supplements [31].

5.9 Case Study: Endurance Runners' Use of Omega-3 Supplements

- **Background**: Omega-3 fatty acids are well-known for their ability to reduce inflammation and speed up healing.
- **Case**: Over the course of 12 weeks, endurance runners who took omega-3 supplements had their performance and recuperation measures monitored.
- **Results**: During training, the supplemented group performed better overall, recovered more quickly, and had less muscular discomfort.
- **Application**: Omega-3 fatty acids can help endurance athletes reduce inflammation and enhance their recuperation [32].

5.10 Case Study: Triathletes' Use of Probiotics

- **Background**: Athletic performance may be impacted by gut health, which is crucial for general well-being.
- **Case**: For eight weeks, probiotics were taken daily by triathletes to examine the impact on their gut well-being and athletic performance.
- **Results**: Better food absorption, less gastrointestinal discomfort during races, and enhanced gut health were all displayed by the probiotic group.
- **Application**: Probiotics can improve performance and promote gut health, especially in endurance-based sports where GI issues are prevalent [33].

6 Future Prospects: Recent Advances and Upcoming Research in Athlete Targeted Delivery

- **Nanotechnology**: Developing methods for nanoencapsulation to improve nutritional absorption and bioavailability. By producing small, protected carriers, these methods improve the transport of nutrients to cells and prevent their destruction. In addition to increasing nutrition absorption efficiency, this invention enables tailored distribution, guaranteeing that the nutrients reach the tissues or organs that require them the most [18].

- **Machine learning and artificial intelligence (AI)**: Applying advanced algorithms to forecast each person's specific nutritional requirements and enhance dietary recommendations. Large-scale datasets from several sources, such as genetic data, lifestyle variables, and performance indicators, are analyzed using AI and machine learning. Personalized nutrition programs that can adjust in real-time to an athlete's changing training schedule, health, and surroundings are made possible by this study [34, 35].
- **Sustainable Practices**: Highlighting the use of biodegradable and environmentally friendly materials in the manufacturing of dietary supplements. This involves using renewable materials and creating packaging solutions that cut down on trash. In addition to helping the environment, sustainable business practices also appeal to the increasing number of customers who give sustainability priority when making purchases [36].
- **Smart Delivery Systems**: Developing wearable and implantable technologies that allow for real-time nutritional status monitoring of athletes and continuous nutrient administration. In addition to tracking biomarkers, wearable technology like smart patches and ingestible sensors can give real-time data on electrolyte balance, hydration levels, and other important parameters. Contrarily, implantable implants provide a more long-term way to administer nutrients straight into the bloodstream, guaranteeing a consistent supply during rigorous training and competition [37, 38].

Sports nutrition is positioned to make major progress in improving athletes' performance, recuperation, and general well-being by adopting these new developments and trends. A comprehensive approach to athlete care is represented by the incorporation of sustainability and technology into nutrition plans, opening the door for further advancements [19, 39].

7 Conclusion

With an emphasis on avoiding disordered eating, this chapter offers suggestions for meeting the nutritional requirements of the active population of athletes. It highlights how crucial it is to customize nutrition regimens to fit the various objectives, body types, metabolic rates, and dietary habits of athletes. In athletic nutrition, macronutrients—carbohydrates, proteins, and fats are essential because they promote energy generation and restore glycogen reserves. Healthy fats support hormone synthesis, long-lasting energy, and general health, while proteins are necessary for muscle growth and repair. For athletes to obtain enough micronutrients and lower their risk of nutritional deficiencies, it is essential to encourage a varied, nutrient-dense diet. Another important strategy for reducing disordered eating in the active population is staying hydrated. Athletes should emphasize whole meals to reduce the danger of disordered eating behaviors, even while sports nutrition supplements are accessible.

Highlighting how important athlete nutrition recommendations are in helping those who participate in sports and physical exercise make the best food choices.

References

1. Martín-Rodríguez A, Belinchón-deMiguel P, Rubio-Zarapuz A, Tornero-Aguilera JF, Martínez-Guardado I, Villanueva-Tobaldo CV et al (2024) Advances in understanding the interplay between dietary practices, body composition, and sports performance in athletes. Nutrients 16(4):571. https://doi.org/10.3390/nu16040571
2. Gonzalez DE, McAllister MJ, Waldman HS, Ferrando AA, Joyce J, Barringer ND et al (2022) International society of sports nutrition position stand: tactical athlete nutrition. J Int Soc Sports Nutr 19(1):267–315. https://doi.org/10.1080/15502783.2022.2086017
3. Cooper C, Heber D (2024) Precision nutrition in exercise and sports performance. In: Precision nutrition. Elsevier, pp 333–354. https://doi.org/10.1016/B978-0-443-15315-0.00004-3
4. Heaney S, O'Connor H, Naughton G, Gifford J (2008) Towards an understanding of the barriers to good nutrition for elite athletes. Int J Sports Sci Coach 3(3):391–401. https://doi.org/10.1260/174795408786238542
5. Exel J, Dabnichki P (2024) Precision sports science: what is next for data analytics for athlete performance and well-being optimization? Appl Sci 14(8):3361. https://doi.org/10.3390/app14083361
6. Jagim AR, Fields JB, Magee M, Kerksick C, Luedke J, Erickson J et al (2021) The influence of sports nutrition knowledge on body composition and perceptions of dietary requirements in collegiate athletes. Nutrients 13(7):2239. https://doi.org/10.3390/nu13072239
7. Kirk D, Catal C, Tekinerdogan B (2021) Precision nutrition: a systematic literature review. Comp Biol Med 133:104365. https://doi.org/10.1016/j.compbiomed.2021.104365
8. Iwasa-Madge K, Sesbreno E (2022) A proposed conceptual sport nutrition approach for athlete development and assessment: the athlete nutrition development approach. Sports Med Open 8(1):142. https://doi.org/10.1186/s40798-022-00532-w
9. Amawi A, AlKasasbeh W, Jaradat M, Almasri A, Alobaidi S, Hammad AA et al (2024) Athletes' nutritional demands: a narrative review of nutritional requirements. Front Nutr 10:1331854. https://doi.org/10.3389/fnut.2023.1331854
10. Bird SP, Rushton BD (2020) Nutritional knowledge of youth academy athletes. BMC Nutr 6:1–8. https://doi.org/10.1186/s40795-020-00360-9
11. Bentley MR, Patterson LB, Mitchell N, Backhouse SH (2021) Athlete perspectives on the enablers and barriers to nutritional adherence in high-performance sport. Psychol Sport Exerc 52:101831. https://doi.org/10.1016/j.psychsport.2020.101831
12. Burke LM, van Loon LJ, Hawley JA (2017) Postexercise muscle glycogen resynthesis in humans. J Appl Physiol. https://doi.org/10.1152/japplphysiol.00860.2016
13. Desbrow B (2021) Youth athlete development and nutrition. Sports Med Open 51(Suppl 1):3–12. https://doi.org/10.1007/s40279-021-01534-6
14. Janiczak A, Devlin BL, Forsyth A, Trakman GL (2022) A systematic review update of athletes' nutrition knowledge and association with dietary intake. Br J Nutr 128(6):1156–1169. https://doi.org/10.1017/S0007114521004311
15. Sasmarianto S, Henjilito R, Zulrafli Z, Kamarudin K, Nazirun N (2021) Understanding the needs of nutrition intake on athletes. J Sport Area 6(2):244–253. http://repository.uir.ac.id/id/eprint/21124
16. Burke L, Eichner E, Maughan R, Montain S, Sawka M, Stachenfeld N (2007) Exercise and fluid replacement. Am Coll Sports Med 39(2):377–390
17. Sawka M, Burke L, Eichner E, Maughan R, Montain S, Stachenfeld N (2007) Exercise and fluid replacement. American college of sports medicine position stand. Med Sci Sports Exerc 39(2):377–390

18. McClements DJ, Xiao H (2012) Potential biological fate of ingested nanoemulsions: influence of particle characteristics. Food Func 3(3):202–220. https://doi.org/10.1039/C1FO10193E
19. Jonvik KL, King M, Rollo I, Stellingwerff T, Pitsiladis Y (2022) New opportunities to advance the field of sports nutrition. Front Sports Active Living 4:852230. https://doi.org/10.3389/fspor.2022.852230
20. Shifflett B, Timm C, Kahanov L (2002) Understanding of athletes' nutritional needs among athletes, coaches, and athletic trainers. Res Q Exer Sport 73(3):357–362. https://doi.org/10.1080/02701367.2002.10609032
21. Firmansyah A, Prasetya MRA (2021) The nutrition needs of adolescent athletes: a systematic review. Jurnal SPORTIF: Jurnal Penelitian Pembelajaran 7(3):400–418. https://doi.org/10.29407/js_unpgri.v7i3.16716
22. Larson-Meyer DE, Woolf K, Burke L (2018) Assessment of nutrient status in athletes and the need for supplementation. Int J Sport Nutr Exer Metab 28(2):139–158. https://doi.org/10.1123/ijsnem.2017-0338
23. Heaney S, O'Connor H, Michael S, Gifford J, Naughton G (2011) Nutrition knowledge in athletes: a systematic review. Int J Sport Nutr Exer Metab 21(3):248–261. https://doi.org/10.1123/ijsnem.21.3.248
24. Smith JW, Holmes ME, McAllister MJ (2015) Nutritional considerations for performance in young athletes. J Sports Med 2015(1):734649. https://doi.org/10.4085/1062-6050-47.2.205
25. Torres-McGehee TM, Pritchett KL, Zippel D, Minton DM, Cellamare A, Sibilia M (2012) Sports nutrition knowledge among collegiate athletes, coaches, athletic trainers, and strength and conditioning specialists. J Athletic Train 47(2):205–211. https://doi.org/10.1155/2015/734649
26. Jäger R, Kerksick CM, Campbell BI, Cribb PJ, Wells SD, Skwiat TM et al (2017) International society of sports nutrition position stands: protein and exercise. J Int Soc Sports Nutr 14:1–25. https://doi.org/10.1186/s12970-017-0177-8
27. Areta JL, Elliott-Sale KJ (2022) Nutrition for female athletes: what we know, what we don't know, and why. Eur J Sport Sci 22(5):669–671. https://doi.org/10.1080/17461391.2022.2046176
28. Ivy J (2004) Nutrient timing: the future of sports nutrition. Basic Health Publications, Inc.
29. Sadowska-Krępa E, Domaszewski P, Pokora I, Żebrowska A, Gdańska A, Podgórski T (2019) Effects of medium-term green tea extract supplementation combined with CrossFit workout on blood antioxidant status and serum brain-derived neurotrophic factor in young men: a pilot study. J Int Soc Sports Nutr 16:1–9. https://doi.org/10.1186/s12970-019-0280-0
30. Kreider RB, Kalman DS, Antonio J, Ziegenfuss TN, Wildman R, Collins R et al (2017) International Society of Sports Nutrition position stand: safety and efficacy of creatine supplementation in exercise, sport, and medicine. J Int Soc Sports Nutr 14(1):18. https://doi.org/10.1186/s12970-017-0173-z
31. Holick MF (2004) Vitamin D: importance in the prevention of cancers, type 1 diabetes, heart disease, and osteoporosis. Am J Clin Nutr 79(3):362–371. https://doi.org/10.1093/ajcn/79.3.362
32. Philpott JD, Witard OC, Galloway SD (2019) Applications of omega-3 polyunsaturated fatty acid supplementation for sports performance. Res Sports Med 27(2):219–237. https://doi.org/10.1080/15438627.2018.1550401
33. West NP, Horn PL, Pyne DB, Gebski VJ, Lahtinen SJ, Fricker PA et al (2014) Probiotic supplementation for respiratory and gastrointestinal illness symptoms in healthy physically active individuals. 33(4):581–587
34. Dunne DM, Lefevre-Lewis C, Cunniffe B, Impey SG, Tod D, Close GL et al (2022) Athlete experiences of communication strategies in applied sports nutrition and future considerations for mobile app supportive solutions. Front Sports Active Living 4:911412. https://doi.org/10.3389/fspor.2022.911412
35. Thomas DT, Erdman KA, Burke LM (2016) Position of the academy of nutrition and dietetics, dietitians of Canada, and the American College of Sports Medicine: nutrition and athletic performance. J Acad Nutr Dietetics 116(3):501–528. https://doi.org/10.1016/j.jand.2015.12.006

36. Birkin F, Polesie T, Lewis L (2009) A new business model for sustainable development: an exploratory study using the theory of constraints in Nordic organizations. Bus Strat Environ 18(5):277–290. https://doi.org/10.1002/bse.581
37. Shi Z, Li X, Shuai Y, Lu Y, Liu Q (2022) The development of wearable technologies and their potential for measuring nutrient intake: towards precision nutrition. Nutrition Bull 47(4):388–406. https://doi.org/10.1111/nbu.12581
38. Yan X, Dunne DM, Impey SG, Cunniffe B, Lefevre CE, Mazorra R et al (2022) A pilot sequential multiple assignment randomized trial (SMART) protocol for developing an adaptive coaching intervention around a mobile application for athletes to improve carbohydrate periodization behavior. Contemp Clin Trials Commun 26:100899. https://doi.org/10.1016/j.conctc.2022.100899
39. Nieman DC (2021) Multi-omics approach to precision sports nutrition: limits, challenges, and possibilities. Front Nutr 8:796360. https://doi.org/10.3389/fnut.2021.796360

Nanotech Innovations in Electrolyte Replacement

Neha Kashyap, Kirti Chhabra, Shikha Singh, and Ashutosh Singh

Abstract Nanotechnology has emerged as a transformative force in the field of electrolyte replacement with exceptional advancements. This chapter explores the juncture between electrolyte therapy and nanotechnology, offering an exhaustive analysis of leading-edge, ground breaking and emergent materials that have revolutionized the critical area of medicine and healthcare. The book begins by explaining the physiological significance of electrolytes and their pivotal role in maintaining homeostasis. It also highlights the drawbacks of traditional electrolyte replacement therapies which includes inadequacy in bioavailability, delivery efficiency and side effects. In light of this, the discussion of how crucial nanotechnology is to overcome these barriers are further discussed. Nanostructured materials like nanoparticles, nano sheets and nano gels which plays imperative role in precise and targeted delivery of electrolytes to body are discussed in this chapter. Improved physiological processes by using artificial membranes or replicating original ones via nano engineered ion channels and artificial membranes are introduced later in chapter.

Keywords Electrolyte · Nanotechnology · Sports science · Nutrition · Nanofuel

N. Kashyap (✉)
SAHS, Galgotias University, Greater Noida, UP, India
e-mail: nehabcip1121@gmail.com

K. Chhabra
Department of Physiotherapy, RPIIMS, Karnal, Haryana, India

S. Singh
Jyotiraophule Subharti College of Physiotherapy, Subharti University, Meerut, India

A. Singh
Department of Physiotherapy, Callingwood Physical Therapy and Sports Injury Clinic, ProActive Physical Therapy and Sports Injury Clinic, Edmonton, Canada

A. Chatterjee et al. (eds.), *Nanofuel: The Future of Sports Nutrition*, Smart Nanomaterials Technology, https://doi.org/10.1007/978-981-96-5471-0_10

1 Introduction

Electrolyte and fluid balance are amongst the fundamental features in maintaining normal homeostasis. Electrolytes are anions and cations carrying electrical charges which are indispensable for various important physiological processes. They are imperative in maintaining uninterrupted smooth functioning of body by regulating countless biological functions, including fluid balance, muscle contraction, relay of nerve impulses and acid- base equilibrium. Sodium (Na^+), Potassium (K^+), Magnesium (Mg^{2+}), Chloride (Cl^-), Calcium (Ca^{2+}), Bicarbonate (HCO_3^-) and phosphate (PO_4^-) are amongst the key electrolytes in human body.

Decrease in electrolyte levels are usually encountered and treated in critical care settings [1–4]. Electrolyte deficiency centrals the reason for severity of primary disorder and often reflects in multiple clinical features like vomiting, nausea, weakness, excessive sleepiness, cardiac arrhythmias, tetany, convulsions, muscle weakness, torsades de pointes, decline in myocardial contractility, impaired diaphragm function, paraesthesia, seizures etc. [3, 4] interventions based on protocols are progressively more utilized in critical care setting to avoid complications of electrolyte imbalance [5].

Electrolyte imbalance/dyselectrolytemia has been associated with numerous performance issues in athletes. It has been reported that electrolyte imbalance leads to muscle cramps, dizziness, nausea and fatigue which in turn increases the probability of stress fractures and infectious diseases [6–10]. The traditional methods like oral rehydration solutions, electrolyte supplements and iv fluids have been in use since ages to address the electrolyte imbalance and dehydration. Even though these methods are effective to some extends but they unveil several physiological incompetencies that bound their ability to re-establish optimal electrolyte imbalance.

Traditional methods like oral solutions rely in passive diffusion or facilitated transport in gastrointestinal tract which again is impacted intestinal permeability, concentration gradient and availability of co-transporters like sodium- glucose co-transporter for Na+ absorption. This whole process is further confronted by factors like competitive absorption which reduces the absorption efficiency and gastrointestinal (GI) issues (bloating, diarrhoea, nausea, gastritis etc.) leading to delay or no absorption rendering the therapy useless. Also traditional methods do not selectively deliver the electrolytes to tissues/organ leading to surplus electrolytre in blood stream misbalancing the osmotic gradients and decrease efficiency in correcting the electrolyte deficiency. Factor like intestinal mobility, electrolyte gradients, intestinal mobility, gastritis etc. affects the utilization as well as absorption of electrolytes and at times leads to quick elimination via kidneys causing spikes and sudden drop of electrolytes in bloodstream. Patients/athletes experiencing compromised health, GI disorders or stress frequently exhibit compromised bioavailability of replacement given orally or intravenously. All this has shaped the need for progressive and innovative approaches such as nanotechnology to overcome these incompetencies. Healthcare has been profoundly impacted by the fusion of nanotechnology, biology, and medicine, as demonstrated by advancements in drug delivery systems, biosensors, and regenerative medicine. Driven by advancements in artificial intelligence, quantum physics, and advanced manufacturing, nanotechnology is still developing today, opening the door to next-generation computing systems, tailored medicine, and sustainable energy sources. Nanotechnology's interdisciplinary character guarantees that it will always be a vital component of scientific and technological advancement.

Nanotechnology addresses the physiological incompetencies by incorporating optimized absorption, innovative delivery, enhanced bioavailability and providing controlled release. Use of nano-encapsulation, nano-hydrogels, nanodrinks, nano formulated transdermal patches, nano-niosensors, nano-enhanced oral rehydration solutions and magnetic nanoparticles in improving electrolyte imbalance has been documented and is gradually becoming popular over the traditional methods.

2 Nano-encapsulation

It's a ground breaking technology that enables encapsulation of active materials like electrolytes within nano-carriers to attain controlled and precise delivery. This method improves bioavailability, efficacy and stability of electrolytes addressing numerous incompetencies of traditional methods.

Nanocarriers like Liposomes, Polymeric nanoparticles.micelles, nano hydrogels etc. are used to encapsulate the electrolytes.

Liposomes: Gathered by virtue of self-organization of phospholipids, these are defined as artificial, round, small vesicles that are both biocompatible and biodegradable [11]. Composed of lipid bilayers, liposomes act like cell membranes enhancing absorption and compatibility. Liposomes are perfect for encapsulating water soluble electrolytes like Na^+ and K^+.

Polymeric Nanoparticles: These are particles made from biodegradable polymers like polylactic acid (PLA) or polylactic-co-glycolic acid (PLGA). These nanoparticles help in sustained release and provides high encapsulation efficiency.

Micelles: These are colloidal particles with spherical shapes and nano-sized diameter with a polar outer surface and non-polar inner surface. These type of nanoparticles are efficient in carrying bioactive agents within their hydrophobic centres or cavalently bound to the surface of it. These are suiotable for encapsulating hydrophilic and hydrophobic ions [11, 12].

Nano Hydragels: These are water soluble yet hydrophilic polymers with a 3D cross linked structure. They have high biocompatibility, high water content, low cytotoxicity, ad small surface to volume ratio. These charactyeristics allow nanohydrogels with multiple privileges like efficient localized sustained release and enhanced cell recognition [13–15].

Nano encapsulation aids in improving solubility and permeability of electrolytes facilitating absorption in GI tract and prevents rapid excretion leading to sustained titres in blood stream. Nano-carriers are engineered to deliver electrolytes to specific tissues making them helpful in targeting muscle cramps during exercise or clearing brain fog in dehydration. Nano carriers are aimed to release the encapsulated substance in response to cues like pH disturbances, temperature or change in ionic concentration, making them an asset in improving electrolyte imbalance.

2.1 Nano Drinks with Pre-dissolved Electrolytes

Drinks with pre-dissolved electrolytes symbolizes a pioneering advancement in hydration and electrolyte replacement. These drinks leverage nanotechnology to improve absorption, efficacy and bioavailability of electrolytes. These are prepared by adding nanoscale particles of electrolytes into liquid formulations, nano-drinks confirms rapid delivery and well-organized use of electrolytes rendering them useful in competitive sports for athletes and clinical settings.

Nano-drinks work by nano-scaling the electrolyte particles (Na^+, K^+, Ca^+, Mg^+) to nanometer size as lesser size increases surface area leading to enhanced solubility and bioavailability. Techniques like emulsification or encapsulation within carriers are used to stabilize electrolyte nanoparticles in water/beverage. These are absorbed rapidly in GI tract bypassing the physiological barriers, providing instant replenishment of lost electrolytes making them ideal for athletes engaging in intense exercise or for quick recovery. Nanoscle electrolytes with higher surface area to volume ratio aids in efficient interaction with absorption sites in GI tract ensures improved absorption. Issues like sedimentation and poor solubility seen in conventional drinks are not seen in nano-drinks ensuring uniform distribution, longer shelf life and steady dosing. These generally comes in ready to use cans providing a quick access solution to athletes. Tolerability of nano-drinks is also commendable as it uses lower concentration of more bioavailable nanoparticles. In traditional therapy higher concentration of salts leads to bloating or diarrhoea due to rerouting of more water to GI tract. As per the needs and severity, nano-drinks can also be personalized to optimize performance and recovery, making it compatible with ongoing special diets of athletes or in case of patients, with metabolic disorders (eg. Diabetes).

2.2 Nano Formulated Transdermal Patches

Transdermal patches laden with electrolytes are an innovative bid of nanotechnology intended to deliver electrolytes and other essential compounds directly through skin. This method provides a barrier free, non-invasive, controlled and sustained delivery system. This technology is particularly beneficial for athletes with GI tract disorders or who needs accurate electrolyte management. Encapsulated carriers like liposomes, polymeric nano-peptides and liposomes are used for electrolytes like Na^+, K^+ and Ca^+ to enhance the permeability and stability of electrolyte across the skin. Nano formulated transdermal patches tackle the biggest barrier to exchange of electrolytes i.e. skin outermost layer (stratum corneum) by using permeation enhancers which incorporates lipophilic substances and electric fields to enhance absorption of skin. Transdermal skin patches progressively release electrolyte maintaining a steady level of plasma electrolyte, attain precise dosing, enhance bioavailability and bypass GI tract making it a preferred choice for selected population. These patches are comfortable to use, non-irritant and adhere to skin snugly providing a reservoir of nanoformulated electrolytes.

2.3 *Nano Enhanced ORS*

Nanoemulsions, dendrimers, micelles, liposomes, solid lipid nanoparticles, polymeric nanoparticles, carbon nanotubes, are amongst the most often utilized nanotechnology-based approaches in drug delivery systems because they offer targeted, regulated, and prolonged drug administration [16, 17]. By using nanoscale formulation, nano enhanced ORS provides efficient and rapid hydration. This technology addresses dehydration caused by a number of causes i.e. intense physical activity, medical conditions, diarrhoea, heat etc. In this formulation nanosized particles or electrolytes encapsulated in nano carriers (liposomes, micelles, polymeric nano particles) are administered for rapid gut absorption, enhanced cellular uptake, controlled and sustained release, rapid hydration, decreased osmotic load and it can be strengthened further by adding additional nutrients e.g. zinc.

2.4 *Nanoemulsions*

Emulsion droplets made with Oil-in-water (o/w) sized between 100 and 500 nm are known as nanoemulsions. They are efficient in providing solubility of hydrophobic substances in oily phase, oil droplets are modified with polymers to extend time of circulation and target specific tissues. Low energy and High energy emulsification are the commonest methods to prepare nanoemulsions [18]. These colloidal systems are extensively used in medicine, cosmetics, food and in industrial application to provide high stability, impactful bioavailability and targeted delivery. Owing to nano size, Nano emulsion possess larger surface area to volume ratio augmenting their interaction with permeable membranes. Due to small droplet size they appear translucent and as they resist separation phase (e.g. sedimentation) rendering them superlative for prolonged use.

As nanotechnology continues to progress and rightfully so as it has the potential to transform the way electrolytic agents, nutrients, and functional chemicals are delivered, improving user experience and efficacy in a variety of applications.

References

1. Hamill-Ruth RJ, McGory R (1996) Magnesium repletion and its effect on potassium homeostasis in critically ill adults: results of a double-blind, randomized, controlled trial. Crit Care Med 24:38–45
2. Hijazi M, Al-Ansari M (2005) Protocol-driven vs. physician-driven electrolyte replacement in adult critically ill patients. Ann Saudi Med 25(March (2)):105–109
3. Kraft MD, Btaiche IF, Sacks GS, Kudsk KA (2005) Treatment of electrolyte disorders in adult patients in the intensive care unit. Am J Health-Syst Pharm 63(August):1663–1682
4. Sedlacek M, Schoolwerth AC, Remillard BD (2006) Electrolyte disturbances in the intensive care unit. Semin Dial 19(Nov–Dec (6)):496–501
5. Zimmerman JE, Alzola C, Von Rueden KT (2003) The use of benchmarking to identify top performing critical care units: a preliminary assessment of their policies and practices. J Crit Care 18(2):76–86
6. Hew-Butler T, Smith-Hale V, Sabou J (2019) Exercise-associated electrolyte disorders. Elsevier Curr Opin Endocrine Metab Res 9:51–55
7. Nikolaidis PT, Veniamakis E, Rosemann T, Knechtle B (2018) Nutrition in ultra-endurance: state of the art. Nutrients 10(12):1995. Published 2018 Dec 16
8. Rehrer NJ (2001) Fluid and electrolyte balance in ultra-endurance sport. Sports Med 31(10):701–715
9. Knechtle B, Chlíbková D, Papadopoulou S, Mantzorou M, Rosemann T, Nikolaidis PT (2019) Exercise-associated hyponatremia in endurance and ultra-endurance performance-aspects of sex, race location, ambient temperature, sports discipline, and length of performance: a narrative review. Medicina (Kaunas) 55(9):537. Published 2019 Aug 26
10. Rogulski M, Patarocha Y, Ślusarska A, Mędyk J, Błasiak P, Wirkijowska M, Mikołajec P, Huk R, Bilecka B, Wirkijowski J (2025) Electrolyte disorders in athletes. Qual Sport 37:57197
11. Crintea A, Dutu AG, Sovrea A, Constantin AM, Samasca G, Masalar AL, Ifju B, Linga E, Neamti L, Tranca RA, Fekete Z, Silaghi CN, Craciun AM (2022) Nanocarriers for drug delivery: an overview with emphasis on Vitamin D and K transportation. Nanomaterials (Basel) 12(8):1376. https://doi.org/10.3390/nano12081376.PMID:35458084;PMCID:PMC9024560
12. Hanafy NAN, El-Kemary M, Leporatti S (2018) Micelles structure development as a strategy to improve smart cancer therapy. Cancers 10:238. https://doi.org/10.3390/cancers10070238
13. Xia L, Cong Z, Liu Z et al (2018) Improvement of the solubility, photostability, antioxidant activity and UVB photoprotection of trans-resveratrol by essential oil based microemulsions for topical application. J Drug Delivery Sci Technol 48:346–354
14. Chatterjee B, Gorain B, Mohananaidu K et al (2019) Targeted drug delivery to the brain via intranasal nanoemulsion: available proof of concept and existing. Int j Pharma 565:258–268
15. Liu Y, Liang Y, Yuhong J, Xin P, Han JL, Du Y, Yu X, Zhu R, Zhang M, Chen W, Ma Y (2024) Advances in nanotechnology for enhancing the solubility and bioavailability of poorly soluble drugs. Drug Des Devel Ther. 1(18):1469–1495. https://doi.org/10.2147/DDDT.S447496.PMID:38707615;PMCID:PMC11070169
16. Boisseau P, Loubaton B (2011) Nanomedicine, nanotechnology in medicine. Journal de Physique Arch 12:620–636
17. Mei L, Zhang Z, Zhao L, Huang L, Yang X-L, Tang J, Feng S-S (2013) Pharmaceutical nanotechnology for oral delivery of anticancer drugs. Adv Drug Deliv Rev 65(6):880–890
18. Solans C, Solé I (2012) Nano-emulsions: formation by low-energy methods. Curr Opin Colloid Interface Sci 17(5):246–254

Smart Nanomaterials for Injury Prevention and Recovery

Rajdip Goswami, Soma Das, Debarati Basu, Wong Tin Wui, Soumik Laha, Sukanta Roy, and Anirbandeep Bose

Abstract The chapter "Smart Nanomaterials for Injury Prevention and Recovery" explores how nanotechnology is used in sports medicine to protect athletes from harm and hasten their recovery. Smart nanomaterials, which are designed at the nanoscale to react actively to physiological changes, present encouraging options for enhancing resilience and athletic performance. Through wearable sensors that track biomechanical loads in real time and notify athletes and trainers of potential hazards before accidents happen, these materials can prevent injury. Furthermore, by maximising the delivery of therapeutic agents precisely where and when needed, smart nanomaterials—such as tailored drug delivery systems and nutrients contained in nanoparticles—can quicken the healing process. Use of nanomaterials is a ground-breaking approach for sports rehabilitation because of its accuracy, which speeds up recovery and lowers the possibility of negative effects. Additionally, protective sports gear made of nanofibers and nanocomposites offers an enhanced layer of protection

R. Goswami · S. Das · S. Laha · A. Bose (✉)
Department of Pharmaceutical Technology, School of Health and Medical Sciences, Adamas University, Barasat, Kolkata, West Bengal, India
e-mail: anirbandeep@gmail.com

R. Goswami
e-mail: rajgoswami4455@gmail.com

S. Das
e-mail: somapharma@gmail.com

S. Laha
e-mail: soumiklaha66@gmail.com

D. Basu
Goenka College of Pharmacy, Lachhmangarh (Sikar), Rajasthan, India
e-mail: bdebarati44@gmail.com

W. T. Wui
Non-Destructive Biomedical and Pharmaceutical Research Centre, Smart Manufacturing Research Institute/Faculty of Pharmacy, Universiti Teknologi MARA, Puncak Alam, Selangor, Malaysia
e-mail: wongtinwui@uitm.edu.my

S. Roy
ABS Clinical Solutions, Garia, Kolkata, India

A. Chatterjee et al. (eds.), *Nanofuel: The Future of Sports Nutrition*, Smart Nanomaterials Technology, https://doi.org/10.1007/978-981-96-5471-0_11

due to their better impact resistance. This chapter demonstrates how smart nanomaterials are changing injury prevention and recovery tactics, opening the door to safer, quicker, and more efficient athletic care through a thorough examination of applications, advantages, and the most recent advancements in nanotechnology for sports.

Keywords Smart nanomaterials · Injury prevention · Sports recovery · Nanodrug delivery · Wearable sensors

1 Introduction to Nanotechnology in Sports Medicine

1.1 The Rise of Nanotechnology in Athletic Care

By manipulating atoms and molecules, nanotechnology reduces the size to less than 100 nm, altering the structure to provide optimal results [1]. Nanotechnology is widely utilized in the sports industry, such as in stadium flooring to enhance cleanliness, physical strength, recovery, antistatic properties, and durability. Waterproof and breathable fabrics protect against moisture, wind, rain, and body heat loss. The use of nanotechnology has improved sports equipment and its durability. Gears have now decreased weight but possess better endurance and are wear-resistant, which increases athletes' performance, safety, and comfort [2]. Nanotechnology has also been applied to address sweating during physical activities. Sportswear can incorporate silver nanoparticles, providing an antibacterial effect by preventing bacterial growth and odors. Similar results are achieved with the addition of organic metals, organic silicones, and zinc oxide. These particles also offer electrical and thermal conductivity advantages. UV protection in sportswear can be implemented using cheaper and chemically stable materials such as aluminum oxide, silicon dioxide, and titanium dioxide, which absorb UV rays. Increasing usage of nano-based medication or nanomedicine for diagnosing and treating cancer, implanting devices, tissue engineering, and surgeries has provided satisfactory results, which can be implemented for any sports injury. Nanomaterials can be utilized to prepare scaffolds to promote the growth of injured tissues or deliver drugs [3, 4]. Thus, nanotechnology can benefit athletes in multiple ways. The selection of athletes with a genome of greater quality using nanotechnology can improve the selection of suitable athletes for each event. Sportswear also possesses additional therapeutic traits, such as recuperating strained or injured muscles from intense physical activity. Fabric materials can be combined with germanium and ceramic nanoparticles such as titanium dioxide and alumina to provide therapeutic purposes through absorbing sun rays from the human body and transforming that energy into far-infrared rays, returning the energy to the body and leading to better blood circulation and metabolism [4].

Nanotechnology holds a promising future for sports injuries and their management and can mark a significant change in the sports field and the performance of athletes. It has significantly contributed to various sports events, such as the Olympics. However, several challenges must be overcome, such as the requirement of clinical studies to establish the safety and efficacy of nanomaterials and the expenses of acquiring nanotechnology and nanomaterials.

2 Mechanisms of Injury Prevention Using Nanomaterials

2.1 Wearable Sensors for Real-Time Monitoring

Chronic diseases such as hypertension, high cholesterol, diabetes, and various other heart diseases have been a primary concern worldwide and require constant monitoring. High-tech devices such as biomedical sensors and wearable sensory devices enable healthcare staff to assess the data whenever needed. These wearable devices constantly track the patient's vitals and alarm the healthcare professional if the patient's vital has reached extreme levels. These wearable devices can monitor athletes' physiological data and physical movement during training and competition to ensure their safety and improve performance. The traditional sensors have a low accuracy rate since they may detach from the body during physical activity. The wearable device must consider the variations that eccrine sweat causes to the sensor output. Therefore, nanotechnology can improve this situation through wearable devices that are much more flexible and can be stretched or bent without compromising their functions, acting almost like a secondary skin on their body, with zero discomfort. Nanosensors can be embedded in wearable devices to monitor athlete performance and detect potential injuries before they become severe damage. Silver or copper nanoparticles and nanowires, carbon nanotubes, graphene two-dimensional layered material, etc., are currently being used to prepare wearable devices [5–7]. **Silver or Copper Nanoparticles and Nanowires**: Enhance electrical conductivity for real-time monitoring and provide antimicrobial properties [5]. **Carbon Nanotubes**: Offer flexibility, mechanical strength, and conductivity for stable and durable sensors [6]. **Graphene and Two-Dimensional Materials**: Provide flexibility, lightweight design, and high sensitivity to biomolecular changes [7].

2.2 Nanomaterials in Injury Risk Assessment

It has been reported that nanomaterials are associated with diseases such as tuberculosis, breast cancer, asthma, respiratory embolism, etc. Therefore, it is necessary to understand nanotoxicity to avoid carcinogenic materials. Toxicity can be studied through in vivo studies using various animal models, which is considered time-consuming, expensive, and requires approval from the ethical committee. In vitro studies are relatively economical and rapid. Risk assessment has four steps:

hazard identification, hazard characterization, exposure assessment, and risk characterization. Substances like copper, silica, titanium dioxide, polystyrene, etc., may accumulate in the body and produce toxicity. It has been reported that copper oxide nanoparticles induce oxidative stress and are genotoxic and cytotoxic, and iron oxide nanoparticles damage the DNA [8–10].

2.3 Early Detection of Biomechanical Stress

Early detection of biomechanical stress is vital in preventing sports-related injuries, mainly muscle strains or ligament sprains, which are common among athletes. Biomechanical stress refers to the force delivered on the body's tissues during physical activity, and when this stress exceeds the tissue's adaptive capacity, injury can occur. Short-term injuries like muscle strains are the result of inadequate recovery or improper movement that place undue stress on specific body parts. Detecting these stresses early can help to prevent more severe injuries [11, 12]. The most significant factor in early detection is monitoring, which observes changes in movement patterns, joint angles, and force distribution during activity. Technologies like motion capture systems, wearable sensors, and pressure sensors are increasingly used to track real-time biomechanical data during competition and training. These tools provide insights into how forces are distributed across muscles, joints, and tendons, which can highlight abnormal stresses on specific body parts [13]. During running or sprinting, abnormal gait patterns, like improper knee alignment, can lead to increased stress on the lower limbs and increase the risk of short-term injuries such as hamstring strains. By analyzing these patterns, early signs of excessive stress can be detected, and corrective actions are taken before inflicting with injury [14, 15]. Biomechanical stress management in athletic performance depends on early detection mechanisms that reduce injury risks. Muscle and tendon adaptations during training are crucial, but fatigue significantly compromises tissue stress tolerance. Innovative wearable technologies and physiological monitoring systems enable real-time tracking of muscle activity, force output, and movement deviations, providing critical insights into an athlete's biomechanical limits [16, 17]. Personalized training and recovery programs, enhanced by artificial intelligence and machine learning, offer tailored interventions that address individual biomechanical variations. By integrating advanced monitoring technologies with data-driven strategies, athletes can identify and mitigate potential injury risks. These sophisticated approaches prevent short-term injuries and optimize performance through proper treatment, representing a revolution in sports biomechanics and athlete health management. The continuous evolution of these technologies promises more effective individualized injury prevention strategies [18, 19].

3 Nanotechnology-Enhanced Sports Gear

3.1 Impact-Resistant Nanofibers

In sports technology, nanotechnology has emerged as a revolutionized approach to enhancing protective gear through impact-resistant nanofibers. Nanofibers can be formed using polymerization, electrospinning, and electrostatic spinning techniques. They possess high mechanical strength, are water resistant, and are lightweight in texture. Carbon nanotubes and advanced polymer nanofibers are at the core of this technological revolution. Because of their exceptional strength-to-weight ratios and distinct molecular structure, manufacturers can produce sporting equipment that is both remarkably durable and lightweight [20]. Traditional protective gear frequently compromises mobility and protection, but nanofiber-enhanced materials overcome this long-standing issue. Athletes across various disciplines are getting benefits from these types of products [21]. Compared to traditional designs, helmets made of nanofiber composites are more effective in absorbing and redistributing impact forces, which drastically minimizes the risk of traumatic brain injuries. Similarly, protective padding used in sports now improves shock absorption without the excessive weight of traditional materials [22].

3.2 Nanocomposites for Improved Protection

Nanocomposites have emerged as advanced materials for enhancing protection in a wide range of applications due to their unique features [23]. nanocomposites offer superior mechanical strength, thermal stability, and resistance to wear and impact by integrating nanoparticles into traditional matrix materials. These attributes make them particularly suitable for applications in industries such as aerospace, automotive, defense, and healthcare. For example, in personal protective equipment, nanocomposites can significantly improve the performance of helmets, body armor, and gloves by providing lightweight yet robust protection [24, 25]. The rigidity of the composite material is improved using nanoscale fillers like silica nanoparticles, graphene, or carbon nanotubes. Better distribution of energy and stress transfer mechanisms—less efficient in traditional composites, are used to accomplish this. Moreover, the multifunctionality of nanocomposites—such as improved flame retardancy, thermal insulation, and antibacterial qualities—increases their use in stressful environments [25, 26]. Considering their advantages, challenges such as cost-effective manufacturing, uniform dispersion of nanoparticles, and scalability are the major obstacles to their widespread adoption. However, ongoing research and advancements in processing techniques address these issues, paving the way for broader implementation. The development of nanocomposites holds great promise for improving protection across diverse areas contributing to both safety and performance [27–29].

3.3 Advances in Protective Gear Using Nanotechnology

Comfort and performance are the key factors that nanotechnology sportswear wants to achieve. It should be able to manage heat, odor, and moisture, be flexible, light in weight, and breathable, protect against UV rays, and resist bacteria. Phase change materials are kept in the core surrounded by a polymer to absorb, store, and liberate latent heat. Phase change materials are extensively used in sportswear for colder environments. Nanofibrous materials have a large surface area and are lightweight. Polyurethane, polyacrylonitrile, and polyamide are commonly used to produce nanofibers by electrospinning separate layers. Polyurethane and nylon 66 are used to waterproof clothes [30]. During skiing, ski goggles made of polyurethanes or polycarbonates are used to provide an antifogging effect or protect the eyes from UV rays. Graphene-based sportswear displays high thermal conductivity and also generates infrared rays, therefore they can increase temperature and boost blood circulation [6]. Antimicrobial properties are exhibited by silver, titanium dioxide, zinc oxide, and copper II oxide. These metals will kill microbes and, through nanotechnology, become small particles with very high specific surface area [31]. Figure 1 shows that nanotechnology sportswear enhances athlete performance, durability, moisture-wicking properties, and temperature regulation.

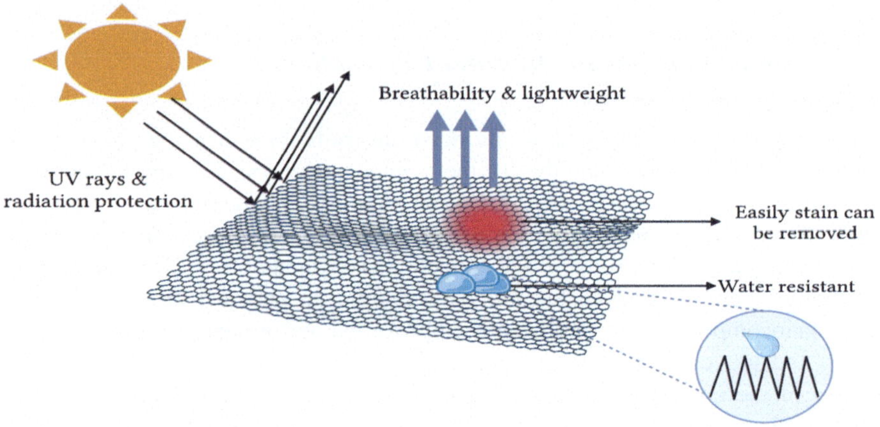

Fig. 1 Nanotechnology sportswear improves athlete performances, durability, moisture-wicking properties, and temperature regulation

4 Smart Nanomaterials for Accelerated Recovery

4.1 Nano-encapsulation of Nutrients and Supplements

The nanoencapsulation of nutrients increases their intracellular uptake by improving bioavailability, extending shelf life, and enhancing controlled dissociation to release nutrients at the required site. It also targets specific areas, enhancing therapeutic efficacy. Nanoencapsulation materials include carbohydrate-based biopolymers such as starch, cyclodextrins, gum arabic, alginate, and ethylcellulose; lipid-based nanomaterials; porous inorganic material; and clay-based materials. Nanoparticles used for encapsulation include nanospheres, nanocapsules, liposomes, solid lipid nanoparticles, and nanogels. A nanocapsule encapsulates a bioactive compound within a cavity surrounded by a polymer membrane, while a nanosphere disperses the bioactive compound uniformly within the polymer matrix. Nanoencapsulation techniques such as spray drying, coacervation, supercritical fluid methods, nanoemulsification, electrospinning, and solvent evaporation increase the solubility of hydrophobic substances like omega-3 fish oil. Lipid-soluble antioxidants or phenolic compounds like carotenoids can be encapsulated in nanoemulsions, while curcumin, quercetin, and resveratrol are incorporated in nanoliposomes or solid lipid nanoparticles for enhanced bioavailability [32–35].

4.2 Targeted Drug Delivery Systems

Targeted drug delivery systems provide site-specific therapeutic action, ensuring high drug concentration at the wound area while minimizing systemic distribution and side effects. While traditional dressings offer general wound protection, targeted systems are designed to release therapeutic agents only where needed, addressing the specific requirements of different wound types. In sports-related injuries, targeted drug delivery maximizes drug efficacy, such as delivering nonsteroidal anti-inflammatory drugs (NSAIDs) directly to inflamed tissues. This reduces systemic effects while enhancing localized healing. Similarly, growth factors, antibiotics, and other agents are efficiently administered to promote tissue repair and prevent infections. Nanoparticles, liposomes, and hydrogels are commonly used as carriers for targeted delivery. These carriers release therapeutic agents based on environmental stimuli like pH or temperature, enabling controlled and sustained drug release. For example, nanoparticles can penetrate deep tissues to deliver drugs precisely, while liposomes and hydrogels maintain a moist wound environment to promote healing [36–43].

4.3 Role of Nanocarriers in Enhanced Healing

Chronic wounds often experience prolonged inflammation, infection, and impaired vascularization, hindering healing. Nanocarriers improve drug pharmacokinetics by efficiently delivering bioactive compounds to specific wound sites, reducing toxicity, and enhancing bioavailability. Metal-based nanoparticles, such as silver, copper, and zinc, exhibit antimicrobial properties and are particularly useful for treating infectious wounds. Gold nanoparticles promote tissue regeneration and provide targeted drug delivery, often modified with crosslinked collagen for enhanced integration with cell adhesion molecules. Silver nanoparticles crosslinked with collagen enhance antibacterial activity. Carbon-based nanomaterials like nanotubes, graphene, fullerenes, and nanodiamonds exhibit antibacterial properties, while dendrimers independently exert antibacterial effects and can be combined with silver nanoparticles for improved efficacy. Incorporating microRNA with carriers such as dendrimers and poly(lactic-co-glycolic acid) provides a novel approach to treating chronic wounds [44–47].

5 Bio-responsive Nanomaterials in Rehabilitation

5.1 Adaptive Nanoparticles for Controlled Therapy

Adaptive nanoparticles are designed to release therapeutic agents in response to physiological stimuli such as pH, enzyme activity, or temperature changes. For instance, pH-sensitive nanoparticles deliver drugs to inflamed areas, while enzyme-responsive systems release agents only in the presence of overexpressed enzymes in damaged tissues. These systems reduce systemic drug administration requirements, providing controlled, localized therapy. Additionally, adaptive nanoparticles can be engineered to mimic the natural healing process by releasing multiple therapeutic agents in sequence. However, ensuring biocompatibility, stability, and scalability remains a challenge. Ongoing advancements are paving the way for personalized, precise treatments using adaptive nanoparticles [48–50].

5.2 Thermo-Responsive and pH-Responsive Systems

Thermo-responsive materials release drugs upon detecting temperature changes, such as inflammation-induced heat, facilitating recovery by ensuring precise delivery. Similarly, pH-responsive systems release therapeutic agents in inflamed or damaged tissues, where the pH differs from healthy tissue. By targeting these unique conditions, healthcare professionals can enhance therapeutic efficacy, ensuring treatment reaches the intended site [51–53].

5.3 Biocompatibility and Safety Considerations

Ensuring the biocompatibility and safety of bio-responsive nanoparticles is critical, as these materials interact directly with biological tissues. Researchers evaluate toxicity, inflammation, and allergic responses to confirm short-term safety and study long-term effects, such as breakdown and potential harmful byproducts. By prioritizing safety and biocompatibility, researchers can develop innovative therapies that enhance patient recovery without compromising health, enabling safe and effective rehabilitation treatments [54–56].

6 Case Studies and Applications in Sports

6.1 Injury Prevention in High-Impact Sports

Preventing injuries during high-impact sports is significant to making sure that athletes are always fit to perform. Cumulatively, they receive strains and fractures with joint damages in between. The most dramatic revolution in personal protection gears in the sports area comes from smart nanomaterials: impact absorption, self-heal, and even antimicrobial features. For example, when fused with carbon nanotubes (CNTs), helmets and pads can be engineered to absorb impacts better and disperse forces evenly through the entire material. As demonstrated by research performed by Smith et al., materials based on CNTs will reduce the head injury ethos to even 30% of that using any other conventional equipment. Nanosensors embedded in clothing and equipment is another technological breakthrough. Biomechanical stress will be monitored on a real-time basis, allowing earlier warnings for the athletes on possible threats before injuries actually occur. Devices using gold nanoparticle-based transducers, for instance, have been noted to accurately show the impact of muscle fatigue as quoted by Jones et al. Moreover, any of the following techniques, such as nanocoating with zinc oxide or silver nanoparticles, is widely, preferably applied for the prevention of microbial growth in order to ensure hygiene in the shared athletic environment. Nanomaterials also parameterize unique designs extrapolated for ankle braces and knee supports. These are materials like graphene-reinforced composites that provide dynamic stiffness as they adapt toward a threshold level of stress that changes with movement. Likewise, bioengineered hydrogels with nano-fibrillar structures with immense cushioning and injury resistance in high-impact zones have been developed. These are but some examples of innovations that would eventually result in safer athletic experiences while maintaining optimum performance [57–61].

6.2 Enhanced Recovery Protocols for Athletes

It is smart nanomaterials that have changed the face of rehabilitation therapy for athletes injured. Drug delivery systems based on nanotechnology, like liposomes and polymeric nanoparticles, have proven themselves effective in targeted therapy. Patel et al. using liposomal ibuprofen, which ensured an anti-inflammatory lasting action with diminished gastrointestinal effects, has pointed out this orientation. Hydrogel-based wound dressings incorporating silver nanoparticles have been the latest in treating abrasions and lacerations. These offer controlled drug release, antibacterial activity, and moisture retention in the matrix that hastens wound healing. Nanofibers of polycaprolactone (PCL) are also used in soft tissue regeneration, as was demonstrated in trials by Zhang et al., where injury recovery, particularly tendon injuries, was reduced by 40%. For example, using nanocapsules as delivery systems for growth factors such as vascular endothelial growth factor (VEGF), they are likely to improve angiogenesis and repair of tissue damage. Additionally, combined with physical therapy, magnetic nanoparticles stimulate deeper tissues and induce quicker muscle regeneration. These also benefit athletes from ligament injuries as they provide mechanical support while promoting cell proliferation and integration. This is actually a paradigm shift as regards the role of nanomaterials in nutritional supplements, where bioavailability is higher and oxidative stress lower with encapsulated nutrients like omega-3 fatty acids or curcumin. All these applications make it possible for athletes to get back into action very quickly [62–65].

6.3 Real-World Applications and Success Stories

Smart nanomaterials have achieved remarkable success in developing injury prevention and recovery measures in sports. A case in point is the partnership of leading sports brands with nanotechnology firms to produce graphene-enhanced running shoes. As validated by Brown et al., these shoes are said to be much more durable and energy-retaining, thus reducing stress injuries among marathon runners. Another similar example is that of professional football clubs that use nanoscale biosensors to monitor hydration and electrolyte levels. Real-time data collected during events have proven valuable in preventing hydration-related muscle injuries. In cycling, air gel nanotechnology is used in helmets to reduce drag while delivering unparalleled impact resistance. Success stories at this stage also reach into the post-injury phase. The example is the case of a famous basketball player treated using nano-scaffold implants, as reported by Green et al. The player regain full mobility six months following acute tearing of the Achilles tendon. Similarly, elite swimmers have also made gains on account of the application of swimming costumes with nanocoatings that have been used to improve swimming dynamics while reducing muscle fatigue. Chronic injuries have also demonstrated considerable improvement with drug-loaded nanoparticles. Athletes with recurring arthritis could now find relief from pain with such nano-encapsulated corticosteroids, thus extending their careers. These examples in real life indicate the application of smart nanomaterials and connect the gap between science and performance improvement [65–69].

7 Challenges and Future Perspectives

7.1 Technical and Regulatory Hurdles

Smart nanomaterials have enormous potential, but there are obstacles even in their application. In fact, scaling is a fundamental issue from that technical perspective. The production of quality nanomaterials such as carbon nanotubes or quantum dots requires, as Lee et al. have mentioned, complex and cost-intensive processes. The assurance of homogeneity with regard to material properties over significantly big batches remains equally daunting. The frameworks of regulation in nanotechnology for sports are underdeveloped. For example, there are no comprehensive guidelines from the European Medicines Agency (EMA) and the FDA for the approval of nanoparticle-based therapies. This fact acts as a deterrent for both widespread adoption and innovation. Studies on biocompatibility and environmental issues for nanomaterials are also relatively rare but raise ethical issues regarding their safety; moreover, the long-term studies are limited. Another challenge is connecting this nanotechnology to existing equipment. For example, embedding nanosensors into protective gear needs futuristic manufacturing techniques, which have yet to evolve. Besides this, issues like data security and privacy arise by use of biosensors, as Martinez et al. point out. These technical and regulatory hurdles indicate the need for interdisciplinary collaboration to tap the full potential of smart nanomaterials in sports [70–72].

7.2 Ethical and Safety Concerns

There are ethical and safety issues around the application of nanotechnology in sports. One of these is toxicity due to nanoparticles. The work of Wang et al. shows cytotoxicity and oxidative stress from silver-nanoparticle prolonged exposure. These remain as priorities for proving biocompatibility and non-toxicity. Another ethical issue concerns the expectation of misuse for performance enhancement, which blurs the line between fair play and doping. For example, wide-spread application of nanoparticle-delivered drugs could give competitive advantages to athletes and undermine the principles of sporting integrity. WADA has flagged the issue and need for rigorous follow-up on nanotechnology application in sports. More so, accessibility and affordability issues would worsen the inequalities that exist in sports. Advanced nanotechnology gear or treatment may only be available to affluent athletes or organizations, which creates an uneven field. These ethical and safety issues will extensively require stringent policy frameworks, public awareness campaigns and equitable access strategies [1, 73–75].

7.3 Future Directions in Smart Nanomaterials for Sports Medicine

The future of smart nanomaterials in sports medicine is both promising and expending source's development in innovations like self-healing polymers and bio-responsive nanomaterials. One such study by Chen et al. explored shape-memory alloys at nanoscale, which may help in revolutionizing braces and splints since they would be able to adapt dynamically to athletes' movements. AI-integrated nanosensors in the future will allow far greater personalization at the hyper level of individuals' training regimens. For example, biomechanical data collected with the immediacy of a clinical examination will feedback to tailoring exercise protocols to reduce injury risk. Advances made in bio-nanotechnology may also lay the foundation for smart implants, like nanoscale electrodes for providing neural rehabilitation of injured athletes. Future focus research will be mainly on green nanotechnology that shall limit environmental impacts of manufacturing processes and modify regulation governing them for clear ways of approval and commercialization. As these miracles come to real-time innovations, their fusion promises a new dimension to the future of athletic performance and safety [59, 62, 67, 76, 77].

8 Conclusion

The future that is being created for sports medicine will surely be an incredible one. With smart nanomaterials, it is possible to take prevention, healing, and performance enhancement to completely new levels. These very smart nanomaterials such as graphene gear and nanoparticle-based drug delivery systems resolve some of the most crucial issues an athlete faces. All promises have technical, regulatory, and ethical barriers to breach before reality can be made to their full potential. Nanotechnology in sports have to work hand in hand and it will take material scientists, medical professionals, and regulators together to make that fruitful outcome. The case studies make a very good point about how nanomaterials have changed the game. Innovations around AI-integrated nanosensors and self-healing materials will further revolutionise sports medicine practice. Safety, economy, and accessibility create a future where smart nanomaterials in sports can be inclusive and sustainable. Thus, this chapter highlights the importance of nanotechnology in moulding the next generation of athletic performance and healthcare solutions.

References

1. da Costa LP (2020) Engineered nanomaterials in the sports industry. In: Handbook of nanomaterials for manufacturing applications. Elsevier, pp 309–320
2. Ansari AA (2023) Nanotechnological advancements in sports rehabilitation to elevate athletic performance levels, pp 377–381
3. Tang M, Yang L, Zhou H (2013) Applications and safety of nanotechnology and nanomaterials in sports, pp 337–343
4. Abbasi S, Peerzada MH, Nizamuddin S, Mubarak NM (2020) Functionalized nanomaterials for the aerospace, vehicle, and sports industries. In: Handbook of functionalized nanomaterials for industrial applications. Elsevier, pp 795–825
5. Brothers MC, DeBrosse M, Grigsby CC, Naik RR, Hussain SM, Heikenfeld J et al (2019) Achievements and challenges for real-time sensing of analytes in sweat within wearable platforms. Acc Chem Res 52(2):297–306
6. He C, Cao M, Liu J, Ge Z, Zhou R, Xu H (2022) Nanotechnology in the olympic winter games and beyond. ACS Nano 16(4):4981–4988
7. Azeem M, Shahid M, Masin I, Petru M (2024) Design and development of textile-based wearable sensors for real-time biomedical monitoring; a review. J Textile Inst 20:1–16
8. Maurer-Jones MA, Bantz KC, Love SA, Marquis BJ, Haynes CL (2009) Toxicity of therapeutic nanoparticles. Nanomedicine 4(2):219–241
9. Ganguly P, Breen A, Pillai SC (2018) Toxicity of nanomaterials: exposure, pathways, assessment, and recent advances. ACS Biomater Sci Eng 4(7):2237–2275
10. Savolainen K, Alenius H, Norppa H, Pylkkänen L, Tuomi T, Kasper G (2010) Risk assessment of engineered nanomaterials and nanotechnologies—a review. Toxicology 269(2–3):92–104
11. Fisher H, Gittoes MJ, Evans L, Bitchell CL, Mullen RJ, Scutari M (2020) An interdisciplinary examination of stress and injury occurrence in athletes. Front Sports Act Living 14:2
12. Mancino F, Kayani B, Gabr A, Fontalis A, Plastow R, Haddad FS (2024) Anterior cruciate ligament injuries in female athletes: risk factors and strategies for prevention. Bone Jt Open 5(2):94–100
13. De Fazio R, Mastronardi VM, De Vittorio M, Visconti P (2023) Wearable sensors and smart devices to monitor rehabilitation parameters and sports performance: an overview. Sensors 23(4):1856
14. Liu Y, Sun Y, Zhu W, Yu J (2017) The late swing and early stance of sprinting are most hazardous for hamstring injuries. J Sport Health Sci 6(2):133–136
15. Bramah C, Mendiguchia J, Dos'santos T, Morin JB (2024) Exploring the role of sprint biomechanics in hamstring strain injuries: a current opinion on existing concepts and evidence. Sports Med 54(4):783–793
16. Seçkin AÇ, Ateş B, Seçkin M (2023) Review on wearable technology in sports: concepts, challenges and opportunities. Appl Sci 13(18):10399
17. Preatoni E, Bergamini E, Fantozzi S, Giraud LI, Orejel Bustos AS, Vannozzi G et al (2022) The use of wearable sensors for preventing, assessing, and informing recovery from sport-related musculoskeletal injuries: a systematic scoping review. Sensors 22(9):3225
18. Yang R, Yuan Q, Zhang W, Cai H, Wu Y (2024) Application of Artificial Intelligence in rehabilitation science: a scientometric investigation utilizing citespace. SLAS Technol 29(4):100162
19. Musat CL, Mereuta C, Nechita A, Tutunaru D, Voipan AE, Voipan D et al (2024) Diagnostic applications of AI in sports: a comprehensive review of injury risk prediction methods. Diagnostics 14(22):2516
20. Malik S, Muhammad K, Waheed Y (2023) Nanotechnology: a revolution in modern industry. Molecules 28(2):661
21. Ali T, Najam-ul-Haq M, Mohyuddin A, Musharraf SG, Hussain D (2023) Next-generation functional nanotextiles—prospects and challenges. Sustain Mater Technol 36:e00640
22. Bottlang M, DiGiacomo G, Tsai S, Madey S (2022) Effect of helmet design on impact performance of industrial safety helmets. Heliyon 8(8):e09962

23. Singh NB, Agarwal S (2016) Nanocomposites: an overview. Emerging Mater Res 5(1):5–43
24. Aththanayaka S, Thiripuranathar G, Ekanayake S (2022) Emerging advances in biomimetic synthesis of nanocomposites and potential applications. Mater Today Sustainability 20:100206
25. Rashid A Bin, Haque M, Islam SMM, Uddin Labib KMR (2024) Nanotechnology-enhanced fiber-reinforced polymer composites: recent advancements on processing techniques and applications. Heliyon 10(2):e24692
26. Papageorgiou DG, Li Z, Liu M, Kinloch IA, Young RJ (2020) Mechanisms of mechanical reinforcement by graphene and carbon nanotubes in polymer nanocomposites. Nanoscale 12(4):2228–2267
27. Macías-Silva MA, Cedeño-Muñoz JS, Morales-Paredes CA, Tinizaray-Castillo R, Perero-Espinoza GA, Rodríguez-Díaz JM et al (2024) Nanomaterials in construction industry: an overview of their properties and contributions in building house. Case Stud Chem Environ Eng 10:100863
28. Rahman M, Islam KS, Dip TM, Chowdhury MFM, Debnath SR, Hasan SMdM et al (2024) A review on nanomaterial-based additive manufacturing: dynamics in properties, prospects, and challenges. Progr Addit Manuf 9(4):1197–224
29. Shahbazi M, Jäger H, Ettelaie R, Chen J, Kashi PA, Mohammadi A (2024) Dispersion strategies of nanomaterials in polymeric inks for efficient 3D printing of soft and smart 3D structures: a systematic review. Adv Colloid Interface Sci 333:103285
30. Bissessur R (2020) Nanomaterials applications. In: Polymer science and nanotechnology. Elsevier, pp 435–453
31. Hardin IR, Kim Y (2016) Nanotechnology for antimicrobial textiles. In: Antimicrobial textiles. Elsevier, pp 87–97
32. Jafarizadeh-Malmiri H, Sayyar Z, Anarjan N, Berenjian A (2019) Nano-encapsulation for nutrition delivery. In: Nanobiotechnology in food: concepts, applications and perspectives. Springer International Publishing, Cham, pp 95–114
33. Katouzian I, Jafari SM (2016) Nano-encapsulation as a promising approach for targeted delivery and controlled release of vitamins. Trends Food Sci Technol 53:34–48
34. Tahir A, Shabir Ahmad R, Imran M, Ahmad MH, Kamran Khan M, Muhammad N et al (2021) Recent approaches for utilization of food components as nano-encapsulation: a review. Int J Food Prop 24(1):1074–1096
35. Assadpour E, Mahdi JS (2019) A systematic review on nanoencapsulation of food bioactive ingredients and nutraceuticals by various nanocarriers. Crit Rev Food Sci Nutr 59(19):3129–3151
36. Kolimi P, Narala S, Nyavanandi D, Youssef AAA, Dudhipala N (2022) Innovative treatment strategies to accelerate wound healing: trajectory and recent advancements. Cells 11(15):2439
37. Landén NX, Li D, Ståhle M (2016) Transition from inflammation to proliferation: a critical step during wound healing. Cell Mol Life Sci 73(20):3861–3885
38. Pereira RF, Bártolo PJ (2016) Traditional therapies for skin wound healing. Adv Wound Care (New Rochelle) 5(5):208–229
39. Saghazadeh S, Rinoldi C, Schot M, Kashaf SS, Sharifi F, Jalilian E et al (2018) Drug delivery systems and materials for wound healing applications. Adv Drug Deliv Rev 127:138–166
40. Rathna RP, Kulandhaivel M (2024) Advancements in wound healing: integrating biomolecules, drug delivery carriers, and targeted therapeutics for enhanced tissue repair. Arch Microbiol 206(4):199
41. Mirhaj M, Labbaf S, Tavakoli M, Seifalian AM (2022) Emerging treatment strategies in wound care. Int Wound J 19(7):1934–1954
42. Li J, Wang Q, Xia G, Adilijiang N, Li Y, Hou Z et al (2023) Recent advances in targeted drug delivery strategy for enhancing oncotherapy. Pharmaceutics 15(9):2233
43. Karthikeyan E, Sivaneswari S (2024) Advancements in transdermal drug delivery systems: enhancing medicine with pain-free and controlled drug release. Intelligent Pharmacy
44. Bernal-Chávez S, Nava-Arzaluz MG, Quiroz-Segoviano RIY, Ganem-Rondero A (2019) Nanocarrier-based systems for wound healing. Drug Dev Ind Pharm 45(9):1389–1402

45. Medina-Cruz D, Saleh B, Vernet-Crua A, Ajo A, Roy AK, Webster TJ (2020) Drug-delivery nanocarriers for skin wound-healing applications. In: Wound healing, tissue repair, and regeneration in diabetes. Elsevier, pp 439–488
46. Toczek J, Sadłocha M, Major K, Stojko R (2022) Benefit of silver and gold nanoparticles in wound healing process after endometrial cancer protocol. Biomedicines 10(3)
47. Deka Dey A, Yousefiasl S, Kumar A, Dabbagh Moghaddam F, Rahimmanesh I, Samandari M et al (2023) MiRNA-encapsulated abiotic materials and biovectors for cutaneous and oral wound healing: biogenesis, mechanisms, and delivery nanocarriers. Bioeng Transl Med. 8(1):e10343
48. Khan I, Saeed K, Khan I (2019) Nanoparticles: properties, applications and toxicities. Arab J Chem 12(7):908–931
49. Cao Z, Li W, Liu R, Li X, Li H, Liu L et al (2019) PH- and enzyme-triggered drug release as an important process in the design of anti-tumor drug delivery systems. Biomed Pharmacother 118:109340
50. Rahman MdM, Islam MdR, Akash S, Harun-Or-Rashid Md, Ray TK, Rahaman MdS et al (2022) Recent advancements of nanoparticles application in cancer and neurodegenerative disorders: at a glance. Biomed Pharmacother 153:113305
51. Salahshoori I, Yazdanbakhsh A, Namayandeh Jorabchi M, Kazemabadi FZ, Khonakdar HA, Mohammadi AH (2024) Recent advances and applications of stimuli-responsive nanomaterials for water treatment: a comprehensive review. Adv Colloid Interface Sci 333:103304
52. Bellotti E, Schilling AL, Little SR, Decuzzi P (2021) Injectable thermoresponsive hydrogels as drug delivery system for the treatment of central nervous system disorders: a review. J Control Release 329:16–35
53. Verkhovskii R, Ivanov A, Lengert E, Tulyakova K, Shilyagina N, Ermakov A (2023) Current principles, challenges, and new metrics in pH-responsive drug delivery systems for systemic cancer therapy. Pharmaceutics 15(5):1566
54. Kyriakides TR, Raj A, Tseng TH, Xiao H, Nguyen R, Mohammed FS et al (2021) Biocompatibility of nanomaterials and their immunological properties. Biomed Mater 16(4):042005
55. Xuan L, Ju Z, Skonieczna M, Zhou P, Huang R (2023) Nanoparticles-induced potential toxicity on human health: applications, toxicity mechanisms, and evaluation models. MedComm (Beijing) 4(4)
56. Ankit, Saha L, Kumar V, Tiwari J, Sweta, Rawat S et al (2021) Electronic waste and their leachates impact on human health and environment: global ecological threat and management. Environ Technol Innov 24:102049
57. Anusiya G, Jaiganesh R (2022) A review on fabrication methods of nanofibers and a special focus on application of cellulose nanofibers. Carbohyd Polym Technol Appl 4:100262
58. Kheyrandish M, Bafande F, Sheikh Arabi M (2022) Features and methods of making nanofibers by electrospinning, phase separation and self-assembly. Jorjani Biomed J 10(1):13–25
59. Patnaik A (2024) Recent developments in application of nanofibers. Processes 12(9):1894
60. Fierascu I, Fierascu IC, Dinu-Pirvu CE, Fierascu RC, Anuta V, Velescu BS et al (2019) A short overview of recent developments on antimicrobial coatings based on phytosynthesized metal nanoparticles. Coatings 9(12):787
61. Abtew MA, Boussu F, Bruniaux P (2021) Dynamic impact protective body armour: a comprehensive appraisal on panel engineering design and its prospective materials. Defence Technol 17(6):2027–2049
62. Farhaj S, Conway BR, Ghori MU (2023) Nanofibres in drug delivery applications. Fibers 11(2):21
63. Wan LY, Wang H, Gao W, Ko F (2015) An analysis of the tensile properties of nanofiber mats. Polymer (Guildf) 73:62–67
64. Hiwrale A, Bharati S, Pingale P, Rajput A (2023) Nanofibers: a current era in drug delivery system. Heliyon 9(9):e18917
65. Jeong GJ, Castels H, Kang I, Aliya B, Jang YC (2022) Nanomaterial for skeletal muscle regeneration. Tissue Eng Regen Med 19(2):253–261

66. Kang MS, Yu Y, Park R, Heo HJ, Lee SH, Hong SW et al (2024) Highly aligned ternary nanofiber matrices loaded with MXene expedite regeneration of volumetric muscle loss. Nanomicro Lett 16(1):73
67. Nathani K, Damani M, Mangrulkar S, Sawarkar S (2024) Polymeric nanofibers derived therapeutic approach for skeletal muscle regeneration: the promising future of muscle repair. J Drug Deliv Sci Technol 94:105503
68. Langwald SV, Ehrmann A, Sabantina L (2023) Measuring physical properties of electrospun nanofiber mats for different biomedical applications. Membranes (Basel) 13(5):488
69. Liu J, Zhang L, Wang N, Zhao H, Li C (2022) Nanofiber-reinforced transparent, tough, and self-healing substrate for an electronic skin with damage detection and program-controlled autonomic repair. Nano Energy 96:107108
70. Subhan MA, Subhan T (2022) Safety and global regulations for application of nanomaterials. In: Nanomaterials recycling. Elsevier, pp 83–107
71. Oksel Karakus C, Bilgi E, Winkler DA (2021) Biomedical nanomaterials: applications, toxicological concerns, and regulatory needs. Nanotoxicology 15(3):331–351
72. Ragelle H, Danhier F, Préat V, Langer R, Anderson DG (2017) Nanoparticle-based drug delivery systems: a commercial and regulatory outlook as the field matures. Expert Opin Drug Deliv 14(7):851–864
73. Wang E, Huang Y, Du Q, Sun Y (2017) Silver nanoparticle induced toxicity to human sperm by increasing ROS(reactive oxygen species) production and DNA damage. Environ Toxicol Pharmacol 52:193–199
74. Overbye M (2018) An (un)desirable trade of harms? How elite athletes might react to medically supervised 'doping' and their considerations of side-effects in this situation. Int J Drug Policy 55:14–30
75. Inbathamizh L, Harsha Varthan MK, Kumar RSR, Rohinth M, Tawfeeq Ahmed ZH (2023) Nanotechnology: ethical impacts, health issues, and safety issues. In: Modern nanotechnology. Springer Nature Switzerland, Cham, pp 455–477
76. Frączek W, Kotela A, Kotela I, Grodzik M (2024) Nanostructures in orthopedics: advancing diagnostics, targeted therapies, and tissue regeneration. Materials 17(24):6162
77. Sharma P, Singh V, Thakur A, Malik AK (2024) Nanotechnology, artificial intelligence, and the future of manufacturing, pp 307–332

Synergic Effects: Combining Nano-nutrients for Maximum Impact

Nafih Cherappurath, Mohammed Navaf, Kappat V. Sunooj, Masilamani Elayaraja, Dilshith A. Kabeer, K. K. Shareef, M. K. Aravind, A. K. Safad, and Borem Siri Reddy

Abstract Nanotechnology advanced sports nutrition through nano-nutrients, a tiny, highly bioavailable particle that overcomes the limits of the traditional supplement. Nano-nutrients differ from traditional supplements mainly due to high precision targeting and enhanced bioavailability. The scientific formulations are optimized for athletic performance and recovery with precision delivery, improved absorption, and targeted physiological benefits. This really happens in the synergetic combinations of the nano-nutrient itself, whereas traditional nutrients come in two or more side by side. Therefore, for example, nano-curcumin was enhanced in combination with nano-piperine in synergy to produce anti-inflammatory results, whereas nano-iron may be combined with vitamin C to optimize oxygen transport. Such synergies allow maximal effectiveness with minimal dosing requirements. This chapter digs into the

N. Cherappurath (✉)
Department of Physical Education, Amal College of Advanced Studies, Nilambur, India
e-mail: nafihch@amalcollege.ac.in

M. Navaf · K. V. Sunooj
Department of Food Science and Technology, Pondicherry University, Puducherry, India
e-mail: navafmak@gmail.com

K. V. Sunooj
e-mail: sunooj4u@gmail.com

M. Elayaraja · D. A. Kabeer (✉) · K. K. Shareef · M. K. Aravind · A. K. Safad
Department of Physical Education and Sports, Pondicherry University, Puducherry, India
e-mail: dilshith.dpe@pondiuni.ac.in

M. Elayaraja
e-mail: elaya.cricket@gmail.com

K. K. Shareef
e-mail: shareefkkvkn@gmail.com

M. K. Aravind
e-mail: mkaravindgk@pondiuni.ac.in

A. K. Safad
e-mail: isafadak@gmail.com

B. S. Reddy
Conestoga College Institute of Technology, Kitchener, ON, Canada

A. Chatterjee et al. (eds.), *Nanofuel: The Future of Sports Nutrition*, Smart Nanomaterials Technology, https://doi.org/10.1007/978-981-96-5471-0_12

science behind nano-nutrients: how they work, how they can work synergistically, and actual applications in sports and health. Several groundbreaking innovations thrusting this field forward are analyzed with special attention to how these novelties are transforming nutrition for athletes and healthy individuals. The chapter further explores evidence-based insights on the transformative potentiality of nano-nutrients for improving performance and recovery.

Keywords Nano-nutrients · Sports nutrition · Bioavailability · Targeted delivery · Nutrient absorption · Synergistic combinations · Athletic performance · Recovery optimization

1 Introduction

In modern civilisation, the current lifestyle routine is combined with more health disorders like diabetes, obesity, cancer and many other diseases. The market of health functioning focuses on food nutritional values, and there is a requirement for bioactive nutrition elements and nutraceuticals, like vitamins, minerals, fatty acids, amino acids and essential oils [1]. However, aiming to elevate bioactive components only through food invention is not rational. Therefore, the best strategy for delivering nutraceutical elements is to use Nano-nutrient compounds through Nanotechnology. The function of nanotechnology in nutrition is to progress the delivery, bioavailability, absorption, solubility, and efficiency. Nano-nutrients play a significant role and contribute to uncountable opportunities for good food quality and safety in human nutrition. It is also progressively included in various sports nutrition fields [2]. Nutritional nano-compounds offer numerous benefits over the standard nutritional supplement; in the context of exercise performance, the benefits of nano-nutritional compounds influence the understanding of new strategies to elevate the action of nutritional supplements and improve exercise performance. Nano-nutritional supplementation focuses on physical capabilities, exercise performance, recovery improvement between workouts, oxidative stress and inflammation reduction, and increasing mitochondrial function and quality [3]. This chapter aims at the synergic effect or influence of Nano-nutrients enhancing and elevating athletes' performance, recovery, and nutritional effectiveness. The combination of nutrients is developing more effectiveness in absorption, target delivery, and bioavailability. Additionally, covering the safety, delivery methods, future and emerging innovations, challenges, and scientific ways of nano-nutrition in sports through encapsulation, nano-scale, and nano-nutrients. The chapter will provide awareness to every individual, especially sports athletes, about their overall health and performance enhancement through the advanced technology of nano-nutrition and its synergic impact.

While nano-nutritional compounds represent several advantages in effectiveness and efficiency in delivering nutrition [3], the specific advantages of nano-nutritional compounds are listed as follows:

- Nano-nutritional compounds are absorbed more effectively into the bloodstream and targeted tissues with higher availability.
- Reducing the side effects and attaining the required effect with smaller amounts due to higher effectiveness.
- Improving energy availability, enhancing muscle recovery and reducing muscle fatigue.
- Customising nutritional resolutions to athletes' insufficiencies with the needed supplements.
- Decreasing the wastage and increasing the impact of nutritional compounds directed to the cells or tissues.

1.1 Definition and Significance of Nano-nutrition

Nano-nutrition means to develop efficacy, bioavailability, and delivery of nutrition with nanotechnology. Similarly, nano-nutrients are substances of specific nutrients, such as vitamins, minerals, and amino acids [4]. It indicates nutrient compounds designed at the nano-scale to enhance their bioavailability, stability and solubility in living systems. The bioavailability of nutritional compounds converting to nano-particles for effective absorption into living tissues and cells [5]. Nano-nutritional compounds are also the latest and most effective channel for nutrient delivery to improve digestion and absorption for enhanced nutrient metabolism. On the other hand, conventional supplements make tasks in the focus delivery to body tissues, lesser solubility, and degradation in the gastrointestinal system.

Nutrient compounds designed at the nano-scale for better nutrient delivery, consuming nano-nutrients improve passage through the mucus layers of the gastrointestinal (GI) tract. The nutrients will first pass through the mucus lining, then cross the cells in the intestine, and thus into the bloodstream. The small scale of the nano-nutrients helps them to pass through these barriers, making them better at absorption. Ensuring the effectiveness of nano-capsules, food compounds, and wall materials is crucial [6]. Figure 1 depicts Nanotechnology for nutrients delivery.

1.2 Historical Evolution of Nano-Nutrients in Nutrition Science

Nano-nutrients started with nanotechnology in the mid-20s, which is the science of working with substances of the nano-scale size range. The famous talk by Richard Feynman in 1959 called "There's Plenty of Room at the Bottom" discusses how materials can be manipulated at their atomic and molecular levels so that they can extended into manufacturing materials on a small scale [7]. Even though the idea was not initially applied to nutrition, it inspired the delivery of better solutions in this area. The beginning of "nanotechnology" was marked by Japanese scientist Norio

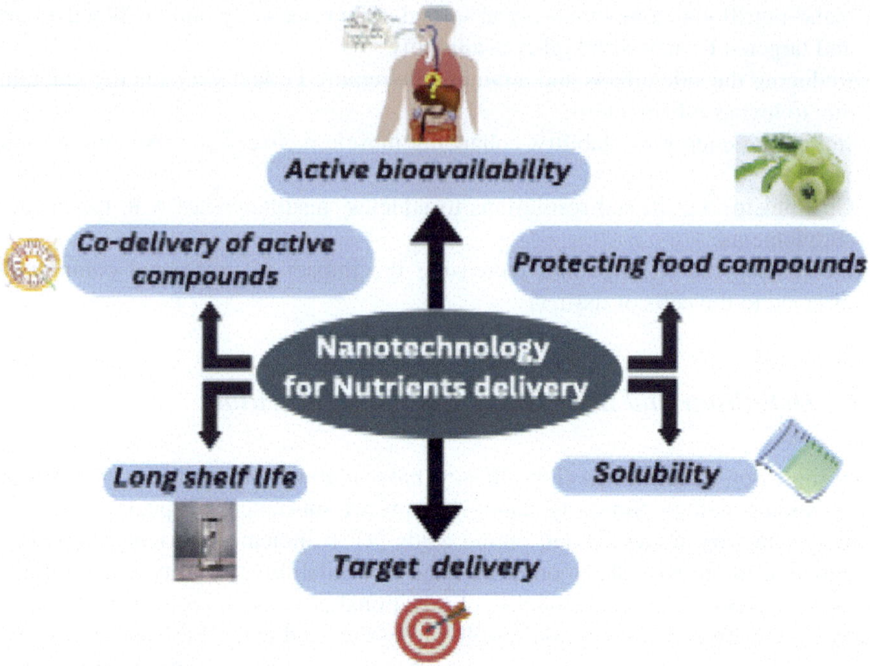

Fig. 1 Nanotechnology for nutrients delivery

Taniguchi in 1974. He focuses on research in nano-scale materials which eventually helped a breakthrough into the effectiveness of nutrients [8].

By the 1990s, nanotechnology was introduced into food science, which paved the way for Nano-encapsulation, which involves enclosing nutrients such as omega-3 fatty acids and vitamins into tiny carriers called nano-capsules. Another milestone in food science was nano-emulsions, tiny drops of fat-soluble vitamins, making them easy to dissolve [9]. By 2010, nano-nutrients started using functional foods, dietary supplements and sports nutrition products, and by the mid-2020s, nano-nutrients started to focus on personalisation, efficiency, and sustainability. From the early twentieth century to this twenty-first century, the evolution of nano-nutrients gives us an insight into how nanotechnology has paved a significant way to nano-nutrients. Nano-nutrients address low absorption, nutrient breakdown, and individual dietary needs. Nano-nutrient redefines the perspective of nutrition in every aspect.

1.3 The Science Behind Nano-nutrition

The science of nano-nutrition is how nutrients are delivered through mini particles to the body. The drawbacks of traditional nutrition delivery were overcome by the concept of nano-nutrition, which focuses on giving unique properties of nano-scale particles to enhance absorption, stability, and targeting deliveries, thus optimising health and performance. These conventional programs changed through nanotechnology like nano-scale, nano-encapsulation, emulsion, etc.

In the background of nano-nutrients, the word Nano-scale refers to a size between 1 to 100 nm (nano-metres) of materials presenting rare biological, chemical and physical things in tiny dimensions [10]. These dimensions create a large percentage of chemical and physical changes in the particle structure and arrangement of electrons in the outer energy level. Nano-compounds are classified through nano-scale, based on size and shape (e.g., rod, globular, conical, hollow, coiled, plane, cylindrical, or asymmetrical).

1.4 Synergetic Effect of Nano-nutrition on Sports Performance

The synergetic effect of nano-nutrition plays a very significant role in improving athletic performance. Combining Iron and vitamin C is the most common example of the synergetic effect. Iron absorption increases with the consumption of vitamin C. These effects create better delivery and absorption for the body. Similarly, athletes get fast recovery, endurance, and good energy levels [11]. More about this topic will be discussed in the upcoming section.

2 Nano-vitamins in Nutrition

Nano-vitamins are vitamins that are encapsulated or delivered using nanotechnology. Traditional vitamins are susceptible to environmental factors such as heat and oxidants, which can degrade their effectiveness. Thus, they must be protected. Nanotechnology and its advanced encapsulation and nano-emulsification techniques enhance stability, bioavailability, and targeted delivery and prevent them against environmental factors, making them highly effective in sports nutrition [12]. Few studies have concluded that using nano-vitamins is beneficial for enhanced physiological outcomes related to exercise performance and general health.

2.1 Vitamin-A in Nano-nutrition

Nano-vitamin-A refers to the formulation of vitamin A within the nano-particles to enhance its stability, bioavailability, and efficacy through encapsulation. This technique is capable of addressing numerous issues arising from the solubility and instability of the vitamin [13]. It is a fat-soluble vitamin found in many foods. This vitamin is not synthesized in the human body and must be taken through food only. It can contribute to various body functions like maintaining healthy vision [14], immune function [15], skin health [16], and wound healing [17] which are essential for athletes who endeavour for peak performance and the ordinary population's overall well-being. Nano-vitamin A formulations address critical issues such as stability and poor bioavailability and ensure optimal release using formulations. This innovation makes it very useful in the prevention of deficiencies and in meeting the enhanced nutrient requirements of athletes and all other individuals.

2.2 Nano-vitamin B Complex

There are eight water-soluble vitamins in the group of B vitamins. Each of them has various critical roles in the human body. These are water-soluble vitamins, and our body does not reserve them. It can perform as coenzymes in a significant role in the enzymatic mechanism that supports every component of cellular functional physiology and plays an essential role in improving brain function [18]. All B vitamins are vital in the harmonious functioning of the energy system as well as performance. Lack of any of these vitamins will decrease energy levels, affect mental health, and have adverse effects on the body [19]. Vitamin B is deficient in a large population due to their dietary choices like minimal intake of animal-source food, alcohol consumption, and lack of nutritional literacy [20]. However, the excess amount of consumption of these vitamins will not have a more positive impact on the human body. It will excrete through urine [21]. From one perspective, it is safe to have more than the suggested daily amount. Five of the eight B vitamins don't have an upper limit for regular intake. However, unlike fat-soluble vitamins, they need more regular intake [18]. Due to the enhanced availability, effectiveness of nano-vitamin B complex, our body can use its benefits more efficiently.

2.3 Nano-vitamin C

It is a water-soluble vitamin and has great antioxidant properties. Our body cannot synthesize or reserve vitamin C, so it should be consumed through food. However, exposure to heat, light, moisture, high-level oxygen concentration, and metal ions can make the vitamin unstable, making its preservation a major challenge [22]. An option to resolve the challenge is implementing nano-encapsulation technology to improve the stability and maximize the benefits of vitamin C. The antioxidant properties of vitamin C can fight against free radicals and other chemicals inside the body. It is also evident that it can help the process of collagen synthesis, and the simultaneous intake of vitamin C can contribute to improved muscle endurance, reduce the chance of injury during training, and speed up the recovery process of the human body.

A deficiency of this vitamin can compromise the immune system, causing muscle fatigue and bone weakness [23]. There are different types of encapsulation methods and agents to stabilize vitamin C. Rice starch and gum Arabic are used as wall material in the spray drying process for the encapsulation of Ascorbic Acid (vitamin C) [24]. Pierucci et al. [25] confirmed pea protein can also serve as the wall material in the same spray drying process. There are more methods to encapsulate vitamin C, such as liposomes [26], spray chilling [27], microfluidic devices [28], emulsification [29].

2.4 Nano-vitamin D

Vitamin D (V-D) is a fat-soluble vitamin. Unlike other vitamins, they can be synthesized in the human body when the body is exposed to UV B radiation from direct sunlight [30]. And can naturally achieve the recommended dietary allowance of V-D through a few foods [31]. V-D has several vital roles in the human physiological system, mainly in improving bone health and the metabolism process of calcium and phosphorus. The latest studies have further examined its vital role in cancer prevention, cardiovascular problems, the growth of cells, the immune system, liver health, the reproductive system, the nervous system, and the respiratory system [32]. In terms of sports, the supplementation of V-D can improve muscular and power [33], muscle function [34], and recovery [35] in individuals. Also, the concentration of V-D in the bloodstream is linked with bone mineral content, which plays a beneficial role in physical exercise and athleticism [36].

However, vitamin D is deficient in most of the world's population. It is because of minimal exposure to sunlight, aerial impurities, use of sun protection factors, the angle of the sunlight penetration, use of clothing for sun protection, skin nature, and limited food intake [36]. The process of synthesis of this vitamin in the human body is a complex process. It is also very difficult to get V-D from the food. Many external factors can affect its potency since it is sensitive to moistness, light, temperature, and a high concentration of oxygen. Nano-encapsulation techniques could be a game changer to protect the nutrients from such external factors [32]. Nano-encapsulation is widely done by formulation approaches like nano-emulsions, nano-particles, lipid nano-capsules, complex coacervate structures, and casein micelles, which can improve the effectiveness of V-D [37].

2.5 Nano-vitamin E

Vitamin E (V-E) is a fat dissolvable vitamin. It is not synthesized in the human body. Photosynthetic organisms and plants are the major sources. V-E has excellent antioxidant properties that fight free radicals. It consists of two main groups, specifically, tocopherols and tocotrienols. These groups are further categorized into four distinct isomers. In humans, the most active form of V-E is tocopherol. The daily required amount of V-E is achievable through a natural diet [38]. Normally, it is stable at room temperature, but it is prone to oxidation in hot conditions and exposed to sunlight.

Due to so many factors in the environment, the bioavailability and absorption amount of V-E in the gastrointestinal tract may be reduced. Nano-encapsulation effectively contributes to preserving its properties and helps to deliver the greatest value to the body [39]. This enhances the solubility and slow and sustained release of V-E and council delivery at the point of need. Nano-liposomes, nano-gels, and nano-fibers are the modes of nano-encapsulation adopted for the modification of solubility and bioavailability of fat-soluble vitamins such as V-E [40].

2.6 Nano-vitamin K

It is an oil-dissolvable nutrient that plays a significant role in assisting many physiological functions, like blood coagulation, supporting skeletal mineralization, improves vascular health [41]. There are a few challenges that can reduce its absorption rate and effectiveness, such as low dissolvability in water. It is hard to absorb the vitamin without the presence of fats since it is lipid-soluble. Environmental factors such as high temperature and sunlight exposure may weaken its effectiveness [42]. There are different nano-carriers to improve the efficiency of V-K, such as lipid-based nano-particles [43], Nano-emulsion [44], Polymeric nano-particles, and Liposomes [45].

3 Nano-minerals

The primary source of minerals is soil. Plants absorb the minerals from there to sustain their development. Humans get the minerals when they consume the plants. Other sources are animal products (milk and meat products) [46] and water sources (nutrient-dense and seawater) [47, 48]. Minerals are essential requirements to perform several operations in a physiological system. They protect our body against nutrient deficiency, enhance the metabolism process, and strengthen bones and teeth [49]. The efficiency of these minerals, their absorption, and bioavailability can be enhanced by incorporating nanotechnology [50].

3.1 Nano-calcium

Due to their significantly smaller size, the calcium nano-particles offer greater surface area compared to the traditional supplements. Therefore, a larger surface area can boost the solubility and absorption amount in the digestive tract, which improves the effectiveness of the mineral [51]. There are several methods for calcium encapsulation, such as lipid-encapsulated systems and polymer-based carriers, which may give an advantage in terms of efficiency and prevent degradation of the mineral. Nano-encapsulation also makes the nutrients more accessible to the body by releasing them in gradual, prolonged release [52]. This mineral may help minimize the chances of osteoporosis by increasing skeletal mineral density [53], as well as by lowering LDL, improving HDL, and developing blood lipid profiles. They may also have a positive impact on the cardiovascular system [54].

3.2 Nano-zinc

The use of nano-zinc allows the body to maximize the usage of its efficiency as compared to conventional zinc supplements, resulting in better absorption by the body [55]. It can guard the human body against cellular damage by lowering oxidative stress due to its great antioxidant qualities. Additionally, it is very effective in safeguarding the human and animal body from various diseases because of its antibacterial properties [56].

3.3 Nano-iron

Nano-iron supplements can be composed of mixed iron compounds, like iron oxides, iron hydroxides, and iron phosphates, to increase bioavailability and boost their efficiency in the biological system [57]. The recent studies on iron supplements confirm that they are very safe to consume and did not find any toxicity in the body when the iron supplements were fed to rats at 25 mg Fe/kg (many times higher RDA than of a human). After contact with iron phosphate, body cells and tissues can survive and maintain regular function [51]. Nowadays, there are several methods to encapsulate iron to improve its effectiveness and bioavailability in the digestive tract, such as the Ion gelation method [58], Spray drying [59], Nano CaO matrix (calcium oxide), Emulsification [60], Noisomal encapsulation [61] These different methods have diverse advantages and disadvantages like bioavailability, stability, cost. However, Iron consumption improves many physiological advantages, which can improve physical efficiencies and day-to-day activities, such as improvement of hemoglobin and myoglobin function [62], fight against anemia [63], and energy metabolism [64].

3.4 Nano-selenium

Selenium and nano-selenium supplements are commonly consumed to support human health. It is confirmed that selenium has anticancer and antimicrobial properties that can significantly support the body in fighting against disease [65]. It also has anti-viral and antioxidant properties, so it is very effective in treating wound healing [66]. Compared to inorganic selenium, nano-selenium has lower toxicity, higher levels of bioavailability, is very safe, and has higher potential in preventing diseases related to selenium deficiency [67].

3.5 Nano-magnesium

After calcium, potassium, and sodium, magnesium is the fourth most available element in the human body. The second most positively charged ion in the body after potassium. This mineral contributes to more than 300 metabolic activities in the human body and plays many vital roles in many physiological systems. Mostly found in bones (60%), soft tissues (40%), and blood (1%) [68]. Magnesium also has many key functions, such as strengthening the skeletal tissue, boosting the immune system, blood pressure regulation, maintaining heart function, and regulating normal muscular and nervous activities that may help individuals improve day-to-day life and physical activities [69]. Taking magnesium supplements also helps athletes recover faster, regulate energy regulation, and enhance performance. Synthesis of nano-magnesium can be done through different physiochemical methods, such as hydrothermal, sonochemical, electrochemical synthesis, template-free synthesis, sol–gel, and precipitation methods [70].

3.6 Nano-manganese

For almost all living organisms, manganese performs a key role as one of the five crucial trace minerals. It is widely found in Earth's outermost layer. It is the most needed mineral that is involved in numerous biochemical functions in the human body, like assisting various metabolic reactions because of its role as a cofactor for a number of enzymes. These enzymes help in proper energy production by the efficient breakdown and synthesis of carbohydrates, lipids, protein, and amino acids [71]. Manganese also acts as an antioxidant, supports bone health and the nervous system, and has more potential advantages for living organisms. However, excessive manganese consumption can lead to toxicity, a condition called manganism, which is similar to Parkinson's disease, cognitive impairment [72], and motor impairment [73].

3.7 *Nano-chromium*

Chromium is a cofactor that plays a key role in supporting insulin's mechanism. It can reduce the risk of insulin resistance because of its capacity to regulate glucose metabolism and balance blood glucose levels [74]. This trace mineral may facilitate the breakdown of carbohydrates, protein, and lipids to improve energy and support overall metabolic health.

4 Nano-proteins

Nano-proteins are protein-based nano-particles that are progressively being investigated for their possible benefits in nutrition and food science. Nano-proteins can improve the solubility, bioavailability, and stability of bioactive compounds, which are very prone to degradation or hard to dissolve, through nano-carriers that deliver the nutrients directly to the targeted tissues in the body [75]. There are some advanced technologies that are used to improve the safety, shelf life, and efficiency of these particles, such as nano-coating, nano-encapsulation, nano-emulsions, and nano-hydrogels. Compared to whole proteins, nano-proteins are much faster in digestion and absorption in the gastrointestinal tract. Subsequently, it can improve the process of muscle recovery and repair shortly after consumption. Because of its quality, it can be the best pick after post-workout [76].

5 Nano-omega 3 Fatty Acid

Nano-omega 3 fatty acids are the encapsulated version of omega 3 to optimize their stability, absorption ability, and controlled delivery. Nano-emulsion, solid lipid nano-particles, and nano-structured lipid carriers are the best nano-delivery systems used to improve the availability and enhance the effectiveness of omega-3 [77]. Omega 3 has many functions in the human body to support better health, such as individual growth and development, enhance visual health, lower blood pressure, provide anti-inflammatory and immunomodulatory benefits, and play a vital role in cardio-protection [78]. However, it cannot be produced by the body and should be consumed through food. In the context of sports nutrition, it is necessary for athletes to reduce inflammation in the body and muscle soreness, enhance muscle recovery, and support joint health due to its anti-inflammatory properties. It can also help muscle functions and maintain muscle mass, which offers benefits for athletes [79].

6 Mechanism of Nano-nutrient Absorption

The significant advantage of nano-nutrients is their proficiency in absorbing nutrients compared to traditional supplement forms. Due to its increased surface area and small size, nano-nutrients demonstrate exceptional solubility. Compared to traditional nutrient forms, nano-nutrient characteristics improve the dissolution efficiency in the digestive tract and increase their bioavailability, making them readily available for the body and utilizing their benefits [2]. There are several mechanisms by which the nano-nutrients interact with the intestinal wall, including passive diffusion, endocytosis, and active transport. Passive diffusion permits the nano-nutrients to pass through the cell membrane without using energy. Endocytosis is when cells surround the particles to form vesicles. Meanwhile, active transport needs to use energy to transport the nutrients from lower concentrated areas to higher ones. These mechanisms also improve the bioavailability of the nutrients [80]. Primarily, nano-nutrients interact with the intestinal wall through epithelial cells. Due to their small size, they penetrate the mucosal layer more effectively. This felicitates efficient absorption by epithelial cells. After that, the nano-nutrients travel through the cytoplasm, involving numerous cellular activities and directing them to diverse cellular regions for further processing. Then, the nutrients are delivered outside the cell via exocytosis from the basal part of the epithelial cells and enter the bloodstream to be carried throughout the body [81].

7 Benefits of Nano-nutrients in Sports Performance

It has been widely proven that individualized nutritional plans can enhance exercise performance and help support a better lifestyle. Nano-nutrients could boost the effectiveness of these bio compounds due to their encapsulated structure. The advanced formulation of nano-nutrients shows immense promise as a workout enhancer. Nanotechnology may also provide a transformative influence on modern sports nutrition and may help to build strategies according to individual demands.

8 Nutrient Synergism: The Combined Effects of Nutrients

Nutrient synergy can be described as a process whereby if two or more nutrients are taken, the outcome or impact on the body is improved. The combinations allow for a better chance at enhancing the body's capacity for assimilating, utilizing, and getting value out of these nutrients. This interaction can lead to multiple advantages for the physiological system [82]. The concept of nano-nutrient synergy has excellent potential in sports nutrition by optimizing the body's ability to absorb

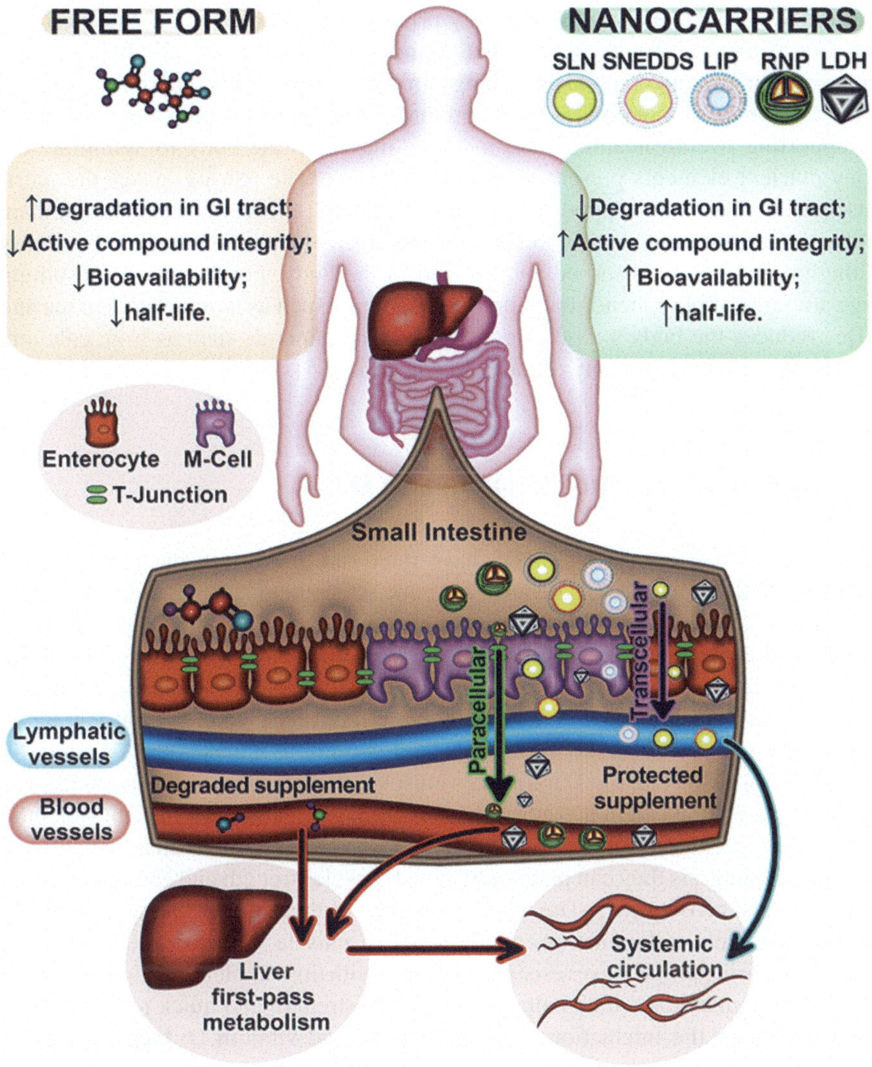

Fig. 2 Comparison between nano-encapsulated active compounds and non-nano-encapsulated active compounds during gastrointestinal transit and absorption (Fig. 2). Solid lipid nanoparticles (SLN), self-nano-emulsifying drug delivery systems (SNEDDS), and liposomes (LIP) are nano-carriers that are directly absorbed in the lymphatic vessels, thereby avoiding the liver's first-pass metabolism. Redox-active nanoparticles (RNP) and layered double hydroxide nano-particles (LDH) are taken up in the blood vessels, meaning that first-pass metabolism occurs before they enter the systemic circulation [3]

and utilize nutrients and improve energy production, endurance, and muscle function. For example, it has been seen that taking a combination of carbohydrates and proteins after exercise replenishes energy in the muscle faster than only carbohydrates and proteins alone [83]. The combinations of nutrients play beneficial roles in supporting overall health. These methods may help the body to assimilate the many nutrient elements present in this combination. The pairing makes the nutrients work together and support health and well-being [84]. On the other hand, there is also antagonistic interaction in some of the nutrients and this hinders the ability of the nutrients to be absorbed and utilized in the body. For example, consuming large amounts of polyphenol-containing compounds such as tannin found in tea and coffee reduces the body's absorption of necessary minerals such as iron, calcium, and copper, resulting in many diseases like anemia [85].

8.1 Nano-calcium and Nano-vitamin D3

Nano-calcium and nano-vitamin D3 have many advantages in the physiological system since they have special benefits for bone strength, muscle function and quick recovery. Vitamin D3 helps to improve the bioavailability and absorption level of calcium in the intestines. This combination improves bone mineral density, lowers the risk of bone injuries like stress fractures, strengthens muscles, and enhances immune function, which is beneficial for athletes more effectively than either nutrient alone [86]. Calcium is also necessary for muscle contractions, as it makes cross-bridges by binding to troponin. Vitamin D influences muscle cell growth and enhances muscular strength by assisting calcium absorption. Combining both nutrients has the potential to improve muscular power and physical performance in adults, even in the absence of exercise. Suggests they can potentially boost muscle strength and endurance while doing physical activity [87]. The interdependence of these nutrients also effectively balances systemic inflammation, which leads to muscle hypertrophy. Researches reveal that pairing progressive resistance training with this supplementation decreases inflammatory markers, which helps boost muscle strength and quick recovery [88]. Figure 3 depicts the interaction between calcium and vitamin D: Increasing bone mass mineralization and calcium bioavailability and absorption.

8.2 Nano-vitamin C and Nano-iron

To date, there are promising breakthroughs in nutritional science regarding the synergic interaction between nano-vitamin C and nano-iron. Vitamin C helps to improve the amount of iron absorption in the body and boosts the iron uptake in the intestinal cells. These nutrients may enhance the iron content in the body and prevent the development of anemia.

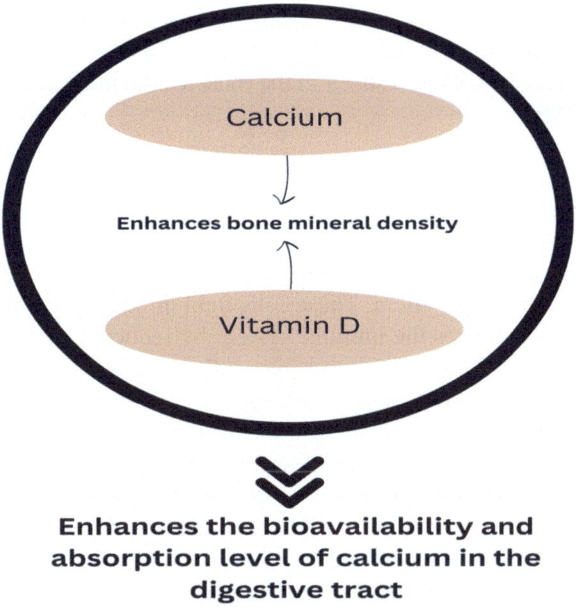

Fig. 3 Interaction between calcium and vitamin D: Increasing bone mass mineralization and calcium bioavailability and absorption

8.2.1 Effects on Hemoglobin Synthesis and Oxygen Transport

According to the discussion above, vitamin C has a role to play in the absorption of iron in the body, control of the synthesis of hemoglobin, and utilization of iron. An increased property of hemoglobin allows the blood to transport more of the oxygen needed for muscle during physical activity. For sustainable performance and optimal exercise volume endurance and intensity of the athletes, it is helpful in high-intensity endurance by delaying the onset of fatigue [89].

8.2.2 Antioxidant Activities and Muscle Recovery

The combination effects of nano-vitamin-C and nano-iron promote the Fenton reaction and increase the ROS (reactive oxygen species) generation. This effect is remarkably useful in therapeutic applications where balanced oxidative stress is needed, like when aiming at cancer cells [90]. Vitamin C can effectively modulate the generation of reactive oxygen species (ROS) caused by the presence of nano-iron. This interaction intensifies the oxidative stress response, leading to selective damaged or stressed cell death and protecting healthy cells because of the antioxidant properties of Vitamin C. The combination can also control the oxidative environment, stimulate cellular repair, and help lower overall oxidative damage to the cells [90]. It is also evident that they can enhance mitochondrial functions and reduce damage to biomolecules, thereby boosting cell health and efficiency [91].

They also help muscle functioning in many ways. Nano-iron stimulates blood circulation and helps oxygen delivery to muscles. Simultaneously, nano-vitamin C reduces oxidative stress in the muscles, facilitating muscle repair and inflammation [92].

8.2.3 Immune System

Vitamin C supports the production and functions of white blood cells, strengthening the immune system. These cells fight infections and improve the immune system. It also regulates the immune system by reducing the production of pro-inflammatory cytokines and helping to reduce inflammation in the body, which helps to support the immune system and protect tissues from damage [93]. Nano-iron also plays a role in the immune system by increasing the number of lymphocytes and T cells, which are essential for effective immune response [94]. Several researches revealed that vitamin C and iron support the immune system. A good immune response will help athletes recover faster from an injury or after exercise, reduce the risk of injuries, and enhance performance.

8.2.4 Benefits of Endurance Exercises

As already explained before, vitamin C enhances the absorption of iron and thus may influence oxygen delivery to the muscle tissue, thereby delaying the onset of fatigue and enhancing exercise performance [95]. Further research is needed to explore the relationship between the combination of these nutrients and the endurance of athletes.

8.2.5 Practical Implications of Nano-vitamin C and Nano-iron

Numerous studies suggest that an optimum dose of nano-iron (e.g., 0.6 mg/kg) may improve the level of hemoglobin level, leading to better exercise performance without causing any side effects [96]. Multiple research studies have suggested that alternate days in the morning are the preferred time to take iron supplements. In the case of vitamin C, moderate intake (0.2 g/day) is beneficial to reduce oxidative stress in the body. Consuming nano-vitamin C before training is preferable to reduce oxidative stress and muscle damage during training (Fig. 4) [97].

8.3 Nano-protein and Nano-carbohydrate

Consuming both nano-proteins and nano-carbohydrates together can improve the carbohydrate-protein interaction (CPI). Using carbohydrates and protein to make glycol-nano-structures can enhance the strength of their interaction, help the body

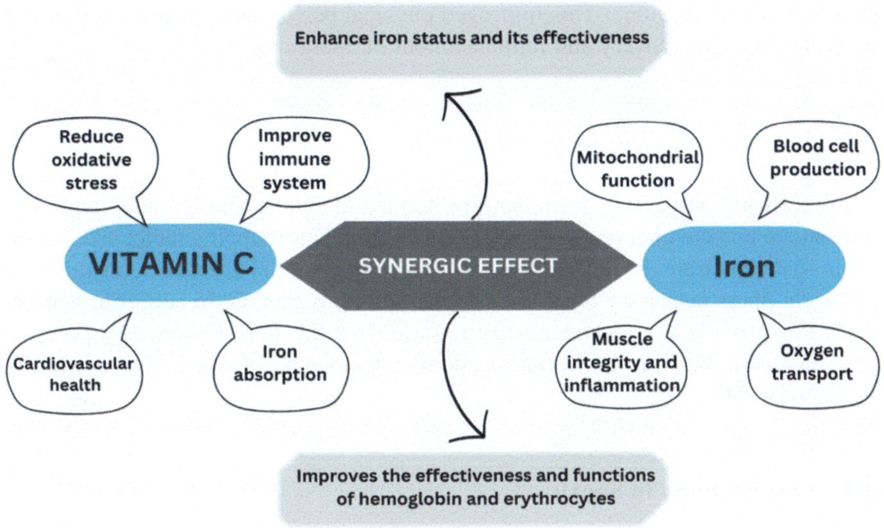

Fig. 4 Synergic effects of nano-vitamin-C and nano-iron on bioavailability and therapeutic outcomes

to utilize its benefits faster, and give greater biological outcomes than consuming alone.

8.3.1 Effects of Nano-carbs and Nano-proteins During Exercise

Taking carbohydrates and proteins during physical activity helps balance glucose levels and slow down glycogen depletion from the muscles, which helps to engage in physical activity for an extended period. This interaction can also help to reduce the onset of fatigue during physical activities, allowing athletes to maintain optimum intensity for a longer duration. It also proved that this nutrient combination can reduce the markers of muscle damage and reduce the breakdown of protein. This pairing helps support muscular functioning during exercise [98].

8.3.2 Nano-carbs and Nano-proteins Post-exercise

Co-ingestion of protein and carbohydrates after physical activity improves the process of muscle glycogen resynthesis and facilitates muscle protein synthesis more efficiently than the nutrients consumed alone. These effects stimulate the recovery process, refill the body's energy stores, and rejuvenate muscle tissues [99]. They also help to reduce the recovery time after vigorous physical activity and assist athletes in training regularly. Several research studies demonstrated that pairing these nutrients assists athletes to perform better in subsequent exercise sessions than those who ingest carbohydrates alone. This improved physical performance is probably linked with better muscle recovery and replenishment of energy stores.

8.3.3 Impact on the Physiological System

As it was mentioned above, both nano-proteins and nano-carbohydrates familiarly contribute to muscle fiber recovery and regeneration. Due to their smaller size, nano-proteins can penetrate the muscle tissues, providing essential amino acids straight to the muscle fibers necessary for muscle repair and rejuvenation. In addition, noncarbohydrates provide the instant energy required for the muscle recovery process. They can also reduce muscle soreness and damage, such as Delayed Onset Muscle Soreness (DOMS).

8.3.4 Practicability in Sports

Several studies showed that the consumption of nano-carbohydrates 30–60 min before exercise ensures enough glycogen and proper energy throughout the activity [100]. Also, the inclusion of small quantity of nano-protein can help the physiological system through muscle preservation, lesser muscle breakdown, and regulate muscle functions [101]. Consuming a blend of protein and carbohydrates during physical activities also enhances performance further. The combination of nano-protein and nano-carbohydrates can also be taken within half hour after the exercise to increase glycogenesis deposit, muscle healing as well as muscle recovery time [102]. The preferred amount of the combination is approximately 30–60 g of carbohydrates and 10–20 g of protein per hour to get the maximum benefits of the nutrients to the body and promote recovery [103]. These studies suggest that consuming a combination of nano-protein and nano-carbohydrate can improve the effects of these nutrients in the physiological system that supports exercise performance before, during, and after a workout.

8.4 Nano-omega-3 and Nano-antioxidant

The combined effect of nano-omega-3 fatty acids and nano-antioxidants has been explored in many studies due to the enhancement in the effectiveness when these nutrients are consumed together rather than when taken alone. Pairing omega-3 and polyphenols (a type of antioxidant) can significantly improve antioxidant enzyme function and plasma antioxidant efficiency, which leads to minimizing oxidative stress in the physiological system. It has also demonstrated that combining these nutrients helps improve the quality of life of patients with challenges like meibomian gland impairment. There are many other physical and mental outcomes of combining these nutrients [104].

8.4.1 Impact on Inflammation and Muscle Repair

Omega-3 fatty acids are quite well known for their antioxidants and anti-inflammatory properties. The combination of these fatty acids with vitamin C has antioxidant properties that may guard against oxidative stress and inflammation, thereby minimizing inflammation-related damage in the tissues of the body [105]. Supplementation in omega-3 has been found to significantly decrease the biomarkers of muscle damage, such as creatine kinase (CK), lactate dehydrogenase (LDH), and myoglobin. That may assist in reducing the incidence of muscle injuries and the time required for muscle repair following exercise [106]. On the other hand, no significant difference in the relationship between antioxidants and muscle recovery is found. Several studies reveal that there is little or no difference in the muscle damage markers or muscle economy [107]. On the other hand, some research studies pointed to a molecule known as a polyphenols-a well-known antioxidant found in fruits, vegetables, and herbs- can help recover muscle damage and lessen the impact of oxidative stress by stimulating exercise-induced cell signalling [108].

8.4.2 Impacts on Muscle Function and Performance

Combining both nano-omega-3 and nano-antioxidants has been shown to improve antioxidant efficiency, which helps to reduce oxidative stress and works as an anti-inflammatory agent in the body. Nano-omega-3 supports decreased reactive oxygen species (ROS) production by developing an antioxidant defence mechanism and supports inflammatory mediators during exercise-related stress. In the same way, nano-antioxidants enhance the presence of electron transport chain proteins. That leads to improving the mitochondria's respiration efficiency and muscle contraction. Research suggests pairing nano-omega-3 and nano-antioxidants can enhance cellular integrity and positively affect athletic muscle functions.

8.4.3 Impacts on the Cardiovascular System

Omega-3 fatty acids, particularly eicosapentaenoic acid (EPA) and docosahexaenoic acid (DHA), are widely known for their impact on cardiovascular health. They can potentially lower systemic inflammation and triglyceride amounts and regulate endothelial functions, contributing to heart health [109]. Additionally, nano-antioxidants also play a crucial role in heart health. A formulation of antioxidants, such as coenzyme 10 and vitamin E, has impressive potential in reducing oxidative stress and inflammation, which may influence cardiovascular wellness. Combining these nutrients can lead to reduced cardiovascular injuries induced by oxidative stress and inflammation [110].

8.5 Nano-prebiotics and Nano-probiotics

The combination of prebiotics and probiotics is called synbiotics. These pairing of nutrients can significantly improve the viability and effectiveness of probiotics by providing them with a proper environment for growth. This synergy can help probiotics maintain viability in the gut for longer, making them offer their benefits more efficiently [111]. They can play a pivotal role in improving gut health by actively supporting gut-friendly beneficial bacteria like Lactobacillus and Bifidobacterium. Prebiotics are characteristically indigestible fibers that selectively feed this beneficial microorganism, which enhances the diversity and functioning of the gut microbiome.

Many studies have demonstrated having a healthy gut microbiome can enhance the absorption level of essential nutrients. Synbiotics can improve the function of the digestive barriers and help to increase nutrient bioavailability. For instance, probiotics can regulate the production of small-chain fatty acids, which are beneficial in nutrient absorption and maintaining a healthy digestive system [112]. In the context of sports, a healthy, balanced gut health is crucial for better athletic performance. Enhancement in the level of absorption of essential nutrients like amino acids facilitates muscle repair and growth, vitamins improve energy metabolism, and minerals enhance muscle performance and recovery [113]. However, traditional microencapsulation has many challenges. To overcome these challenges, cytoprotective nanocoating for probiotics has emerged. This method improves particle size control and enhances its effectiveness in vivo.

8.5.1 Nano-synbiotics in Gut Health and Performance

Nano-prebiotics, such as Phthalyl pullulan nano-particles (PPN) combined with nano-probiotics like Lactobacillus Plantarum (LP) have proven higher efficacy in defending against harmful pathogens and help to restore gut dysbiosis than consuming probiotics alone. These combinations significantly enhance the development of beneficial bacteria like LP and Bifidobacterium and enrich microbial diversity, which contributes significantly to maintaining a healthy gut environment [114]. They also have potential antimicrobial, anti-cancerous, and antioxidant properties, making them valuable across various therapeutic applications [115]. This research reveals that diverse gut microbiota is beneficial in the production of short-chain fatty acids (SCFA), which enhances energy metabolism and lowers inflammation in the body. Improved gut health through consuming synbiotics can significantly enhance immune responses, prevent infections, and improve overall well-being, which is essential for maintaining peak performance [116].

8.6 *Nano-curcumin and Nano-piperine*

Nano-curcumin significantly enhances the bioavailability and therapeutic potential of curcumin, thereby increasing its beneficial features in the body [117]. The synergic effects of curcumin and piperine have the potential to reduce exercise-induced inflammation, joint pain, and muscle soreness. This pairing of bioactive components enhances curcumin's anti-inflammatory and antioxidant features, maximized by piperine, which improves the bioavailability of curcumin by 2000% [118]. As outlined before, the anti-inflammatory and antioxidant features of curcumin help to reduce inflammation and oxidative stress caused by physical activity. It also minimizes muscle tears associated with exercise by preventing free radical formation in the body and stabilizing cell membranes. On the other hand, piperine increases curcumin absorption level and bioavailability of curcumin and helps the body to utilize its benefits.

8.6.1 Synergic Effects of Nano-curcumin and Nano-piperine Benefits

Nano-curcumin improves the antioxidant mechanism in the body and helps to neutralize free radicals formed during intense exercise. This function can protect muscle cells from oxidative stress and encourage better recovery and holistic performance. Due to its anti-inflammatory and antioxidant properties, it can play a crucial role in stabilizing muscle integrity and function. It also aids in maintaining muscular strength and power and supports muscle repair between training sessions [119]. Curcumin is also beneficial for supporting joint health. It helps to prevent joint injuries and situations like joint arthritis by lowering elevated levels of pro-inflammatory cytokines. This may reduce the severity of joint discomfort and aid mobility and rejuvenation [120]. When curcumin is paired with piperine, it plays a significant role in increasing the bioavailability of curcumin in the intestinal line. They may also effectively reduce pain and discomfort from joint injuries and guarantee better joint comfort.

8.6.2 Practicability of the Synergy of Nano-curcumin and Nano-piperine

Several studies suggest that a daily dose of 80 mg of nano-curcumin can significantly optimize metabolic efficiency and calm inflammation in the body [121]. In addition, a higher dose of 180 mg daily intake can offer a meaningful improvement in reducing muscle damage and inflammation. These compounds may also promote exercise recovery after intense exercise [122]. Curcumin paired with piperine at a 100:1 weight ratio can greatly enhance the bioavailability and effectiveness of curcumin in the digestive tract [123]. Consuming these bioactive components prior to exercise can diminish acute inflammation related to exercise [122]. Alternatively, consuming them after 24 h post-exercise drastically reduces serum creatine kinase (CK) levels, signalling reduced muscle damage and better muscle recovery [124]. Pairing these components is usually safe and easily tolerated at recommended doses with minimal side effects. Excessive or prolonged consumption may have health risks like toxicity in the liver. Intake of these bioactive components can cause gastrointestinal problems in some individuals [125].

8.7 Nano-creatine and Nano-carbohydrate

Nano-creatine, when encapsulated in nano-liposomes, can improve the penetration ability of creatine through the blood–brain barrier. Compared to traditional creatine supplements, nano-creatine is more efficient in delivering creatine to the digestive wall and central nervous system. It can contribute to various physiological activities like improved enzymic activity and antioxidant activity [126]. It is very common to take nano-creatine to boost exercise performance. Because it allows athletes to perform short, high-intensity activities such as sprinting and weight lifting. This combination may also help the athletes to perform more repetitions and aid in building mass [127].

Nano-particles improve the biological properties of nano-carbohydrates. The nano-particles deliver the nutrients to the targeted cells and enhance their bioavailability in the digestive tract. They interact with receptors on the cell surface and help target specific cells, and they have great potential in targeted treatment, great drug delivery efficiency, and more practical applications [128]. It is evident that carbohydrates are a primary energy source for athletes. Maintaining a balanced amount of carbohydrates is beneficial, especially in endurance sports and prolonged physical activities. Its major role is to produce ATP (adenosine triphosphate), which the body can use as a primary source of energy.

8.7.1 Combined Effects of Nano-creatine and Nano-carbohydrate on Physical and Physiological Performance

Both nano-creatine and nano-carbohydrates are used by athletes as an ergogenic aid. Consuming creatine along with carbohydrates can significantly improve the creatine uptake in the muscle more efficiently than creatine consuming alone. Several studies demonstrated that the combination of creatine and carbohydrates assists in performing high-intensity activities like sprints and high-intensity interval training due to their sustained release of energy. This synergy also improves muscular strength and muscular endurance and helps with muscle functions. In addition, these compounds can speed up the recovery process, lower muscle damage, and prevent muscular injuries [129]. Nano-creatine particles have the quality of involving ATP rejuvenation by boosting the Phosphocreatine (PC) level during any physical activities. On the other hand, nano-carbohydrates help enrich glycogen storage in the muscles and supply sustained energy throughout prolonged exercise. Consuming these compounds prior to high-intensity sports like football and basketball is beneficial for releasing sustainable energy.

8.7.2 Practicability of Nano-creatine and Nano-carbohydrate

There are some protocols that need to be followed during the initial loading phase of creatine. A dosage of 0.3 g per kilogram of an individual's body weight is preferable to consume for a period of 5 days. This is followed by a maintenance phase in which the daily can dosage is reduced to 0.03 g per kilogram. Which effectively stores the phosphocreatine in the skeletal muscles [130]. Consumption of creatine prior to physical activities can promote blood flow to the skeletal muscles and aid in better muscle contractions. In addition, emerging studies reveal that intake of creatine after exercise promotes increased absorption levels in the muscles and helps the body produce more glycogen, which can speed up the recovery process. It is generally safe and comfortable in healthy individuals and doesn't harm any physiological system like renal, hepatic, cardiac, or muscle function in the body.

In the case of nano-carbohydrates, consuming them before physical activities can release sustained energy and help the body mitigate fatigue while doing any physical activity. These components may help reduce fatigue by stabilizing blood glucose levels during any physical activity. Moreover, an intake of 1.2 g per kilogram of body weight within the 4 h of post-workout is suggested to boost glycogen levels in the skeletal muscle and help to accelerate the recovery process [131]. It is very safe to consume in moderate amounts. However, taking excessive amounts can cause gastrointestinal problems that can affect exercise performance. Table 1 depicts Effects of single and combined nano-nutrients on physiological system.

9 Mechanisms of Synergic Actions in Nano-nutrients Combination

Combining some nano-nutrients can improve their stability and bioavailability of the nutrients. Nanotechnology enables the controlled and targeted release of bio compounds by encapsulating them into nano-hydrogels, nano-particles, nano-emulsions, or nano-capsules. These techniques boost overall nutritional benefits by encapsulating and supplying multiple nutrients in a single form. Pairing multiple nutrients in a single form may significantly enhance the nutrient's effectiveness compared to consuming it alone. For example, taking nano-encapsulated omega-3 fatty acids and antioxidants may elevate antioxidant efficiency and enhance nutrient absorption in the digestive tract. Similarly, combining nano-creatine with nano-carbohydrate may aid better ATP production and glycogen synthesis and improve

Table 1 Effects of single and combined nano-nutrients on physiological system

Nano-nutrients	Physiological effects	Combined effects	References
Vitamin D3	Enhances bone mineral density, reduces pro-inflammatory markers, lowers the risk of rheumatoid arthritis, reduces the risk of fractures	Enhances bioavailability and absorption level of calcium in the digestive tract	[132–136]
Calcium	Improves bone mineral density, reduces inflammatory markers, reduces the progression of rheumatoid arthritis, and reduces the chance of fracture and osteoporosis		
Vitamin C	Reduce oxidative stress and inflammatory markers, improving nutrient absorption and bioavailability. Mixed results in the case of performance enhancement, some studies show positive outcomes, and others show neutral effects	May enhance iron status and its effectiveness in the body and blood elements. Improves the effectiveness and functions of hemoglobin and erythrocytes. Enhances the anti-oxidant level and also improves the biomarkers of oxidative stress	[2, 3, 96, 137–140]
Iron	Maintains muscle integrity and inflammation status, Improves mitochondrial functions and quality. Increases hemoglobin and red blood cell count		
Protein	Improves muscle recovery and performance, reduces markers of muscle damage, reduces fat mass when combined with resistance training,	Improve performance and muscle damage and leverage the benefits of both macronutrients	[98, 141–143]
Carbohydrates	Post-exercise consumption could enhance muscle glycogen synthesis. Promote muscle recovery and are beneficial in endurance activity		
Omega 3	Beneficial in post-exercise inflammatory responses and can help lower the muscle damage and oxidative stress caused by exercise. Omega 3 with resistance training could enhance muscle strength	Significant improvement in antioxidant enzyme activities and oxidative stress markers in the body more efficiently than either take alone	[104, 144–146]

(continued)

Table 1 (continued)

Nano-nutrients	Physiological effects	Combined effects	References
Antioxidants	Enhance antioxidant activities and reduce oxidative stress caused by physical activities. Reduce fatigue and help in prolonged activities by enhancing antioxidant defence		
Prebiotics	Stimulate the immune system and improves overall health. Helps to improve gut bacteria and prevent gastrointestinal infections	Helps to elevate the antimicrobial activities of probiotics. Increases the effectiveness of gut microbiomes and may lead to athletic performance. Synbiotics also aid in barrier functions of the digestive tract and reduce systemic inflammation	[114, 116, 147–153]
Probiotics	Reduce gastrointestinal distress and enhances recovery from fatigue by improving the absorption level of nutrients. improved exercise endurance performance. Probiotics may enhance gut microbiota, which may be beneficial in immune response and nutrient absorption. They also improve overall health and recovery by reducing oxidative stress and inflammation		
Curcumin	Reduces inflammatory markers and oxidative stress. shows toxic effects against cancer cells and it has antibacterial properties	Piperine significantly improves the bioavailability of curcumin. Enhanced anti-inflammatory activity. Better antioxidant effects than individual components	[121, 154–160]
Piperine	Anti-inflammatory effects. Antioxidants. Antibacterial		
Creatine	Beneficial for muscle strength and efficiency. Decreases muscle damage and helps immediate recovery. It may help to improve high-intensity endurance	Significantly enhances glycogen resynthesis. Speed up recovery and supports both aerobic and anaerobic systems more effectively than either take alone	[161, 162]
Carbohydrates	Supply instant energy for high-intensity exercise and endurance activities		

overall energy metabolism in the body. Researchers are actively studying the mechanism behind nano-nutrient synergy, which offers great promise in revealing more effective nutrient combinations. This type of nano combination may be able to meet specific nutritional demands and offer different therapeutic needs.

10 Practical Application in Sports Nutrition

In nutrition, the practical application of sports nutrition through nanotechnological systems like nano-encapsulation and nano-emulsions, the nutritional compounds engineered to improve absorption, stability, and target delivery efficiently with the physiological needs of athletes. The synergetic effect of nutrients enhances energy production, muscle recovery, and endurance. This technique improves the direct application of nutritional compounds, such as amino acids, vitamins, and minerals, with maximum delivery of nutrition throughout physical activities. Preparation of quicker muscle recovery, nano-electrolytes for hydration, and nano-calcium for bone strength [92]. These advanced nanotechnologies in nutrition help during sports but also develop permanent health. Additionally, nano-based supplements, such as nano-iron for resisting exercise-induced anemia and nano-zinc for enhancing the immune system, report certain physiological challenges athletes face.

11 Delivery Methods of Nano-nutrients

11.1 Nano-encapsulation

The encapsulation is a protective procedure for bioactive materials compounds from other extraneous factors with a coating shell of organic or inorganic material. Similarly, the process of nano-encapsulation carries small nutrients (1 to 100) in nano-scale size. The human body uses these engineered nano-materials to target areas absorbed by the encapsulation method. Nano-encapsulation techniques include coacervation, emulsion polymerisation, solvent evaporation, and naturalised participation.

11.2 Nano-emulsification

The Nano-emulsification process typically involves the creation of stable emulsions. The size range of the droplets can range between 20 and 200 nano-metres. This size range enhances the delivery of fat-soluble nutrients. An emulsifier or surfactant like lecithin or polysorbates, water and oil are mixed thoroughly to form a preliminary emulsion that undergoes different high-energy methods such as ultrasonication, high-pressure homogenization or micro fluidisation. This will break down the oil droplets into nano-sized particles. Emulsifiers are used to stabilise the resulting nano-emulsion, which prevents the droplet coalescence and maintains its structure. Nano-emulsification is very effective for nutrients like vitamins A, D, E and K because of its properties like improved solubility, bioavailability and stability towards hydrophobic compounds [163].

11.3 Nano-liposomes

Liposomes are formed by the interaction between phospholipids and water, which gives them the benefit of controlling the quick release of materials. Liposome encapsulation uses a lipid layer to trap, deliver and release different types of materials. The materials can be either fat-soluble or water-soluble or can have both properties. This property of liposomes ensures that they reach the right place at the right time [164]. In nano-nutrition, liposomes encapsulate hydrophilic and lipophilic elements, which are also important to sports nutrition.

12 Preparation Techniques of Nano-nutrient Delivery Methods

The following techniques, Coacervation, solvent evaporation, electrospinning, electrospraying, spray drying, and freeze drying, are the process of creating Nano-capsules, Nano-liposomes, and Nano-emulsions to help target the delivery and effectiveness of the final product (Fig. 5).

- ***Coacervation***: The process of coacervation in encapsulation works by separating numerous hydrocolloids from the initial solution to create a new layer around the active substance. This method encapsulates nutritional compounds like vitamins and minerals to protect from other living substances [165]. These protective layers enhance bioavailability and stability and also shield from environmental components such as heat, light and oxygen.
- ***Evaporation***: The evaporation process can break down pre-synthesized polymer or a mixture of separate polymers dissolved in an appropriate solvent and blended with a non-mixable liquid containing a surfactant. After that, using a low vacuum, remove the solvent by heating the mixture. This evaporation method can create small capsules and particles in pharmaceutical science to casing drugs in a biodegradable polymer [166]. This will be useful in delivering nutritional or medical compounds in a suitable way.

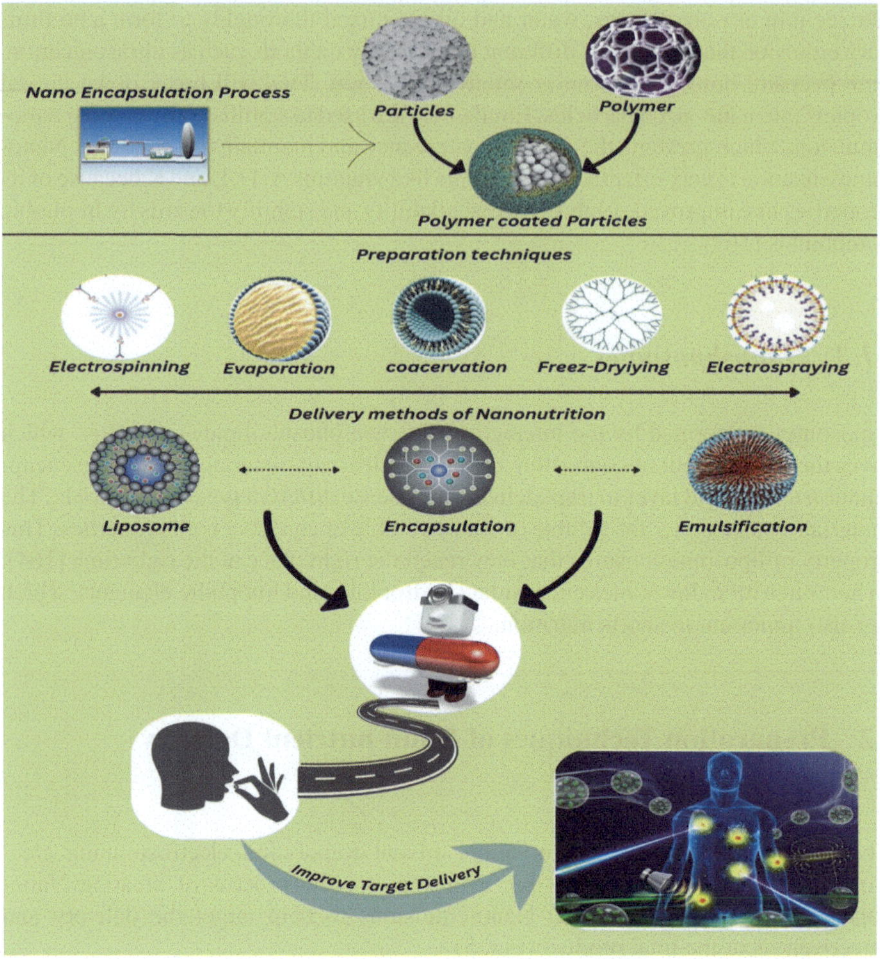

Fig. 5 Nano-nutrient preparation techniques and delivery methods

- ***Electrospinning***: Nano-fibers are created using a process called electrospinning, which is usually done by applying a high-voltage electric field to a polymer solution. A polymer solution mix is prepared by using bioactive compounds such as nutrients or drugs that need to be encapsulated. This compound is loaded into a syringe fitted with a fine needle, and a strong electric field is applied between this needle tip and a collector. The electrostatic force overcomes the surface tension of the solution, which leads to a charged jet being ejected. A nano-fibrous mat is created as the solvent evaporates and by the deposition of the ultrafine fibre on the collector. Fibers are integrated with these encapsulated bioactive compounds, which protect them and allow them to be released in a controlled way.

- ***Electrospraying***: The electrospraying device consists of a spinneret and an electric field. A spinneret is a device that collects particles, and a high-voltage electric field is connected to the spinneret and collector. Recently, electrospraying devices have been modified to produce particles with various functions. This will further improve the encapsulation efficiency of the bioactive compounds and thus expand the scope of electrospraying [167].
- ***Spray drying***: A nutrient in a liquid form is turned to powder for fast drying with hot air. This technique supports the creation of stability and easy transportation of the nutrient. The process of Spray drying prevents the moistening effects from destroying the nutrients, increasing shelf life, making it easy to handle, and protecting all compounds [168]. As an economical process, spray drying is a suitable choice for food and pharmaceutical products and also meets the essential nutrients of athletes' demands effectively and efficiently.
- ***Freeze Drying/Lyophilization***: Freeze drying is a method to prevent variability by delay enhancing the stability of nano-particles for target delivery. The process of lyophilisation freezes the working materials and turns them into ice. It produces a suitable powdered product from the absorbent material. The technique designs a shield to guard nano-particles, building solid quantity forms, keeping them from environmental pressure and lowering exposure to moisture of nano-particles. They also maintain their effectiveness and durability from environmental factors like oxidation, heat, and light [169].

13 Commercial Innovations in Nano-nutrition

Innovation in nano-nutrition has advanced significantly and now focuses on applying nanotechnology to create a product that enhances the bioavailability, stability and targeted delivery of nutrients to offer enhanced health benefits. Recent applications involve nano-encapsulated vitamins and minerals in beverages fortified with nano-calcium for bone health, functional foods, and nano-emulsions of fat-soluble vitamins (A, D, E, K) for better absorption. Energy drinks with nano-formulated ingredients and protein supplements with nano-delivery systems meet the specific needs of athletes in sports nutrition. These innovations are helping to re-engineer dietary supplements and functional foods, and they pave the way for personalized nutrition solutions that transform the area of health and wellness using nanotechnology.

14 Trends in Nano-nutrition for Athletes

Engineered nano-supplements are developing sports nutrition to the particular demands of several sports and specific athletes' physical and physiological needs for their achievement or goal. The required supplements, such as nano-carbohydrate drinks, provide sustained energy for long-duration workouts and reduce fatigue in the body. Athletes who need strength-based supplements can consume nano-enhanced protein for quick absorption, muscle recovery and repair. Similarly, nano-hydration supplements maintain hydration in the body by reducing inflammations and protecting from dehydration quickly also, nano-nutritional supplements make crucial progress in sports nutrition. Nutritional compounds offer specialised solutions to meet several requirements, like improving energy, maintaining hydration, faster recovery and enhancing the whole performance of an athlete. Also, it will help athletes' long-term health other than their athletic performance.

15 Nano-nutrient Products for Athletes

Nano-nutrient products provide developed and innovative contributions to better performance and recovery for athletes in sports nutrition. This section examines several approaches to athletic supplementation of nutritional compounds through nano-enhanced nutrition, like protein delivery, carbohydrate energy sources, and hydration technologies. This technology improves bioavailability, absorption, and target delivery of necessary nutrients, and physical and physiological needs. Nano-nutrient compounds are engineered nano-size for faster muscle recovery, prolonged energy levels, and faster and well-organized application through hydration. Integrating nanotechnology helps achieve nutritional compounds with low wastage and improves the efficiency of using supplements, leading to better performance and quick muscle recovery.

15.1 Protein-Based Nanoparticles

In sports nutrition, athletes' performance enhancement, muscle recovery, and muscle repair are getting through the massive representation of protein-based nanoparticles. These advanced nanotechnology influences overcoming limitations of conventional protein supplements, like lacking digestion, slow absorption, and insufficient application of nutrient compounds [170]. Whey protein and casein supplements are by-products of the dairy industry. These products are rich in essential amino acids like leucine, valine, isoleucine, threonine, and cysteine, significantly improving the body's metabolic and neural functions [171]. Whey protein is processed through nano-encapsulation, such as evaporation, spray-drying, and freeze-drying. These processes alter by-products into concentrated whey protein, isolated whey protein, whey permeate, lactose and other products [170].

The delivery of nano-enhanced whey protein with the essential amino acids in nano-scale structure. The nanoparticles increase the shell area and allow digestion to improve absorption in the gastrointestinal tract quickly. The process of nano-enhanced technology reduces post-workout effects such as repairing and recovering muscles, growth of muscles, and reduced soreness with the help of essential amino acids.

15.2 Nano-carbohydrate Formulations

Carbohydrates are essential for endurance athletes to meet more extended energy requirements. Conventional carbohydrates are fiber, starch, and sugar or essential food nutrients the body turns into energy. Carbohydrates contain single sugar, short chains of sugars, and long chains of sugar. The long chains are the best trapping and bioactive compounds called polysaccharides. Polysaccharide structure in huge and natural polymers is made from single sugar and adjusts in type, quantity, and arrangement [172]. Nano-carbohydrate is an encapsulated delivery substance that creates a product for target delivery with an efficient energy source. Glucose nano-drinks are a type of nano-carbohydrate to perform athletes for their best. These glucose drinks are broken into nanoparticles, which help to better body absorption, faster absorption, and quick energy, so athletes can maintain their energy and reduce fatigue during a workout [173].

15.3 Nano-enhanced Hydration

Advanced nano-enhanced hydration techniques help athletes restore their fluids effectively, and nanotechnology improves absorption and target delivery of electrolytes. These fluid solutions can keep hydration during and after the athlete's activity. Electrolytes are readily available sources in the body, and these solutions enhance performance and recovery [174]. Electrolyte nano-drinks and gels are innovative solutions for hydrating the body with the required compounds like sodium, potassium, magnesium, and calcium in a nanosized structure. As a result of these components, the body maintains the appropriate balanced electrolytes and prevents muscle cramps and fatigue through improved performance. Innovative hydration techniques improve great sports performance using nanotechnology and replace fluids and electrolytes quickly. These techniques also help to better absorption, fast delivery of nutrients, long time availability, and prevention from fatigue.

16 Emerging Innovations and Potential Developments

Nano-nutrition introduces innovative and efficient nutrient delivery and absorption procedures through nanotechnology in sports nutrition. Nanotechnology improves bioavailability, specific target delivery, and the constant release of needed nutrients, focusing on athletes' different and active needs. Innovations through the nano-encapsulation method on vitamins, minerals, and proteins ensure the supply of nutrition directly to organs and muscles, emerging recovery, performance, and total health. In sports science, nanotechnology is creating a new era through potential developments, like the integration of wearable technology and AI (artificial intelligence) [175]. Nanotechnology explored through nanoscience materials acts on a small scale, usually at the nano-scale (between 1 to 100 nm). This technology is important in many science fields, and it can design new technologies and suitable solutions. Artificial intelligence has a developing relationship with these technologies. Nanoscience, nanotechnology and artificial intelligence will work together to help the modern world cultivate medical tests with enhanced robotics and improve the accuracy of medicine, drug delivery efficiency, and treatment techniques.

17 Advanced Delivery Mechanism

Nutrient delivery in sports nutrition using innovative nano-carriers such as liposomes, polymeric nanoparticles, and nano-emulsions. These nano-carriers make the delivery of nutrients more effective in the targeted areas. Liposomes are created from phospholipids, and the structure of the liposome is small and rounded. It will protect water-soluble and fat-soluble from damage and allow for organized release. Polymeric nanoparticles contribute accurate control over the choice of nutrients and are produced from safe polymers. These work well for delicate nutritional compounds like protein and peptides. Nano-emulsions are a constant mix of oil and water at several small scales; they help satisfactory finishing of badly water-soluble. This method will improve the effectiveness of absorption quickly. The target delivery of nutritional compounds is a key development factor in sports nutrition. For direct delivery of nutritional compounds, nanotechnology-designed nano-carriers using environmental triggers like pH changes and biological markers to specific tissues. After a heavy workout, these nanotechnological systems can release amino acids or antioxidants for the required muscle damage. This process will help muscle recovery quickly, and nano-carriers provide balanced nutrient delivery, reduce inflammation, and help muscle repair.

18 Future Directions and Innovations

Sports nutrition is changing quickly due to new technology, scientific discoveries, and a better understanding of individual needs. One key future trend is personalized nutrition, which uses genetic testing, microbiome analysis, and metabolic profiling. This approach tailors nutrition plans to fit each athlete's unique biology. It helps improve performance and addresses individual issues like injury risks or chronic fatigue by adjusting nutrient intake precisely. Recently, AI driven innovations can be found in every sector, thus in nano-nutrition also. AI-based technologies are integrating with nanotechnology, giving innovative solutions in the bioavailability and efficacy of nutrients that provide tailored nutritional solutions. AI plays a significant part in designing, predicting and implementing new bioactive components, providing personalized nutrition plans for athletes, optimising the delivery of nutrients, etc. This interdependence between AI and nano-technology paves the way for efficient and effective sports nutrition strategies [176].

AI basically works by using existing datasets and it identifies patterns and relationships within the data and applies them to make predictions about unknown situations. In the field of nano-nutrition, AI can be used to create models that predictive models to enhance the knowledge about existing nutrients and discover and develop new ones, like new bioactive components [177]. AI can outperform human expertise in complicated fields such as body composition and supplements, sports nutrition knowledge, etc., which can be used to optimize the performance of athletes. As mentioned earlier, AI uses large datasets to make predictions from the patterns obtained from them. This property can be utilised in precision nutrition to provide dietary plans to athletes based on their lifestyle factors and environmental and genetic factors, thus providing an effective nutrition intervention [178]. AI can predict the nutrient needs of an athlete based on these factors so that personalised meal plans and recipes can be provided based on the athlete's nutrition requirements. The most suitable nanomaterials and the appropriate delivery system for specific nutrients can be identified using AI based on their properties and can help in the nano formulation. In order to enhance the stability, targeted delivery and bioavailability, AI can perfect the structure and composition of nano-formulation [179]. By implementing a predictive model that learns from huge datasets, AI can predict the benefit and behaviour of the nano-formulation in the body which helps the athletes in so many aspects. Wearable sensors and devices are another future innovation that can be expected. Data from wearable devices and sensors can be fed into AI models so that the real-time performance, nutrient level, psychological response of the athlete, etc., can be monitored and traced, and patterns can be drawn from this. This data can be used to provide feedback to both athletes and coaches which makes them change their routine nutrition intake and their training methods. Through this, the athletes can do the needful thing to achieve better performance. Research has been going on recently regarding the advantages of integrating nanotechnology with nutrition and its impact on sports. This huge amount of literature can be used to escalate the ideas for the development and discovery of nano-nutrition delivery systems. AI can also be used to provide a vision into the potential effects and the behaviour of nano-nutrients in the environment and the body. It can also provide insights into discovering personalised medicine approaches so that individualized treatment and interventions can be done. AI has a wide range of possible applications in sports nutrition with the help of nanotechnology, it may help athletes to achieve better dietary habits and improve performance.

19 Challenges and Safety Concerns in Nano-nutrition

The effective integration of nano-nutrition into sports exercises ensures this safe, equitable, comprehensive understanding. In inference, we can integrate nanotechnology into safe and fair sports nutrition with exciting changes and innovations in a biocompatible and friendly environment. Nano-compounds can mitigate health possibilities while reducing construction expenses and creating a makeover in sports nutrition, such as enhanced nutrition delivery, absorption, and athletes' performance. However, despite its hopeful potential for nanotechnology application in nutrition, it also comes with challenges and risks. Health concerns and challenges like potential toxicity and long-lasting effects, over-supplementation, imbalance in nutritional ratios and monitoring barriers involving control and approval procedures, ethical problems related to reasonable and availability, and sensible concerns, like high production expenses and incomplete consumer knowledge [180]. Therefore, this section discovers the various risks and challenges facing nano-nutrition in sports.

19.1 Risk of Over-Supplementation

Over-supplementation or consuming more nutrients than the body requires can lead to risks in sports nutrition and dangerous consequences. Nano-nutritional compounds improve the absorption and bioavailability compared to conventional nutritional compounds. Therefore, over-supplementation carries a high level of concentrated doses in a few products. The high consumption of nutrients can cause toxicity, stress, and imbalances in the body's functions, such as the kidneys and liver [181]. Excessive calcium consumption reduces magnesium absorption and is a suitable example of over-supplementation. These problems affect athletes' performance and cause health issues, causing endless injuries. Athletes and every individual need well-defined instruction and supervision of consuming nutrients involving appropriate intake quantities and checking before taking.

19.2 Toxicity Risks Due to Enhanced Bioavailability

Nanotechnology in sports nutrition enhances bioavailability and improves nutrient delivery, absorption, and application. However, this process causes a risk of nutrient toxicity by small doses of significant absorption to the body. Contrasting conventional nano-nutritional supplements can result in specific physiological difficulties in the body. The unpredicted relations with cells and tissues can lead to oxidative stress, cellular damage, or toxicity in the liver and kidney [182]. Everyone lacks awareness about safety restrictions regarding how much nano-nutrient compounds are safe to intake. Radically, the toxicity is when changes in nanomaterials are used in food condiments in a complex system. The combination of synergic mechanisms creates toxicity, and specific causes of the combined nutritional compounds affect athletes' performance and endurance life.

19.3 Imbalance in Nutrient Ratios

In nutrition, the main requirement of the human body is balanced nutrients for proper functioning, especially for athletes' sports performance. They need a massive amount of nutrients for faster recovery. Nanotechnology in sports nutrition enhances benefits such as absorption, bioavailability, and target delivery; however, these benefits create imbalances in nutrient ratio. The nanosized compound formulation of one nutrient is absorbed and accelerated at the rate it can restrict with another nutrient's absorption and acceleration. Massive levels of nano-iron nutrient supplements reduce zinc or calcium absorption, and this imbalanced supplementation causes deficiencies like damage to bone health, immunity, and muscle function [183]. The synergic effects of nutrients are not only a problem for imbalance of nutrition but also a cause of an athlete's overall health and performance. Modified method of supplementation reduces the risk of nutrient imbalance and dangerous consequences. Therefore, the nutrient ratio, maintaining the exact consumption of nutritional supplements, and regular checkups of overall nutritional levels meet the specific requirements of athletes. Sciences and innovations ensure the benefits of evolution in nano-nutrition help athletes' health and improve their performance.

19.4 Mitigating Risks

We need a perfect and active strategy to reduce the risks in sports nutrition. To ensure safety and understand the effect of nanosized materials duration, severe scientific research and toxicity assessment. Educated athletes and coaches are aware of nutritional supplements correctly [184]. Likewise, regulatory agencies must provide clear rules for nanotechnological product quality and proper guidance. That will help to reduce over-supplementation and misuse of nutrients. In other way collaboration between researchers, policymakers, and industry leaders can create innovative products and safety insurance. The collaboration between regulatory bodies around the world improves safety and reduces risks. Similarly, it will implement the same rules in different regions. These collaborations provide campaigns that will increase public awareness about nano-nutrition risks and benefits. Also, it helps consumers choose the most suitable option [185].

19.5 Ethical Consideration and Current Regulatory Frameworks for Nano-nutrients

To ensure nanotechnology's safe and equitable function in sports nutrition, it must be focused on regulatory and ethical considerations. Developed nano-nutritional products with nano-sized, that should face the present techniques, which do not consider the regular type of resources. Individually, long-lasting instructions are needed to ensure safety, effectiveness, and quality through proper authority clearance [186]. They also necessary ethical concerns, such as product access, item expenditure, and probabilities in sports. Integrity and trust are fundamental support for new concepts of nano-nutrients. Therefore, regulatory and ethical concerns are essential in nano-nutrition in nanotechnology.

The nano-nutrients in sports nutrition are still increasing rules and regulations, and the new use of nanotechnology is managing the challenges. Nano-nutrition differs from conventional nutritional supplements because they need particular procedures for their unique qualities, such as small-sized compounds and better absorption. Worldwide, many regulatory frameworks, like FDA (Food and Drug Administration) and EFSA (European Food Safety Authority). We discuss these regulatory frameworks to create safety authorisations, labelling guidelines, and quality criteria in nano-nutrition. Nanotechnology, nanosized materials in foods face various challenges in food safety, finding harmful pathogens, and processing procedures. Regulatory agencies are working to report these problems and ensure safety. In the U.S., the Food and Drug Administration (FDA) documented regulatory challenges issued by nanotechnology and set up a nanotechnology task force. This force supports the safe development of FDA-regulated nanosized products. The European Food Safety Authority (EFSA) is responsible for food protection and safety. The control of nanosized material issued by EFSA [187]. The guidance and control ensure protection from the risk of nanosized products or materials.

19.6 Role of Clinical Trials in Evidence-Based Nutrition

Nano-nutrition in sports, clinical trials and evidence-based recommendations are essential functions for the safety and success of athletes. Integration of nanotechnology in nano-nutrition has extraordinary qualities, such as target delivery of nutrients, absorption, and bioavailability. These features affect athletes' performance and overall health intensely [188]. Therefore, clinical trials are mandatory to recognize the exact quantities, potential hazards, and nutrient synergic effects. Similarly, it will ensure the benefits of products with less damage. Nano-nutrition in sports creates better training and recovery for an athlete through evidence-based guidelines like systematic research and help from healthcare authorities and nutritionists. These scientific systems and recommendations will provide trust in the sports industry and support revolutions.

19.7 Sports Scientists and Nutritionists Role in Monitoring Usage

The programmed dietary guidance of nutritionists supports better health and avoiding diseases of individuals. Therefore, athletes ensure their required macronutrients in the exact balance like protein, fat, and carbohydrates. Similarly, micronutrients are vitamins and minerals suitable for performance enhancement, fast recovery, and better overall health. The influence of nano-enhanced solutions focuses on better target delivery and absorption. Nutritionists can spot nutritional gaps by regularly testing their body composition with food consumption, and they can manage the food ratio with training sessions and competitions, which will help athletes' performance and reduce injury rates. Sports scientists collect data to identify how the body performs and responds to food, which helps nutritionists understand the present levels of athletes. They use tools for testing hydration, energy use, and health through metabolic tests and wearable devices. Therefore, a well-rounded database from the analysis of scientists and nutritionists and this information helps athletes modify their diet plans and supplementation ratio of nutrients.

20 Conclusion

This chapter highlights the game-changing potential of nano-nutrient combinations for maximal impact. It explores how advanced nanotechnology improves the bioavailability of nutrients, enabling better absorption levels and optimizing the body's use of the nutrients. Through the targeted delivery system of nanotechnology, nano-nutrients may aid in maximizing the physiological benefits by delivering essential nutrients to where the body needs them most. This chapter also mentioned nano-encapsulation techniques and methods, such as nano-emulsification, liposome, electrospraying, spray drying, spray freezing, and more. These advanced techniques and features may help athletes to get a sustainable supply of nutrients, targeted delivery, and readily available to the body. The sustainable release of the nutrients makes them suitable for prolonged activities, speed up muscle healing, and peak athletic performance. The field of nutrition is growing widely due to its applicability in different contexts. Nano-nutrition will take sports nutrition science to the next level because of its advanced features.

In sports nutrition, the implementation of nano-nutrient combinations is making a revolution. Pairing some nutrients may improve the bioavailability, absorption level, and effectiveness compared to those consumed alone. These synergic effects of nano-nutrients may meet the specialized demands of athletic performance, such as accelerating energy production, promoting muscle repair, improving bone health, and supporting other physiological processes. Combining nano-creatine and nano-carbohydrates may increase ATP synthesis and help the body replenish glycogen stores more quickly than it consumes alone. At the same time, nano-proteins and nano-carbohydrates aid protein synthesis and speed up the recovery process after exercise. Additionally, pairing nano-omega-3 fatty acids and nano-antioxidants could help lower oxidative stress and anti-inflammatory response and may support heart health. Combining nano-nutrients may elevate athletic performance and promote overall health and well-being. There are more nano-nutrient combinations that may influence the physiological system. Future research on this topic may reveal the efficiency of many nano combinations.

Incorporating these nano-nutrient combinations into an athlete's schedule may give more immediate and long-term performance than traditional nutrient supplements. Athletes can make the best use of these combinations by strategically schedule pre-workout and post-workout combinations. For example, consuming nano-creatine and nano-carbohydrate as pre-workout may increase ATP production and glycogen synthesis and supply sustained energy throughout exercise. In addition, taking nano-proteins and nano-carbohydrates post-workout improves protein synthesis and helps the muscle recover faster, replenishing energy stores. Highlighting the synergic effects of nano-nutrients specific to sports will be more effective when athletes understand what combinations should be consumed, when to take them, how they should be consumed, and the physiological benefits of consuming specific nutrient combinations. These understandings will help athletes to achieve better training and exercise performance and help to maintain optimal physical health over time.

In conclusion, nano-nutrients claim great promise in many therapeutic contexts, especially in sports. Therefore, future research can unlock the unexplored areas related to combinations of nano-nutrients and how those nutrients interact in the human body. Initiating more scientific studies addressing innovative pairing and upgrading current formulas may help athletes design more effective strategies that could optimize physical performance, muscle healing, and overall health and well-being. In addition, future studies also should focus on their long-term effects on the human body to ensure its safety regulations or any side effects. Integrating nano-nutrient synergy with customized nutritional plans and AI-guided dietary planning could shape individualized supplementations, specific needs, genetic profiles, and lifestyle demands. Future research can contribute significant progress and unlock potential benefits of the synergic effects of nano-nutrients that could empower athletes to achieve peak performance.

References

1. Jafari SM, Arpagaus C, Cerqueira MA, Samborska K (2021) Nano spray drying of food ingredients; materials, processing and applications. Trends Food Sci Technol 109:632–646. https://doi.org/10.1016/j.tifs.2021.01.061
2. Altemimi AB et al (2024) Application of nanoparticles in human nutrition: a review. Nutrients 16(5):636. https://doi.org/10.3390/nu16050636
3. Lima MR, Moreira BJ, Bertuzzi R, Lima-Silva AE (2024) Could nanotechnology improve exercise performance? Evidence from animal studies. Brazilian J Med Biol Res 57. https://doi.org/10.1590/1414-431x2024e13360
4. Hilty FM, Zimmermann MB (2014) Nano-structured minerals and trace elements for food and nutrition applications. In: Nano- and microencapsulation for foods. Wiley, pp 199–222. https://doi.org/10.1002/9781118292327.ch8
5. Duragkar A, Mishra S, Pethe AM (2024) An insight into the nanomaterials fabricated nanotechnology in the nutraceutical field
6. Date AA, Hanes J, Ensign LM (2016) Nanoparticles for oral delivery: design, evaluation and state-of-the-art. J Control Release 240:504–526. https://doi.org/10.1016/j.jconrel.2016.06.016
7. Mukherjee A, Das S, Chatterjee S (2017) An architecture of calcium signaling for molecular communication based nano network, pp 165–203. https://doi.org/10.1007/978-3-319-50688-3_8
8. Fulekar MH (2009) Nanotechnology—in relation to bioinformatics. In: Bioinformatics: applications in life and environmental sciences. Springer Netherlands, Dordrecht, pp 200–206. https://doi.org/10.1007/978-1-4020-8880-3_11
9. Katouzian I, Jafari SM (2017) Nanoencapsulation of vitamins. In: Nanoencapsulation of food bioactive ingredients. Elsevier, pp 145–181. https://doi.org/10.1016/B978-0-12-809740-3.00004-0
10. Bhowmick AK (2008) Current topics in elastomers research
11. Fantini AP, Canniatti-Brazaca SG, Souza MC, Mansi DN (2008) Disponibilidade de ferro em misturas de alimentos com adição de alimentos com alto teor de vitamina C e de cisteína. Ciênc Tecnol Aliment 28(2):435–439. https://doi.org/10.1590/S0101-20612008000200026
12. Kakkar V, Kumar M, Saini K (2018) Nanoceuticals governance and market review. Environ Chem Lett 16(4):1293–1300. https://doi.org/10.1007/s10311-018-0754-3
13. Resende D, Costa Lima SA, Reis S (2020) Nanoencapsulation approaches for oral delivery of vitamin A. Coll Surf B Biointerfaces 193:111121. https://doi.org/10.1016/j.colsurfb.2020.111121
14. Koprivica M, Bjelanović J (2021) The importance of vitamin a in the nutrition. Medicinski casopis 55(3):99–103. https://doi.org/10.5937/mckg55-31633
15. Gürbüz M, Aktaç Ş (2022) Understanding the role of vitamin A and its precursors in the immune system. Nutrition Clinique et Métabolisme 36(2):89–98. https://doi.org/10.1016/j.nupar.2021.10.002
16. Roche FC, Harris-Tryon TA (2021) Illuminating the role of vitamin A in skin innate immunity and the skin microbiome: a narrative review. Nutrients 13(2):302. https://doi.org/10.3390/nu13020302

17. Bono M, Tejon G, Flores-Santibañez F, Fernandez D, Rosemblatt M, Sauma D (2016) Retinoic acid as a modulator of T cell immunity. Nutrients 8(6):349. https://doi.org/10.3390/nu8060349
18. Kennedy D (2016) B vitamins and the brain: mechanisms, dose and efficacy—a review. Nutrients 8(2):68. https://doi.org/10.3390/nu8020068
19. Mahmudiono T, Haliman CD (2023) B Vitamins. In: Handbook of food bioactive ingredients. Springer International Publishing, Cham, pp 1209–1239. https://doi.org/10.1007/978-3-031-28109-9_27
20. Allen LH (2008) Causes of Vitamin B12 and folate deficiency. Food Nutr Bull 29(2_suppl1):S20–S34. https://doi.org/10.1177/15648265080292S105
21. Aşkın Ö, Uzunçakmak TKÜ, Altunkalem N, Tüzün Y (2021) Vitamin deficiencies/hypervitaminosis and the skin. Clin Dermatol 39(5):847–857. https://doi.org/10.1016/j.clindermatol.2021.05.010
22. Comunian T et al (2022) Protection and controlled release of vitamin C by different micro/nanocarriers. Crit Rev Food Sci Nutr 62(12):3301–3322. https://doi.org/10.1080/10408398.2020.1865258
23. Chugaeva UY, Raouf M, Morozova NS, Mahdavian L (2023) Effects of l-ascorbic acid (C6H8O6: Vit-C) on collagen amino acids: DFT study. Amino Acids 55(11):1655–1664. https://doi.org/10.1007/s00726-023-03339-5
24. Comunian TA, Thomazini M, Alves AJG, de Matos Junior FE, de Carvalho Balieiro JC, Favaro-Trindade CS (2013) Microencapsulation of ascorbic acid by complex coacervation: protection and controlled release. Food Res Int 52(1):373–379. https://doi.org/10.1016/j.foodres.2013.03.028
25. Pierucci APTR, Andrade LR, Baptista EB, Volpato NM, Rocha-Leão MHM (2006) New microencapsulation system for ascorbic acid using pea protein concentrate as coat protector. J Microencapsul 23(6):654–662. https://doi.org/10.1080/02652040600776523
26. Xiao W et al (2019) Ascorbic acid-modified brain-specific liposomes drug delivery system with 'lock-in' function. Chem Phys Lipids 224:104727. https://doi.org/10.1016/j.chemphyslip.2019.01.005
27. de Matos-Jr FE, Comunian TA, Thomazini M, Favaro-Trindade CS (2017) Effect of feed preparation on the properties and stability of ascorbic acid microparticles produced by spray chilling. LWT 75:251–260. https://doi.org/10.1016/j.lwt.2016.09.006
28. Comunian TA, Abbaspourrad A, Favaro-Trindade CS, Weitz DA (2014) Fabrication of solid lipid microcapsules containing ascorbic acid using a microfluidic technique. Food Chem 152:271–275. https://doi.org/10.1016/j.foodchem.2013.11.149
29. Khalid N, Kobayashi I, Neves MA, Uemura K, Nakajima M (2013) Preparation and characterization of water-in-oil emulsions loaded with high concentration of l-ascorbic acid. LWT Food Sci Technol 51(2):448–454. https://doi.org/10.1016/j.lwt.2012.11.020
30. Sylvester FA, Casserino A (2024) Are there patient populations at risk for vitamin d deficiency? How do i diagnose and treat it? In: Curbside consultation in pediatric GI. CRC Press, Boca Raton, pp 101–102. https://doi.org/10.1201/9781003523635-24
31. Benedik E (2022) Sources of vitamin D for humans. Int J Vitam Nutr Res 92(2):118–125. https://doi.org/10.1024/0300-9831/a000733
32. Maurya VK, Bashir K, Aggarwal M (2020) Vitamin D microencapsulation and fortification: trends and technologies. J Steroid Biochem Mol Biol 196:105489. https://doi.org/10.1016/j.jsbmb.2019.105489
33. Weiss K et al (2022) Performance improvement in sport through vitamin D—a narrative review
34. Książek A, Zagrodna A, Słowińska-Lisowska M (2019) Vitamin D, skeletal muscle function and athletic performance in athletes—a narrative review. Nutrients 11(8):1800. https://doi.org/10.3390/nu11081800
35. Sikora-Klak J, Narvy SJ, Yang J, Makhni E, Kharrazi FD, Mehran N (2018) The effect of abnormal vitamin D levels in athletes. Perm J 22(3). https://doi.org/10.7812/TPP/17-216
36. de la Puente Yagüe M, Collado Yurrita L, Ciudad Cabañas M, Cuadrado Cenzual M (2020) Role of Vitamin D in athletes and their performance: current concepts and new trends. Nutrients 12(2):579. https://doi.org/10.3390/nu12020579

37. Lee J, Duggan E (2022) Improved stability of vitamin D3 encapsulated in whey protein isolate microgels. Int Dairy J 129:105351. https://doi.org/10.1016/j.idairyj.2022.105351
38. Niki E (2019) Vitamin E. Royal Society of Chemistry, Cambridge. https://doi.org/10.1039/9781788016216
39. Fu J-Y, Meganathan P, Gunasegaran N, Tan DMY (2023) Effect of nano-delivery systems on the bioavailability and tissue biodistribution of vitamin E tocotrienols. Food Res Int 171:113048. https://doi.org/10.1016/j.foodres.2023.113048
40. Mishra S et al (2024) Enhancement in biological availability of vitamins by nano-engineering and its applications: an update. Curr Pharm Biotechnol 25(12):1523–1537. https://doi.org/10.2174/0113892010251234231025085759
41. Parama Iswara R, Yunus K, Pati N (2024) Vitamin K deficiency, pp 214–218. https://doi.org/10.4018/979-8-3693-2947-4.ch011
42. Raseetha S, Aida FMNA, Roslan F (2023) "Vitamin K". In: Handbook of food bioactive ingredients. Springer International Publishing, Cham, pp 1149–1185. https://doi.org/10.1007/978-3-031-28109-9_25
43. Ashfaq R, Rasul A, Asghar S, Kovács A, Berkó S, Budai-Szűcs M (2023) Lipid nanoparticles: an effective tool to improve the bioavailability of nutraceuticals. Int J Mol Sci 24(21):15764. https://doi.org/10.3390/ijms242115764
44. Dhaval M et al (2022) Lipid-based emulsion drug delivery systems—a comprehensive review. Drug Deliv Transl Res 12(7):1616–1639. https://doi.org/10.1007/s13346-021-01071-9
45. Biswas J, Datta B (2022) Nanocarriers as drug delivery vectors, pp 49–62. https://doi.org/10.1007/978-981-16-8399-2_2
46. Gupta UC, Gupta SC (2014) Sources and deficiency diseases of mineral nutrients in human health and nutrition: a review. Pedosphere 24(1):13–38. https://doi.org/10.1016/S1002-0160(13)60077-6
47. Rosborg I, Kozisek F, Selinus O, Ferrante M, Jovanovic D (2024) Background. In: Drinking water minerals and mineral balance. Springer International Publishing, Cham, pp 1–24. https://doi.org/10.1007/978-3-030-18034-8_1
48. Yuzawa S et al (2024) Chronic intake of water supplemented with deep-sea water extract reduces body fat percentage and increases basal metabolic rate and muscle mass in humans. Water (Basel) 16(17):2400. https://doi.org/10.3390/w16172400
49. Smita RM et al (2022) The role of mineral deficiencies in insulin resistance and obesity. Curr Diabetes Rev 18(7). https://doi.org/10.2174/1573399818666211117104626
50. Abdelnour SA et al (2021) Nanominerals: fabrication methods, benefits and hazards, and their applications in ruminants with special reference to selenium and zinc nanoparticles. Animals 11(7):1916. https://doi.org/10.3390/ani11071916
51. Knijnenburg JTN, Posavec L, Teleki A (2019) Nanostructured minerals and vitamins for food fortification and food supplementation. In: Nanomaterials for food applications. Elsevier, pp 63–98. https://doi.org/10.1016/B978-0-12-814130-4.00004-X
52. Gupta C, Prakash D (2023) Nanonutraceuticals for drug delivery. In: Advances in novel formulations for drug delivery. Wiley, pp 521–540. https://doi.org/10.1002/9781394167708.ch27
53. Dashnyam K et al (2020) Nanoscale calcium salt-based formulations as potential therapeutics for osteoporosis. ACS Biomater Sci Eng 6(8):4604–4613. https://doi.org/10.1021/acsbiomaterials.0c00219
54. Heshmati J et al (2019) Impact of dietary calcium supplement on circulating lipoprotein concentrations and atherogenic indices in overweight and obese individuals: a systematic review. J Diet Suppl 16(3):357–367. https://doi.org/10.1080/19390211.2018.1440685
55. Singh O et al (2024) Chapter 6 Enhancing crop nutrition with zinc nanoparticles: a novel approach to cereal crop micronutrient management. In: Sustainable agriculture. De Gruyter, pp 85–106. https://doi.org/10.1515/9783111234694-006
56. Singh NB, Agarwal A, Uzo-God OC (2021) Zinc nanomaterials: a safe animal feed supplement. In: Zinc-based nanostructures for environmental and agricultural applications. Elsevier, pp 385–401. https://doi.org/10.1016/B978-0-12-822836-4.00026-4

57. Mohammed NI et al (2023) A novel nano-iron supplement versus standard treatment for iron deficiency anaemia in children 6–35 months (IHAT-GUT trial): a double-blind, randomised, placebo-controlled non-inferiority phase II trial in The Gambia. EClinicalMedicine 56:101853. https://doi.org/10.1016/j.eclinm.2023.101853
58. Gholam Jamshidi E, Behzad F, Adabi M, Esnaashari SS (2024) Edible iron-pectin nanoparticles: preparation, physicochemical characterization and release study. Food Bioproc Tech 17(3):628–639. https://doi.org/10.1007/s11947-023-03156-4
59. Pratap-Singh A, Leiva A (2021) Double fortified (iron and zinc) spray-dried microencapsulated premix for food fortification. LWT 151:112189. https://doi.org/10.1016/j.lwt.2021.112189
60. Gupta C, Chawla P, Arora S (2015) Development and evaluation of iron microencapsules for milk fortification. CyTA—J Food 13(1):116–123. https://doi.org/10.1080/19476337.2014.918179
61. Dehnad D, Emadzadeh B, Ghorani B, Rajabzadeh G, Jafari SM (2024) Assembly of niosomal nanodelivery system for iron: introducing a green-hybrid method to enhance encapsulation efficiency. J Food Measure Characterization 18(12):9793–9804. https://doi.org/10.1007/s11694-024-02918-8
62. Babitt JL, Ganz T (2024) Molecular basis of iron metabolism. In: Molecular hematology. Wiley, pp 175–185. https://doi.org/10.1002/9781394180486.ch13
63. MacGregor E, Conway D, Henderson MA (2024) Iron metabolism. Anaes Intensive Care Med. https://doi.org/10.1016/j.mpaic.2024.11.001
64. Conway D, Henderson MA (2022) Iron metabolism. Anaes Intensive Care Med 23(2):123–125. https://doi.org/10.1016/j.mpaic.2021.10.021
65. El-Ramady H et al (2020) Selenium and nano-selenium biofortification for human health: opportunities and challenges. Soil Syst 4(3):57. https://doi.org/10.3390/soilsystems4030057
66. Fang M, Zhang H, Wang Y, Zhang H, Zhang D, Xu P (2023) Biomimetic selenium nanosystems for infectious wound healing. Eng Regener 4(2):152–160. https://doi.org/10.1016/j.engreg.2023.01.004
67. Zhang T, Qi M, Wu Q, Xiang P, Tang D, Li Q (2023) Recent research progress on the synthesis and biological effects of selenium nanoparticles. Front Nutr 10. https://doi.org/10.3389/fnut.2023.1183487
68. Piuri G et al (2021) Magnesium in obesity, metabolic syndrome, and type 2 diabetes. Nutrients 13(2):320. https://doi.org/10.3390/nu13020320
69. Reddy ST, Soman SS, Yee J (2018) Magnesium balance and measurement. Adv Chronic Kidney Dis 25(3):224–229. https://doi.org/10.1053/j.ackd.2018.03.002
70. Chinthala M, Balakrishnan A, Venkataraman P, Manaswini Gowtham V, Polagani RK (2021) Synthesis and applications of nano-MgO and composites for medicine, energy, and environmental remediation: a review. Environ Chem Lett 19(6):4415–4454. https://doi.org/10.1007/s10311-021-01299-4
71. European Food Safety Authority (EFSA) PI (2013) Scientific opinion on dietary reference values for manganese. EFSA J 11(11). https://doi.org/10.2903/j.efsa.2013.3419
72. Ikeda K, Tachibana T, Manome Y (2017) The applications, neurotoxicity, and related mechanisms of manganese-containing nanoparticles. In: Neurotoxicity of nanomaterials and nanomedicine. Elsevier, pp 205–225. https://doi.org/10.1016/B978-0-12-804598-5.00009-X
73. Bouabid S, Delaville C, De Deurwaerdère P, Lakhdar-Ghazal N, Benazzouz A (2014) Manganese-induced atypical parkinsonism is associated with altered basal ganglia activity and changes in tissue levels of monoamines in the rat. PLoS ONE 9(6):e98952. https://doi.org/10.1371/journal.pone.0098952
74. Hameed HM (2023) Ameliorative effect of nano chromium on some hematological values and GLUT-2 in rabbits fed a high- Fat diet. Iraqi J Vet Sci 37(4):775–780. https://doi.org/10.33899/ijvs.2022.136659.2604
75. Lin Q et al (2023) Advances in preparation, interaction and stimulus responsiveness of protein-based nanodelivery systems. Crit Rev Food Sci Nutr 63(19):4092–4105. https://doi.org/10.1080/10408398.2021.1997908

76. Yuan J, Jiang B, Li K, Shen W, Tang JL (2017) Beneficial effects of protein hydrolysates in exercise and sports nutrition. J Biol Regul Homeost Agents 31(1):183–188
77. Gill R, Al-Badr M, Alghouti M, Mohamed NA, Abou-Saleh H, Rahman MM (2024) Revolutionizing cardiovascular health with nano encapsulated omega-3 fatty acids: a nano-solution approach. Mar Drugs 22(6):256. https://doi.org/10.3390/md22060256
78. Li D, Wahlqvist ML, Sinclair AJ (2019) Advances in n-3 polyunsaturated fatty acid nutrition. Asia Pac J Clin Nutr 28(1):1–5. https://doi.org/10.6133/apjcn.201903_28(1).0001
79. Mesta K, Medithi S (2023) Effect of omega-3 fatty acid supplementation on the delayed-onset muscle soreness (DOMS) and inflammatory response among athletes and non-athletes. Curr Nutr Food Sci 19(8):796–805. https://doi.org/10.2174/1573401318666220413090030
80. Guo S et al (2021) Research on the fate of polymeric nanoparticles in the process of the intestinal absorption based on model nanoparticles with various characteristics: size, surface charge and pro-hydrophobics. J Nanobiotechnology 19(1):32. https://doi.org/10.1186/s12951-021-00770-2
81. He Y et al (2023) Research progress on the mechanism of nanoparticles crossing the intestinal epithelial cell membrane. Pharmaceutics 15(7):1816. https://doi.org/10.3390/pharmaceutics15071816
82. Townsend JR, Kirby TO, Sapp PA, Gonzalez AM, Marshall TM, Esposito R (2023) Nutrient synergy: definition, evidence, and future directions. Front Nutr 10. https://doi.org/10.3389/fnut.2023.1279925
83. Ihsan F, Kozina Z, Sukendro S, Nasrulloh A, Arzhan Hidayat R, Perdana S (2024) Nutritional strategies for rapid recovery in sport: a literature review. Retos 57:153–164. https://doi.org/10.47197/retos.v57.105622
84. Amawi A et al (2024) Athletes' nutritional demands: a narrative review of nutritional requirements. Front Nutr 10. https://doi.org/10.3389/fnut.2023.1331854
85. Zawistowski J, Kopeć A (2022) Effect of functional food ingredients on nutrient absorption and digestion. In: Nutrition and functional foods in boosting digestion, metabolism and immune health. Elsevier, pp 173–189. https://doi.org/10.1016/B978-0-12-821232-5.00003-3
86. Hendy GN (2018) The calcium-sensing receptor and vitamin D. Vitamin D, Elsevier, pp 477–495. https://doi.org/10.1016/B978-0-12-809965-0.00028-8
87. Rathmacher JA et al (2020) Long-term effects of calcium β-Hydroxy-β-methylbutyrate and vitamin D3 supplementation on muscular function in older adults with and without resistance training: a randomized, double-blind, controlled study. J Gerontol: Ser A 75(11):2089–2097. https://doi.org/10.1093/gerona/glaa218
88. Peake JM, Kukuljan S, Nowson CA, Sanders K, Daly RM (2011) Inflammatory cytokine responses to progressive resistance training and supplementation with fortified milk in men aged 50+ years: an 18-month randomized controlled trial. Eur J Appl Physiol 111(12):3079–3088. https://doi.org/10.1007/s00421-011-1942-z
89. Solberg A, Reikvam H (2023) Iron status and physical performance in athletes. Life 13(10):2007. https://doi.org/10.3390/life13102007
90. Pal S, Jana NR (2020) Pharmacologic vitamin C-based cell therapy via iron oxide nanoparticle-induced intracellular fenton reaction. ACS Appl Nano Mater 3(2):1683–1692. https://doi.org/10.1021/acsanm.9b02405
91. He H et al (2018) Dual action of vitamin C in iron supplement therapeutics for iron deficiency anemia: prevention of liver damage induced by iron overload. Food Funct 9(10):5390–5401. https://doi.org/10.1039/C7FO02057K
92. Ansari AA (2023) Nanotechnological advancements in sports rehabilitation to elevate athletic performance levels, pp 377–381. https://doi.org/10.1007/978-981-99-4685-3_54
93. Selvamary VN, Brundha MP, Girija AS (2020) Role of vitamin c in immune function of human body. Indian J Forensic Med Toxicol. https://doi.org/10.37506/ijfmt.v14i4.12429
94. Subroto E, Indiarto R, Andoyo R (2023) Bioavailability of iron and its potential to improve the immune system and ward off COVID-19: a review. Food Res 7(1):76–92. https://doi.org/10.26656/fr.2017.7(1).701

95. Nolte S, Krüger K, Hollander K, Carlsohn A (2024) Approaches to prevent iron deficiency in athletes. German J Sports Med 75(5):195–202. https://doi.org/10.5960/dzsm.2024.607
96. Abdolvand EFAVA (2020) Effects of supplementation cysteine-coated Fe_3O_4 nanoparticles compared to Fe_3O_4, on reproductive performance in male quail
97. Braakhuis AJ (2012) Effect of vitamin C supplements on physical performance. Curr Sports Med Rep 11(4):180–184. https://doi.org/10.1249/JSR.0b013e31825e19cd
98. Qin L, Wong SH, Sun F-H, Huang Y, Sheridan S, Sit CHP (2017) The effect of carbohydrate and protein co-ingestion on energy substrate metabolism, sense of effort, and affective responses during prolonged strenuous endurance exercise. Physiol Behav 174:170–177. https://doi.org/10.1016/j.physbeh.2017.02.035
99. Dahl MA et al (2020) Coingestion of protein and carbohydrate in the early recovery phase, compared with carbohydrate only, improves endurance performance despite similar glycogen degradation and AMPK phosphorylation. J Appl Physiol 129(2):297–310. https://doi.org/10.1152/japplphysiol.00817.2019
100. Dhiman C, Chand Kapri B (2023) Optimizing athletic performance and post-exercise recovery: the significance of carbohydrates and nutrition. Montenegrin J Sci Med 12(2):49–56. https://doi.org/10.26773/mjssm.230907
101. Zhao S, Zhang H, Xu Y, Li J, Du S, Ning Z (2024) The effect of protein intake on athletic performance: a systematic review and meta-analysis. Front Nutr 11. https://doi.org/10.3389/fnut.2024.1455728
102. Kessinger TK (2018) Nutritional recovery considerations for intermittent exercise and sport. Strategies 31(6):26–33. https://doi.org/10.1080/08924562.2018.1515678
103. Kloby Nielsen LL, Tandrup Lambert MN, Jeppesen PB (2021) The effect of ingesting carbohydrate and proteins on athletic performance: a systematic review and meta-analysis of randomized controlled trials. Nutrients 12(5):1483. https://doi.org/10.3390/nu12051483
104. Méndez L, Medina I (2021) Polyphenols and fish oils for improving metabolic health: a revision of the recent evidence for their combined nutraceutical effects. Molecules 26(9):2438. https://doi.org/10.3390/molecules26092438
105. Alorabi M et al (2022) Combination treatment of omega-3 fatty acids and vitamin c exhibited promising therapeutic effect against oxidative impairment of the liver in methotrexate-intoxicated mice. Biomed Res Int 2022:1–11. https://doi.org/10.1155/2022/4122166
106. Xin G, Eshaghi H (2021) Effect of omega-3 fatty acids supplementation on indirect blood markers of exercise-induced muscle damage: systematic review and meta-analysis of randomized controlled trials. Food Sci Nutr 9(11):6429–6442. https://doi.org/10.1002/fsn3.2598
107. Bailey DM, Williams C, Betts JA, Thompson D, Hurst TL (2011) Oxidative stress, inflammation and recovery of muscle function after damaging exercise: effect of 6-week mixed antioxidant supplementation. Eur J Appl Physiol 111(6):925–936. https://doi.org/10.1007/s00421-010-1718-x
108. Volpe-Fix AR, de França E, Silvestre JC, Thomatieli-Santos RV (2023) The use of some polyphenols in the modulation of muscle damage and inflammation induced by physical exercise: a review. Foods 12(5):916. https://doi.org/10.3390/foods12050916
109. Remila L et al (2022) Intake of coated EPA:DHA 6:1 nanoparticles improves age-related endothelial dysfunction by restoring the endothelial formation of NO and improving oxidative stress: role of the local angiotensin system. J Funct Foods 91:105003. https://doi.org/10.1016/j.jff.2022.105003
110. Du X et al (2017) Combined effects of vitamin E and omega-3 fatty acids on protecting ambient PM2.5-induced cardiovascular injury in rats. Chemosphere 173:14–21. https://doi.org/10.1016/j.chemosphere.2017.01.042
111. Habib M, Qureshi I, Jan K, Bashir K (2024) Prebiotics, probiotics, synbiotics, and nutraceuticals. In: Functional foods and nutraceuticals: chemistry, health benefits and the way forward. Springer International Publishing, Cham, pp 199–219. https://doi.org/10.1007/978-3-031-59365-9_9

112. Kumari T, Bag KK, Das AB, Deka SC (2024) Synergistic role of prebiotics and probiotics in gut microbiome health: mechanisms and clinical applications. Food Bioeng. https://doi.org/10.1002/fbe2.12107
113. Mohr AE et al (2020) The athletic gut microbiota. J Int Soc Sports Nutr 17(1). https://doi.org/10.1186/s12970-020-00353-w
114. Hong L et al (2021) Synbiotics containing nanoprebiotics: a novel therapeutic strategy to restore gut dysbiosis. Front Microbiol 12. https://doi.org/10.3389/fmicb.2021.715241
115. Dangi P et al (2023) Nanotechnology impacting probiotics and prebiotics: a paradigm shift in nutraceuticals technology. Int J Food Microbiol 388:110083. https://doi.org/10.1016/j.ijfoodmicro.2022.110083
116. Nami Y, Barghi A, Shahgolzari M, Salehian M, Haghshenas B (2024) Mechanism of action and beneficial effects of probiotics in amateur and professional athletes. Food Sci Nutr. https://doi.org/10.1002/fsn3.4658
117. Hassanizadeh S, Shojaei M, Bagherniya M, Orekhov AN, Sahebkar A (2023) Effect of nano-curcumin on various diseases: a comprehensive review of clinical trials. BioFactors 49(3):512–533. https://doi.org/10.1002/biof.1932
118. Hewlings S, Kalman D (2017) Curcumin: a review of its effects on human health. Foods 6(10):92. https://doi.org/10.3390/foods6100092
119. Liu Q, Wang C, Guo X, Du Q, Keshavarzi M (2024) Curcumin and its nano-formulations combined with exercise: from molecular mechanisms to clinic. Cell Biochem Funct 42(4). https://doi.org/10.1002/cbf.4061
120. Ms SAB, Waldman PHS, Krings PBM, Lamberth PJ, Smith PJW, McAllister PMJ (2020) Effect of curcumin supplementation on exercise-induced oxidative stress, inflammation, muscle damage, and muscle soreness. J Diet Suppl 17(4):401–414. https://doi.org/10.1080/19390211.2019.1604604
121. Shafabakhsh R, Asemi Z, Reiner Z, Soleimani A, Aghadavod E, Bahmani F (2020) The effects of nano-curcumin on metabolic status in patients with diabetes on hemodialysis, a randomized, double blind, placebo-controlled trial. Iran J Kidney Dis 14(4):290–299
122. Tanabe Y et al (2019) Effects of oral curcumin ingested before or after eccentric exercise on markers of muscle damage and inflammation. Scand J Med Sci Sports 29(4):524–534. https://doi.org/10.1111/sms.13373
123. Vecchione R et al (2016) Curcumin bioavailability from oil in water nano-emulsions: in vitro and in vivo study on the dimensional, compositional and interactional dependence. J Control Release 233:88–100. https://doi.org/10.1016/j.jconrel.2016.05.004
124. Kusnanik NW, Ayubi N, Herawati L, Jatmiko T, Muin A, Bird SP (2023) Curcumin reduce creatine kinase (CK) levels without decreasing malondialdehyde (MDA) levels after 24 hours of high-intensity physical exercise. Retos 48:878–882. https://doi.org/10.47197/retos.v48.96966
125. Pancholi V, Smina TP, Kunnumakkara AB, Maliakel B, Krishnakumar IM (2021) Safety assessment of a highly bioavailable curcumin-galactomannoside complex (CurQfen) in healthy volunteers, with a special reference to the recent hepatotoxic reports of curcumin supplements: a 90-days prospective study. Toxicol Rep 8:1255–1264. https://doi.org/10.1016/j.toxrep.2021.06.008
126. Mezzomo NJ et al (2019) Creatine nanoliposome reverts the HPA-induced damage in complex II–III activity of the rats' cerebral cortex. Mol Biol Rep 46(6):5897–5908. https://doi.org/10.1007/s11033-019-05023-y
127. Saunders B, Gough LA, Carvalho A, Torregrosa A, Oliveira TN, Reis C (2024) Creatine supplementation. In: Nutrition and supplements in cycling. Routledge, New York, pp 266–278. https://doi.org/10.4324/9781003375418-15
128. Jebali A et al (2017) Nano-carbohydrates: synthesis and application in genetics, biotechnology, and medicine. Adv Colloid Interface Sci 240:1–14. https://doi.org/10.1016/j.cis.2016.11.002
129. Liao H, Zhu S, Li Y, Huang D (2024) The synergistic effect of compound sugar with different glycemic indices combined with creatine on exercise-related fatigue in mice. Foods 13(3):489. https://doi.org/10.3390/foods13030489

130. Jurado-Castro J, Navarrete-Pérez A, Ranchal-Sánchez A, Mata Ordóñez F (2021) Optimum timing in creatine supplementation for improved sporting performance. Archivos de Medicina del Deporte 38(1):48–53. https://doi.org/10.18176/archmeddeporte.00026
131. Lawler TP, Cialdella-Kam L (2020) Non-carbohydrate dietary factors and their influence on post-exercise glycogen storage: a review. Curr Nutr Rep 9(4):394–404. https://doi.org/10.1007/s13668-020-00335-z
132. Gariballa S, Yasin J, Alessa A (2022) A randomized, double-blind, placebo-controlled trial of vitamin D supplementation with or without calcium in community-dwelling vitamin D deficient subjects. BMC Musculoskelet Disord 23(1):415. https://doi.org/10.1186/s12891-022-05364-z
133. Hopkins MH et al (2011) Effects of supplemental vitamin D and calcium on biomarkers of inflammation in colorectal adenoma patients: a randomized, controlled clinical trial. Cancer Prev Res 4(10):1645–1654. https://doi.org/10.1158/1940-6207.CAPR-11-0105
134. Abdalrida AQHHHASA (2018) Calcium phosphate nanoparticles and vitamin D in sera of rheumatoid patients. Connect Journal
135. Bischoff-Ferrari H, Steahelin H (2010) Vitamin D: update bone and muscle effects. Aktuel Ernahrungsmed 35(01):18–22. https://doi.org/10.1055/s-0029-1223412
136. Verbrugge FH, Gielen E, Milisen K, Boonen S (2012) Who should receive calcium and vitamin D supplementation? Age Ageing 41(5):576–580. https://doi.org/10.1093/ageing/afs094
137. Righi NC et al (2020) Effects of vitamin C on oxidative stress, inflammation, muscle soreness, and strength following acute exercise: meta-analyses of randomized clinical trials. Eur J Nutr 59(7):2827–2839. https://doi.org/10.1007/s00394-020-02215-2
138. Rogers DR, Lawlor DJ, Moeller JL (2023) Vitamin C supplementation and athletic performance: a review. Curr Sports Med Rep 22(7):255–259. https://doi.org/10.1249/JSR.0000000000001083
139. Jiménez-Fernández E, Ponce M, Rodríguez-Rúa A, Manchado M, Fernández-Díaz C (2018) Assessing the role of vitamin C and iron in early larvae stages of Solea senegalensis fed enriched Artemia. Aquaculture 488:145–154. https://doi.org/10.1016/j.aquaculture.2018.01.021
140. Khoshfetrat MR et al (2013) The effect of iron-vitamin C co-supplementation on biomarkers of oxidative stress in iron-deficient female youth. Biol Trace Elem Res 153(1–3):171–177. https://doi.org/10.1007/s12011-013-9695-7
141. Hansen M, Bangsbo J, Jensen J, Bibby BM, Madsen K (2015) Effect of whey protein hydrolysate on performance and recovery of top-class orienteering runners. Int J Sport Nutr Exerc Metab 25(2):97–109. https://doi.org/10.1123/ijsnem.2014-0083
142. Lockwood CM et al (2017) Effects of hydrolyzed whey versus other whey protein supplements on the physiological response to 8 weeks of resistance exercise in college-aged males. J Am Coll Nutr 36(1):16–27. https://doi.org/10.1080/07315724.2016.1140094
143. Alghannam A, Gonzalez J, Betts J (2018) Restoration of muscle glycogen and functional capacity: role of post-exercise carbohydrate and protein co-ingestion. Nutrients 10(2):253. https://doi.org/10.3390/nu10020253
144. Fernández-Lázaro D, Arribalzaga S, Gutiérrez-Abejón E, Azarbayjani MA, Mielgo-Ayuso J, Roche E (2024) Omega-3 fatty acid supplementation on post-exercise inflammation, muscle damage, oxidative response, and sports performance in physically healthy adults—a systematic review of randomized controlled trials. Nutrients 16(13):2044. https://doi.org/10.3390/nu16132044
145. Uchida Y, Tsuji K, Ochi E (2024) Effects of Omega-3 fatty acids supplementation and resistance training on skeletal muscle. Clin Nutr ESPEN 61:189–196. https://doi.org/10.1016/j.clnesp.2024.03.019
146. Qin L et al (2020) In vivo effect of resveratrol-loaded solid lipid nanoparticles to relieve physical fatigue for sports nutrition supplements. Molecules 25(22):5302. https://doi.org/10.3390/molecules25225302
147. Charalampopoulos D, Rastall RA (2012) Prebiotics in foods. Curr Opin Biotechnol 23(2):187–191. https://doi.org/10.1016/j.copbio.2011.12.028

148. Sankarganesh P, Adeyeye SAO, Chakka AK, Babu AS (2024) Prebiotics and probiotics food. In: Food coatings and preservation technologies. Wiley, pp 551–583. https://doi.org/10.1002/9781394237623.ch16
149. Saini K, Minj J (2020) Multifunctional aspects of probiotics and prebiotics in health management: an overview. In: Dairy processing: advanced research to applications. Springer Singapore, Singapore, pp 119–131. https://doi.org/10.1007/978-981-15-2608-4_6
150. Candrawati DPMA, Mahardika IG, Bidura IGNG, Siti NW (2024) Effect of prebiotics-probiotics synbiotic supplementation in feed on performance, serum cholesterol, and pathogenic bacteria in broilers intestines. Adv Anim Vet Sci 12(9). https://doi.org/10.17582/journal.aavs/2024/12.9.1752.1758
151. Lin C-L et al (2020) Bifidobacterium longum subsp. longum OLP-01 supplementation during endurance running training improves exercise performance in middle- and long-distance runners: a double-blind controlled trial. Nutrients 12(7):1972. https://doi.org/10.3390/nu12071972
152. Patel BK, Patel KH, Lee CN, Moochhala S (2024) Intestinal microbiota interventions to enhance athletic performance—a review. Int J Mol Sci 25(18):10076. https://doi.org/10.3390/ijms251810076
153. Zhang J, Chen L, Zhang L, Chen Q, Tan F, Zhao X (2021) Effect of Lactobacillus fermentum HFY03 on the antifatigue and antioxidation ability of running exhausted mice. Oxid Med Cell Longev 2021(1). https://doi.org/10.1155/2021/8013681
154. Zeng X et al (2017) Selective reduction in the expression of UGTs and SULTs, a novel mechanism by which piperine enhances the bioavailability of curcumin in rat. Biopharm Drug Dispos 38(1):3–19. https://doi.org/10.1002/bdd.2049
155. Alam J et al (2022) Curcumin encapsulated into biocompatible co-polymer PLGA nanoparticle enhanced anti-gastric cancer and anti-helicobacter pylori effect. Asian Pac J Cancer Prev 23(1):61–70. https://doi.org/10.31557/APJCP.2022.23.1.61
156. Kumar MN, Kalarikkal SP, Bethi CMS, Singh SN, Narayanan J, Sundaram GM (2023) An eco-friendly one-pot extraction process for curcumin and its bioenhancer, piperine, from edible plants in exosome-like nanovesicles. Green Chem 25(16):6472–6488. https://doi.org/10.1039/D3GC01287E
157. AbouAitah K et al (2021) Anti-inflammatory and antioxidant effects of nanoformulations composed of metal-organic frameworks delivering rutin and/or piperine natural agents. Drug Deliv 28(1):1478–1495. https://doi.org/10.1080/10717544.2021.1949073
158. Shirmohammadli F, Nikzad M, Ghoreyshi AA, Mohammadi M, Poureini F (2021) Preparation and characterization of zein/sodium caseinate/xanthan gum complex for encapsulation of piperine and its in vitro release study. Food Biophys 16(2):254–269. https://doi.org/10.1007/s11483-021-09668-w
159. Ouni M, Ghadami A, Zandi M (2022) Studies on synergistic effect of curcumin, piperine and ellagic acid on antibacterial properties of biocompatible nanofibers based on polycaprolactone. Iranian J Polym Sci Technol
160. Ren T et al (2019) Piperine-loaded nanoparticles with enhanced dissolution and oral bioavailability for epilepsy control. Eur J Pharm Sci 137:104988. https://doi.org/10.1016/j.ejps.2019.104988
161. Forbes SC et al (2023) Creatine supplementation and endurance performance: surges and sprints to win the race. J Int Soc Sports Nutr 20(1). https://doi.org/10.1080/15502783.2023.2204071
162. Wax B, Kerksick CM, Jagim AR, Mayo JJ, Lyons BC, Kreider RB (2021) Creatine for exercise and sports performance, with recovery considerations for healthy populations. Nutrients 13(6):1915. https://doi.org/10.3390/nu13061915
163. Choi SJ, McClements DJ (2020) Nanoemulsions as delivery systems for lipophilic nutraceuticals: strategies for improving their formulation, stability, functionality and bioavailability. Food Sci Biotechnol 29(2):149–168. https://doi.org/10.1007/s10068-019-00731-4
164. Fang Z, Bhandari B (2010) Encapsulation of polyphenols—a review. Trends Food Sci Technol 21(10):510–523. https://doi.org/10.1016/j.tifs.2010.08.003

165. Gouin S (2004) Microencapsulation. Trends Food Sci Technol 15(7–8):330–347. https://doi.org/10.1016/j.tifs.2003.10.005
166. Staff RH, Landfester K, Crespy D (2013) Recent advances in the emulsion solvent evaporation technique for the preparation of nanoparticles and nanocapsules, pp 329–344. https://doi.org/10.1007/12_2013_233
167. Jaworek A, Sobczyk AT (2008) Electrospraying route to nanotechnology: an overview. J Electrostat 66(3–4):197–219. https://doi.org/10.1016/j.elstat.2007.10.001
168. Gohel MC et al (2009) Spray drying: a review. Pharmainfo.net
169. Arora S, Dash SK, Dhawan D, Sahoo PK, Jindal A, Gugulothu D (2024) Freeze-drying revolution: unleashing the potential of lyophilization in advancing drug delivery systems. Drug Deliv Transl Res 14(5):1111–1153. https://doi.org/10.1007/s13346-023-01477-7
170. Manjusha, Singh R, Srivastava A (2023) Protein-based nanostructures: a promising trend in food industry. Res J Chem Environ 27(9):123–133. https://doi.org/10.25303/2709rjce1230133
171. Patel S (2015) Functional food relevance of whey protein: a review of recent findings and scopes ahead. J Funct Foods 19:308–319. https://doi.org/10.1016/j.jff.2015.09.040
172. Fathi M, Martín Á, McClements DJ (2014) Nanoencapsulation of food ingredients using carbohydrate based delivery systems. Trends Food Sci Technol 39(1):18–39. https://doi.org/10.1016/j.tifs.2014.06.007
173. Khanna GL, Manna I (2005) Supplementary effect of carbohydrate-electrolyte drink on sports performance, lactate removal and cardiovascular response of athletes. Indian J Med Res 121(5):665–669
174. Bobbala S, Vincent MP, Scott EA (2021) Just add water: hydratable, morphologically diverse nanocarrier powders for targeted delivery. Nanoscale 13(26):11349–11359. https://doi.org/10.1039/D1NR02188E
175. Kalinina I, Potoroko I, Nenasheva A, Velyamov M, Bagale U (2019) Prospects for the application of taxifolin based nanoemulsions as a part of sport nutrition products. Human Sport Medicine 19(1):100–107. https://doi.org/10.14529/hsm190114
176. Côté M, Lamarche B (2024) Artificial intelligence in nutrition research. In: Artificial intelligence in clinical practice. Elsevier, pp 465–473. https://doi.org/10.1016/B978-0-443-15688-5.00031-0
177. Su Q et al (2022) Nano functional food: opportunities, development, and future perspectives. Int J Mol Sci 24(1):234. https://doi.org/10.3390/ijms24010234
178. de Moraes Lopes MHB, Ferreira DD, Ferreira ACBH, da Silva GR, Caetano AS, Braz VN (2020) Use of artificial intelligence in precision nutrition and fitness. In: Artificial intelligence in precision health. Elsevier, pp 465–496. https://doi.org/10.1016/B978-0-12-817133-2.00020-3
179. Duan Z, Ge N, Kong Y (2024) The factors affecting aerobics athletes' performance using artificial intelligence neural networks with sports nutrition assistance. Sci Rep 14(1):29639. https://doi.org/10.1038/s41598-024-81437-4
180. Colica C et al (2018) The role of nanotechnology in food safety. Minerva Biotechnol Biomolec Res 30(2). https://doi.org/10.23736/S1120-4826.18.02394-7
181. Georgescu L, Ciucurel C, Burcea C-C, Burlibasa M, Armean P, Chivu R-D (2013) Practical approach to dietary supplements used by athletes
182. Luo X et al (2022) Evaluation of the protective roles of alpha-lipoic acid supplementation on nanomaterial-induced toxicity: a meta-analysis of in vitro and in vivo studies. Front Nutr 9. https://doi.org/10.3389/fnut.2022.991524
183. Prasad AS, Beck FWJ (2023) Metal-metal interactions. In: Toxicology of metals, Vol I. CRC Press, Boca Raton, pp 189–197. https://doi.org/10.1201/9781003418917-16
184. Maad AH, Aljerf L, AlJerf A, Rafaat Dayoob C, Jehad Sukkar A, Alajlani M (2024) A cross-sectional study on demographic characteristics, nutritional knowledge, and supplement use patterns. Int J Kinesiol Sports Sci 12(4):21–30. https://doi.org/10.7575/aiac.ijkss.v.12n.4p.21
185. Rawat S et al (2024) Chapter 23 Harvesting nanotech: responsible policies for integrating nanotechnology in agriculture and health. In: Sustainable agriculture. De Gruyter, pp 495–510. https://doi.org/10.1515/9783111369754-023

186. Coles D, Frewer LJ (2013) Nanotechnology applied to European food production—a review of ethical and regulatory issues. Trends Food Sci Technol 34(1):32–43. https://doi.org/10.1016/j.tifs.2013.08.006
187. Bugusu B, Mejia C, Magnlison B, Tafazoli S (2009) Global regulatory food policies on nanotechnology
188. Kong F, Ren S (2022) Analysis on the influence of nanotechnology development on sports health industry. Int J Nanotechnol 19(6/7/8/9/10/11):1034. https://doi.org/10.1504/IJNT.2022.128986

Innovations in Nano-packaging for Nutrient Stability

Shobana Srinivasan, Vinitha Shankar, and Sivakumar Venkatachalam

Abstract Preserving the nutrient stability and improving the shelf life of food products necessitates nanotechnological advancements. In order to maintain the nutrient integrity and extend the shelf life, nanoparticles with antimicrobial activity like zinc and silver oxide, cellulose and lignin are incorporated into packaging films. In addition, barrier properties offered by nanocomposites and nano-laminates prevents the nutrient loss that occurs due to light, moisture and oxygen exposure. The reinforcement of oxygen scavengers in nanoscale dimension and other freshness indicators facilitates the real-time monitoring of nutrient stability and food quality that comes in an active and intelligent food packaging films. These biodegradable materials and edible films, not only preserve the food quality but also reduce the environmental impact and carbon footprint. This chapter explorescutting edge nano-packaging innovations that offer substantial advantage for the food industry, including preservation of nutrients, shelf-life extension and environment friendly packaging solutions.

Keywords Nanotechnology · Sports science · Food packaging · Edible films

1 Introduction

Nanotechnology has emerged as a significant innovation in food packaging, offering new solutions to address both food preservation challenges and environmental concerns [1]. The role of packaging materials serves as essential barriers that protect food products from various degradation factors while maintaining their nutritional quality throughout storage and distribution. The unique characteristics of nanomaterials, including their size, morphology, and distribution patterns, provide enhanced functionality compared to conventional materials [2]. Several types of nanoparticles have shown promising results in food packaging applications [3]. Metal-based nanoparticles (silver, copper) and metal oxide nanostructures (zinc oxide, titanium dioxide) have demonstrated improved barrier properties and antimicrobial

S. Srinivasan · V. Shankar · S. Venkatachalam (✉)
Department of Chemical Engineering, AC Tech Campus, Anna University, Chennai, India
e-mail: drvsivakumar@yahoo.com

A. Chatterjee et al. (eds.), *Nanofuel: The Future of Sports Nutrition*, Smart Nanomaterials Technology, https://doi.org/10.1007/978-981-96-5471-0_13

effects. Additionally, naturally-derived nanomaterials such as cellulose, chitosan, and montmorillonite have enhanced packaging performance through improved water resistance, mechanical strength, and thermal stability, though these properties vary depending on concentration levels [4].

Biopolymer-based nanocomposites represent a particularly interesting development in this field. For example, cellulose nanocrystals (CNC) and cellulose nanofibers (CNF) have successfully improved the mechanical properties and moisture barrier characteristics of natural polymers like chitosan and whey protein isolate [5]. These improvements directly contribute to better food preservation and extended shelf life.

The application of nanomaterials in intelligent packaging systems has enabled real-time monitoring of food quality. Silver nanoparticles combined with specific amino acids can detect lactic acid in dairy products, while titanium dioxide nanoparticles help monitor meat freshness [6]. These developments represent significant progress in food safety monitoring and quality assurance.

As environmental considerations become increasingly important, nanotechnology offers potential alternatives to traditional non-biodegradable packaging materials [7]. The versatility of nanomaterials allows for the development of composite materials that combine organic, inorganic, and hybrid components, creating packaging solutions that balance functionality with environmental responsibility.

This chapter examines current developments in nano packaging technologies with a focus on nutrient stability. It explores how various nanomaterials contribute to food preservation while considering their broader impact on environmental sustainability. The discussion covers recent advances, practical applications, and future directions in this developing field.

2 Challenges in Maintaining Nutrient Stability

The transition from conventional plastic packaging to biodegradable alternatives has gained significant momentum due to rising environmental concerns and the health risks associated with plastic-derived chemicals. However, biodegradable packaging materials, while eco-friendly, present several challenges in maintaining the nutrient stability of packaged foods [8]. These challenges stem from factors such as moisture sensitivity, mechanical weakness, oxygen permeability, and interactions with food components [9]. Addressing these issues is crucial to ensure that biodegradable packaging can effectively preserve food quality and extend shelf life.

One of the primary challenges in biodegradable packaging is its susceptibility to moisture. Biopolymers such as polysaccharides (e.g., pectin, starch) and proteins (e.g., whey, soy) tend to absorb water due to their hydrophilic nature [10]. This increased water activity can lead to nutrient degradation, particularly for water-soluble vitamins like vitamin C and B-complex vitamins Excess moisture also creates an environment conducive to microbial growth, accelerating spoilage [11].

Biodegradable films often exhibit poor mechanical strength compared to plastics, making them prone to tearing and puncturing during handling and storage.

Such damage compromises the protective barrier, exposing foods to oxygen, light, and contaminants that accelerate nutrient oxidation and degradation [12]. Mechanical reinforcement through the addition of cellulose or chitin nanoparticles has been shown to improve tensile strength and durability. Additionally, multilayer film technologies, where biodegradable layers are combined with stronger bio-based polymers, have demonstrated enhanced structural integrity [13, 14].

While biodegradable materials such as starch and protein films can provide moderate oxygen barriers, their performance declines under humid conditions. Increased oxygen permeability facilitates oxidative reactions, degrading sensitive nutrients like unsaturated fatty acids, vitamin E, and carotenoids. Strategies to mitigate oxygen permeability include the use of layered nanocomposites or blending biodegradable films with materials like chitosan, which has excellent oxygen barrier properties [15]

Light exposure can degrade light-sensitive nutrients, such as riboflavin (vitamin B2) and β-carotene, through photochemical reactions. Biodegradable films often lack the opacity required to block light effectively. The addition of UV-blocking agents or the incorporation of plant pigments and anthocyanins into the film matrix has shown promise in reducing light-induced nutrient degradation [16].

Biodegradable packaging materials may interact with the food they are designed to protect. For instance, protein-based films can bind to lipids, affecting the bioavailability of fat-soluble vitamins. Similarly, the migration of active compounds (e.g., antimicrobial agents or antioxidants) from the packaging into the food can alter nutrient composition. The present chapter highlight the importance of selecting compatible biopolymers and limiting the migration of reactive substances to maintain nutrient stability.

Temperature fluctuations, humidity, and prolonged storage significantly impact the performance of biodegradable packaging. High temperatures accelerate nutrient degradation and compromise the structural integrity of biodegradable films [13]. Controlled atmosphere packaging (CAP) and modified atmosphere packaging (MAP) techniques, combined with biodegradable materials, have been explored to counteract these effects and preserve nutrient quality.

3 Role of Nanotechnology in Food Packaging

Nanotechnology has transformed food packaging by enhancing functional properties, improving food safety, and extending shelf life. The integration of nanomaterials at the molecular or atomic scale (typically below 100 nm) into packaging systems results in superior barrier performance, mechanical strength, antimicrobial properties, and real-time monitoring capabilities [17]. These advancements address key challenges associated with conventional and biodegradable packaging. Nanomaterials significantly improve the barrier properties of packaging films by reducing the permeability of gases, moisture, and UV light [18, 19]. The incorporation of nano-clays, silica nanoparticles, and graphene forms tortuous pathways that hinder the diffusion of oxygen and water vapor. This minimizes oxidative degradation and moisture absorption, crucial for preserving the stability of sensitive nutrients and extending shelf life. In biodegradable packaging, where moisture sensitivity is a common drawback, the addition of nano-fillers helps mitigate water uptake, enhancing the performance of starch, pectin, and chitosan films.

Biodegradable films often lack the mechanical strength necessary for practical packaging applications. The addition of cellulose nanocrystals (CNC) and nanofibrillated cellulose (NFC) reinforces the polymer matrix, improving tensile strength, puncture resistance, and elasticity [20]. This prevents damage during handling and storage, reducing the risk of nutrient degradation through exposure to oxygen and contaminants. Similarly, nanoclay and graphene oxide improve the flexibility and durability of biopolymer films, making them suitable for demanding food packaging applications.

The integration of antimicrobial and antioxidant nanomaterials into packaging films actively inhibits microbial growth and oxidative processes that lead to spoilage [21]. Table 1 illustrates the role of inorganic nanoparticles in food packaging. Silver, zinc oxide, and titanium dioxide nanoparticles exhibit potent antimicrobial activity against a broad spectrum of bacteria and fungi. These nanoparticles prevent the proliferation of spoilage microorganisms, extending the shelf life of packaged foods. Antioxidant nanoparticles, such as green tea polyphenols and curcumin, are incorporated to protect against lipid oxidation, preserving the quality and nutritional value of food products. Active packaging systems that gradually release antimicrobial agents help maintain food safety over extended periods.

Nanosensors embedded within packaging materials enable real-time monitoring of food quality and safety. These sensors detect changes in temperature, humidity, gas composition, and microbial contamination. For example, nanosensors that respond to the release of ammonia or carbon dioxide provide visual indicators of spoilage, enhancing consumer confidence and reducing food waste. Intelligent packaging systems utilizing colorimetric sensors or fluorescence-based indicators are increasingly being developed for meat, dairy, and seafood products, where spoilage detection is critical [14].

Nanotechnology plays a vital role in developing sustainable packaging solutions by reinforcing biodegradable polymers with bio-based nanomaterials. Nanocellulose, starch nanoparticles, and lignin nanoparticles improve the mechanical, barrier, and antimicrobial properties of biodegradable films without compromising their compostability. Hemicellulose and lignin nanoparticles possess good antioxidant and antimicrobial properties [30, 31]. These bio-nanocomposites address the performance limitations of plant-based polymers, facilitating their broader application in food packaging [32, 33]. The use of renewable nanomaterials aligns with global sustainability goals by reducing reliance on petroleum-based plastics and promoting circular economy practices.

By enhancing packaging performance, nanotechnology contributes to the reduction of food waste across supply chains. Prolonged freshness, delayed spoilage, and real-time spoilage detection prevent premature disposal of food products. This not only improves food security but also minimizes environmental impacts associated

Table 1 Role of inorganic nanoparticles in food packaging

Inorganic nanoparticles	Food packaging material	Main findings	References
Silver nanoparticles	Cellulose	The composite films demonstrated antimicrobial efficacy against *A. hydrophila* and prolonged the shelf life of cabbages and tomatoes	[22]
	Polyvinyl alcohol	The silver/PVA composite effectively preserved salmon fillets for up to 7 days of storage and exhibited antimicrobial efficacy against *Staphylococcus aureus*	[23]
	Polylactic acid	Silver nanoparticle incorporation improved the film's mechanical strength and barrier properties against water vapor and oxygen, while its antimicrobial activity contributed to prolonged strawberry preservation	[24]
Zinc oxide nanoparticles	Polyvinyl alcohol	The film incorporating ZnO nanoparticles demonstrated antimicrobial efficacy against both S. aureus and E. coli, effectively prolonging cherry tomato storage stability	[25]
	Polylactic acid	The films exhibited improved water vapor barrier and mechanical properties, including elongation and elasticity, while effectively preserving fresh-cut apples for 14 days	[26]
	Chitosanand cellulose acetate phthalate	ZnO nanoparticle incorporation enhanced the film's thermal stability and barrier characteristics, successfully extending black grape storage stability to 9 days	[27]
Titanium dioxide nanoparticles	Polyvinyl alcohol	PVA films incorporating TiO_2 nanoparticles showed antimicrobial activity against *Shewanella spp.*, *P. putida*, and A. hydrophilia, effectively prolonging *M. rosenbergii* preservation by 1–2 days	[28]
	Chitosan	The film demonstrated improved mechanical strength and barrier characteristics, while its enhanced ethylene photodegradation capability successfully delayed the ripening process of tomatoes	[29]

with food production and waste management. The integration of intelligent packaging technologies ensures that food is consumed within safe limits, preventing unnecessary discarding of products still fit for consumption [11].

3.1 Nanomaterials for Nutrient Protection

Nanomaterials play a critical role in preserving the nutritional quality of packaged foods by enhancing barrier properties, preventing oxidative degradation, and providing active protection against environmental stressors. The integration of nanoparticles into biopolymer matrices addresses key issues such as nutrient loss due to oxygen exposure, light, and moisture, contributing to extended shelf life and improved food quality.

Mahajan et al. [34] developed a moisture absorber using bentonite/sorbitol/$CaCl_2$ (0.55:0.25:0.2 g/g) that achieved 0.91 ± 0.02 g H_2O/g absorption in 5 days, significantly reducing mushroom browning. In oxygen scavenging research, Gomes et al. [35] demonstrated that iron-containing laminates reduced oxygen from 67.44% to 1% within 24 h, maintaining these levels for 1 year. For antimicrobial packaging, Higueras et al. [36] showed chitosan films with LAE reduced Listeria monocytogenes by 2 log cycles over 60 days in chicken fillets. In ethylene scavenger studies, Bodaghi et al. [37] developed TiO_2 nanocomposite films that effectively reduced microbial growth through photocatalytic effects. Das et al. [38] found blueberry extract had a minimum inhibitory concentration of 50 mg/mL against foodborne pathogens, being more effective against Gram-positive bacteria. For antioxidant packaging, Barbosa-Pereira et al. [39] created films with natural phenolics that significantly improved oxidative stability of beef, while López-de-Dicastillo et al. [40] developed EVOH films with flavonoids that effectively reduced lipid oxidation in brined sardines.

Oxygen exposure is a primary factor in nutrient degradation, particularly for vitamins (e.g., vitamin C, vitamin E) and unsaturated fatty acids. Nanoclays, graphene oxide, and layered silicates are widely used to enhance oxygen barrier properties in packaging films [5]. These nanomaterials create labyrinth-like structures that reduce oxygen permeability by forcing gas molecules to traverse longer, more complex paths. This significantly lowers the oxidation rate of sensitive nutrients, maintaining the nutritional value of packaged foods over time.

Moisture absorption can accelerate the degradation of water-soluble vitamins and promote microbial growth. Hydrophilic biopolymers such as starch and pectin are prone to moisture uptake, compromising their protective capabilities. The addition of nanocellulose, silica nanoparticles, and chitosan nanofibers enhances the water vapor barrier of biodegradable films by filling polymer voids and reducing water permeability [41, 42]. This moisture control is essential for preserving vitamins B and C, which are highly susceptible to hydrolysis in humid environments.

Light exposure triggers photodegradation of nutrients such as riboflavin (vitamin B2), carotenoids, and polyphenols. Nanomaterials with UV-blocking properties, such as titanium dioxide (TiO_2) and zinc oxide (ZnO), are incorporated into packaging films to protect light-sensitive nutrients. These nanoparticles absorb and scatter UV radiation, preventing photochemical reactions that degrade vitamins and antioxidants. Additionally, bio-based pigments and anthocyanins are used to impart natural UV-blocking characteristics to biodegradable films.

Antioxidant nanoparticles are directly incorporated into packaging materials to inhibit lipid oxidation and protect fat-soluble vitamins (e.g., vitamins A, D, E, K). Polyphenols, quercetin, and curcumin are nano-encapsulated within biodegradable films to provide localized antioxidant activity [43]. These nanocomposites gradually release antioxidants, neutralizing free radicals and delaying the oxidative degradation of packaged foods. This technique is particularly beneficial for protecting oils, dairy products, and snack foods rich in polyunsaturated fatty acids.

Nanocellulose and lignin nanoparticles are increasingly explored for their ability to act as carriers for bioactive compounds, including antioxidants and antimicrobial agents [16, 21]. These nanoparticles provide a controlled release mechanism, ensuring sustained nutrient protection throughout the storage period. Cellulose nanofibers on its incorporation in guar gum-based films exhibited better film forming properties making it suitable for packaging of food products like cheese [44, 45]. Nanohemicellulose also plays a key role in preserving the nutrient stability of packaged foods due to its antioxidant properties [20, 46]. Organic nanoparticles like almond shells are more effective in improving the packaging film properties and its potentiality of active packaging [47]. By embedding essential oils or plant extracts within nanocellulose, packaging films can continuously prevent oxidation and microbial contamination, preserving both nutrient integrity and food safety.

Edible nanocoatings made from polysaccharides, proteins, and lipids form an additional protective layer around food surfaces. These coatings, enriched with nanoparticles such as nano-emulsified essential oils, enhance nutrient stability by reducing oxygen and moisture diffusion. Edible coatings provide a direct barrier that minimizes nutrient loss, particularly for fresh produce, meat, and dairy products [12, 17, 48]. The use of edible nanocoatings extends shelf life while maintaining the sensory and nutritional attributes of packaged foods.

Nanotechnology-driven advancements in food packaging continue to play a pivotal role in protecting nutrients, reducing food waste, and enhancing the overall quality of packaged products. The strategic incorporation of nanomaterials ensures that foods retain their nutritional value and freshness, aligning with the growing demand for healthier and more sustainable packaging solutions.

3.2 Mechanisms of Nutrient Preservation in Nanopackaging

Nanopackaging employs a range of physical, chemical, and biological mechanisms to protect nutrients, ensuring the stability and quality of packaged foods over extended periods. The nanoscale incorporation of active and passive elements into packaging materials enhances their ability to mitigate nutrient degradation caused by oxygen, moisture, light, and microbial activity. Nanomaterials improve the barrier properties of packaging films by creating dense, tortuous pathways that impede the movement of gases, water vapor, and light. This physical barrier minimizes the penetration of oxygen and moisture—key factors in nutrient degradation. Layered nanoclays, graphene oxide, and silicon oxide nanoparticles disrupt the free flow of oxygen by forming impermeable structures within the polymer matrix. By preventing oxidation, these materials help preserve sensitive nutrients like vitamin C, vitamin E, and polyunsaturated fatty acids. Nanocellulose, chitosan nanofibers, and silica nanoparticles fill the pores of biodegradable films, reducing water vapor transmission rates. This protects water-soluble vitamins (e.g., B-complex, vitamin C) from hydrolysis and nutrient leaching. UV-blocking nanoparticles such as zinc oxide (ZnO) and titanium dioxide (TiO_2)[49] scatter and absorb harmful radiation, protecting light-sensitive nutrients like riboflavin (vitamin B2) and carotenoids from photodegradation.

Active nanopackaging involves the integration of nanoparticles that actively interact with food or the surrounding environment to prevent nutrient loss [50]. This includes antioxidant, antimicrobial, and oxygen-scavenging systems. Various researches has been carried out towards it. Han et al. [51] demonstrated that active packaging reduced food spoilage by 45% compared to traditional petroleum-based packaging through controlled gas and moisture barriers. Lee et al. [52] developed a pH-responsive packaging system that showed distinct color changes from red (pH 3) to purple (pH 7) to yellow (pH 11), accurately indicating bacterial growth in stored foods. Zheng et al. [53] combined anthocyanins with polymer matrices to create dual-functional packaging that exhibited 82% antioxidant activity and reduced microbial growth by 65% during a 14-day storage period. Miao et al. [54] successfully incorporated anthocyanins into nanofiber films that showed color transitions within 2 min of pH changes and maintained 78% of their activity after 30 days of storage. Min et al. (2020) enhanced chitosan's antimicrobial properties through quaternization, achieving a 95% bacterial growth inhibition rate against E. coli and S. aureus while maintaining film flexibility with an elongation at break of 30%Nano-encapsulated antioxidants, such as quercetin, polyphenols, and tocopherols, are embedded within packaging films. These antioxidants neutralize free radicals, preventing lipid oxidation and protecting fat-soluble vitamins. Iron oxide and palladium nanoparticles serve as oxygen scavengers by chemically binding oxygen molecules that penetrate the packaging [55]. This reduces oxidative degradation of nutrients and preserves the sensory attributes of food. Silver, copper oxide, and zinc oxide nanoparticles exhibit antimicrobial properties that prevent the growth of spoilage microorganisms, indirectly preserving the nutrient profile of foods by reducing microbial-induced degradation [56].

Nanopackaging can act as a carrier for bioactive compounds, offering controlled release systems that gradually dispense antioxidants, antimicrobials, and moisture absorbers [57, 58]. This ensures sustained nutrient protection throughout the storage period. Bioactive compounds such as essential oils, vitamins, and phenolic compounds are encapsulated within nanocarriers (e.g., nanoliposomes, nanocellulose) and released gradually to counteract oxidation and microbial growth. Some nanomaterials are designed to release active agents in response to specific environmental triggers, such as changes in humidity or pH [59, 60]. For example, nano-encapsulated citric acid can release antioxidants when the food's pH drops, preventing further nutrient degradation.

Edible nanocoatings applied directly to the surface of food create an additional layer of protection. These coatings, often composed of polysaccharides, proteins, and lipid nanoparticles, reduce the diffusion of oxygen and moisture while enhancing the retention of nutrients [60]. Nanolipids and protein-based nanoparticles create a semi-permeable layer that regulates gas exchange, maintaining the nutrient integrity of fresh produce and dairy products [61]. Edible coatings infused with antimicrobial nanoparticles further prevent spoilage.

3.3 Applications of Nano-packaging in Food Industries

The various types of packaging involving nanoparticles are active and intelligent. Active packaging has good antibacterial, antioxidant, and UV absorption properties. Intelligent packaging gives information about the internal environment of the food product, like pH and temperature.

Teymouri and Shekarchizadeh monitored fish spoilage using intelligent packaging, spoilage of fish leads to the release of sulfur compounds. The Cu nanoparticle in the packagingis a selective and sensitive indicator of sulfur compounds. Simple, low-cost, and rapid detection of fish spoilage by film color variation was not affected by parameters like moisture, pH, and time. The film color variations are white, orange, and brown, indicating the freshness of fish, semi-freshness of stored fish, and spoiled fish [62]. Kwon and Ko studied chicken spoilage by preparing an indicator film made of cellulose nanocrystals incorporated with silver nanoparticles. The variation of film color occurred when silver nanoparticles reacted with hydrogen sulfide released during the chicken spoilage. The film color changes from yellowish/dark wine-red color to colorless/metallic grey based on the silver nanoparticle's presence in the film which shows the spoilage of chicken [63].

Ligaj et al. studied the antimicrobial properties of active packaging prepared using the casting method. The zero-valent iron (ZVI) was incorporated in the PLA (Polylactide) matrix to improve the shelf life of packed food. The PLA films and the PLA film added with ZVI at different concentrations (1%, 3%, and 5%) are compared. The addition of ZVI improves the film's appearance, optical properties, and 3% of ZVI exhibited better antimicrobial properties [64]. Ahmadi et al. prepared eco-friendly active packaging with antimicrobial properties. The film consists of gelatin, cellulose nanofiber (CNF) at varying concentrations (2.5%, 5%, and 7%), and ZnO nanoparticles at varying concentrations (1%, 3%, 5%, and 7%). The 5% CNF in the film improves strength, stiffness, whereas decreases flexibility, water vapor permeability, and moisture absorption. The 5% ZnO shows a better antibacterial effect on gram-positive bacteria than that of gram-negative bacteria and prevents the growth of bacteria on packed chicken to improve the shelf life [65]. Sani et al.

studied the shelf life of lamb meat packed using active packaging. The film consists of cellulose nanofiber (CNF) to improve mechanical properties, essential oil (2% rosemary oil) to enhance antimicrobial and antioxidant activity, and 1% titanium dioxide to improve antimicrobial and antioxidant activity. The prepared film was used to store the lamb meat under refrigerated conditions for 15 days and monitored for microbial spoilage [66]. This film decreased microbial growth, lipid oxidation, and lipolysis during storage. The eco-friendly film enhanced the shelf life of lamb meat and prevented food wastage.

3.4 Merits and Demerits of Nano-packaging

The nano-packaging material is prepared by incorporating the nanoparticle in the packaging material or by coating the particles to create functional and resistant food packaging. These nanoparticles improve the physical and mechanical properties of the packaging by enhancing the barrier properties, thermal stability, strength, durability, flexibility, biodegradability, UV absorptivity, water vapor impermeability, and oxygen impermeability properties [8]. The nano-packaging material's antibacterial properties help kill the bacteria by degrading bacterial cell walls, damaging intracellular components, and affecting bacterial metabolism. This property maintains the quality of the food products during storage, transportation, and consumption [67]. Nano-packaging material prevents food from spoilage, prevents the active compounds of food from external environmental conditions, and improves the shelf life of the food products. Thus, helps in reducing food waste generation and prevents the harmful effects of non-biodegradable packaging [8, 68].

Nguyen et al. prepared polyvinyl alcohol (PVOH)/agar/maltodextrin and silver nanoparticles coating to extend the shelf life of bananas. The silver nanoparticles improve the ductility and plasticity of the film. The coating acts as the physical barrier preventing the physiochemical properties and is effective against bacteria like E. coli and S. aureus [48]. Mathew et al. developed a packaging film using PVA/clay and silver nanoparticles to improve the shelf life of packed chicken sausage. The silver nanoparticles are effective against bacteria like S. Typhimurium and S. aureus by breaking their cell wall [6]. The film has good mechanical properties, light barrier properties, water resistivity, and biodegradability. Shankar et al. developed a polylactic film incorporated with zinc oxide nanoparticles to store minced fish paste. The Zinc oxide nanoparticles in the film improved the tensile strength, smoothness, and compactness of the film. The incorporation of ZnO in film enhances the UV-light barrier, mechanical, and antibacterial properties [15]. Hezma et al. prepared the polyvinyl alcohol/chitosan and different concentrations of zinc oxide nanoparticles incorporated film. The 10 and 15 wt% zinc oxide nanoparticles improved the thermal stability, mechanical properties, and antibacterial effects of the film [69]. Qu et al. developed a zein/chitosan (cs) film incorporated with titanium dioxide to improve the mechanical properties, thermal stability, tensile strength, and antibacterial properties.

The film is more effective against gram-positive bacteria (S. aureus) than gram-negative bacteria (E. coli and S. enteritidis) [70].

The commercialization of nano-packaging was difficult due to limited knowledge about its toxic effects and health impacts. The hydrophilic properties of the starch and cellulose packaging led to poor mechanical strength. Replacing non-biodegradable plastic packaging was difficult because of poor marketing and public awareness about biodegradable packaging. Large-scale packaging production was challenging due to cost and poor properties like durability, thermal capacity, and moisture resistance [4]. Incorporating nanoparticles in the matrix alter the physiochemical properties (viscosity, tensile strength, elastic modulus, water solubility, thermal conductivity, thermal stability) and intramolecular interactions [67]. The nanoparticles released from the packaging material into the food products have toxic effects on humans during consumption. The harmful effects of nanoparticles depend on their properties, concentration, and exposure time. These nanoparticles are carcinogenic and genotoxic, have cardiovascular effects when they enter the bloodstream, and affect the brain by crossing the blood–brain barrier [71]. Reactive oxidative species released by nanoparticles activate the cell-damaging reactions, inhibit DNA replication, ATP formation, and cause cell death [72]. The release of the reagents and chemicals used in the extraction of nanoparticles also has toxic effects on humans and the environment [3]. The metal oxide nanoparticle decreases cell viability and damages the DNA [73]. Various nanoparticles like silver, zinc oxide, titanium dioxide, copper oxide, carbon nanotubes, graphene oxide, and cellulose nanocrystals are used in packaging applications. The research and development regarding the migration rate, toxic effects, and health and environmental effects of these nanoparticles are to be investigated further for better understanding [61].

4 Regulatory and Safety Considerations

Analyzing nanoparticles' migration, toxicity, permissible limit, and interaction with a matrix is important before developing the nano-packaging material [8]. The cytotoxicity of nanoparticles is studied by exposing the cell to nanoparticles in a medium [73]. The migration test is to study the release of nanoparticles from the packaging material into the food product [8]. The physicochemical properties such as absorption, biodistribution, metabolism, and excretion of nanoparticles are studied for nanotoxicity [72]. It is important to know the nanoparticle's size, shape, surface charge, solubility, and surface chemistry to determine their toxicity [68].

Each country prepares its list regarding the use and trade of materials such as REACH (Registration, Evaluation, Authorization, and Restriction of Chemicals) by European Chemicals Agents (ECHA) for chemical solvents used by industries, Food and Drugs Administration (FDA) by the United States gives information about the risk of nanomaterials used in nano-packaging, Inventory of Existing Chemical Substances in China (IECSC), Agency of National Health Surveillance (ANVISA) in Brazil [67].

The producer should get permission from the FDA to produce nanomaterials through a Food Additive Petition (FAP) or by Food Contact Notification (FCN). The manufacturer can market the product directly without getting approval from the FDA if they have published a scientific journal about the product risk assessment [61]. The International Organization for Standardization (ISO) provides the standards for the analyses of nanomaterials. The proper investigation regarding the toxic effects of nanoparticles migrating into food products from the packaging material has to be studied further. The potential risk of nanoparticles (additives) and matrix in the human body is unclear, and a detailed view of it is important [3].

5 Sustainability and Future Perspectives

The integration of nanotechnology into food packaging presents a transformative opportunity to address both food preservation challenges and the environmental impact of conventional packaging. As global concerns surrounding plastic pollution and food waste intensify, nanotechnology offers innovative solutions that enhance packaging performance, improve food safety, and align with sustainability goals. This section explores the sustainable aspects of nanotechnology in food packaging and outlines future directions that will shape the industry. Nanomaterials enable the development of thinner, stronger, and more durable packaging films, reducing the overall material required for packaging without compromising performance. For instance, the addition of nanoclays or cellulose nanofibers enhances the mechanical strength and barrier properties of biodegradable polymers, allowing for reduced film thickness while maintaining product integrity [74]. This contributes to lower raw material consumption and decreased environmental footprint.

Nanotechnology is increasingly integrated with biodegradable polymers such as starch, pectin, polylactic acid (PLA), and chitosan. The addition of nanocellulose, lignin nanoparticles, and silica-based fillers improves the mechanical and barrier properties of these materials, making them competitive with traditional plastics. Unlike conventional plastics, nanocomposite films made from biodegradable matrices decompose naturally, contributing to reduced plastic waste and aligning with circular economy principles.

Nanotechnology plays a pivotal role in extending the shelf life of food products by enhancing packaging properties. Nanoclays, metal oxide nanoparticles, and antioxidant-releasing nanocomposites minimize oxygen and moisture penetration, delaying spoilage and nutrient degradation. Intelligent packaging equipped with nanosensors provides real-time indicators of food quality, reducing unnecessary food disposal and ensuring optimal consumption periods [22]. This directly addresses global food waste, which contributes significantly to environmental degradation and resource depletion.

The future of nanotechnology in food packaging lies in the expansion of smart and active packaging systems. Smart packaging, embedded with nanosensors, will continue to evolve, offering enhanced monitoring capabilities that detect microbial contamination, gas composition, and temperature fluctuations. This technology will enable dynamic responses, such as antimicrobial or antioxidant release when food quality begins to deteriorate. Active packaging will shift toward controlled-release systems, where bioactive nanoparticles gradually release compounds that prevent spoilage and oxidation, further reducing food waste [17].

Edible nanocoatings represent a rapidly growing area of research, providing a sustainable solution for protecting perishable foods. Future edible coatings will incorporate nanolipids, nanoemulsions, and biopolymer-based nanomaterials, creating barriers against moisture and oxygen while enhancing the sensory attributes of food. These coatings will be particularly beneficial for fresh produce, dairy, and meat products, offering direct protection without generating additional waste.

Nanotechnology-driven packaging will increasingly align with circular economy principles by focusing on recyclability, compostability, and reusability. Research is underway to develop nanocomposites that retain their functional properties during recycling processes. The integration of bio-based nanomaterials will ensure that packaging decomposes safely, returning valuable nutrients to the environment and closing the material loop [4]. The environmental footprint of nanomaterial synthesis is a growing area of focus. Future efforts will prioritize green synthesis methods, utilizing plant extracts, bio-waste, and renewable precursors for nanoparticle production. Green chemistry approaches will reduce energy consumption, eliminate hazardous by-products, and ensure the scalability of sustainable nanomaterial production.

As nanotechnology adoption grows, regulatory frameworks will need to evolve to ensure consumer safety and environmental protection [75]. Governments and international bodies are actively developing guidelines for the use of nanomaterials in food contact applications, with a focus on migration studies and long-term health assessments. Collaboration between researchers, policymakers, and industry stakeholders will drive the creation of standardized protocols, fostering public trust and facilitating market entry for novel packaging technologies.

Consumer perception and acceptance of nanotechnology in food packaging will shape its future trajectory. Transparent communication regarding the benefits, safety, and sustainability of nanopackaging will be essential to address potential concerns. Educational initiatives and labeling strategies highlighting the eco-friendly and health-preserving aspects of nanotechnology will enhance consumer confidence and drive demand for sustainable packaging solutions.

6 Conclusion

Food packaging has seen notable improvements through recent technological developments, responding to the growing need for better food preservation and sustainability. Active and intelligent packaging systems show promising potential in enhancing food quality and extending shelf life, though current challenges include production costs and regulatory requirements. The emergence of nanotechnology offers viable solutions in this field, particularly through the development of bio-based nanocomposites and organic nanoparticles derived from cellulose, lignin, and hemicellulose. Advancements in environmentally friendly synthesis methods and balanced regulatory approaches may help establish more practical food packaging systems. Research efforts focusing on optimizing production costs combined with nanomaterial innovations, could contribute to commercial applications in food packaging.

Acknowledgements The author (Shobana Srinivasan) acknowledges Anna University for supporting with Anna Centenary Research fellowship (Ref. No. CFR/ACRF-2021/AR1).

AuthorContribution CRediT Authorship Contribution Statement
Shobana Srinivasan: Conceptualization, Methodology, Writing—originaldraft. **Vinitha Shankar**: Conceptualization, Methodology, Writing. **Sivakumar Venkatachalam**: Supervision, Conceptualization, Visualization, Writing—review and editing.

Data Availability No data was used for the research described in the article.

Declaration of Competing Interest The authors declare that they have no known competing financialinterests or personal relationships that could have appeared to influencethe work reported in this paper.

References

1. D'Almeida AP, de Albuquerque TL (2024) Innovations in food packaging: from bio-based materials to smart packaging systems. Processes 12
2. Anjum A, Garg R, Kashif M, Eddy NO (2023) Nano-scale innovations in packaging: properties, types, and applications of nanomaterials for the future. Food Chem Adv 3
3. Manikandan V, Min SC (2023) Roles of polysaccharides-based nanomaterials in food preservation and extension of shelf-life of food products: a review. Int J Biol Macromol 252. https://doi.org/10.1016/j.ijbiomac.2023.126381
4. Khandeparkar AS, Paul R, Sridhar A, et al (2024) Eco-friendly innovations in food packaging: A sustainable revolution. Sustain Chem Pharm 39
5. Palanisamy S, Selvaraju GD, Selvakesavan RK et al (2024) Unlocking sustainable solutions: nanocellulose innovations for enhancing the shelf life of fruits and vegetables—a comprehensive review. Int J Biol Macromol 261. https://doi.org/10.1016/j.ijbiomac.2024.129592
6. Mathew S, Snigdha S, Mathew J, Radhakrishnan EK (2019) Biodegradable and active nanocomposite pouches reinforced with silver nanoparticles for improved packaging of chicken sausages. Food Packag Shelf Life 19:155–166. https://doi.org/10.1016/j.fpsl.2018.12.009
7. Wang M, Li S, Chen Z et al (2021) Safety assessment of nanoparticles in food: current status and prospective. Nano Today 39
8. Ashfaq A, Khursheed N, Fatima S et al (2022) Application of nanotechnology in food packaging: pros and cons. J Agric Food Res 7. https://doi.org/10.1016/j.jafr.2022.100270
9. Kuai L, Liu F, Chiou BS et al (2021) Controlled release of antioxidants from active food packaging: a review. Food Hydrocoll 120
10. Kuswandi B (2016) Nanotechnology in food packaging, pp 151–183

11. Ahmed MW, Haque MA, Mohibbullah M et al (2022) A review on active packaging for quality and safety of foods: Current trends, applications, prospects and challenges. Food Packag Shelf Life 33
12. Qian M, Liu D, Zhang X et al (2021) A review of active packaging in bakery products: applications and future trends. Trends Food Sci Technol 114:459–471
13. Gaikwad KK (2021) Recent developments in intelligent packaging systems for food processing industry: a review
14. Mirza Alizadeh A, Masoomian M, Shakooie M et al (2021) Trends and applications of intelligent packaging in dairy products: a review. Crit Rev Food Sci Nutr 62:383–397
15. Shankar S, Wang LF, Rhim JW (2018) Incorporation of zinc oxide nanoparticles improved the mechanical, water vapor barrier, UV-light barrier, and antibacterial properties of PLA-based nanocomposite films. Mater Sci Eng, C 93:289–298. https://doi.org/10.1016/j.msec.2018.08.002
16. Jacob Rani BS, Venkatachalam S (2024) Biomass-derived nanoparticles reinforced chitosan films: as high barrier active packaging for extending the shelf life of highly perishable food. J Food Sci Technol 61:990–1002. https://doi.org/10.1007/s13197-023-05896-9
17. Wu D, Zhang M, Chen H, Bhandari B (2021) Freshness monitoring technology of fish products in intelligent packaging. Crit Rev Food Sci Nutr 61:1279–1292
18. Zhang X, Zhang B, Mao R et al (2024) A novel multilayer film based on sodium alginate/k-carrageenan-gelatin incorporated with ZnO nanoparticles and oregano essential oil for active food packing. Prog Org Coat 187. https://doi.org/10.1016/j.porgcoat.2023.108170
19. Yadav N, Kaur R (2024) Innovations in packaging to monitor and maintain the quality of the food products. J Packag Technol Res 8:15–50. https://doi.org/10.1007/s41783-024-00163-4
20. Louis ACF, Venkatachalam S, Gupta S (2022) Innovative strategy for rice straw valorization into nanocellulose and nanohemicellulose and its application. Ind Crops Prod 179. https://doi.org/10.1016/j.indcrop.2022.114695
21. Jacob Rani BS, Venkatachalam S (2023) Unleashing the potential of bio-nanoparticles from invasive cattail as reinforcement in packaging film to extend the shelf life of poultry meat. Packag Technol Sci 36:1021–1036. https://doi.org/10.1002/pts.2773
22. Singh M, Sahareen T (2017) Investigation of cellulosic packets impregnated with silver nanoparticles for enhancing shelf-life of vegetables. LWT 86:116–122. https://doi.org/10.1016/j.lwt.2017.07.056
23. Kavakebi E, Anvar A, Ahari H, Motalebi A (2020) The effect of PVA coated film containing silver nanoparticles synthesized from aqueous Satureja rechingeri extract on shelf life of rainbow trout)oncorhynchus mykiss) fillet. https://doi.org/10.22059/jfabe.2021.325907.1094
24. Cheng J, Lin X, Wu X et al (2021) Preparation of a multifunctional silver nanoparticles polylactic acid food packaging film using mango peel extract. Int J Biol Macromol 188:678–688. https://doi.org/10.1016/J.IJBIOMAC.2021.07.161
25. Yang L, Ren L, Zhao Y et al (2023) Preparation and characterization of PVA/arginine chitosan/ZnO NPs composite films. Int J Biol Macromol 226:184–193. https://doi.org/10.1016/J.IJBIOMAC.2022.12.020
26. Li W, Li L, Cao Y et al (2017) Effects of PLA film incorporated with ZnO nanoparticle on the quality attributes of fresh-cut apple. Nanomaterials 7. https://doi.org/10.3390/nano7080207
27. Indumathi MP, Saral Sarojini K, Rajarajeswari GR (2019) Antimicrobial and biodegradable chitosan/cellulose acetate phthalate/ZnO nano composite films with optimal oxygen permeability and hydrophobicity for extending the shelf life of black grape fruits. Int J Biol Macromol 132:1112–1120. https://doi.org/10.1016/J.IJBIOMAC.2019.03.171
28. Tang ZP, Chen CW, Xie J (2018) Development of antimicrobial active films based on poly(vinyl alcohol) containing nano-TiO_2 and its application in macrobrachium rosenbergii packaging. J Food Process Preserv 42. https://doi.org/10.1111/jfpp.13702
29. Siripatrawan U, Kaewklin P (2018) Fabrication and characterization of chitosan-titanium dioxide nanocomposite film as ethylene scavenging and antimicrobial active food packaging. Food Hydrocoll 84:125–134. https://doi.org/10.1016/J.FOODHYD.2018.04.049

30. Srinivasan S, Venkatachalam S (2024) One pot green process for facile fractionation of sorghum biomass to lignin, cellulose and hemicellulose nanoparticles using deep eutectic solvent. Int J Biol Macromol 277. https://doi.org/10.1016/j.ijbiomac.2024.134295
31. Srinivasan S, Venkatachalam S (2024) An energy-efficient process for enhanced production of bioethanol from sorghum biomass: a futuristic approach towards circular economy. Biomass Convers Biorefin. https://doi.org/10.1007/s13399-024-06083-2
32. Jacob Rani BS, Venkatachalam S (2022) Cleaner approach for the cascade production of nanocellulose, nanohemicellulose and nanolignin from Prosopis juliflora. Carbohydr Polym 294:119807. https://doi.org/10.1016/j.carbpol.2022.119807
33. Jacob Rani BS, Venkatachalam S (2022) A neoteric approach for the complete valorization of Typha angustifolia leaf biomass: a drive towards environmental sustainability. J Environ Manage 318:115579. https://doi.org/10.1016/j.jenvman.2022.115579
34. Mahajan PV, Rodrigues FAS, Motel A, Leonhard A (2008) Development of a moisture absorber for packaging of fresh mushrooms (Agaricus bisporous). Postharvest Biol Technol 48:408–414. https://doi.org/10.1016/J.POSTHARVBIO.2007.11.007
35. Gomes C, Castell-Perez ME, Chimbombi E et al (2009) Effect of oxygen-absorbing packaging on the shelf life of a liquid-based component of military operational rations. J Food Sci 74:E167–E176. https://doi.org/10.1111/j.1750-3841.2009.01120.x
36. Higueras L, López-Carballo G, Hernández-Muñoz P et al (2013) Development of a novel antimicrobial film based on chitosan with LAE (ethyl-Nα-dodecanoyl-l-arginate) and its application to fresh chicken. Int J Food Microbiol 165:339–345. https://doi.org/10.1016/J.IJFOODMICRO.2013.06.003
37. Bodaghi H, Mostofi Y, Oromiehie A et al (2013) Evaluation of the photocatalytic antimicrobial effects of a TiO2 nanocomposite food packaging film by in vitro and in vivo tests. Food Sci Technol 5:702–706
38. Das Q, Islam MR, Marcone MF et al (2017) Potential of berry extracts to control foodborne pathogens. Food Control 73:650–662. https://doi.org/10.1016/J.FOODCONT.2016.09.019
39. Barbosa-Pereira L, Aurrekoetxea GP, Angulo I et al (2014) Development of new active packaging films coated with natural phenolic compounds to improve the oxidative stability of beef. Meat Sci 97:249–254. https://doi.org/10.1016/J.MEATSCI.2014.02.006
40. López-De-Dicastillo C, Gómez-Estaca J, Catalá R et al (2012) Active antioxidant packaging films: development and effect on lipid stability of brined sardines. Food Chem 131:1376–1384. https://doi.org/10.1016/J.FOODCHEM.2011.10.002
41. Ezati P, Rhim JW (2020) pH-responsive pectin-based multifunctional films incorporated with curcumin and sulfur nanoparticles. Carbohydr Polym 230. https://doi.org/10.1016/j.carbpol.2019.115638
42. Yang W, Zhang S, Hu Y et al (2024) Pectin-based film activated with carboxylated cellulose nanocrystals-stabilized oregano essential oil Pickering emulsion. Food Hydrocoll 151. https://doi.org/10.1016/j.foodhyd.2024.109781
43. Almasi H, Azizi S, Amjadi S (2020) Development and characterization of pectin films activated by nanoemulsion and Pickering emulsion stabilized marjoram (Origanum majorana L.) essential oil. Food Hydrocoll 99. https://doi.org/10.1016/j.foodhyd.2019.105338
44. Palanichamy P, Venkatachalam S, Gupta S (2023) Improved recovery of cellulose nanoparticles from printed wastepaper and its reinforcement in guar gum films. Biomass Convers Biorefin 13:14113–14125. https://doi.org/10.1007/s13399-022-02516-y
45. Palanichamy P, Venkatachalam S, Gupta S (2023) Tough, flexible and oil-resistant film from sonicated guar gum and cellulose nanofibers for food packaging. Food Packag Shelf Life 40. https://doi.org/10.1016/j.fpsl.2023.101189
46. Louis ACF, Venkatachalam S (2020) Energy efficient process for valorization of corn cob as a source for nanocrystalline cellulose and hemicellulose production. Int J Biol Macromol 163:260–269. https://doi.org/10.1016/j.ijbiomac.2020.06.276
47. Edward MSG, Louis ACF, Srinivasan H, Venkatachalam S (2022) A mechanochemical approach for synthesizing almond shell nanoparticles and their potential application on the enhancement of polylactic acid film properties. Iranian Polym J (English Edition) 31:1523–1535. https://doi.org/10.1007/s13726-022-01099-9

48. Nguyen TT, Huynh Nguyen TT, Tran Pham BT et al (2021) Development of poly (vinyl alcohol)/agar/maltodextrin coating containing silver nanoparticles for banana (Musa acuminate) preservation. Food Packag Shelf Life 29. https://doi.org/10.1016/j.fpsl.2021.100740
49. Zhang W, Rhim JW (2022) Titanium dioxide (TiO_2) for the manufacture of multifunctional active food packaging films. Food Packag Shelf Life 31
50. Roy S, Rhim JW (2021) Preparation of pectin/agar-based functional films integrated with zinc sulfide nano petals for active packaging applications. Colloids Surf B Biointerfaces 207. https://doi.org/10.1016/j.colsurfb.2021.111999
51. Han J-W, Ruiz-Garcia L, Qian J-P, Yang X-T (2018) Food packaging: a comprehensive review and future trends. Compr Rev Food Sci Food Saf 17:860–877. https://doi.org/10.1111/1541-4337.12343
52. Lee K, Park H, Baek S et al (2019) Colorimetric array freshness indicator and digital color processing for monitoring the freshness of packaged chicken breast. Food Packag Shelf Life 22:100408. https://doi.org/10.1016/J.FPSL.2019.100408
53. Zheng L, Liu L, Yu J et al (2023) Intelligent starch/chitosan-based film incorporated by anthocyanin-encapsulated amylopectin nanoparticles with high stability for food freshness monitoring. Food Control 151:109798. https://doi.org/10.1016/J.FOODCONT.2023.109798
54. Miao Y, Chen Z, Zhang J et al (2023) Exopolysaccharide riclin and anthocyanin-based composite colorimetric indicator film for food freshness monitoring. Carbohydr Polym 314:120882. https://doi.org/10.1016/J.CARBPOL.2023.120882
55. Riaz A, Mostafa H, Lawal KG et al (2024) A novel source of lignin from date palm leaves as a reinforcing agent for fabrication of carboxymethyl cellulose-based active food packaging film. Food Biophys 19:334–346. https://doi.org/10.1007/s11483-024-09826-w
56. Tiwari K, Singh R, Negi P et al (2021) Application of nanomaterials in food packaging industry: a review. Mater Today: Proc:10652–10655
57. Yang L, Sheng W, Yang Y et al (2025) Bacterial cellulose and photothermal material from biomass-based zero-waste biorefinery functionalized chitosan film for sustainable active food packaging. Food Hydrocoll 162. https://doi.org/10.1016/j.foodhyd.2024.110977
58. Sun F, Zhao J, Shan P et al (2025) Gelatin-based composite film integrated with nanocellulose and extract-metal complex derived from coffee leaf for sustainable and active food packaging. Food Hydrocoll 159. https://doi.org/10.1016/j.foodhyd.2024.110610
59. Xu M, Fang D, Shi C et al (2025) Anthocyanin-loaded polylactic acid/quaternized chitosan electrospun nanofiber as an intelligent and active packaging film in blueberry preservation. Food Hydrocoll 158. https://doi.org/10.1016/j.foodhyd.2024.110586
60. Fu X, Chang X, Xu S et al (2024) Development of a chitosan/pectin-based multi-active food packaging with both UV and microbial defense functions for effectively preserving of strawberry. Int J Biol Macromol 254. https://doi.org/10.1016/j.ijbiomac.2023.127968
61. Herrera-Rivera M del R, Torres-Arellanes SP, Cortés-Martínez CI et al (2024) Nanotechnology in food packaging materials: role and application of nanoparticles. RSC Adv 14:21832–21858
62. Teymouri Z, Shekarchizadeh H (2022) A colorimetric indicator based on copper nanoparticles for volatile sulfur compounds to monitor fish spoilage in intelligent packaging. Food Packag Shelf Life 33. https://doi.org/10.1016/j.fpsl.2022.100884
63. Kwon S, Ko S (2022) Colorimetric freshness indicator based on cellulose nanocrystal–silver nanoparticle composite for intelligent food packaging. Polymers (Basel) 14. https://doi.org/10.3390/polym14173695
64. Ligaj M, Tichoniuk M, Cierpiszewski R, Foltynowicz Z (2020) Efficiency of novel antimicrobial coating based on iron nanoparticles for dairy products' packaging. Coatings 10. https://doi.org/10.3390/coatings10020156
65. Ahmadi A, Ahmadi P, Ehsani A (2020) Development of an active packaging system containing zinc oxide nanoparticles for the extension of chicken fillet shelf life. Food Sci Nutr 8:5461–5473. https://doi.org/10.1002/fsn3.1812
66. Alizadeh-Sani M, Mohammadian E, McClements DJ (2020) Eco-friendly active packaging consisting of nanostructured biopolymer matrix reinforced with TiO_2 and essential oil: application for preservation of refrigerated meat. Food Chem 322. https://doi.org/10.1016/j.foodchem.2020.126782

67. de Sousa MS, Schlogl AE, Estanislau FR et al (2023) Nanotechnology in packaging for food industry: past, present, and future. Coatings 13
68. Sagar NA, Kumar N, Choudhary R et al (2022) Prospecting the role of nanotechnology in extending the shelf-life of fresh produce and in developing advanced packaging. Food Packag Shelf Life 34
69. Hezma AM, Rajeh A, Mannaa MA (2019) An insight into the effect of zinc oxide nanoparticles on the structural, thermal, mechanical properties and antimicrobial activity of Cs/PVA composite. Colloids Surf A Physicochem Eng Asp 581. https://doi.org/10.1016/j.colsurfa.2019.123821
70. Qu L, Chen G, Dong S et al (2019) Improved mechanical and antimicrobial properties of zein/chitosan films by adding highly dispersed nano-TiO_2. Ind Crops Prod 130:450–458. https://doi.org/10.1016/j.indcrop.2018.12.093
71. Dimitrijevic M, Karabasil N, Boskovic M et al (2015) Safety aspects of nanotechnology applications in food packaging. Procedia Food Sci 5:57–60. https://doi.org/10.1016/j.profoo.2015.09.015
72. Sahani S, Sharma YC (2021) Advancements in applications of nanotechnology in global food industry. Food Chem 342
73. Nikolic MV, Vasiljevic ZZ, Auger S, Vidic J (2021) Metal oxide nanoparticles for safe active and intelligent food packaging. Trends Food Sci Technol 116:655–668
74. Biswas R, Alam M, Sarkar A et al (2022) Application of nanotechnology in food: processing, preservation, packaging and safety assessment. Heliyon 8
75. Istiqola A, Syafiuddin A (2020) A review of silver nanoparticles in food packaging technologies: regulation, methods, properties, migration, and future challenges. J Chin Chem Soc 67:1942–1956

Post-workout Recovery Nano Enhancement Solution

W. Vinu, Dilshith A. Kabeer, Ketheeswaran, G. Manibharathi, and Gnanaratnam Giritharan

Abstract The pursuit of optimal post-workout recovery is central to enhancing athletic performance, reducing fatigue, and promoting muscle regeneration. Traditional recovery methods, while effective, are often limited by issues such as nutrient bioavailability and delayed recovery times. Nanotechnology, particularly through the use of nano-enhancements, is revolutionizing the sports nutrition field by providing innovative solutions that accelerate recovery processes. We delve into various nanoparticle systems, including liposomes, micelles, and polymeric nanoparticles, which enhance the bioavailability of proteins, amino acids, and antioxidants essential for recovery. By improving cellular uptake and targeted delivery, these nanomaterials optimize the absorption of key recovery nutrients and reduce muscle damage. This chapter examines the role of nano-enhancements in post-workout recovery, focusing on their ability to improve nutrient delivery, support muscle repair, and reduce inflammation. It highlights recent research, clinical trials, and real-world applications, demonstrating the effectiveness of nano-enhanced solutions in accelerating recovery. Additionally, the chapter addresses safety, regulatory, and ethical considerations surrounding the integration of nanotechnology into sports nutrition. By exploring the challenges and future potential, it offers insights into how these innovations could reshape post-workout care, offering athletes, trainers, and healthcare professionals powerful solutions for optimizing performance and well-being.

Keywords Post-workout recovery · Nanotechnology · Nano-enhancements · Muscle repair · Nano carriers · Bioavailability

W. Vinu (✉) · D. A. Kabeer (✉) · G. Manibharathi
Department of Physical Education and Sports, Pondicherry University, Puducherry, India
e-mail: vinu@pondiuni.ac.in

D. A. Kabeer
e-mail: dilshith.dpe@pondiuni.ac.in

Ketheeswaran
Science Unit, Allied Health Sciences, University of Jaffna, Jaffna, Sri Lanka

G. Giritharan
Nevada Center for Reproductive Medicine, Reno, NV, USA

A. Chatterjee et al. (eds.), *Nanofuel: The Future of Sports Nutrition*, Smart Nanomaterials Technology, https://doi.org/10.1007/978-981-96-5471-0_14

1 Introduction

1.1 Post-workout Recovery

Recovery or recovery period means the process of bringing the body of a performer, or athlete to a steady state after exerting strenuous physical effort. It is the time when energy is resynchronized, muscle tissues rebuilt, fatigue lessened, and risks of injuries lowered. Thus, recovery is a multiple biological event combined with glycogen synthesis, muscle protein synthesis, replenishment of water-potassium balance, and anti-inflammatory processes.

What are the Benefits of Staying Sober During Exercise?

Athletes or anyone into exercise perform challenging training programs that demand a lot of stress. Applying the suggestions of this article in practice will contribute to avoiding possible muscle damage, further overtraining, and a corresponding decrease in performance if there is no efficient recovery strategy. The recovery phase is also important since, during the intense exercise training, small breaks or damage in the muscle fiber need to recover and grow Glycogen depletion is another issue that also plays its part in the energy- storing macromolecule, and post exercise, the glycogen reserves need to be refueled and restored to their ideal capacity a gain to serve as a source to support the following training sessions [1, p. 201].

One of the most common issues athletes face is delayed onset muscle soreness (DOMS), which occurs due to microscopic muscle damage. Research indicates that proper recovery strategies, including hydration, stretching, and nutrient intake, can significantly reduce soreness and enhance recovery speed [2]. Additionally, hydration plays a crucial role in post-workout recovery, as excessive sweating leads to electrolyte loss, which can impair muscle function and prolong fatigue. Studies emphasize the importance of fluid balance for performance and recovery [3].

Beyond physical recovery, immune system support is also essential for athletes, as intense workouts can temporarily suppress immune function, making athletes more susceptible to illness. Antioxidant-rich foods and probiotics have been shown to help maintain immune resilience [4]. Furthermore, exercise influences the body's hormonal balance, with cortisol levels rising post-workout. Proper rest, sleep, and nutrition help regulate these hormones, preventing overtraining syndrome and promoting mental well-being [5, 6]

1.2 Key Components of an Effective Post-workout Recovery Plan

A well-structured post-workout recovery plan consists of several critical elements. Proper nutrition is fundamental, as consuming an adequate balance of proteins, carbohydrates, and healthy fats aids muscle repair and energy restoration. Hydration is equally vital, with electrolyte-infused beverages restoring fluid balance and preventing dehydration. Additionally, rest and quality sleep play a crucial role in muscle regeneration, allowing for physiological repair. Stretching and active recovery techniques help reduce stiffness and improve circulation, further accelerating the recovery. Some athletes also benefit from cold therapy and compression techniques, which minimize muscle inflammation and soreness.

With advancements in nanotechnology, post-workout recovery has become more efficient. Nano-protein shakes containing nano-encapsulated BCAAs and creatine enhance muscle repair by improving nutrient bioavailability. Similarly, nano-electrolyte drinks provide faster hydration due to their increased solubility and absorption properties. Nano-antioxidant formulations help reduce oxidative stress and support overall recovery. These innovations demonstrate the potential of nanotechnology in optimizing athletic performance and recovery strategies.

1.3 Overview of Nanotechnology in Sports

Nanotechnology is the use of equipment, instruments, and systems to control matter at atomic and molecular levels, namely 1 to 100 nm is transforming many sectors, and possibly the biggest influence is transports, foods, and recreational activities such as sports. Nanotechnology, which employs the science of manipulating materials at the atomic level, has great potential in stimulating the improvement of athletic performance, duration of performance, and health. It is used in a number of applications; however, none is as fascinating as its place in the creation of the new generation of sport drinks, and functional food. These ideas are revolutionizing not only how athletes nourish themselves but are now setting benchmarks in targeted nutrition and ergogenic enhancement.

The load put into the athletes during the training and even during the competition is very high, and this often calls for precise and timely delivery of nutrients for energy reproduction and replenishing other parts of the body that may be used. Though useful in improving nutrient bioavailability, current sports drinks and the general food products may pose shortcomings like low solubility, poor nutrient uptake or delivery, and a very short shelf life. Nanotechnology thus attempts to overcome these shortcomings through the fabrication of nanotherapeutic carriers, oil-in-water emulsions, and encapsulation systems that improve bioavailability, delivery, and functionality of nanocarriers.

For example, nanoencapsulation helps vitamins, minerals, amino acids, and other super nutrients delivered to the body as and when required in their most potent form. Not only does this process increase solubility, but it also increases the stabile time of these nutrients, making them more active for longer hours. In the same way, in sports drinks, nanoemulsions enhance the dissolution of the ingredients, such as caffeine or specific vitamins, so the body easily assimilates them. These developments allow the athletes to enjoy increased energy combustion, better osmotic reabsorption, and increased healing capacities.

2 Elements Necessary for Athletic Performance and Recovery

Finding better ways to boost recovery after a workout is essential for athletic performance. The process ensures that the body gets the necessary power to recover from the physical activity applied to it. Athletes who recover properly may avoid the effects of exercise-induced muscle damage (EIMD), use the maximum muscle repair features, and replenish the energy stores. In athletic care, traditional methods used techniques such as hydration with carbohydrates for replenishment and proteins and antioxidant supplements. However, these methods often have many shortcomings, especially due to factors like poor nutritional bioavailability, absorption ineffectiveness, and delayed recovery times [7]. Nanotechnology has recently been seen as a promising solution to these problems, and interest in creative solutions has arisen.

Food is one of the sectors of the economy that have been transformed by nanotechnology, which is the construction of nanoscale instruments. These materials of the same coldness, smaller than the rest of the bulk, can significantly extend surface areas and time decay of gas rings in the short term, making them higher solutes. Nutrition provided on a nanoscale can easily cross physiological barriers that usual supplements cannot. This means the digestive limitations of substances typically given orally digestion are overcome by the nutrients encapsulated in nanocarriers such as liposomes or micelles. Thus, through this, the delivery system is targeted to the tissues where specific recovery support is needed, and the bioavailability is enhanced [8].

The main factors critical for after-exercise regeneration are decreasing the lumens, reforming muscles, and replenishing the glycogen. Reactive oxygen species (ROS), produced due to strenuous physical exercise, are the main factors affected by inflammation and oxidative stress [9]. To deal with these processes, antioxidants such as curcumin, vitamin E, and polyphenols are essential; nevertheless, the hydrophobic property limits their adsorptions in grooming samples [10].

The synthesizing of more stable and soluble substances achieved by nanotechnology enables these drugs to contribute to reducing oxidative damage and inflammation, thus being more effective [11].

Another setback to recovery is exercise-induced muscle soreness, sometimes called delayed-onset muscle soreness or DOMS. Over the years, it has been established that the faster proteins and amino acids are absorbed, which are crucial for muscle regeneration, the quicker muscle repair can be attained with Nano delivery systems. These progresses contribute substantially to an athlete's readiness for the next training session or the game and faster recovery.

3 The Science of Muscle Growth and Repair in Post-workout Recovery

Restoring muscles and their growth is one of the most important parts of rest after the exercises. This particular aspect makes the person more resilient, stronger, and a better sports performer. Muscle damage during exercise (exercise-induced muscle damage—EIMD) results from microscopic tears that occur at a high intensity, especially in resistance or endurance training. This is everyday life, but the perfection of the muscles' adaptation makes it possible for the body to heal these microtears and have the muscles ready for the next round. The body must comply with the recovery process protocol to make the microtears in the muscles disappear and fiber sustainably to maintain life's usual functioning. The process should stimulate physiological adaptation, such as microtears, for the body to respond by producing and repairing damaged cells, thereby maintaining homeostasis, which is the balance between the internal and external factors of the body.

3.1 The Physiology of Muscle Repair

Muscle repair is a complex interaction between cellular and molecular mechanisms. It starts with an inflammatory phase that causes the attraction of immune cells to the injured tissue, particularly neutrophils and macrophages. In addition to the disposal of cellular debris, immune cells are necessary, and cytokines are the attachment molecules that stimulate tissue repair. The problem is that even when it is essential to cure a disease, an inflamed condition that is either prolonged or too severe might inhibit such a cure, causing prolonged muscle soreness. This underscores the necessity of carefully managing this phase. The repair phase is driven by satellite cells, a kind of stem cell found on the surface of skeletal muscle fibers, which can convert into muscle cells after being triggered. In addition to these cells, cytokines play an essential role in the proliferation, differentiation, and amalgamation of these stem cells with the existing satellite cells. Consequently, improving muscle functioning and enhancing muscle integrity are hinged on this cellular mission. The quality and quantity of the cells in the tissue decide the effectiveness and speed of the tissue repair process and the consequent standard functionality of the organs and systems. Abilities such as regeneration and the repair of tissues continue to be increasingly important, both for the prevention of tissue loss and the healing and restoration of tissue and organ functions in various disease conditions (Fig. 1).

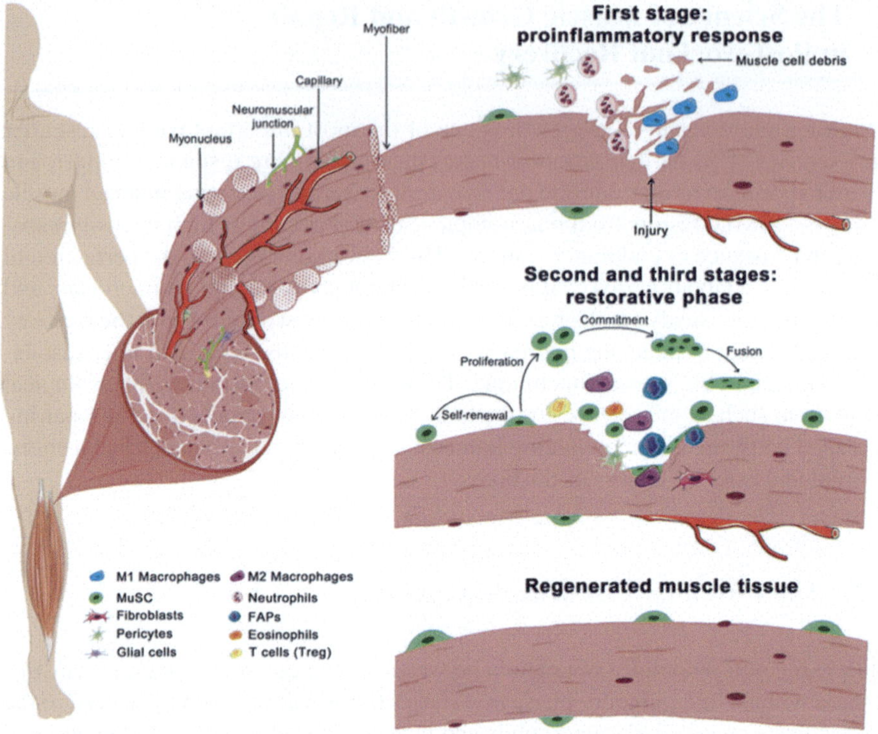

Fig. 1 Schematic overview of muscle repair, showing immune-mediated debris clearance, satellite cell activation, and tissue regeneration [12]

3.2 *Nutrients' Role in Muscle Growth and Repair*

Correct fuel is the cornerstone of muscle growth and recovery malady, so intake of essential substances is crucial. The process owes its importance to protein. Protein is a key trigger of muscle mass synthesis by supplying essential amino acids for muscle protein synthesis (MPS). Among them, the branched-chain amino acids, mainly valine, isoleucine, and leucine, are the ones accountable for regulating MPS. For instance, the mTOR (mammalian target of rapamycin) signaling pathway is activated by leucine and is essential for synthesizing muscle proteins [13].

Protein is a critical macronutrient in muscle hypertrophy and remodeling and also in the capacity of an organism for muscle adaptation to training stimuli. Optimal macronutrients, specifically proteins, must meet acute muscle repair, Growth and Adaptation. Also, adequate energy substrate is required for metabolic replenishment, and some micronutrients have a role in skeletal muscle building (Holick, n.d.). EAAs, especially leucine, are key nutrients that significantly influence MPS. Regarding muscles, leucine is a great activator of the mTOR which plays a role in building muscle proteins. Several investigations have shown that pre-exercise supplementation with leucine-fortified products facilitates post-workout recovery and remodeling Other micronutrients, such as vitamins, repair muscles engaged in macronutrition. For example, vitamin D plays a central role in muscle tissue remodeling and repair, and lack of it appears to result in muscle atrophy and a slower recovery rate [14]. Furthermore, Table 1 shows that creatine, selenium, calcium, and vitamins D, C, and E all promote the rate of muscle protein synthesis and depress the rate of proteolysis [15].

Another factor about how food is consumed impacts muscles is that the timing of nutrient intake affects muscle recovery. The metabolic window suggests that post-exercise food consumption alters body metabolism from catabolism to anabolism to support muscle repair and growth [16]. Research shows immediate and optimal post-workout protein and carbohydrate consumption enhances muscle glycogen and protein synthesis [17].

3.3 NanoNutrients' Role in Muscle Growth and Repair

Muscle tissue engineering has remarkably grown through nanotechnology due to nano-biomaterials and various fabrication methods. Current research has shown great promise for bioengineered 3D tissue scaffolds that can be constructed from multiple bio-nano-micro scale techniques and would remain fertile in resemblance to native muscular tissues. Such constructs have improved muscle tissue regeneration and function [18].

Structural scaffolds have been developed from nanotechnology and have supported muscle regeneration through molecular launch pads. For instance, when MXene nanoparticles have been integrated into nanofiber structures, it has been revealed that the enhanced proliferation and differentiation of muscle cells significantly improve the regeneration of volumetric muscle loss. It must also be noted that this approach holds promise in using nanomaterials to promote conditions favorable for muscle regeneration [19]. In addition, using nanomaterials as a tool in skeletal muscle regeneration has received more attention, especially recently. Zero-dimensional nanoparticles, including quantum dots, one-dimensional nanorods/ nanotubes, and two-dimensional nanofilms /nanolayers, have been deployed to support the formation of new muscle tissue. Such an application provides opportunities to use nanomaterials with overall physical and biochemical characteristics that can be optimized for tissue engineering of skeletal muscle tissue [20].

Not only does using nanotechnology in muscle repair provide new structural support for regenerating tissues, but it also helps deliver therapeutic agents. By controlling the size of particles involved, drugs, growth factors, or genes can be delivered at targeted sites of tissue injury, thus making treatments more effective and less hazardous to the patient. This targeted delivery system is particularly advantageous for targeting the multilayered microenvironment of traumatized muscles.

4 Nanoparticles: Types Related to Post-workout Recovery

4.1 Nano-particles Materials Used in Sports Drinks, Food, and Medicines

Nanoemulsions: Employed to deliver lipophilic nutrients, including omega-3 fatty acids and fat-soluble vitamins. Enhance bioavailability and guarantee long-lasting release.

Solid Lipid Nanoparticles (SLNs): Perfect for encapsulating such active antioxidants as anti-inflammatory compounds, accelerating healing by minimizing oxidisation stress and inflammation.

Polymeric Nanoparticles: Coating agents encapsulate proteins, amino acids, or other progressive active ingredients delivered to muscles to support repair and build-up functions. Be able to meter release so that nutrients are released on time during the critical post-workout recovery window.

Nanostructured Lipid Carriers (NLCs): A lipid-based system that could release a range of nutrients, such as one that could deliver a single nutrient conveniently, has multiple recovery benefits that can be obtained through a single formulation.

Cubosomes: Cubosomes are nanostructured liquid crystal particles that act as carriers made of amphiphilic lipids. It is a very stable nanoparticle made of a honeycomb structure. This is used in the delivery of carrier particles to all the organs.

Metallic Nanoparticles: Nanosilver and Gold Nanoparticles: These are employed due to their anti-microbial and anti-inflammatory effects, providing immune rebound and suppressing muscle soreness.

Water Soluble Carbohydrate Based Nanoparticle: Water-soluble carbohydrate-based nanoparticles have been developed to enhance biological activity. Based on natural polymers such as chitosan or starch, these particles enhance the structural features involving stability and absorption of bioactive compounds. Approximately efficient vehicles for prebiotics and probiotics promoting gut health and overall recovery.

5 Approaches to Include Nanoparticles in Recovery Products

5.1 Nanoencapsulation

Ingredients in Nano are protected within tiny spheres so that only the active compounds come out. Degradation and controlled release are essential characteristics that can be incorporated into a polymer matrix. It is popular for amino acids, creatine, and boosting recovery nutrients.

5.2 *Nano-protein Supplements*

Nano-Whey Protein: The encapsulation of whey protein hydrolysates assures improved digestion and absorption rate and faster muscle repair. mNano-Casein Proteins: In the body, they provide a slow release, allowing amino acids to be in the bloodstream for longer recovery.

5.3 *Nano-amino Acids and Peptides*

Nano-BCAA: They work to stimulate muscle protein synthesis by reducing muscle fatigue.

*Nano-glutamine is helpful for muscle repair, promoting immune func*tion, and lowering exercise-induced oxidative stress.

5.4 *Nano-antioxidants for Muscle Recovery*

Nano-Curcumin: A potent anti-inflammatory agent that affects muscle soreness and oxidative stress.

Nano-Resveratrol: Enhances mitochondrial function and decreases post-exercise fatigue.

Nano-Quercetin: Supports endurance and decreases exercise-induced oxidative damage.

5.5 *Nano-energy and Rehydration Solutions*

Nano-Caffeine: Aids muscle glycogen replenishment and makes them less fatigued after a workout. Nano-Electrolytes: Improve hydration through mineral supply, namely sodium, potassium, and magnesium, to the cellular medium.

5.6 *Nano-adaptogens and Anti-inflammatory Compounds*

Nano-Ashwagandha: Reduces cortisol levels and improves post-exercise recovery. Nano-Ginger Extract: Aids muscle recovery by decreasing inflammation caused by exercise.

5.7 Mechanisms of Nanoencapsulation for Recovery

Liposomal Encapsulation: Lipid-based vesicles enhance the absorption of fat-soluble nutrients like curcumin and resveratrol. Polymeric Nanoparticles: Biodegradable polymers allow the controlled release of amino acids and proteins. Nano-Emulsions: Increase the bioavailability of hydrophobic bioactives such as omega-3 fatty acids.

6 Clinical and Performance Implications

Fast Muscle Repair: Nano-proteins and BCAAs quicken muscle regeneration. Reduced Inflammation: Curcumin and resveratrol in nano-encapsulated preparations lower oxidative stress.

Improved Hydration: Nano-electrolytes ensure better fluid retention post-exercise. Sustained Energy Release: Nano-caffeine and adaptogens assist with long-term energy balance.

6.1 Nanoemulsification

Transforms hydrophobic materials into water-soluble materials, enhancing their solubility and absorption. It is used in most sports drinks to provide necessary vitamins and energy-boosting compounds. Nanoemulsification refers to manufacturing an ultrafine emulsion in which one liquid (oil) is dispersed in another liquid (water) by using surfactants and high-energy processing techniques. They will facilitate solubility and absorption of bioactives and thus are very efficacious in post-workout recovery solutions.

6.2 Mechanisms of Nanoemulsification in Recovery

Higher solubility: Optimize the spreading of lipophilic compounds in water-based environments for better absorption. Enhanced permeability: Nano-sized particles cross cell membranes more efficiently.

Faster cellular uptake: Absorption occurs quicker, leading to quicker physiological results.

Sustained Energy: Nano-MCT and nano-caffeine improve metabolic recovery.

7 Nanocarriers

Liposomes, dendrimers, or micelles transport bioactives to targeted cells or tissues. Optimise nutrient utilization within muscles after exercising, which would shorten recovery time.

7.1 Understanding Nanocarriers in Post-workout Recovery

Nanocarriers are administered for encapsulation and delivery of bioactive compounds to increase stability and improve bioavailability. They may be lipid-based, polymer-based, or hybrid systems designed for the prompt delivery of nutrients.

Nanoemulsions: Liquid-based nanocarriers that enhance the solubility and absorption of hydrophobic compounds.

Solid Lipid Nanoparticles: Stable lipid-based nanocarriers that offer protection of bioactive compounds from degradation.

Nanomicelles: Lead to improved water solubility and absorption of poorly soluble compounds.

7.2 Key Nanocarrier-Based Products for Post-workout Recovery

7.2.1 Nanocarrier-Enhanced Proteins and Amino Acids

Liposomal Whey Protein: Ensures fast absorption and a good muscle-building effect. Polymeric Nano-BCAA: Allows long-lasting presence in the body of essential amino acids necessary for muscle repair.

Nanoencapsulated Creatine*:* Assures better solubility, ATP production improvement, and better endurance.

7.2.2 Nanocarrier Delivered Antioxidants and Anti-inflammatory Agents

Nano-Curcumin: Improved bioavailability reduces inflammation and stress from oxidation. Nano-Liposomal Resveratrol: Increased mitochondrial function and muscle recovery. Solid Lipid Quercetin Nanoparticles: Better endurance with decreased oxidative stress caused by exercise.

7.2.3 Carrying in Nanoform for Post-workout Recovery Agents

Nano-Omega-3 (EPA/DHA): Lowers inflammation, promotes joint health, and enhances cardiovascular function. Nano-MCT: Gives a very fast energy refuel-up and enhanced metabolism recovery.

7.2.4 Nanocarrier-Based Rehydration and Electrolyte Solutions

Liposomal Electrolytes: Improves hydration efficiency and prevents cramps. Nano-Caffeine *Delivery*: Supports faster glycogen replenishment and post-workout alertness.

7.2.5 Nanocarrier-Encapsulated Adaptogens and Herbal Extracts

Polymeric Ashwagandha Nanoparticles: Decreases cortisol levels and recovers performance.

Nano-Ginger Extract: Moderates muscle soreness and exercise-induced inflammation. Nano carriers are reviewed in this section based on different mechanisms in recovery.

Increased solubility: Enhances dispersion of lipophilic compounds in aqueous environments for enhanced absorption.

Targeted delivery: Delivers nutrients to specific muscle tissues to enhance recovery.

Controlled release: Tethered nutrients result in optimized long-term recovery.

Enhanced cellular uptake: Nano-size particles penetrate cell membranes more effectively.

8 Nano Coatings

Coating of nutrients avoid denial from environment by enclosing them in coat factors. To enhance the shelf-life of sports drinks and foods. Nano coatings are a new technology getting much attention in sports science. They give athletes the possibility of improving post-workout recovery and athletic performance. Applying ultra-thin nanometer-scale coatings to sportswear, recovery devices, nutritional supplements, and rehabilitation tools could introduce properties such as durability, antimicrobial protection, temperature regulation, and better nutrient delivery.

Nanocoatings are crucial for muscle work recovery after a workout to lower fatigue, increase comfort, enhance bioavailability in active supplementation, and prevent the spread of sweat that can cause infections. The application of nano-coating technology in sports is multidisciplinary, encompassing materials science, nanotechnology, biomedical engineering, and sports medicine. Thus, one will first require an understanding nano coatings in sports and recovery. Nano coatings are made using nanomaterials, and the surfaces are modified at the molecular scale, contributing to better quality. These coatings can be hydrophobic, antimicrobial, thermoregulating, or biofunctional, contingent upon their applications in sports and post-exercise recovery.

8.1 Types of Nano Coatings Used in Sports and Recovery

8.1.1 Nano-coating Applications on Sports Attire for Restoration

Compression Garments Using Nano-Coated Fabrics:

- Promote blood circulation and oxygen delivery to muscles.
- Reduce muscle soreness and inflammation post-workout.
- Infuse antimicrobial properties to prevent bacterial growth.

Smart Textiles with Nano-Coated Bioactive Agents:

- Dissolves magnesium, menthol, or curcumin to block muscle fatigue.
- Gradually releases bioactive compounds to boost recovery.

8.1.2 Nano Coatings for Sports Equipment and Gym Surfaces

Antibacterial Nano Coatings on gym equipment (dumbbells, yoga mats, machines) prevent microbial contamination. Anti-Sweat & Water-Resistant Nano Coatings enhance grip and durability of sports gear.

8.1.3 Nano Coatings in Supplements and Nutritional Products

Nano-Coated Protein Supplements:

- Protects proteins, BCAAs, and creatine from degradation.
- Enhances absorption and bioavailability for muscle recovery.

Liposomal Nano Coatings for Curcumin and Omega-3:

- Improves solubility and anti-inflammatory benefits post-exercise.
- Ensures slow and sustained release for extended muscle recovery.

8.1.4 Nano Coatings in Wearable Recovery Devices

- Electrotherapy and Massage Devices with Nano-Coated Electrodes
- Enhances conductivity and efficiency of recovery devices.
- Prevents sweat-induced degradation for long-term durability.

8.2 Mechanisms of Nano Coatings in Recovery Enhancement

Prevention of Microbial Infections and Moisture Retention: Hydrophobicity and Antimicrobial Activity. Controlled release of bioactive compounds-Nano coatings allows gradual release of nutrients, amino acids, and antioxidants from recovery-based supplements in sportswear.

Thermoregulation: Keeping in mind the surrounding conditions, certain nano coatings allow for adjustment of heat dissipation and insulation to promote muscle relaxation.

Biocompatibility: Nanocoatings develop permeability of nutrients, allowing for quick muscle repair and enhanced immune function [21].

8.3 Clinical and Performance Benefits of Nano Coatings

Faster Muscle Recovery: Nano-coated textiles and supplements reduce soreness and enhance recovery.

Enhanced Comfort and Hygiene: Antibacterial coatings prevent odors and infections.

Improved Endurance and Hydration: Sweat-resistant nano coatings regulate body temperature and moisture control.

Extended Product Lifespan: Nanocoatings protect equipment and clothing, increasing durability.

9 Bioavailability in Post-workout Recovery and Nanotechnology

Bioavailability is the proportion of a nutrient, supplement, or pharmaceutical active that passes into the systemic circulation and becomes available for use by the body or the tissues. For efficient absorption and muscle repair, along with much-needed anti-inflammatory activity and glycogen synthesis in post-workout recovery, there is a great need for high bioavailability. Conventional supplements suffer a lot in bioavailability due to discarding enzymatic digestion, metabolic breakdown, or limited absorption in the GI tract Nanotechnology presents novel applications in accelerating bioavailability, such as enhancing solubility, preventing nutrient degradation, and promoting cellular uptake [22].

9.1 Factors Affecting Bioavailability

The factors influencing bioavailability of nutrients and agents of recovery are as follows

Solubility and absorption efficiency

For this reason, they have low solubility in water, such as compounds such as curcumin, resveratrol, and omega-3 fatty acids. Water-soluble compounds tend to have a greater absorption level but get rapidly excreted before showing their full effectiveness [23].

Gastrointestinal barriers

Enzymatic digestion and metabolism in the liver can take some nutrients down before reaching their tissues of action. pH changes in the stomach and intestines can degrade sensitive compounds and thus lessen their efficacy.

Transport and Cellular Uptake

Some compounds require dedicated transporters to pass cell membranes, limiting their absorption. Larger molecules, i.e., proteins and peptides, may penetrate the cell less efficiently.

Dosage and Timing

Certain nutrients in very high doses have to be used to compensate for the very low bioavailability, though it could lead to adverse effects or useless waste from not being absorbed. Timing of the nutrients after exercise will influence the uptake since, after workouts, the muscle is still absorptively receptive. Nanotechnology enhances bioavailability by modifying nutrients at the nanoscale for improved solubility, absorption, and cellular uptake. The following nano-based strategies adapt to surmount the challenges of the conventional delivery systems:

9.2 Nanoencapsulation for Enhanced Solubility

Hydrophobic nutrients like curcumin and omega-3 fatty acids can be encapsulated within nanoparticles, such as liposomes or micelles, to render them water-dispersible and enhance their absorption ability inside the body. Nanoemulsions-in-the-water or water-in-oil formulations are substrates that enhance the solubility of fat-soluble vitamins and recovery agents.

9.3 Liposomal Delivery for Stability and Protection

Liposomes are made from phospholipid bilayers and shield the nutrients from digestion in the stomach, enhancing absorption in the intestines. Liposomal vitamin C and curcumin are absorbed more significantly than their non-liposomal equivalents.

9.4 Nanocarriers for Targeted Delivery

Polymeric nanoparticles can deliver nutrients directly to muscle tissues or an area of inflammation for maximum effectiveness. Stimuli-responsive nanoparticles release their contents when exposed to specific biological signals, including pH changes or oxidative stress [24].

9.5 Controlled and Sustained Release

Slow-release nanocarriers enable the release of nutrients such that they remain bioavailable in the system for a longer time; this minimizes the times of supplementation. Polymeric protein nanoparticles slowly release amino acids to continuously support muscle protein synthesis after exercising.

9.6 Enhanced Cellular Uptake via Nanoparticles

Additionally, because they are nanoscale-sized, they can penetrate cellular membranes and allow for enhanced concentrations of essential recovery agents within the cell. Gold and silver nanoparticles have been investigated for their antioxidant delivery capabilities and protection against cellular-onset oxidative stress. By improving the bioavailability of essential nutrients, nanotechnology accelerates recovery, reduces fatigue, and enhances performance. Table 1 shows Benefits of Nano-Enhanced Bioavailability on Post-Workout Recovery.

10 Nanotechnology's Role in Muscle Repair

After exercising, focus on that food that is tailored based on the workout and repair requirements, which athletes need to encourage muscle hypertrophy and restitution. Utilize recovery products that incorporate nanotechnology, such as protein

Table 1 Benefits of nano-enhanced bioavailability on post-workout recovery

Nutrient/Compound	Conventional bioavailability (%)	Nano-enhanced bioavailability (%)	Recovery benefits	References
Curcumin (Anti-inflammatory)	~ 5%	Up to 40% (liposomal)	Reduces muscle soreness, speeds up healing	[25]
Omega-3 fatty acids	~ 10–15%	~ 50% (Nano emulsified)	Reduces inflammation, supports joint health	[26]
Protein (Whey, BCAAs)	~ 60–70%	~ 90% (nanoparticles)	Enhances muscle repair, sustains protein synthesis	[27]
Vitamin C (Antioxidant)	~ 20–30%	~ 85% (liposomal)	Neutralizes oxidative stress, boosts immunity	[28]
Electrolytes (Sodium, Potassium)	~ 50–60%	~ 95% (nanoelectrolytes)	Improves hydration, maintains electrolyte balance	[29]
Carbohydrates (Glycogen replenishment)	~ 60–80%	~ 98% (nanocarriers)	Enhances energy restoration post-workout	[30]

nanoparticles and liposomes containing antioxidants, thereby having a considerable influence on the process of the growth of the wound. The healing process is completed more quickly through innovative recovery products like protein nanoparticles and liposomes loaded with antioxidants. These are combined with the more traditional recovery techniques like stretching and drinking sufficient water, which, in turn, improves the whole recovery strategy, optimizes immediate performance, and extends the long-term adaptations. Athletes can only consistently achieve their performance goals, recover more quickly, and decrease the time they are out of action because they have grasped the progress of nanotechnology and the whole muscle repair bodily functions.

10.1 Practical Applications for Athletes

Athletes should prioritize post-workout nutrition tailored to their training intensity and recovery needs to optimize muscle repair and growth. Using nano-enhanced recovery products, such as protein nanoparticles and antioxidant-loaded liposomes, can provide significant advantages in accelerating the repair process. Combining these with traditional recovery strategies, such as stretching and adequate hydration, creates a comprehensive recovery plan that maximizes both short-term performance and long-term adaptations.

By leveraging advancements in nanotechnology and a deep understanding of the physiology of muscle repair, athletes can recover more efficiently, minimize downtime, and achieve their performance goals with greater consistency.

10.2 Inflammation and Oxidative Stress in Post-workout Recovery

Mild to moderate levels of inflammation and oxidative stress are side effects produced naturally after exercise, a necessary part of the body's healing process. Inflammation and oxidative stress, which are too much or too long and can hinder the recovery process, can also add to muscle soreness and impair performance because they are the key to the beginning of muscle repair and adjustment. In the end, to provide a proper recovery, it is paramount to be thoroughly acquainted with the underlying mechanisms of these processes and to be able to come up with effective management techniques.

10.2.1 Inflammation

Muscle damage triggered by exercise is mostly a body's first reaction to it, especially when it comes to exercise with long duration and high intensity, which is inflammation, the positive and the negative, the two-sided sword. The immune system is the first to get activated when the body detects the pain.

- *Emergency waste collection*:

Neutrophils and macrophages are then guided to the damaged area to start the immune response. They are the ones that, among other things, disintegrate the organism's waste. By phagocytosis operation, neutrophils destroy out-of-order cellular elements.

- *Communications that Signal Regeneration*:

Macrophages secrete cytokines, stimulating satellite cells and muscle repair, such as interleukin-6 (IL-6) and tumor necrosis factor-alpha (TNF-α).

Inflammation, on the one hand, is a sign of the start of the healing while there's a danger of producing too much inflammation, thus making the body recovery slow and swelling to extend the duration of Neutrophils use phagocytosis, a standard process to eliminate debris from injured cellular parts. Macrophages can also release certain types of cytokines like TNF-α and interleukin-6 to activate the muscle repair process of the satellite cells.

10.2.2 Oxidative Stress

The Function of Reactive Oxygen Species

Oxidative stress results from the body producing more reactive oxygen species (ROS) than it can neutralize with antioxidants. As a consequence of energy metabolism, ROS are produced by mitochondria during exercise, a vigorous workout with an intensity close to maximum. Even though ROS at normal levels can contribute to the signaling mechanisms that favor adaptation, excessive ROS can damage proteins, lipids, and DNA, which may lead to cell dysfunction. Inflammation, muscle fatigue, and delayed recovery are all issues that are directly associated with oxidative stress [31]. The athlete's performance in the long term may be decreased, and untreated may lead to chronic muscle damage.

How our body manage inflammation and oxidative stress?

To reach the best point of recovery, it is essential to let the body's natural inflammatory response take its course while minimizing oxidative stress's harmful effects. Some of the most important strategies will be:

- *Anti-Inflammatory Nutrients*

Nutrients such as curcumin, omega-3 fatty acids, and polyphenols are well-known for their anti-inflammatory effects. Curcumin, a compound that is isolated from turmeric, acts by preventing the release of inflammatory enzymes and cytokines, which in turn reduces the occurrence of muscle soreness and swelling. Omega-3 fatty acids are an example of these, which are genuinely effective in fish oil because they will cause the production of pro-inflammatory cytokines to go down, and correspondingly, there will be quicker tissue regeneration

- *Antioxidants*

Antioxidants like vitamin C, E, and flavonoids cancel the ROS, slowing the oxidative damage and helping in fast recovery. Post-workout recovery is a common choice for foods full of antioxidants, e.g., strawberries, spinach, and green tea

- *Adequate Rest and Sleep*

Inflammation and oxidative stress need rest and sufficient good quality sleep to be cured as they are controlled by a surge of your hormones only during the deep sleep phase.

10.3 Management of Inflammation and Oxidative Stress by Nanotechnology

Nanotechnology is changing how inflammation and oxidative stress are managed in the recovery process after exercise. A major drawback of traditional antioxidant and anti-inflammatory supplements is their low bioavailability, and are easily broken down in the digestive tract. To overcome these limitations, nanotechnology establishes the delivery and efficiency of these compounds by using tiny particles at large scales.

- *Nano-Antioxidants*

Nanoparticles could help grow food derived from plants, using nano-curcumin, for example, which would be up to 40 times more bioavailable than ordinary curcumin and alleviate inflammation more effectively [9]. A case in point, the introduction of nano-encapsulated curcumin has enabled a dramatic increase in absorption rate of about 40 times more than the traditional curcumin, and thus, more effective inflammation management has been realized

- *Liposomal Delivery System*

Liposomal technology is more effective at delivering antioxidants like vitamin E to their destined cells, thereby ensuring their stability and their transfer to the site of injury [32]. Their effectiveness in providing medicine to the desired areas and, hence, their viability have been raised by this

- *Controlled-ReleaseNanocarriers*

The drug delivery platform fabricated from synthesizing existing polymers into nanoparticles is being developed to minimize the release of anti-inflammatory drugs, thereby managing therapeutic levels in the bloodstream for prolonged periods. This kind of drug delivery system ensures the presence of the drug in the blood for a prolonged period, which is helpful for long-term inflammation management, especially during intense exercise.

11 Muscle Growth and Recovery: The TimeLine of Nanotechnology (Fig. 2)

Step 1: Delivery of Nutrients via Nanocarriers

Once exercising is conducted, nanoparticles from the bloodstream immediately deliver essential nutrients like proteins, amino acids, and minerals.

Step 2: Improved Absorption and Bioavailability

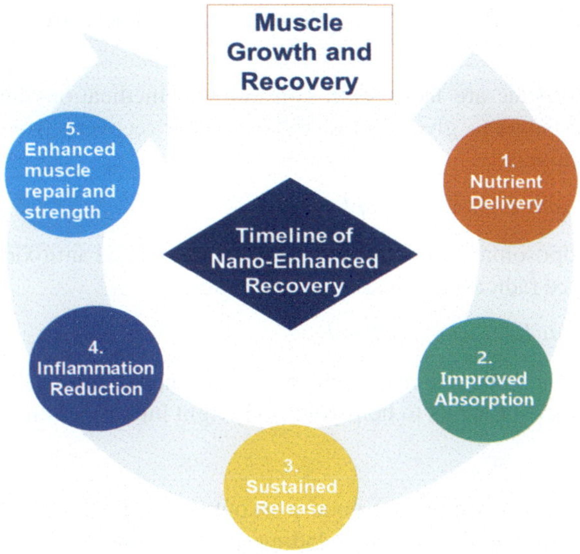

Fig. 2 Timeline showing nanotechnology's role in muscle growth and recovery: nutrient delivery (Step 1), improved absorption (Step 2), sustained release (Step 3), inflammation reduction (Step 4), and enhanced muscle repair and strength (Step 5)

Traditional supplementation suffers from one of the worst items: low absorption rates. The digestive system would destroy many nutrients once ingested before they reach muscle cells. By improving bioavailability, nanotechnology seeks to get around that very benefit: more nutrients absorbed efficiently [33].

Step 3: Controlled, Sustained Release

Thus, instead of a huge nutrient spike with depletion thereafter, nanocarriers slowly release nutrients over time, consequently allowing protein synthesis and muscle repair to occur for many hours after exercise [34].

Step 4: Reduction of Inflammation and Oxidative Stress

Following intensive exercise, muscle tissue experiences inflammation, resulting in soreness. For the reduction of stress oxides, metallic nanoparticles and lipid-based carriers are already available with faster recovery in such a way as to prevent further injury to the muscle [35].

Step 5: Muscle Fiber Repair and Strength

When there is sufficient and continuous nutrient delivery to the muscle fibers, they will repair more effectively and be stronger, allowing better endurance, strength, and performance for future workouts.

11.1 Practical Applications for Athletes

Athletes are using both conventional medications and the newer nano-enhanced versions to give a bit of relief in their recovery from the hurting aspects of their training:

Nano-Enhanced Supplements:

Liposomal curcumin or the nano-encapsulated antioxidants can be used to deactivate free radicals and lessen inflammation.

Anti-Inflammatory Foods:

Choose food that is rich in natural antioxidants and omega-3 fats for your post-workout meals to help your body fight inflammation.

Recovery:

The recovery after a good workout should be built on a combination of sleep, flexibility, and fluid intake. Leading sportspersons are likely to experience higher productivity, shorter recovery time, and lower severity of muscle aches by taking advantage of the benefits of proper pain. They can achieve that by controlling the processes of inflammation and oxidation with the appropriate use of the respective functional and efficacious recovery agents and by combining their benefits with the most recent nanotechnological treatments, which are important components of the latest recovery therapies (Table 2).

Recovery necessary for the body includes rehydration and electrolyte balance. In workouts, the body may suffer from dehydration and the lessening of water and minerals through sweating. Besides getting dehydrated, a lack of electrolytes may happen in some areas of the body. Dehydration that gets to the level of heat exhaustion or heat stroke to a person is the perfect element for those who prefer heat to the winter season's difficulties. Still, one should be aware that such a state can negatively affect physical performance and delay the cure of physical illness. Consequently, aside from restoring electrolytes, drinking fluid and thermal equilibrium restoration are crucial steps for making an organism return to the norm.

11.2 The Value of Rehydration

The body comprises approximately 60 to 70 percent of the water needed for performing different types of exercise, and the bodily structure and processes depend on it quite a lot. e.g., during exercise, the body releases the heat it produces through sweating. However, this cooling mechanism comes at the cost of losing fluids. The lifespan of thinking processes, cardiac power, and muscle efficacy have consistently been implicated in the impairment of dehydrated people, a dehydration volume of 2% of the body weight. In this way, dehydration will be more drastic, and thus, the

Table 2 Nano-enhanced sports nutrition products benefits and delivery mechanism

Product type	Key ingredients	Benefits	Delivery mechanism	References
Nano-protein shake	Nano-encapsulated Whey/Plant Protein, BCAAs, Creatine	Enhances muscle repair and growth, improves recovery speed	Nanoencapsulation for better protein absorption	[25]
Nano-hydration drink	Nano-emulsified Electrolytes, Vitamins, Minerals	Rapid hydration, better electrolyte absorption	Nanoemulsification for enhanced solubility	[26]
Nano-energy bar	Nano-structured Carbohydrates, Omega-3, Fiber	Sustained energy release, digestive support	Nanostructured carbohydrates for sustained release	[21]
Nano-recovery gel	Nano-encapsulated L-Glutamine, Curcumin, Beta-Alanine	Reduces muscle inflammation, improves endurance	Liposome-based delivery for controlled nutrient absorption	[28]
Nano-electrolyte drink	Nano-formulated Sodium, Potassium, Magnesium	Prevents dehydration, enhances muscle contraction	Nano-electrolyte dispersion for rapid rehydration	[29]
Nano-probiotic yogurt	Nano-coated Probiotics, Prebiotic Fiber, Vitamin D	Supports gut health, boosts immunity	Nano-coating to enhance probiotic survival	[36]
Nano-antioxidant smoothie	Nano-infused Antioxidants (Vitamin C, E, Polyphenols)	Fights oxidative stress, promotes overall recovery	Nano-infusion for higher bioavailability	[30]
Nano-amino acid capsules	Nano-delivered Essential Amino Acids, Collagen Peptides	Optimizes muscle protein synthesis, strengthens joints	Nanoparticle-based transport for targeted muscle repair	[37]

endurance ability of the athlete will be lessened, and fatigue and recovery will be slower. The major idea behind post-workout rehydration is to replace the lost fluids.

11.3 Electrolyte Balance

11.3.1 Muscle Contraction

An Essential Part of Recovery Electrolytes are substances that dissolve in body fluids to form electrically charged ions. The minerals include sodium, potassium, calcium, magnesium, and chloride. Some of the ions are needed for muscles to be able to contract. Nerve signals are sent to muscles via the central nervous system, controlling the heart and lungs. Some of the biotin is also helpful to the skin. Blood clotting is also a function of calcium. Both calcium and magnesium muscles accelerate and relax.

11.3.2 Nerve Function

The electrical gradients required for nerve impulse transmission are maintained by sodium and potassium.

11.3.3 Fluid Balance

The balancing of chloride and sodium is a factor that helps control water distribution in blood and cells. Various electrolytes, of which sodium is the main, are lost through perspiration during intense exercise. Pain in the muscles, little strength, and sometimes, heart irregularities might accompany these electrolyte deficiency states. As a result, permanent electrolyte loss is more relevant to post-exercise than only refueling fluids.

12 Techniques for Electrolyte Replacement and Effective Rehydration

It cannot be emphasized enough just how vital proper hydration and electrolyte balance are for post-workout recovery since strenuous bouts of exercise cause a huge fluid loss through sweat and are known to include key electrolytes like sodium, potassium, and magnesium that are critical to muscle function, nerve signaling, and overall hydration. Traditional means of rehydration, such as water or electrolyte-based sports drinks, do not often reach optimal bioavailability, which prolongs recovery times. Nanotechnology, however, has sparked an advantageous revolution in restoring hydration and electrolyte balance, increasing the speed and efficiency of nutrient delivery at the cellular level.

Such nanoparticles play a significant role in enhancing the absorption and retention of electrolytes. Nanoemulsions and lipid bilayer systems facilitate enhanced absorption and metabolic action of fat-soluble and water-soluble electrolytes. They enhance bioavailability, via which more significant amounts of the electrolyte are absorbed, with lesser portions lost during digestion. In addition, with the advent of polymeric nanoparticles, it became possible to control what to incorporate and the release of nourishing compounds over longer timescales- a feature very close to the hydration imprinted into food supplements for athletes and endurance activities.

Other than that, another considerable advantage of this developed technology in rehydration is spectacular in that it can promise rapid hydration to organs by targeted delivery of nano-supplementation. Micron-sized nano-enhanced electrolyte solutions work much faster than regular drinks because they fall through the cell walls quicker, returning muscle cells to complete well-being within the shortest time. Metallic nanoparticles, including nano-zinc and nano-magnesium, fight oxidative stress and inflammation that may arise from altered electrolyte composition-induced exercise stress.

Nanotechnology ultimately incorporates faster absorption, greater bioavailability, stable electrolyte levels, and lowered fatigue. This novel technology is changing the business side of sports and fitness by optimally allowing post-workout recovery to boost athletic performance and avert complications of dehydration. As research continues, nanotech-based hydration becomes an expected necessity in the present and extended future, the gold standard in sports nutrition and recovery, offering a more efficient, targeted, and scientifically advanced method of maintaining electrolyte balance.

13 Nanotechnology in Electrolyte Balance and Rehydration

Nanotechnology plays a crucial role in electrolyte replenishment and rehydration by boosting Lithgow infusion and electrolyte efficiency. Traditional rehydrating methods often have to deal with issues like irregular electrolyte intake or delayed absorption. These limitations are abolished by nano-enhanced conveyance, which tends to evolve to the stage of even faster and more precise results.

Nano-enhanced Drinks: Hydration aids that feature nanoemulsions can incorporate electrolytes more efficiently and give complete assimilation. For one, sodium nanoparticles found in sports drinks, mainly when dehydration arises, quickly re-establish the necessary ions, which is the root of prompt recovery.

Liposomal delivery systems: Coping with electrolyte supplementation interruptions and boosting targeted delivery to cells can now happen with liposomes by encapsulating electrolytes. As opposed to traditional technologies, this approach gives patients more efficient and targeted treatment; therefore, the electrolyte imbalance is resolved faster nanocarriers made of hydrogel. Loaded hydrogels with water and electrolytes can serve as long-acting hydration supplements. Long-term rehydration and maintaining an electrolyte balance are assured through the slow release of contents of such systems.

Athletes could use nanoproducts such as nano-based electrolyte drinks and capsules for quicker and more effective recovery. The athletes could build fluid rehydration and electrolyte balance, demining between old working out and nano-enhanced strategies, thus achieving optimal performance. Measuring the weight of one's body before and after exercise may promote efficient fluid loss and guide rehydration efforts. A liquid intake regimen may be employed throughout the workout and recuperation periods to prevent dehydration. Athletes can sustain the highest performance levels, minimize cramps and exhaustion, and recover faster if adequately replenished and balanced with electrolytes. The advancements in nanotechnology have brought about a new way of effective electrolyte transmission and opened experimental questions in post-exercise initiatives.

14 Disadvantages of Traditional Ways of Recuperating After a Workout

Rest, nutritionally dense food, and hydration have been the common methods adopted while treating injured athletes for years, but traditional post-workout recovery techniques sluggish the recovery process. They do not always cover the specific recovery requirements. Some editions of them are particularly troublesome in the case of elite athletes and individuals who do regular intensive exercise.

14.1 Nutrient Absorption Delays

Most previous approaches to nutrient replenishment involve ingesting micronutrients through the mouth, which must travel through the gastrointestinal tract before getting to the bloodstream. This process can often be slow and ineffective for several reasons, such as the action of the digestive enzymes, which may prevent the absorption of many nutrients. Further, the acidic pH of the stomach may cause the degradation of some sensitive components, affecting the active software. After absorption through the intestine, the first-pass hepatic metabolism of ingested nutrients significantly reduces their bioavailability. For instance, curcumin, omega-3, and some polyphenols are poorly soluble and have low bioavailability due to their inability to dissolve in water, making them weak fighters of inflammation and oxidative stress.

14.2 Key Nutrients with Low Bioavailability

For instance, vitamins, antioxidants, and hydrophobic compounds, common in many recovery nutrients, are characterized by low bioavailability. Fat-Soluble Nutrients: Vitamins A, d, E, and K need to absorb fat, and sometimes post-workout meals may not contain sufficient fat. Hydrophobic Compounds: Some compounds, such as resveratrol curcumin, cannot quickly assimilate into the body because of their poor aqueous solubility. When absorption is compromised, their impact on recovery is reduced, resulting in suboptimal delivery into target tissues.

14.3 Ineffective Recovery Agent Targeting

Such delivery of nutrients is inadequate when using conventional methods, especially when targeting tissue or inflammation areas. These factors suggest that a systemic distribution of nutrients can decrease their effectiveness for suppressing inflammation or localized muscle repair. For example, the anti-inflammatory nutrients like omega-3 fatty acids and curcumin become metabolized before they get to the repair site and hence are of little benefit.

14.4 Excessive Dosage Needs and Adverse Effects

The first identified medical need was an excessive dosage need or a need for frequent dosage, and its associations are presented in Table 3.

Several conventional recovery boosters require that more amounts of the drugs be taken to compensate for the poor absorption and metabolic capacity. The following adverse effects are more likely to occur with this method:

Gut Issues: Since rich sources of protein or specific antioxidants such as vitamin C can cause inflammation of the gastrointestinal tract, bloating, nausea, or diarrhea may occur.

14.5 Long-Hour Techniques of Recovery

To ensure the best recovery, it is customary that many conventional techniques demand extended periods of rest and consecutive applications, such as several meals or hydration plans. Depending on the extent of the injury, the time required for recovery may not allow the athletes who have other events to block for or may be involved in other consecutive events.

14.6 Limited Nutrient Stabilit

Some nutrients are incorporated in traditional recovery techniques, and when exposed to environmental components such as light, heat, or oxygen, they decompose at a very high speed. For example, Vitamin C is a delicate vitamin that readily undergoes oxidation and is deactivated. Omega-3 Fatty Acids: When stored inappropriately, they can go bad. This instability may temper the efficacy of food-based recovery solutions or supplements.

14.7 Inadequate Handling of Individual Variability

One-day diet meals vary not only by athletes' age, genetics, and training regimen but also by athletes' metabolism, digestive system, or recuperation period. Many traditional recovery approaches often rely on a conventional approach in handling patients that may not cater to the patient's individual differences. For example, the delayed gut runners may not be able to assimilate nutrients at the same rate as they extra digest their unraveled macronutrients. Moreover, some athletes may require more antioxidant capacity to combat osteoblastic activity than other athletes.

15 The Need for Sophisticated Services

The limitations of traditional approaches emphasize the need for innovative approaches- thus achieving these complexities. Possible enabling technologies by nanotechnology include enhancement of bioavailability, ability to deliver nutrients selectively, and controlled release of nutrients. By doing so, they surpass conventional methods in the time taken to heal, inflammatory response, and performance enhancement. Nano-enhanced recovery techniques offer a new solution to persisting challenges such as poor targeting, slow uptake, and poor bioavailability of nutrients. Nanoparticles, therefore, remove these limitations and increase the accuracy and efficiency of post-workout recovery plans by utilizing nanocarriers like liposomes, micelles, and polymeric nanoparticles.

16 Nano-enhanced Recovery

In view of that, nanotechnology has been proven to bring a new paradigm for post-workout recovery over existing traditional methods. Nanotechnology enhances the recovery rate through muscle repair, inflammation relief, and glycogen replenishment through improved nutrient uptake, using tissue selectivity to deliver nutrients and rate controlled release. Below is the list showing how Nano-enhanced systems improve outcomes of recovery:

16.1 Increased Bioavailability

Bioavailability, in simple terms, is the proportion of a nutrient or other medicinal compound available to do its intended job at the site where it is required. Many potential recovery agents such as omega-3 fatty acid based agents, resveratrol and curcumin possess low bioavailability because they are poorly absorbed in the gastrointestinal tract, have poor water solubility, or are vulnerable to degradation. Nanoparticles Increase Solubility: Compounds with hydrophobic properties become soluble and can be absorbed easily when encased in a liposome or micelle formation. To illustrate, while conventional curcumin can be absorbed at 1%, nano-curcumin can be absorbed at up to 40%, making inflammation management more effective Protection Against Degradation: To preserve the therapeutic capacity of temperature-sensitive nutrients, nanocarriers shield them from the activities of the GI tract acids and enzymes.

16.2 Targeted Delivery

One of the most significant advantages is its potential to put the nutrients and the therapeutic agents where needed, for example, the detrimental muscle tissues or the inflamed region. Functionalized Nanoparticles: It is possible to incorporate surface ligands that recognize and bind specifically to receptors on target tissues into nanoparticles. To enhance the limited side effects, it is possible to localize the anti-inflammatory agent-carrying nanoparticles, for example, to the inflammatory muscles. Enhanced Efficiency: Site-specific delivery reduces the required amount of the drug and enhances the therapeutic effect by focusing the recovery agents on the affected part.

16.3 Enhanced Cellular Uptake

Given the nature of the particles involved, nanoscale particles are far more likely to be transported across cellular membranes than the larger molecules. This enhanced uptake ensures that recovery agents get to the intracellular structures where they are most wanted. Endocytosis Pathways: Nanoparticles can penetrate cellular barriers that otherwise would not allow the access of specific nutrients to the cell straight through endocytosis, among others. Intracellular Delivery: For nanocarriers to be most effective, they should be able to release their payload directly into the cytoplasm or other organelles after the nanocarrier has entered the cell.

16.4 Delivery Based on Controlled and Sustained Release

To ensure the formulation has a long-term therapeutic effect, the nanocarriers are deliberately designed to release their contents at a controlled and slow rate. Those nutrients and agents that are required to work for extended periods for optimal performance are particularly benefited by this flexibility. One of the most frequent uses of such polymeric nanoparticles is what is referred to as a controlled release system. For example, to sustain MPS for a long time, polymeric particle-stabilized proteins or amino acids ensure constant nutrient delivery. Hydrogel Systems: In the recovery stages, electrolytes and hydration are sustained by nanohydrogels characterized by the slow discharge of these values.

16.5 Decreased Systemic Side Effects

Conventional recovery agents often exert system effects on enzymes, which may lead to side effects, for example, immunosuppression or gut irritation. These dangers are minimized by nanotechnology because it ensures that the delivery of the treatment is local and requires a lower quantity of the solution. Localized Action: Instead of quelling the immune system, which is how most diseases begin, nano-curcumin specifically calms inflammation-ridden areas. Reduced Dosage Requirements: When the nutrient's bioavailability is enhanced, it needs a small quantity to bring about therapeutic effects, thereby reducing the probability of side effects.

16.6 Stability and Improving the Shelf-Life of Food

Improving the shelf-life of food is a priority in the food industry for biological and economic reasons. Many recovery agents, including antioxidants and some vitamins, are inversely affected by oxygen, heat, or light. These substances are more stable thanks to nanotechnology. Observations in Nanocarriers: Nanoparticles protect vulnerable content from the environment, thus enhancing their longevity. Liposomes function as protective barriers, ensuring nutrients do not lose efficacy before arriving at the right place.

16.7 Multi-functionality

The applicability of post-workout recovery agents is available by virtue of multifunctional nanotechnology that enables the delivery of several agents. For example, Combination Systems can include antioxidant ingredients to combat the effects of oxidation stress and carbohydrates for glycogen replenishment in nanoemulsions. Compared to short-term recovery, layered nanocarriers can also provide a sustained delivery system where a burst-release carbohydrate is followed by a slow-release protein.

16.8 Examples of the Type of Nano-Enhanced Recovery Mechanisms

Nano-liposomal curcumin: Incorporating curcumin in liposomes significantly enhances the compound's bioavailability. This further improved the capability to decrease inflammatory markers such as IL 6 & TNF compared to conventional formulations. Nanoemulsion Omega-3 Fatty Acids Nanoemulsions enhance the anti-inflammatory effects, tissue healing, solubility, and absorption of omega-3-polyunsaturated fatty acids. Polymer Protein Nanoparticles In the recuperation phase, muscle protein synthesis is optimized by the Sustained release of protein nanoparticles to provide a continuous supply of amino acids.

Opportunities and Risks of Nanotechnology for Post-Workout Recovery: Future Perspective

Despite the huge potential for nanotechnology in post-workout recovery, this technology has several barriers. If the kindred issues, in addition to the posited future courses, are approached, it can optimize its ability to improve athletic efficiency, diminishing fatigue and general human health.

Applied to nutrition and medicine, nanotechnology makes people recover faster by improving the process of taking nutrients and therapeutic agents and using these effectively. Nanoscale material properties differ from more extensive materials; hence, smaller materials possess higher stability, better solubility, and larger surface area. Due to these properties, nanocarriers can cross such barriers, including cell membranes and gastrointestinal (GI) tracts, which commonly decline the effectiveness of ordinary supplements.

16.9 Nanotechnology in Muscle Repair

Among the rest periods, muscle repair is considered one of the most significant factors that nanotechnology improves. Extended MPS is achieved by controlling the release kinetics of EAA from the polymeric nanoparticles based on proteins such as whey or casein. Muscle tissues utilize branch-chain amino acids (BCAAs) more efficiently when these amino acids are incorporated in nanocarriers. Hydrolyzed Collagen Nanoparticles: These assist connective tissues in reinforcing faster, reducing the incidence of joint pains and recurring injuries characteristic of high-impact activities.

While assessing normal curcumin supplements, nano-encapsulated curcumin has a better anti-inflammatory property to minimize muscle pain and the time required for reparation.

The present invention relates to the topical application of nanotechnology whereby anti-inflammatory and antioxidant potential can be controlled. Soreness and increased oxidation are regular events in people's training process, but if they are not counteracted, they cause a slow recovery process. Liposomal antioxidants are one of the accurate treatments nanotechnology has offered to reduce these impacts. Vitamin C, E, polyphenols, and others are incorporated in liposomes, which can protect them from degradation in the GI tract and ensure their delivery to the target tissue. Different from the general antioxidants, these enzymes reduce oxidative damage and improve faster healing by efficiently neutralizing ROS. Free omega-3 fatty acids with nanoencapsulation or nanoformulation show higher bioavailability and reduced inflammation profiles. It also modifies cytokine release and has anti-inflammatory effects on any cellular level. Anti-inflammatory active ingredients with *controlled release:* Various polymeric nanoparticles, including active substances like ibuprofen or curcumin, provide an extended-release for the sustained concentration that does not cause consistent inflammation.

17 Application of Nanotechnology in Electrolyte Balance and Rehydration

A prime application of nanoemulsions: enhanced mineral replenishment through optimized hydration. Nanoemulsions bring improved solubility and absorption kinetics in electrolytes like sodium, potassium, and magnesium lost through perspiration in our bodies. The formulations ensure athletes replenish their bodies more than traditional electrolyte beverages do. Hydrogel-based nanocarrier delivers water and electrolytes slowly for a long duration, thus maintaining athletes' water balance during training or competition to avoid dehydration (Fig. 3).

18 Realization and Practical Use

Recent clinical investigations and field deployments demonstrate how nanotech will play an essential role in recuperating injured athletes. For example, nano-antioxidant-aged athletes kept the view that they sort out and comparatively lowered the markers of oxidative stress and muscle soreness among marathon runners compared to conventional forms of antioxidants, as ascribed by Goel and Aggarwal in their findings. Liposomal Protein Supplements: A study done with athletes who participated

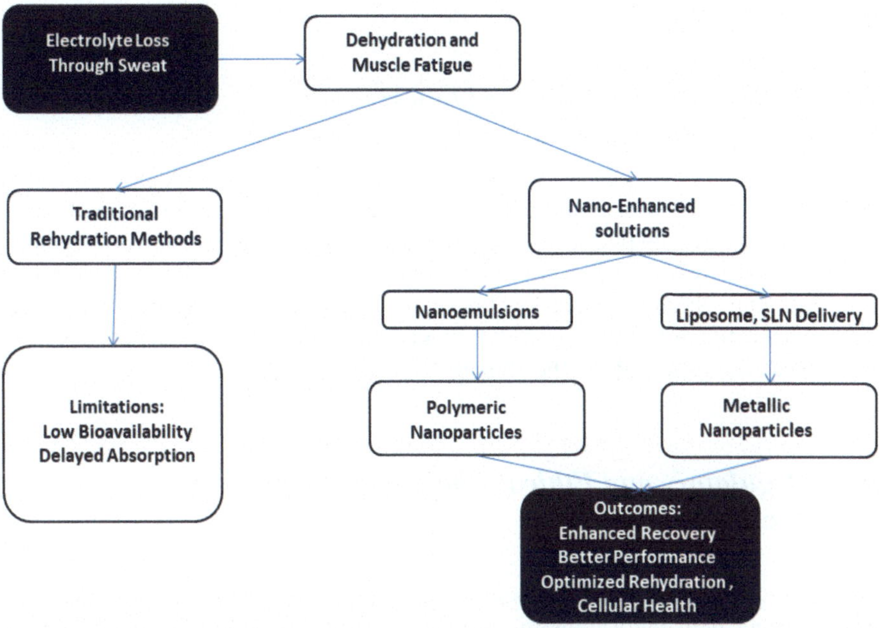

Fig. 3 Nano-enhanced electrolyte balance and rehydration

in acute high-intensity resistance training showed that using commercially available protein supplements manufactured employing nanotechnology enhanced muscle recovery by 30%.

19 Challenges and Ethical Considerations

19.1 Safety Concerns

Generally, there is a need for more research to establish the toxicity profiles and other chronic effects of specific nanoparticles, notwithstanding the promising capability of the technology. *Regulatory Compliance:* This means using nanotechnology to prepare sports supplements must be closely regulated for safety and fair play in competitive sports.

Cost and Accessibility: Nano-enhanced products may be expensive, especially for amateur athletes or people in areas with restricted access.

19.2 Challenges

Unintended Side Effects and Safety and Toxicity Issues

Due to their size and reactivity, there are concerns related to nanoparticles, such as their ability to induce oxidative stress, their tendency to accumulate in tissues, and their toxicity to cells. Acute and chronic health consequences are linked with these effects. Immune System Interactions: The advantage of nanoparticles in recovery may be accompanied by immune reactions leading to inflammation or allergy. Absence of Thorough Safety Studies: The lack of long-term nanoparticle exposure studies makes it somewhat difficult to fully grasp the weekly cumulative derivational changes due to nanoparticles, especially in the athletes who often use them.

19.3 Regulatory and Ethical Challenges Regulatory Frameworks

The failure concerning the elaboration of rigorous regulatory norms has been found wanting with the astounding pace with which the field of nanotechnology is being innovated. This leads to several questions on nano-technology product utilization, classification, and authorization. Equity in Access: The higher cost of research, production, and advertisement accompanying products involving nanotechnology means that the elites may not access the products, thus the performance disparity. Ethical Issues in Sports: The predicate between performance-enhancing interventional and legal recovery assistances might be blurred by the nano-enhanced solutions, which would complicate fairness issues in competitive sports.

19.4 Production and Cost Issues High Manufacturing Costs

Nanotechnology products entail complex modes of operation and ingredients that explain why they are expensive to produce and optimize. Complex Production Procedures: Nanoparticle synthesis, functionalization, and stabilization include multistep processes that require special experience and equipment.

19.5 Stability and Scalability Issues

Because nanoparticles are influenced a lot by light, temperature, and pH levels, they may be less effective and have less shelf life than their larger counterparts. Scalability Issues: However, manufacturing high-quality, credible Nano-enhanced products that can be scaled up is still challenging and expensive.

19.6 Lack of Awareness and Acceptance

Customers often have a sort of mistrust as to the workings of this nanotechnology and the benefits that come along with it. Misconceptions Sometimes, when you present products that are safe and effective, acceptance is not forthcoming due to anxiety and efforts to spread fake news.

20 Future Perspectives Conclusion

The following processes have been observed as part of developing biodegradable nanoparticles. Due to such safety issues, the subsequent studies focus on fabricating ecologically degradable and tissue-friendly nanoparticles. Biodegradable promising substitutes chitosan, PLA, and PLGA reduce the risk of toxicity and bioaccumulation.

Smart Nanocarriers for Precision and Personalized Recovery: Through nanoparticles designed to disintegrate in response to stimuli as specific as temperature and pH, recovery agents can be made available as specialized solutions. Wearable Technology Integration: Integrating nanotechnology with biosensors and wearables provides data on real-time oxidative stress, hydration, nutrient needs, and deficiencies to enable ideal supplementation.

Better Production Methods Cost-Effective Manufacturing: These green chemistry processes and other advancements in the synthesis of nanoparticles are supposed to reduce cost and its impact on the community.

Scalable Nanoparticle Systems: It is possible to achieve reliable quality and reasonable prices for such products by progressing in large-scale manufacturing methods such as microfluidics. The development of nanoparticles that can deliver anti-inflammatory, glycogen synthesis, and water delivery functions can contribute to the simplification of the recovery procedures and an overall increase in efficacy.

Global Standards for Regulatory Development: Developing clear and standard regulatory frameworks for sports recovery products with nanotechnology features will ensure effectiveness, safety, and consumer satisfaction. Post-Market.

Surveillance: Continuous surveillance of off-the-shelf nanoproducts helps identify and mitigate emergent risks.

Environmental Sustainability Eco-Friendly Nanomaterials: Separation of nanoparticles from their matrices, as well as methods of disposal and waste management, also have an ecological impact implication to nanotechnology: this can be mitigated through the use of renewable and biodegradable materials for the production of nanoparticles. Safe Disposal Procedures: In the case of products containing nanoparticles, generalised ecological risks can be mitigated by observing proper disposal procedures as prescribed.

Research and Collaboration Multi-disciplinary Research: Multi-disciplinary nanotechnologists, sports scientists, and clinicians working side by side can help advance the identification of new interventions. Clinical Trials: Subsequent clinical trials on a diverse sample of athletes will establish nano-enhanced therapy's efficacy, safety, and recovery protocols.

Applications of Nanotechnology in Recovery Beyond Sports Rehabilitation: This is another field where nanotechnology applications can help the other populace other than athletes regain their fitness after surgeries, accidents, or extended illness. Enhancing the repair and healing activities of older people or those involved in strenuous jobs is called general well-being.

Nanotechnology promises to revolutionize the science of recovery, providing precise, efficient, and tailored answers for athletes and non-athletes alike. While research takes care of current issues, nanotechnology in sports recovery can be furthered by improvements in performance and well-being in the wide field of medicine and wellness. By cultivating these innovations responsibly and collaboratively with stakeholder members, the potentiality of nanotechnology toward a cause to help shape the future of recovery and rehabilitation can be fully availed.

As discussed under some challenges that traditional forms of recovery encounter, nanotechnology has emerged as a tool that breaks the molds in post-workout and injury/recovery periods. Nanotechnology enhances nutrient delivery by increasing the rate at which nutrients are assimilated and brings this delivery right to the area to be healed to form a therapy that only lasts for the shortest time possible with long-term therapeutic effects. It remains adaptable enough to meet muscle repair, inflammation control, glycogen restoration, hydration, and infection prevention needs, making it variable enough to solve several recovery needs. The use of nanotechnology in sports recovery has been shown to yield significant improvements in one of the following ways: improving muscle regeneration with protected proteins and growth factors, delivering antioxidants and Growth factors Through liposomes that assist in inflammation and oxidative stress reduction, replenishing lost electrolytes and carbohydrates with the help of nanotechnology-based delivery systems, and enhancing the healing process, painless recovery, and anti-bacterial property. Notwithstanding the apparent benefits of nanotechnology, the problems include low consumer awareness, high production costs, safety and legislation concerns, and others. Further work is required to ensure that nanomaterials are properly utilized without any health harm since their toxicity, accumulation in living organisms, and some other long-lasting impacts have been scientifically established. Nevertheless, there is room for more comprehensive research and technical advancements that may place nanotechnology on the map for sports physiotherapy and athletic recovery.

References

1. Ivy JL (1998) Glycogen resynthesis after exercise: effect of carbohydrate intake. Int J Sports Med 19(Suppl 2):S142-145. https://doi.org/10.1055/s-2007-971981
2. Cheung K, Hume P, Maxwell L (2003) Delayed onset muscle soreness: treatment strategies and performance factors. Sports Med 33(2):145–164. https://doi.org/10.2165/00007256-200333020-00005
3. Shirreffs SM, Sawka MN (2011) Fluid and electrolyte needs for training, competition, and recovery. J Sports Sci 29(Suppl 1):S39-46. https://doi.org/10.1080/02640414.2011.614269
4. Gleeson M, Nieman DC, Pedersen BK (2004) Exercise, nutrition and immune function. J Sports Sci 22(1):115–125. https://doi.org/10.1080/0264041031000140590
5. Kageta T, Tsuchiya Y, Morishima T, Hasegawa Y, Sasaki H, Goto K (2016) Influences of increased training volume on exercise performance, physiological and psychological parameters. J Sports Med Phys Fitness 56(7–8):913–921
6. Mennitti C et al (2024) How does physical activity modulate hormone responses? Biomolecules 14(11). https://doi.org/10.3390/biom14111418
7. Vitale K, Getzin A (2019) Nutrition and supplement update for the endurance athlete: review and recommendations. Nutrients 11(6). https://doi.org/10.3390/nu11061289
8. Adetunji CO et al (2023) Biological nanocarriers for nutrient delivery. In: Nanobiotechnology for the livestock industry: animal health and nutrition, pp 99–110. https://doi.org/10.1016/B978-0-323-98387-7.00009-4
9. Gomes EC, Silva AN, Oliveira MRD (2012) Oxidants, antioxidants, and the beneficial roles of exercise-induced production of reactive species. Oxidative Med Cell Longevity. https://doi.org/10.1155/2012/756132
10. Li J, Xiong K, Gong Y, Yu X (2016) Advances in studies on anti-oxidant mechanism of curcumin based on signal transduction pathway. Chinese Traditional Herbal Drugs 47(13):2373–2380. https://doi.org/10.7501/j.issn.0253-2670.2016.13.028
11. Singh S (2024) Antioxidant nanozymes as next-generation therapeutics to free radical-mediated inflammatory diseases: a comprehensive review. Int J Biol Macromole 260. https://doi.org/10.1016/j.ijbiomac.2024.129374
12. Loreti M, Sacco A (2022) The jam session between muscle stem cells and the extracellular matrix in the tissue microenvironment. NPJ Regen Med 7(1):1–15. https://doi.org/10.1038/s41536-022-00204-z
13. Gorissen SHM, Phillips SM (2018) Branched-chain amino acids (leucine, isoleucine, and valine) and skeletal muscle. In: Nutrition and skeletal muscle, pp 283–298. https://doi.org/10.1016/B978-0-12-810422-4.00016-6
14. Holick MF (2011) Vitamin D: evolutionary, physiological and health perspectives. Current Drug Targets 12(1):4–18. https://doi.org/10.2174/138945011793591635
15. Katsanos CS, Kobayashi H, Sheffield-Moore M, Aarsland A, Wolfe RR (2005) Aging is associated with diminished accretion of muscle proteins after the ingestion of a small bolus of essential amino acids 2. Am J Clin Nutr 82(5):1065–1073. https://doi.org/10.1093/ajcn/82.5.1065
16. Aragon AA, Schoenfeld BJ (2013) Nutrient timing revisited: is there a post-exercise anabolic window? J Int Soc Sports Nutrition 10. https://doi.org/10.1186/1550-2783-10-5
17. Kerksick CM et al (2018) ISSN exercise & sports nutrition review update: research & recommendations. J Int Soc Sports Nutr 15(1):38. https://doi.org/10.1186/s12970-018-0242-y
18. Kim JH et al (2018) 3D bioprinted human skeletal muscle constructs for muscle function restoration. Scientific Reports 8(1). https://doi.org/10.1038/s41598-018-29968-5
19. Liu Y, Zhao X, Gan F, Chen X, Deng K, Crowe SA, ... Keasling JD (2024) Complete biosynthesis of QS-21 in engineered yeast. Nat 629(8013):937–944
20. Wang L et al (2022) Enhanced oral absorption and liver distribution of polymeric nanoparticles through traveling the enterohepatic circulation pathways of bile acid. ACS Appl Mater Interfaces 14(37):41712–41725. https://doi.org/10.1021/acsami.2c10322

21. Ranjha MMAN et al (2022) Biocompatible nanomaterials in food science, technology, and nutrient drug delivery: recent developments and applications. Front Nutr 8. https://doi.org/10.3389/fnut.2021.778155
22. Altemimi AB et al (2024) Application of nanoparticles in human nutrition: a review. Nutrients 16(5). https://doi.org/10.3390/nu16050636
23. Lu Y, Qi J, Wu W (2012) Absorption, disposition and pharmacokinetics of nanoemulsions. CDM 13(4):396–417. https://doi.org/10.2174/138920012800166544
24. Cheng R, Meng F, Deng C, Klok H-A, Zhong Z (2013) Dual and multi-stimuli responsive polymeric nanoparticles for programmed site-specific drug delivery. Biomaterials 34(14):3647–3657. https://doi.org/10.1016/j.biomaterials.2013.01.084
25. Amna T, Hassan MS, Gharsan FN, Rehman S, Sheikh FA (2022) Nanotechnology in drug delivery systems: ways to boost bioavailability of drugs. In: Hameed S, Rehman S (eds) Nanotechnology for infectious diseases. Springer, Singapore, pp 223–236. https://doi.org/10.1007/978-981-16-9190-4_10
26. O'Connor E, Mündel T, Barnes MJ (2022) Nutritional compounds to improve post-exercise recovery. Nutrients 14(23), Art. no. 23. https://doi.org/10.3390/nu14235069
27. Morgan PT, Breen L (2021) The role of protein hydrolysates for exercise-induced skeletal muscle recovery and adaptation: a current perspective. Nutr Metab (Lond) 18(1):44. https://doi.org/10.1186/s12986-021-00574-z
28. Zhuo Y, Zhao Y-G, Zhang Y (2024) Enhancing drug solubility, bioavailability, and targeted therapeutic applications through magnetic nanoparticles. Molecules 29(20):4854. https://doi.org/10.3390/molecules29204854
29. A. C. Tesseract Founder | Enhancing bioavailability of supplements through nanotechnology. Tesseract. Accessed 7 Jan 2025. [Online]. Available: https://insights.tessmed.com/bioavailability-of-supplements-new-technology-tsr/
30. Verma S, Pandey AK (2021) Improving bioavailability of nutrients through nanotechnology. In: Maurya VK, Gothandam KM, Ranjan S, Dasgupta N, Lichtfouse E (eds) Sustainable agriculture reviews 55: micro and nano engineering in food science, vol 1. Springer International Publishing, Cham, pp 135–170. https://doi.org/10.1007/978-3-030-76813-3_6
31. Niess AM, Simon P (2007) Response and adaptation of skeletal muscle to exercise—the role of reactive oxygen species. Front Biosci 12(13):4826–4838. https://doi.org/10.2741/2431
32. Battistelli M, Salucci S, Falcieri E (2018) Morphological evaluation of liposomal iron carriers. Microsc Res Tech 81(11):1295–1300. https://doi.org/10.1002/jemt.23137
33. Leng J, Li L, Zhao W, Bian Y (2025) Enhancing lipid digestion and absorption rate: the role of whey protein isolate short fibrils in emulsions. Int J Biol Macromole 285. https://doi.org/10.1016/j.ijbiomac.2024.138351
34. Boostani S, Jafari SM (2020) Controlled release of nanoencapsulated food ingredients. In: Release and bioavailability of nanoencapsulated food ingredients, pp 27–78. https://doi.org/10.1016/B978-0-12-815665-0.00002-3
35. Wardani G, Nugraha J, Mustafa MR, Kurnijasanti R, Sudjarwo SA (2022) Antioxidative stress and antiapoptosis effect of chitosan nanoparticles to protect cardiac cell damage on streptozotocin-induced diabetic rat. Oxidative Med Cellular Longevity 2022. https://doi.org/10.1155/2022/3081397
36. Shin J-S, Kim B-H, Baik M-Y (2021) Applicable plant proteins and dietary fibers for simulate plant-based yogurts. Foods 10(10), Art. no. 10. https://doi.org/10.3390/foods10102305
37. Sjögren E, Tarning J, Barnes KI, Jonsson EN (2021) A physiologically-based pharmacokinetic framework for prediction of drug exposure in malnourished children. Pharmaceutics 13(2), Art. no. 2. https://doi.org/10.3390/pharmaceutics13020204

Immune Support for Athletes: Nanotech Approaches

Debajit Karmakar, Anindita Das, Karuppasamy Govindasamy, and Koulla Parpa

Abstract To that extent, this chapter presents an account of the innovative application of nanotechnology in sports medicine and probably the most effective nanonutrients for athletes craving for superior performance as invincible immune shields. This leads to an increase immunosuppression and susceptibility to diseases owing to rigorous physical exerted put on athletes. Nano-nutrients, that is nutrients that are nanoscaled, will present enhanced immunology performance and bioavailability, site of action targeting the cells of interest and up to 90% absorption rates; this is a novel strategy towards enhancing immune strength. Something that will be seen as one of the main top line benefits of nanotechnology in relation with nutrient delivery has to do with supporting the cell, repairing and protecting it from oxidative damage. This chapter also discusses the progress achieved in the past few years in the encapsulation of vitamins, minerals and bioactive compounds into nanoparticles for the ability of the later to affect immune responses and inflammation as well as oxidation stress. Nano enhanced antioxidants as described in the previous chapters are beneficial in reversing oxidative damage that is caused by exercises which help to maintain muscle strengths and shortened recovery time. The incorporation of nanotech into sports nutrition provides one of the best ways toward supporting the athlete's immune system and maintain the body and mind's consistently in great performance conditions. Based on the findings of the latest scientific advances from the field of sports

D. Karmakar (✉) · A. Das (✉)
Department of Physical Education Pedagogy, Lakshmibai National Institute of Physical Education, Gwalior, Madhya Pradesh, India
e-mail: debajitkarmakar2200@gmail.com

A. Das
e-mail: anindita416@gmail.com

K. Govindasamy
Department of Sports, Recreation and Wellness, Symbiosis International (Deemed University), Hyderabad Campus, Modallaguda (V), Nandigama (M), Rangareddy, Telangana 509217, India
e-mail: gowthamadnivog@gmail.com

K. Parpa
Faculty of Sports and Exercise Science, UCLan University of Cyprus, Pyla 7080, Cyprus
e-mail: kparpa@uclan.ac.uk

A. Chatterjee et al. (eds.), *Nanofuel: The Future of Sports Nutrition*, Smart Nanomaterials Technology, https://doi.org/10.1007/978-981-96-5471-0_15

immunology and nanotechnology this chapter offers a critical analysis of the opportunities for the use of nanotechnology as a key to enhance the health, longevity and performance of the athletes.

Keywords Nanocarriers · Nanoemulsions · Antioxidants · Oxidative stress · Immunomodulation · Athletic performance · Bioactive compounds

1 Introduction

Anti-stress protection or defence of athletes constitutes one of the vital ways of boosting performance and health in demanding sports disciplines. Physical activity especially at the professional level exerts much pressure on the body and this leads to many changes. Some of the changes that occur as a result of the impact of these genetic factors include; metabolic changes, inflammation and immune function changes [1]. In many cases over the strongly catabolic stimuli, the strength of immune system is reduced because of the elevation of levels of pro-inflammatory cytokines and the suppression of immune-surveillance. This has been termed exercise-induced immunosuppression, where athletes are at high risk of falling sick due to reduction in immunity for some time [2]. Discussing the physiological concept of the stress of athletic performance as multiple stressors including mechanical, metabolic, and psychological stress indicators and all of them can result in temporary negative impact on the immunological parameters. A series of high-intensity exercises, which are usually prescribed for athletes, give a cumulative effect that triggers chronic low-grade inflammation and hence immune dysfunction, which is not conducive for an athlete in terms of consistent and progressive performance improvements [3].

Research on applications of nanotechnology in sports medicine has been identified to have developed into a rapidly growing tradition, primarily holding potential in regulating of immune response. Due to the versatility of nanotechnology and control of material at the nano level, scientists have come up with sophisticated techniques in the targeted release of bioactive compounds [4]. Nanotech approach shows promising prospectus for improving immune status and avoiding the effect of exercise-induced decline of immune function through targeted delivery of immunomodulatory agents. Nanoparticles are incorporated in delivery systems to achieve controlled release of immune-active molecules; to increase their bioavailability, and to target the required tissues and cells. In addition, it has been possible to design new types of immune stimulating agents that could help an athlete adapt to various loads implied by training and competitions [3]. The advances these nano-carriers portray in the field of nanomedicine present a bright future for athletes and their immune system, and therefore their competitive endowments in their respective games.

Nevertheless, the interest in enhancing some aspects of immune status by approximately different approaches, as the current traditional interventions for increasing immunocompetent are quite restricted. Traditional practices that have for example involved the use of vitamins, minerals, and other products, which are usually considered as dietary supplements also show a lot of promise that however doubles up in providing the kind of impact required for uplifting athletes' performance to peak levels [5]. Further, a significant number of these approaches operates under the premise of high dosage and hence can complicate matters by provoking toxicity or side effects or influencing other bodily functions. Immunomodulation via these traditional approaches is rather crude and usually nonspecific compared to what can be accomplished with nanoscale approaches, so it is less suitable for meeting the multifaceted challenges that the immune system presents under physical stress [2]. In addition, the usual immune enhancement practices do not cover the specificities necessary for athletes, including the promotion of recovery time and the prevention of overtraining syndrome. As such, there remains a great void for most present immunological support and this is where nano-technology comes into the frame with new improved and personalized strategic immunological support. Therefore, there is much missing from the current Immune support methodology and this is where nanotechnology steps in with more tailored approaches suited for the immune needs of athletes [4].

2 Immunological Challenges in Athletic Performance

There is a form of immunosuppression commonly arising from exercise and especially those athletes who involve themselves in rigorous or for a prolonged period. This condition occurs due to immunological reaction to the physical stress linked with rigorous exercise [5]. By far the greatest impact on immune function during exercise comes from the stress, high temperature and shifts in hormonal balances which are the ingredients of mechanical stress. The underlying mechanism of exercise-induced immunosuppression is primarily caused by Cortisol stress hormone elevations during exercise, thereby reducing most immunological processes [2]. Also, during heavy training or exercise activity, there is an immediate inflammatory response, and the immune system capacity to mount a fitness response to organisms that may cause infection is reduced temporarily, thus exposing the athlete to infection particularly upper respiratory tract infections in the post-training period [3].

In athletes, immune response can be defined on the basis of acute and chronic immunology and the two has physiological significance difference. Said to be acute, inflammation is a precisely orchestrated physiology that accompanies and regulates the exercise-induced external cellular disruption following tissue injury. They are usually short lived but during stress to any bodily tissue the process can get amplified. Chronic inflammation can therefore be defined as the prolonged and more or less pathological reaction to regular physical stress [6]. Prolonged sub-clinical inflammation is invariably seen in athletes who undergo intense continuous training because the successive bouts of acute inflammation do not completely return to baseline, keeping the immune machinery continuously active. Such a condition means that healing, performance reduction, and increase vulnerability to overtraining syndrome occurs because the immune system becomes chronically weakened [2].

Oxidation stress is also pin pointed as another factor implicated in modulating immune responses in athletes. High-intensity exercise causes an acceleration of delivery as a natural consequence of aerobic metabolism primarily within muscle cells [7]. Even though reactive oxygen species (ROS) are fundamental to cellular signalling and allowing cells to adapt to exercise, if allowed to accumulate too much, these molecules burden the body's antioxidant armour and cause oxidative destruction of tissues including the immune cells [8]. The chronic process of oxidative damage to immune cells may result in reduced capacity to eliminate pathogens which weakens immune reserve and increases an athlete's vulnerability to illnesses and slow recovery [1].

Another factor which impairs the immune competence of athletes is nutrient deficiencies—particularly micronutrient ones. Vitamin A, C, D and E and minerals including zinc and selenium are important in immune system health and efficiency. Lack of these micronutrients, which are often low in athletes because of low calorie diets or even standard dietary advises, result in a suboptimal immune status [6]. For instance, vitamin D is well known to play an important role in the proper functioning of innate and adaptive immune system, and deficiency of this vitamin is linked to higher rate of infections and longer times to recovery. Likewise, vitamin C, an antioxidant, preserves the immune structure and directs cytokine manufacturing processes that refine immune reactions [9]. Zinc is another micronutrient that boosts the activity of the immune cells including T-Lymphocytes, and macrophage. Lack of these micronutrients to the recommended nutritional need significantly may enhance the vulnerability of athletes to infections due to a poor immune system though training [2].

3 Nanomaterial-Based Immune Modulators

Far reaching power has been paid to nanomaterial founded immune-modulators in the recent past due to their possible use in fun performance improvement at the Olympic level. These sophisticated materials provide the means by which the shape and function of the immune system can be manipulated, revealing the potential for enhanced tissue healing, increased and more frequent remission from illness and inflammation [5]. Through utilizing nanotechnology, immune stimulus is achieved on target muscles; thus enhancing athlete immunity and overall performance. Due to their synergistic interaction with immune system at molecular level, nanomaterials can the targeted at specific immune pathways providing fresh approach in enhancing sports performance and recovery [10].

Designing nanoparticles for specific interferences with the immune system is a core aspect of nanotechnology-based immunomodulatory agents on-targeted immune system modulation. By controlling the size, surface charge and further functionalization these particles may be tailored for targeting selectively immune cells like macrophage or dendritic cells involved in the innate and adaptive immunity response respectively [11]. Thus, the specificity of nanoparticle targeting is of paramount importance to the success of the nanoparticle intervention on the one hand and its safety on the other. Some uses of cytokines, peptides, or bioactive compounds in immunomodulatory nanoparticles will be loaded in a manner that boosts the immune system to help in responding to stress factors such as strenuous activities or infections [10]. Since the active agents delivered by nanoparticles are specifically targeted towards Immune cells, the transport of the agents to the site of action is efficient than the conventional delivery systems which blesses the immune system and athletic recovery [8].

Lipid based nanocarriers are being identified with the prospective of reinforcing immune-nutrients delivery to the athletes in a systematic way. Lipid based particulate systems, including solid lipid nanoparticles and liposomes, provide numerous advantages of being biocompatible carriers for the delivery of active molecules, including vitamins, minerals, and polyunsaturated fatty acid is crucial for immune support [12]. Such lipid based nanocarriers are particularly useful in immune-nutrition because they can be designed to transport nutrients across various barriers like the blood brain barrier or the epithelial lining of the intestine so that delivery is directly to the target site [8]. Also, by controlled release lipid nanoparticles, it is practically possible to have sustained release of the contents for a longer duration for immune modulation which is important for athletes who need immune support during trainings and various competitions.

Metallic and metal oxide nanoparticles all display varied immunomodulatory characteristics which make them suitable for boosting athlete immunity. In the present invention, metallic nanoparticles including gold, silver and copper and metal oxide nanoparticles including zinc oxide and titanium dioxide have been reported to have antioxidant, antimicrobial and immunomodulatory properties [11]. These nanoparticles may modulate the release of cytokines from immune cells, increase the process of phagocytosis, and facilitate pathogen elimination. These nanoparticles are relatively larger than other nanomaterials and have certain surface characters that enable them to cross the cell membrane and interact with intracellular structures to trigger an early immune response [13]. To the athletes this is a way of accelerating their healing time in regard to inflammation, decreasing their possibilities of getting an infection, and improving their immune system. Furthermore, the introduction of metallic nanoparticles into different forms of supplement or topical applications come with the easy method of transporting the immune-modulating properties to athletes with minimal side effects [1].

Another emergent strategy achieved by polymer-based nanostructures is related to immune response regulation. Polymeric systems including poly (lactic-co-glycolic acid) (PLGA), chitosan, polyethylene glycol (PEG) where some of the polymers that have been put into practice for nanocarriers for immune-modulating agents delivery [2]. These bioactive molecules incorporated into the polymer based nanostructures afford the opportunity of controlled and maintainable release of the respective compounds thus ensuring a constant and durable immune state in the athletes. These features of polymers also include the ability to adjust the properties of the polymers including degradation profiles and surface chemistry to meet specific therapeutic requirements [14]. Specifically for the purpose of immune modulation, it is possible to encapsulate a variety of immune-nutrients, including antioxidants, anti-inflammatory agents or immune stimulating molecules into the polymer matrix and then control the rate of their release to aid the immune system during both training and recovery periods. It means that the proposed approach excludes such dangers associated with immunomodulation strategy, such as over stimulation or under stimulation of the immunity that results in the organism's ability to deliver improved athletic performance that is free from negative health implications [15].

4 Advanced Nanodelivery Systems for Immune Support

The integration of the user-friendly nanodelivery systems concering immune boosting in elite athletes has taken a new idea on immune boosting and reactivity. These innovative technologies represent a radical improvement for the delivery of nutrients, especially antioxidants, inside the body, while guaranteeing the highest impact for health and performance. Antioxidants are one of the more promising applications of the recently emerged nanotechnology known as nanoemulsions [13]. Nanoemulsions simply refer to formulations containing droplets of active substances less than 1 μm distributed in water with the aid of surfactants. These emulsions created a large surface area for absorption of active ingredients like vitamin C, vitamin E, and polyphenols, which are used in fighting oxidative stress or inflammation in athletes [15]. This nanotechnology can be achieved in way to target antioxidants directly to the immune cells where it will be effectively absorbed and utilized by the body most especially in post exercise sessions where oxidative stress and immune cells suppression are at their highest [2].

Smart nanocarriers are another example of the rapidly progressing field of nanotechnology that offers athletes micronutrient delivery in detail. Such as liposomes, dendrimers or nanoparticles incorporated nanocarriers that have enhanced targeting ability to deliver micronutrients like vitamins, minerals and amino acid selectively to the cells that need them [16]. By the overall organization of these smart carriers that could be modified on the surface with ligands or antibodies, those micronutrients can get through other biological barriers and go directly to reach the specific cells or tissues like muscle cells or immune system, or the gut [8]. Such controlled delivery system can further intensify the therapeutic value of micronutrients and exclusively targets its immune modulating effect without the necessity of exposing high doses that can produce side effects. The enhancement of these nanocarriers to target tissues/receptors enhances the general well being of athletes by focusing on areas that trigger significant immune support, especially when recovering or in periods of training intensification [17].

Prolonged release nanotechnology has once again enhanced the capability of immune hosting and healing matrices for elite sportspersons. This is different from conventional supplementation techniques where there is high concentrations for a while before the concentration of nutrients drops drastically sustained-release systems deal with the constant discharge of bioactive compounds within a particular timeframe [12]. These systems can use polymeric nanoparticles or hydrogels, so an extended and stable source of nutrients can be maintained in the bloodstream or target tissues, excluding the necessity of boosters during the resting or easy training sessions [17]. Regular release of these key immune supportive nutrients such as zinc, selenium and probiotics can assist in building a stable immunity system to fight infections and inflammation which might be contracted from constant athletic training. These nanotechnology systems give a slow and steady release so that athletes receive what they need without interruptions due to depot effects, while at the same time maintaining adequate immune function during their training and periods of rest when the immune system is most susceptible to being suppressed [11].

Such sophisticated nanocarriers have paved way to enhancement of the health status of athletes and correspondingly performance enhancement while excluding chances of disease and immune detriment [18]. Incorporation of antioxidants, micronutrient delivery vesicles and systems, smart nanocarriers, and sustained-release systems deliberate a quad- dimensional modality of immune support contextualized for optimum utility in elite athletes. These systems do not only augment bioavailability and effectiveness of critical immune supporting nutrients but also introduce another layer of sophistication to athletic health optimization [8].

5 Cutting-Edge Nanotech Immune Interventions

New developments in nanotechnology have made it possible for athletes to access immune support that takes performance to another level by equally boosting their immune system in preparation and response to the stress of training and competing [16]. One of the most creative solutions is the application of broadband nanotechnology for shaping the immune genetic code based on the CRISPR platform. Since this therapy work at the genetic level, the CRISPR-Cas9 can be use to alter specific genes involved in immune responses in athletes directly thus addressing affected immune circuits. These genetic changes are aimed at enhancing the immune system's sensitization to pathogens, decreasing inflammation responses and shortening post exercise recovery times [14]. CRISPR based treatment interventions present an ability to develop a unique fix, which can help construct or amplify athlete prowess by strengthening their natural immunity against high pressure performances. This new gene-editing process could transform the science of restoring an athlete and substituting the optimal physiologic state at the cellular and molecular levels [19].

Other exciting advancement in nanotechnology for immune support is the utilization of exosome nanocarriers that help the immune cells to communicate. Exosomes are small, membrane-bound vesicles, released by cells and contain proteins, lipids and RNA that mediates the cellular signalling [18]. In the case of the immune system, molecules are provided to those particular immune cells through exosome nanocarrier to increase their efficiency. With respect to athletic performance, the present strategy entails the ability to fine-tune immune responses by augmenting anti-inflammatory cytokine secretion while increasing immune cell vigilance [14]. Exosome based delivery systems provide a novel non-invasive and efficient method of modulating immune cell function, which will enhance the healing process, overall health and thus performance of the athletes. These nanoscale vesicles can also be designed to promote tissue repair and remodelling, which will be of added value to athletes who often undergo micro traumas in training [19].

Other technological development includes nanorobotic platforms that have also been used to enhance germinal immune performance in athletes. These microscopic robots are intended to communicate with the immune system at a molecular level hence allowing for the feedback and control of the immune system [18]. Allergic reactions, infections, fatigue and inflammation affecting the immune cells can be observed and diagnosed by these nanorobots without causing harm on the athlete [19]. In addition, these platforms can be made to release therapeutic agents at immune cells only in cases of imbalance thereby making the system very sensitive to the needs of the athletes. In this manner, by allowing the constant and precise monitoring of immune performance, nanorobots can play a critical role in the goal of managing immunity to help athletes avoid diseases and injuries, and to optimise training capacity [14].

Personalized nanomedicine for immune profiling is yet another technological advancement that may greatly advance the process of maintaining the immune status of athletes. Personalized nanomedicine makes interventions depend on an athlete's immune response, and these can be obtained through genomics, proteomics, and metabolomics. It provided for a very directed immune plan, that is to say it would be a best fit immune strategy for the athlete in question based on their individual physiological attributes and their training regimen [20]. Consequently, by discovering athlete's immune profiles or susceptibilities to diseases, researchers can design immunologically optimized nanotech systems that increase immunity's effectiveness, shield athletes from specific pathogens, and quicken recovery. Such care guarantees the best possible protection to athletes from diseases and enhanced vulnerability to injuries given that every athlete receives the best care suited to their body and needs. Applying both diagnosis and treatment at the nanoscale level into athletic health plans entails a massive advancement in both sports physiology as well as immunity enhancement [19].

6 Molecular Mechanisms of Nanotech Immune Enhancement

This approach to immune enhancement by specific interactions of nanotechnology with immune cell receptors has brought new life to this field. At the molecular level, it is possible to design nanoparticles that could selectively accumulate in immune cells including macrophages, dendritic cells, and T lymphocyte. These nanoparticles are to work with receptors present in the surface of immune cells in order to facilitate the control of these immune responses [21]. The ensuing acknowledgments clarify that whenever nanoparticles get bound to a specific immune cell receptor, they can cause various biological outcomes as they stimulate or inhibit the activity of targeted immune pathways. For instance, nanoparticles can behave as Pathogen-Associated Molecular Pattern imitating pathogen associated molecular patterns and stimulate effector proteins such as toll like receptors to improve immunological responses. It is beneficial in enhancing the body's immune system hence the need to engage in such an interaction especially athletes who are under high pressure of training and competition in their training process leading to enhancement of antibodies production [22].

This approach to immune enhancement by specific interactions of nanotechnology with immune cell receptors has greatly rejuvenated this area. Molecularly, there are opportunities of creating nanoparticles with the ability to focus themselves in immune cells such as macrophages, dendritic cells and T lymphocytes cell [23]. These nanoparticles are to interact with receptors that are found at the surface of immune cells in order to help regulate such immune responses [24]. The subsequent acknowledgments stress that in case nanoparticles bind to a definite immunocyte receptor, it will lead to diverse biological effects because nanoparticles stimulate or suppress the activity of affected immunologic pathways [21]. For example, nanoparticles can mimic PAMPs, the pathogen associated molecular patterns and activate effector proteins including the toll like receptors in an effort to enhance immunological outcomes. It is effective in improving immune system hence such an interaction should be done including athletes most of whom are under pressure in terms of training and competition which makes them produce more antibodies [20].

Exercise-induced inflammation is important to the immune response, but it can become detrimental when constantly recurring and chronic. The nanotechnology has presented enhanced approaches towards resolving inflammation via delivering systems [25]. A wide variety of drugs that possess anti-inflammatory properties can be administered at the Nanoparticles targeting the affected tissues so as to enhance the quick and healthy healing processes without unnecessary tissue deterioration. These nano-carriers can also enhance the activity of molecules like resolvins and protectins that are, in any case, engaged in the anti-inflammatory process [7]. Therefore, nanotechnology helps an athlete to recover quicker from inflammation making him/her suitable for other events, or competitions since the injury free period would be shorter and therefore improving an athlete's performance.

Immuno-metabolism means the combination of immunology and metabolism to describe immune cell function. This paper seeks to establish the extent to which nanotechnology is used in the regulation of these processes in order to enhance athlete performance [25]. Nanoparticles are also capable of modulating the immune cells metabolic activity as well as the metabolic pathways involved in the catabolism of energy molecules. Specifically, the increase in mitochondrial efficiency or the modulation of cytoplasmic metabolic pathways by nanomaterials can promote efficient functionality of immune cells. Athletes can particularly benefit from it since training and increasing the immune response will improve endurance and reduce the time required to recover from intense training [5]. For instance, Nano technology can enhance energy requirements of glucose metabolism of immunological cells to enable them get adequate power to respond proactively to infections and injuries [7]. Moreover, nanotechnology has the ability to propagate the metabolic pathways of the immune responses permitting a more moderate reaction. Such a fine-tuning of immune-metabolism not only supports the enhanced immune strategy but also contributes to an athletic body by preserving their well-being as athletes to make highly immune athletic demanding sports [24].

7 Safety and Toxicological Considerations

The introduction of nanotechnology in to immune support mechanisms for athletes has opened great possibilities for performance improvement but a huge bubble of safety issues comes along with it. Some properties of nanomaterials are different from those larger materials because of their small size, which may cause unforeseen effects in biological systems [26]. Such materials may have potential to get across the cellular membranes, be deposited within tissues, and produce reactive oxygen species or inflammation if their concentration is not tightly controlled. More recently, possible long-term effects of the toxicity of nanomaterials have been considered that may destabilise the immune system, affect cell membranes or even worsen pre-existing pathologies [27]. Despite many nanomaterials qualifying for enhancing the capabilities of the immune system, a close attention should be paid to the dangers of their application.

Besides, there is a considerable concern about delayed immune response to nanomaterial exposure in athletes. As athletic performance partially depends on the immune system, and provided that nanotechnology based therapies may alter immune tolerance, cytokines, and also functioning of immune cells, these phenomena need to be studied in the long-term [28]. However, continued exposure to specific types of nanomaterials could slowly disrupt the immune balancing mechanism making the body prone to autoimmune diseases or chronic inflammation [26]. In addition, the build-up of nanomaterial at the target organs like liver, lungs or kidneys can lead to sequestration toxicity and may have an impact on health and performance of the system. As such, there is hope for nanotech in increasing immunity but one must practice caution as to where these long-term effects put the positive affirmatives [19].

The regulatory environment of nanomedicine in sport remains in the developmental stages. The reason is that the present legal frameworks primarily deal with pharmaceutical and medical devices, not technologies which improve athletic performance [18]. The frameworks by bodies like the FDA and EMA do not focus on issues connected to the use of nanomaterials in sport. Now that nanomaterials are introduced to sports medicine, there is a need for particular guidelines in the regulation that will be able to evaluate these technologies regarding safety and efficacy in athletic populations [28]. It would ensure rigorous pre-market testing as well as post-market surveillance of adverse effects that may surface later. In this potential performance enhancement, regulatory bodies must ensure that such interventions do not give some athletes an unfair advantage while making sure that all those concerned remain safe and well [1].

The ethical issues related to nanotechnology in improving immunity for athletes are complicated and multifaceted. With new technologies, the line becomes further blurred between legitimate optimization of performance and ethically unacceptable enhancement [29]. For instance, if nanotech interventions significantly enhance immune function, they might end up giving athletes using such treatments an unfair competitive advantage and thus create a wedge between those who can have access to such technologies and those who cannot [19]. The ethical issues relating to this include changing the immune system of an athlete to increase performance, and all these questions raise further queries on the issues of consent, long-term risks on health, and preserving the fair game [22]. The idea of remaining competitive pushes some athletes to undergo nanotech treatments with unknown side effects. It is not just the athletes but also medical professionals and governing bodies that should be concerned about these technologies and how they will be used, in complete transparency and maintaining the integrity of sports.

8 Future Perspectives

The future of immune support for athletes through nanonutrient based approaches is extremely promising with emerging research directions shaping a new paradigm in sports science. Advances in nanotechnology are offering the possibility to revolutionize how athletes maintain and optimize their immune function [16]. As more focus is laid on the improvement of performance and recovery from injury, it becomes a vital factor for the recovery capacity of the athlete in general. Scientific research will further advance nanoscale interventions targeted at immune cells and pathways in more specificity [1]. This may involve designing immunomodulatory nanoparticles in a way that infections are prevented right from the onset or minimize inflammation to yield more effective tools for exercise recovery and performance in athletic activities [30]. It might be possible to discover more research with nano-delivery systems, where the immunomodulatory compounds could be precisely released at the site of inflammation or stress, that would result in optimum recovery time while keeping the risk of disease or injury as low as possible [22].

Among the promising breakthrough technologies that can shape the future of immune support for athletes, there is a potential application of targeted nanomaterials in order to modulate the gut microbiome [31]. Gut microbiota significantly participates in immune regulation; an imbalance in its components may weaken an athlete's immune defence and health. Nanoparticles carrying prebiotics, probiotics, or bioactive compounds may be directly delivered to the gut, resulting in enhanced growth of good microbes, a better immune response, and prevention of gastrointestinal disorders among athletes [32]. Improvement in techniques of nano-encapsulation further may lead to the efficient uptake of critical nutrients such as vitamins and minerals, boosting immunity and overall resilience among athletes [33]. Other areas of innovation include nano-based vaccines or immune enhancers tailored to the individual athlete's immune profile, thus offering personal support to an athlete's immunity and decreasing chances of side effects.

The integration of AI and nanotechnology will expedite the rate at which immune support technologies are developed for athletes [26]. It may improve the design, synthesis, and application of nanomaterials by faster simulations and predictions of how they would interact with biological systems. Machine learning algorithms may be used to study huge datasets of athlete health, performance, and immune responses to offer insight into optimal nutrient delivery strategies [22]. This could help researchers determine patterns in immune response that have specificity to certain athletic populations or training regimens that would be helpful in building more customized nanonutrient interventions. Furthermore, AI-enabled nanosensors embedded in wearable devices could allow for continuous monitoring of an athlete's immune status, thus enabling real-time adjustments in an athlete's nutritional regimen or training protocols to maximize the support to the immune system [33]. This nanotechnology-AI fusion will herald a new frontier in sports science, where precision and personalization will be the driving forces behind more effective immune support strategies for athletes [34].

Another promising direction for the future of immune support in athletes is predictive modeling of immune responses. Advances in computational biology and systems immunology help in the development of models that predict how an athlete's immune system will respond to different training loads, recovery periods, and environmental conditions [34]. Models based on genetic predispositions, training history, and nutritional intake would have allowed for personal recommendations on immune system modulation [35]. Predictive models, for instance, could pinpoint early warning signs of immunosuppression or inflammation before this condition has developed into full-blown illness or injury by incorporating data from many sources, including wearable sensors, genetic testing, and laboratory analyses [12]. Second, through this modeling, researchers would also be able to predict what effect some new nanonutrient interventions might have and would then use this to predict how efficacious these interventions would prove in promoting immune function under various circumstances. This means that with such models, there is likely to be more pro-active and targeted immune-support strategies, meaning improved performances and health among athletes [34].

9 Conclusion

Nanotechnology has been a revolutionary means of delivering immune support to athletes with unprecedented precision in targeting immune cells and delivering bioactive compounds. Advanced nanodelivery systems, including nanoemulsions, smart carriers, and sustained-release technologies, have modified the delivery and usage of immunonutrients within the body [12]. Whereas the conventional immune-boosting methods have yielded very limited success, the nanotechnology approaches have delivered superior bioavailability, specific targeting of cells, and controlled release of immune-modulating compounds to meet rather complex demands of athletic performance [15]. The advent of newer platforms, including CRISPR-based approaches, exosome carriers, and nanorobotic systems, presents new avenues for personalized enhancement of immunity but these should be weighed against safety aspects and the requirements of the regulatory agencies. As the field continues to advance, artificial intelligence will be incorporated with nanotechnology to develop these interventions further, and they will become even more precise and personalized, allowing for better immune support of athletes [35]. The future of immune support in sports depends on these technological advancements that will provide athletes with new tools that can help them maintain their optimal immune function when working out and competing at their best [22]. Still, the research should go on and all the safety, ethical, and regulatory considerations should be made judiciously for the proper execution of these innovative approaches.

References

1. Suzuki K (2019) Chronic inflammation as an immunological abnormality and effectiveness of exercise. Biomolecules 9:223. https://doi.org/10.3390/biom9060223
2. Woolf K, Manore MM (2006) B-vitamins and exercise: does exercise alter requirements? Int J Sport Nutr Exerc Metab 16:453–484. https://doi.org/10.1123/ijsnem.16.5.453
3. Close GL, Russell J, Cobley JN, Owens DJ, Wilson G, Gregson W, Fraser WD, Morton JP (2013) Assessment of vitamin D concentration in non-supplemented professional athletes and healthy adults during the winter months in the UK: implications for skeletal muscle function. J Sports Sci 31:344–353. https://doi.org/10.1080/02640414.2012.733822
4. Rumpf C, Proschinger S, Schenk A, Bloch W, Lampit A, Javelle F, Zimmer P (2021) The effect of acute physical exercise on NK-cell cytolytic activity: a systematic review and meta-analysis. Sports Med 51:519–530. https://doi.org/10.1007/s40279-020-01402-9
5. Shephard RJ, Shek PN (1999) Effects of exercise and training on natural killer cell counts and cytolytic activity: a meta-analysis. Sports Med 28:177–195. https://doi.org/10.2165/00007256-199928030-00003
6. Khakroo Abkenar I, Rahmani-Nia F, Lombardi G (2019) The effects of acute and chronic aerobic activity on the signaling pathway of the inflammasome NLRP3 complex in young men. Medicina (Kaunas) 55:105. https://doi.org/10.3390/medicina55040105
7. Nieman DC, Wentz LM (2019) The compelling link between physical activity and the body's defense system. J Sport Health Sci 8:201–217. https://doi.org/10.1016/j.jshs.2018.09.009
8. Simpson RJ, Campbell JP, Gleeson M, Krüger K, Nieman DC, Pyne DB, Turner JE, Walsh NP (2020) Can exercise affect immune function to increase susceptibility to infection? Exerc Immunol Rev 26:8–22
9. Jamurtas AZ, Fatouros IG, Deli CK, Georgakouli K, Poulios A, Draganidis D, Papanikolaou K, Tsimeas P, Chatzinikolaou A, Avloniti A, Tsiokanos A, Koutedakis Y (2018) The effects of acute low-volume HIIT and aerobic exercise on leukocyte count and redox status. J Sports Sci Med 17:501–508
10. Kakanis M, Peake J, Hooper S, Gray B, Marshall-Gradisnik S (2010) The open window of susceptibility to infection after acute exercise in healthy young male elite athletes. J Sci Med Sport 13:e85–e86. https://doi.org/10.1016/j.jsams.2010.10.642

11. Ahmadinejad Z, Alijani N, Mansori S, Ziaee V (2014) Common sports-related infections: a review on clinical pictures, management and time to return to sports. Asian J Sports Med 5:1–9. https://doi.org/10.5812/asjsm.34174
12. Brolinson PG, Elliott D (2007) Exercise and the immune system. Clin Sports Med 26:311–319. https://doi.org/10.1016/j.csm.2007.04.011
13. Collins J, Maughan RJ, Gleeson M, Bilsborough J, Jeukendrup A, Morton JP, Phillips SM, Armstrong L, Burke LM, Close GL, Duffield R, Larson-Meyer E, Louis J, Medina D, Meyer F, Rollo I, Sundgot-Borgen J, Wall BT, Boullosa B, Dupont G, Lizarraga A, Res P, Bizzini M, Castagna C, Cowie CM, D"Hooghe M, Geyer H, Meyer T, Papadimitriou N, Vouillamoz M, McCall A (2021) UEFA expert group statement on nutrition in elite football. Current evidence to inform practical recommendations and guide future research. Br J Sports Med 55:416–416. https://doi.org/10.1136/bjsports-2019-101961
14. Micheletti A, Rossi R, Rufini S (2001) Zinc status in athletes: relation to diet and exercise. Sports Med 31:577–582. https://doi.org/10.2165/00007256-200131080-00002
15. Duggal NA, Niemiro G, Harridge SDR, Simpson RJ, Lord JM (2019) Can physical activity ameliorate immunosenescence and thereby reduce age-related multi-morbidity? Nat Rev Immunol 19:563–572. https://doi.org/10.1038/s41577-019-0177-9
16. Wardenaar F, Brinkmans N, Ceelen I, Van Rooij B, Mensink M, Witkamp R, De Vries J (2017) Micronutrient intakes in 553 Dutch elite and sub-elite athletes: prevalence of low and high intakes in users and non-users of nutritional supplements. Nutrients 9:142. https://doi.org/10.3390/nu9020142
17. Yamauchi R, Shimizu K, Kimura F, Takemura M, Suzuki K, Akama T, Kono I, Akimoto T (2011) Virus activation and immune function during intense training in rugby football players. Int J Sports Med 32:393–398. https://doi.org/10.1055/s-0031-1271674
18. Wu D, Lewis ED, Pae M, Meydani SN (2019) Nutritional modulation of immune function: analysis of evidence, mechanisms, and clinical relevance. Front Immunol 9. https://doi.org/10.3389/fimmu.2018.03160
19. Walsh NP (2019) Nutrition and athlete immune health: new perspectives on an old paradigm. Sports Med 49:153–168. https://doi.org/10.1007/s40279-019-01160-3
20. Manore MM (2000) Effect of physical activity on thiamine, riboflavin, and vitamin B-6 requirements. Am J Clin Nutr 72:598S-606S. https://doi.org/10.1093/ajcn/72.2.598S
21. Shaw DM, Merien F, Braakhuis A, Dulson D (2018) T-cells and their cytokine production: the anti-inflammatory and immunosuppressive effects of strenuous exercise. Cytokine 104:136–142. https://doi.org/10.1016/j.cyto.2017.10.001
22. Pedersen BK, Toft AD (2000) Effects of exercise on lymphocytes and cytokines. Br J Sports Med 34:246–251. https://doi.org/10.1136/bjsm.34.4.246
23. Hernández-Camacho JD, Vicente-García C, Parsons DS, Navas-Enamorado I (2020) Zinc at the crossroads of exercise and proteostasis. Redox Biol 35:101529. https://doi.org/10.1016/j.redox.2020.101529
24. Braakhuis AJ (2012) Effect of vitamin C supplements on physical performance. Curr Sports Med Rep 11:180–184. https://doi.org/10.1249/JSR.0b013e31825e19cd
25. Steensberg A, Toft AD, Bruunsgaard H, Sandmand M, Halkjær-Kristensen J, Pedersen BK (2001) Strenuous exercise decreases the percentage of type 1 T cells in the circulation. J Appl Physiol 91:1708–1712. https://doi.org/10.1152/jappl.2001.91.4.1708
26. Carins J, Booth C (2002) Salivary immunoglobulin-A as a marker of stress during strenuous physical training. Aviat Space Environ Med 73:1203–1207
27. Robson PJ, Blannin AK, Walsh NP, Castell LM, Gleeson M (2007) Effects of exercise intensity, duration and recovery on in vitro neutrophil function in male athletes. Int J Sports Med 20:128–130. https://doi.org/10.1055/s-2007-971106
28. Steensberg A, van Hall G, Osada T, Sacchetti M, Saltin B, Pedersen BK (2000) Production of interleukin-6 in contracting human skeletal muscles can account for the exercise-induced increase in plasma interleukin-6. J Physiol 529:237–242. https://doi.org/10.1111/j.1469-7793.2000.00237.x

29. Tirla A, Islam F, Islam MR, Ioana Vicas S, Cavalu S (2022) New insight and future perspectives on nutraceuticals for improving sports performance of combat players: focus on natural supplements, importance and advantages over synthetic ones. Appl Sci 12:8611
30. Fahlman MM, Engels H-J (2005) Mucosal IgA and URTI in American college football players: a year longitudinal study. Med Sci Sports Exerc 37:374–380. https://doi.org/10.1249/01.mss.0000155432.67020.88
31. Kerksick CM, Wilborn CD, Roberts MD, Smith-Ryan A, Kleiner SM, Jäger R, Collins R, Cooke M, Davis JN, Galvan E, Greenwood M, Lowery LM, Wildman R, Antonio J, Kreider RB (2018) ISSN exercise and sports nutrition review update: research and recommendations. J Int Soc Sports Nutr 15:38. https://doi.org/10.1186/s12970-018-0242-y
32. Friman G, Wesslén L (2000) Infections and exercise in high-performance athletes. Immunol Cell Biol 78:510–522. https://doi.org/10.1111/j.1440-1711.2000.t01-12-.x
33. Müller SM, Dawczynski C, Wiest J, Lorkowski S, Kipp AP, Schwerdtle T (2020) Functional biomarkers for the selenium status in a human nutritional intervention study. Nutrients 12:676. https://doi.org/10.3390/nu12030676
34. Bian L, Wang X, Liu D (2024) Enhanced detection of vitamin C in sports beverages using electrochemical sensors with nano-gold particles. Food Measure 18:3820–3826. https://doi.org/10.1007/s11694-024-02457-2
35. Kistler W, Villiger M, Villiger B, Yazici D, Pat Y, Mitamura Y, Ardicli S, Skolnick S, Dhir R, Akdis M, Nadeau K, Ogulur I, Akdis CA (2024) Epithelial barrier theory in the context of nutrition and environmental exposure in athletes. Allergy 79:2912–2923. https://doi.org/10.1111/all.16221

The Future of Personalized Sports Nutrition with Nanotech

Bitu Gogoi, Akramul Ansary, Deijy Choudhury, Babita Deka, Juti Rani Devi, Pooja Patowary, Prakash Kishore Hazam, and Koushik Nandan Dutta

Abstract Nanotechnology has transformed the landscape of personalized sports nutrition, providing novel tools for nutrient delivery, bioavailability enhancement, and recovery acceleration. Nanotechnology enables the targeted delivery of nutrients based on a specific athlete's metabolic status, genetic background, exercise regimen, and exposure to demanding environmental conditions. Techniques like nanoencapsulation can protect labile nutrients (vitamins, omega-3 fatty acids) against thermal or oxidative degradation and simultaneously improve their oral absorption. Wearable nanosensors enable real-time monitoring of multiple biomarkers in body fluids/sweat, allowing immediate modifications in nutrient intake patterns to optimize performance and recovery time. Also "phytochemical" nanoformulations (curcumin nanoparticles; resveratrol nanoparticles) show superior effects against exercise-induced stress/inflammation-mediated detrimental changes than their conventional counterparts in simpler formulations due to an enhanced therapeutic index. The overall availability/uptake of plant-origin proteins/nutrients is also augmented by nanotechnology in view of a global upsurge in vegan consumers who face substantial nutritional challenges during muscle repair after physical assault/exercise overexertion. However the applications still remain far from realization owing to high production costs associated with most nanopreparations/formulations; safety concerns particularly arising from nanoemulsions/nanomicelles that become bioactive upon gastrointestinal uptake; compliance issues connected with unobtrusive wearable technology-based biosensors; apprehensions regarding privacy breaches when browsing dietary/sports-nutrition-habit databases via portable technology platforms without appropriate security safeguards in place; lack or lagging behind strict legislative regulations surrounding manufacture/approval/marketing-an essential prerequisite for ensuring widespread legal marketing/uniform access at affordable prices globally as well as enforcing systematic post-market surveillance compliances relating to use dosage

B. Gogoi · A. Ansary · D. Choudhury · B. Deka · J. R. Devi · P. Patowary · K. N. Dutta (✉)
NETES Institute of Pharmaceutical Science, NEMCARE Group of Institutions, Mirza, Kamrup, Assam, India
e-mail: koushik5dutta@gmail.com

P. K. Hazam
Marine Research Station, Institute of Cellular and Organismic Biology, Academia Sinica, 23-10 Dahuen Rd., Jiaushi, Ilan 262, Taiwan

A. Chatterjee et al. (eds.), *Nanofuel: The Future of Sports Nutrition*, Smart Nanomaterials Technology, https://doi.org/10.1007/978-981-96-5471-0_16

efficient dosing schedules during development/marketability patent protection procedures products' registration/nutraceutical approval benefits/risks analyses toxicity/adverse-effect evaluations.

Keywords Nanotechnology · Nutrient delivery · Bioavailability · Wearable nanosensors · Phytochemical nanoformulations

1 Introduction

Implementation of personalized strategies is becoming more widespread in sports nutrition as it helps to cater for the peculiarities of individual athletes. While the general methods are somewhat sufficient, they are not fully effective in providing specific nutritional plans which cater for the diverse physiological and metabolic needs of athlete personnel. There have been widespread changes in particular areas such as medicine, energy, and materials science by the use of nanotechnology and now those changes are expected to impact sports nutrition [1]. Athletes can now combine personalized nutrition with nanotechnology in a bid to achieve optimum performance, faster recovery among other possible health benefits. Nanotechnology characterized by manufacture and manipulation of materials ranging between (1–100) nanometers is at best the solution at a time of great possibility of access to nutrients. For example, nanoparticles, liposomes, nanoemulsions and dendrimers have been able to serve as forms of nutrient nano delivery system which has enabled protecting sensitive nutrients from degradation, targeted delivery and increased absorption in the gastrointestinal tract [2]. To the athletes these are great advances because it means efficient use of the nutrients which in turn determines the performance and recovery rates. The tailor-made nutrition of an athlete starts with looking deeply into the individual athlete's profile such as genetic profile, metabolic profiling, and microbiome composition. These insights can be garnered via advanced diagnostic tools such as genomic analysis/metabolomics/microbiome sequencing etc. looking at the individual amenable nutrition guidelines can be developed. Nanotechnology improves on this model as it provides precision in supplementation delivery and application thereby addressing the challenges of conventional supplementation. For example, placing Omega 3 fatty acids or further branched chain amino acids in nanoparticles would ensure durability through digestion and functioning whereby these are efficiently delivered to tissues attached to muscles increasing the recovery rate and aiding in reducing inflammation [3].

One of the greatest issues encountered regarding sports nutrition, especially for individuals who are lipophilic which include fat soluble vitamins, coenzyme Q10 and curcumin is bioavailability. Nanoemulsions and nanoencapsulation technologies promise to boost the evidence that these compounds can be efficiently absorbed by improving the solubility and stability of the compounds. Curcumin, a natural compound well-known for its anti-inflammatory properties is hydrophobic and their bioavailability is poor. Different Nanoformulations of curcumin have given significantly better absorption, thus they are highly beneficial in the management of exercise caused inflammation and oxidative stress [4]. Likewise, far- and fat-soluble vitamins which are essential for bone density, immunity, and blood coagulation, namely, D, E, and K, can also gain from nano-based drug delivery systems. Wearables are now being integrated with nanotechnology to create advanced, real-time nutritional monitoring and management solutions. For example, biosensors housed in wearables can now measure blood markers like hydration, electrolyte balance and oxidative stress, and reports an athlete's current state in real time. Using these data-driven insights, dietary intake, and nutritional strategies can be adjusted in real-time to maximize [5].

Including nutrient delivery and monitoring, nanotechnology enables detailed food formulations to the dietary demand of the increasing preference for plant-based and sustainable diets among athletes. Plant-based proteins are environmentally friendly but can be deficient in certain essential amino acids and have lower bioavailability than animal-based proteins. Nanotechnology overcomes these challenges by improving the digestibility and bioavailability of plant proteins through nanoencapsulation and fortification with vitamins, minerals and essential amino acids. For example, nanotechnology formulations of plant-based proteins from legumes and grains exhibited a superior amino acid profile and absorption rate, indicating their efficacy in muscle building and recovery [6]. Intensive sports often cause oxidative stress and inflammation of athletes, muscle damage, and delayed recovery. Phytochemicals are bioactive compounds derived from plants known to halt such outcomes, as they exert antioxidant and anti-inflammatory activities. However, many phytochemicals have poor bioavailability, which restricts their effectiveness. Phytochemical nanoformulations, including quercetin and resveratrol, improve solubility, stability, and absorption, thereby optimizingtherapeuticefficacy. As an illustration, quercetin-loaded nanoparticles have outperformed classical supplements in minimizing oxidative stress from exercise [7].

The role of nanotechnology in sports nutrition goes beyond that of improving the delivery of individual nutrients and encompasses more sophisticated therapeutic uses. Injuries are a recurring problem for most athletes, and nanotechnology provides new solutions to injury prevention and recovery. Nanoengineered molecules can be used to enhance the effect of anti-inflammatory drugs and antioxidants on specific body structures or tissues and thus minimizing the systemic effects typically associated with conventional drug administration in other applications. For example, curcumin nanoparticles applied directly to muscle-injury sites were shown to enhance anti-inflammatory activity resulting in faster healing. Another example is the use of nanofibrous scaffolds for tissue engineering applications (e.g., ligaments and tendons) [8].

The prohibitive cost associated with the development and production of nano-based nutritional products represents a major barrier to their wider use. As things stand, it is only the elite athletes and professional teams who can afford these cutting-edge performance solutions. However, in order to make these products readily accessible to amateur athletes and society more generally, it will be necessary to inject much capital into R&D. To this end, multi-stakeholder collaboration between higher education institutions, industry leaders and policy makers will be required in a bid to collectively address the economic challenges posed by nanotechnology that need overcoming so as to ensure equal access by all [9]. Ethics also poses an important challenge—indeed obstacle—to wider exploitation of nanotechnology in personalized sports nutrition. A major concern relates to data privacy when dealing with personalized approaches which require harvests of genetic and metabolic data for subsequent analysis. Assigning athletes control over their own data usage must happen if trust is the goal; this would facilitate engagement with emerging paradigm shifts. Secondly, but equally important is ensuring equal access for all should nanotech-enabled personalized nutrition strategies become mainstream methods, so as not to widen further any performance/health gaps between better-off and worse-off athletes [10].

2 Nanotechnology-Based Solutions in Personalized Sports Nutrition

The quest for optimal athletic performance has always been intertwined with nutrition. Athletes meticulously plan their diets, supplementing with specific nutrients to enhance energy levels, muscle growth, and recovery. However, traditional dietary approaches often rely on generalizations and overlook individual variations in metabolism, genetics, and training regimes. This is where nanotechnology steps in, offering the potential to deliver targeted, personalized nutrition based on an individual's specific needs [11].

Nanotechnology, at its core, deals with matter on an atomic and molecular scale (1–100 nm). At this scale, materials exhibit unique properties that can be harnessed for various applications, including drug delivery, diagnostics, and in this case, nutritional supplementation. The tiny size of nanoparticles allows them to interact with biological systems in novel ways, improving bioavailability, targeting specific tissues, and providing controlled release of nutrients [12].

In the context of sports nutrition, nanotechnology offers solutions that go beyond conventional dietary supplements. It enables the development of smart, responsive nutritional strategies that are precisely tailored to an athlete's [13]:

Individual Needs: Based on their specific sport, training intensity, goals, and physiological parameters.

Real-time Monitoring: Using nanosensors to track biomarkers and adjust nutritional intake accordingly.

Efficient Delivery: Enhancing the bioavailability and effectiveness of crucial nutrients.

Improved Recovery: Accelerating muscle repair and reducing inflammation.

2.1 Benefits of Nanotechnology in Personalized Sports Nutrition [14]

The adoption of nanotechnology in sports nutrition offers a multitude of benefits:

Personalization: Tailoring nutritional strategies to individual needs based on real-time data and physiological responses.

Improved Bioavailability: Enhancing the absorption and utilization of crucial nutrients.

Targeted Delivery: Ensuring nutrients reach specific tissues and cells for maximum efficacy.

Enhanced Recovery: Accelerating muscle repair, reducing inflammation, and minimizing downtime.

Real-Time Monitoring: Tracking physiological parameters and adjusting nutrition plans proactively.

Performance Optimization: Maximizing athletic potential through evidence-based nutritional approaches.

Convenience: Delivering nutrients in easy-to-consume formats.

2.2 Key Applications of Nanoparticles in Supplement Design

Here are some of the most promising applications of nanotechnology in this burgeoning field.

2.2.1 Targeted Nutrient Delivery

Nanoencapsulation: This is a technique where nutrients are encapsulated within nanoscale carriers, such as liposomes, polymeric nanoparticles, or nanoemulsions. These carriers can protect nutrients from degradation in the digestive tract, enhancing their absorption and bioavailability. For instance, highly unstable vitamins or omega-3 fatty acids can be delivered effectively to specific tissues, ensuring maximum utilization [15].

Ligand-Mediated Targeting: Nanoparticles can be functionalized with ligands (molecules that bind to specific receptors) to target specific cells or tissues, like muscles or liver. This ensures that nutrients are delivered directly to the areas where they are needed most, minimizing off-target effects. For example, nanoparticles carrying amino acids could be targeted to muscle cells during recovery for efficient protein synthesis [16].

Controlled Release: Nanotechnology allows for the precise control of nutrient release over time. This can be beneficial for athletes requiring sustained energy levels during endurance activities or for delayed delivery of recovery nutrients. For example, a nanoparticle containing a slow-release carbohydrate could provide a gradual supply of glucose during a marathon [17].

2.2.2 Nano-enabled Nutritional Supplements

Enhanced Bioavailability: Many essential nutrients, such as iron, calcium, and certain vitamins, have poor bioavailability in their traditional forms. Nanoparticles can improve the solubility and absorption of these nutrients, leading to better utilization by the body. For instance, iron-oxide nanoparticles have been explored to enhance iron uptake, addressing iron deficiency in athletes [18].

Improved Flavor and Texture: Nanoemulsions can be used to encapsulate flavor compounds, resulting in enhanced taste and texture of nutritional products, making them more appealing for athletes. This is crucial for ensuring adherence to nutrition plans [19].

Multifunctional Supplements: Nanoparticles can be designed to carry multiple nutrients simultaneously, creating synergistic effects and providing a comprehensive approach to supplementation. For instance, a single nanoparticle could deliver a combination of amino acids, electrolytes, and antioxidants [20].

2.2.3 Nanosensors for Real-Time Monitoring

Wearable Sensors: Nanomaterials, such as carbon nanotubes, graphene, and metal nanoparticles, can be used to create wearable sensors that can continuously monitor an athlete's physiological parameters, including:

Hydration Levels: By detecting changes in electrolytes in sweat.
Glucose Levels: To optimize carbohydrate intake.
Lactate Levels: To monitor muscle fatigue and performance.
Cortisol Levels: To assess stress and recovery status [21].

Implantable Sensors: While more invasive, implantable nanosensors offer the potential for even more precise, real-time data on metabolic activity and nutrient levels directly within the bloodstream [22].

Data Integration: The data collected from these nanosensors can be transmitted wirelessly to mobile devices or cloud-based platforms. This allows for real-time monitoring and personalized adjustments to nutritional strategies, creating a closed-loop system for optimized performance [23].

2.2.4 Nano-enhanced Recovery Strategies

Anti-Inflammatory Delivery: Nanoparticles can deliver anti-inflammatory agents, such as curcumin or omega-3 fatty acids, to targeted areas to reduce muscle soreness and inflammation after intense workouts, accelerating recovery [24].

Antioxidant Delivery: Exercise-induced oxidative stress can lead to muscle damage and fatigue. Nanoparticles carrying antioxidants, like vitamin C or E, can neutralize free radicals and enhance recovery [25].

Muscle Repair and Regeneration: Nanoparticles can deliver specific growth factors or amino acids to muscle tissue to promote repair and regeneration. This can lead to faster recovery and reduced downtime between training sessions [26].

2.2.5 Specific Examples of Nanoparticle Applications in Sports Supplements

Protein Supplements: Nano-encapsulated whey protein with controlled release can provide a sustained amino acid supply for muscle protein synthesis.

Electrolyte Supplements: Nano-encapsulated electrolytes can offer more efficient hydration by improving absorption and reducing gastrointestinal distress.

Vitamin and Mineral Supplements: Liposomes and polymeric nanoparticles can improve the bioavailability of poorly absorbed micronutrients like iron, calcium, and vitamin D.

Creatine Supplements: Nanoparticle formulations of creatine can enhance its absorption and uptake by muscles, leading to improved performance benefits.

Anti-inflammatory Supplements: Targeted delivery of curcumin, omega-3 fatty acids, and other anti-inflammatory compounds using nanoparticles can accelerate muscle recovery and reduce soreness.

Antioxidant Supplements: Nanoparticles can deliver antioxidants like vitamin C, vitamin E, and glutathione directly to cells, protecting them from oxidative stress.

Nitric Oxide Boosters: Nanoparticles can improve the delivery and effectiveness of nitric oxide precursors, enhancing blood flow and muscle performance [27, 28]. Figure 1 shows Applications of nanoparticles in supplement design.

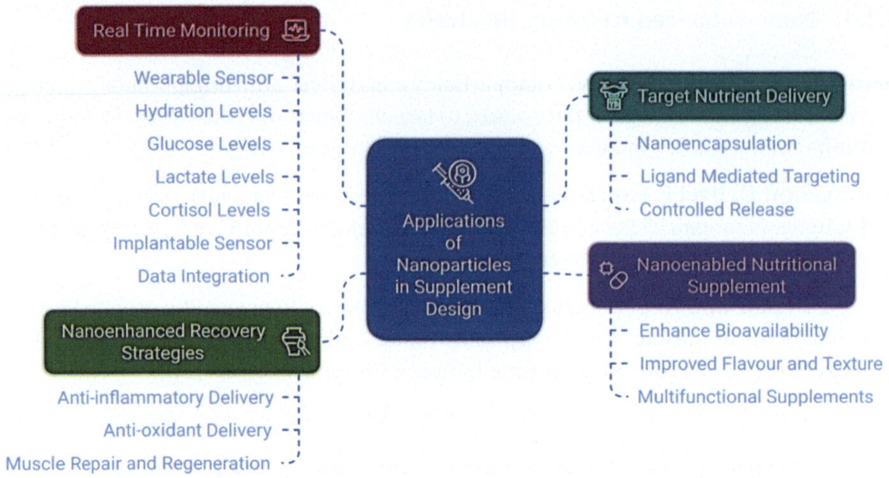

Fig. 1 Applications of nanoparticles in supplement design

2.3 Nanoparticles for Supplement Design

2.3.1 Enhancing Bioavailability Through Nanoencapsulation

Liposomes: These are spherical vesicles with a lipid bilayer membrane, similar to cell membranes. They can encapsulate water-soluble or fat-soluble nutrients, protecting them from degradation and enhancing their absorption across the intestinal lining. Liposomes are particularly effective for delivering vitamins (like C and B), antioxidants (like glutathione), and omega-3 fatty acids.

Example: Encapsulating curcumin, a poorly absorbed anti-inflammatory compound, in liposomes significantly enhances its bioavailability, making it more effective in reducing muscle soreness and inflammation [29].

Polymeric Nanoparticles: These are made from biocompatible polymers and can be engineered with varying sizes, shapes, and surface properties. They can encapsulate a wide range of nutrients and offer controlled release capabilities.

Example: Encapsulating branched-chain amino acids (BCAAs) in polymeric nanoparticles can improve their absorption and delivery to muscle cells, enhancing protein synthesis and recovery [30].

Nanoemulsions: These are dispersions of one liquid in another, where at least one liquid is in the nanoscale range. They are particularly effective for delivering fat-soluble vitamins (A, D, E, K) and hydrophobic compounds like CoQ10.

Example: Nanoemulsified vitamin D is more readily absorbed compared to traditional forms, helping athletes maintain optimal levels for bone health and muscle function [31].

Solid Lipid Nanoparticles (SLNs): These are made from biocompatible lipids and are solid at room temperature. They are effective for delivering lipophilic drugs and nutrients, offering good stability and controlled release.

Example: SLNs can be used to deliver vitamin E, a potent antioxidant, protecting it from degradation and enhancing its absorption [32].

2.3.2 Targeted Delivery of Nutrients to Specific Tissues

Ligand-Mediated Targeting: Nanoparticles can be functionalized with specific molecules (ligands) that bind to receptors on target cells. This approach ensures that nutrients are delivered precisely where they are needed.

Example: Folate receptors are overexpressed on muscle cells after exercise. Nanoparticles carrying amino acids or growth factors can be functionalized with folate ligands to target muscle tissue for efficient repair and recovery.

pH-Sensitive Nanoparticles: These nanoparticles release their payload in response to changes in pH, such as the acidic environment of inflamed tissues.

Example: Nanoparticles carrying anti-inflammatory agents could be designed to release their cargo specifically in inflamed muscles after intense training.

Magnetic Nanoparticles: These can be guided to specific locations using an external magnetic field, allowing for localized drug or nutrient delivery.

Example: Magnetic nanoparticles carrying recovery nutrients could be directed to injured muscle groups to accelerate healing [33, 34].

2.3.3 Controlled Release of Nutrients for Sustained Effects

Diffusion-Controlled Release: The nutrient is released slowly over time as it diffuses through the nanoparticle matrix. This approach is suitable for providing sustained energy levels during endurance events.

Example: Encapsulating carbohydrates in polymeric nanoparticles with a diffusion-controlled release mechanism can provide a gradual and sustained release of glucose to the bloodstream.

Erosion-Controlled Release: The nanoparticle degrades over time, releasing its contents. This method is suitable for delivering nutrients that are needed at specific times during the recovery process.

Example: Nanoparticles designed to degrade slowly can release amino acids or growth factors over several hours after a workout, maximizing muscle protein synthesis.

Stimuli-Responsive Release: Nanoparticles can be designed to release their contents in response to a specific stimulus, such as temperature, light, or enzyme activity.

Example: Nanoparticles that release electrolytes upon sensing changes in sweat can help maintain hydration levels during intense workouts [35, 36].

2.3.4 Improving the Taste and Stability of Supplements

Flavor Encapsulation: Nanoemulsions and other nanoparticles can encapsulate flavor compounds, masking unpleasant tastes and improving the palatability of nutritional supplements.

Stabilization of Nutrients: Nanoencapsulation can protect sensitive nutrients from degradation due to light, heat, or oxidation, extending their shelf life [37].

2.3.5 Targeted Nutrient Delivery

Nanoencapsulation: This is a technique where nutrients are encapsulated within nanoscale carriers, such as liposomes, polymeric nanoparticles, or nanoemulsions. These carriers can protect nutrients from degradation in the digestive tract, enhancing their absorption and bioavailability. For instance, highly unstable vitamins or omega-3 fatty acids can be delivered effectively to specific tissues, ensuring maximum utilization.

Ligand-Mediated Targeting: Nanoparticles can be functionalized with ligands (molecules that bind to specific receptors) to target specific cells or tissues, like muscles or liver. This ensures that nutrients are delivered directly to the areas where they are needed most, minimizing off-target effects. For example, nanoparticles carrying amino acids could be targeted to muscle cells during recovery for efficient protein synthesis.

Controlled Release: Nanotechnology allows for the precise control of nutrient release over time. This can be beneficial for athletes requiring sustained energy levels during endurance activities or for delayed delivery of recovery nutrients. For example, a nanoparticle containing a slow-release carbohydrate could provide a gradual supply of glucose during a marathon [38].

2.3.6 Nano-enabled Nutritional Supplements

Enhanced Bioavailability: Many essential nutrients, such as iron, calcium, and certain vitamins, have poor bioavailability in their traditional forms. Nanoparticles can improve the solubility and absorption of these nutrients, leading to better utilization by the body. For instance, iron-oxide nanoparticles have been explored to enhance iron uptake, addressing iron deficiency in athletes.

Improved Flavor and Texture: Nanoemulsions can be used to encapsulate flavor compounds, resulting in enhanced taste and texture of nutritional products, making them more appealing for athletes. This is crucial for ensuring adherence to nutrition plans.

Multifunctional Supplements: Nanoparticles can be designed to carry multiple nutrients simultaneously, creating synergistic effects and providing a comprehensive approach to supplementation. For instance, a single nanoparticle could deliver a combination of amino acids, electrolytes, and antioxidants [39, 40].

2.3.7 Nanosensors for Real-Time Monitoring

Wearable Sensors: Nanomaterials, such as carbon nanotubes, graphene, and metal nanoparticles, can be used to create wearable sensors that can continuously monitor an athlete's physiological parameters, including:

Hydration Levels: By detecting changes in electrolytes in sweat.
Glucose Levels: To optimize carbohydrate intake.
Lactate Levels: To monitor muscle fatigue and performance.
Cortisol Levels: To assess stress and recovery status.

Implantable Sensors: While more invasive, implantable nanosensors offer the potential for even more precise, real-time data on metabolic activity and nutrient levels directly within the bloodstream.

Data Integration: The data collected from these nanosensors can be transmitted wirelessly to mobile devices or cloud-based platforms. This allows for real-time monitoring and personalized adjustments to nutritional strategies, creating a closed-loop system for optimized performance [41, 42].

2.3.8 Nano-enhanced Recovery Strategies

Anti-inflammatory Delivery: Nanoparticles can deliver anti-inflammatory agents, such as curcumin or omega-3 fatty acids, to targeted areas to reduce muscle soreness and inflammation after intense workouts, accelerating recovery.

Antioxidant Delivery: Exercise-induced oxidative stress can lead to muscle damage and fatigue. Nanoparticles carrying antioxidants, like vitamin C or E, can neutralize free radicals and enhance recovery [43].

2.3.9 Muscle Repair and Regeneration

1. Nanoparticles can deliver specific growth factors or amino acids to muscle tissue to promote repair and regeneration. This can lead to faster recovery and reduced

downtime between training sessions [44]. Table 1 shows Nutritional components with formulation for sports personnel

2.4 Advantages of Nanoparticle-Based Supplement Design

Increased Bioavailability: Greater absorption and utilization of nutrients.

Argeted Delivery: Precisely delivering nutrients to specific tissues and cells.

Controlled Release: Sustained delivery or specific timing of nutrient release.

Enhanced Effectiveness: Maximizing the therapeutic potential of supplements.

Improved Stability: Protecting nutrients from degradation and prolonging shelf life.

Better Taste: Masking unpleasant flavors and enhancing palatability [45].

2.5 Challenges and Considerations

Poor Bioavailability: Many nutrients are not easily absorbed by the body, reducing their effectiveness.

Degradation: Some nutrients are unstable in the digestive tract and are broken down before they can be utilized.

Slow Absorption Rates: Delaying the delivery of crucial nutrients to muscles or other tissues can limit their impact on performance and recovery.

Off-Target Effects: Some nutrients may have unintended side effects if not delivered to the right tissues at the right time.

Safety: Comprehensive toxicological studies are necessary to ensure the safety of nanoparticles used in supplements.

Regulatory Approval: Clear guidelines and regulations are needed for the production and sale of nano-based supplements.

Cost: Nanoparticle production can be expensive, potentially increasing the price of supplements.

Scalability: Transitioning from laboratory-scale to mass production of nano-based supplements is necessary for widespread availability.

Public Perception: Educating the public about the safety and benefits of nanotechnology in supplements is essential.

Nanotechnology offers a powerful toolbox to overcome these limitations by engineering nanoparticles that can act as:

Table 1 Nutritional components with formulation for sports personnel

No	Supplement name	Formulation	Primary use
1	Whey protein	Powder, bar, shake	Muscle repair and growth
2	Casein protein	Powder	Sustained-release protein for overnight recovery
3	Creatine monohydrate	Powder, capsule	Boosts strength and performance
4	Beta-alanine	Powder, capsule	Enhances endurance by buffering lactic acid
5	BCAAs (branched-chain amino acids)	Powder, capsule	Muscle recovery and reduced soreness
6	L-Glutamine	Powder, capsule	Supports muscle recovery and immune health
7	L-Carnitine	Capsule, liquid	Fat metabolism and energy production
8	Multivitamin	Tablet, capsule	Overall nutrient balance
9	Fish oil (Omega-3)	Capsule, liquid	Reduces inflammation and supports cardiovascular health
10	Vitamin D	Tablet, capsule	Bone health and muscle function
11	Zinc	Tablet, capsule	Supports immune function and protein synthesis
12	Magnesium	Tablet, capsule, powder	Muscle relaxation and energy production
13	Iron	Capsule, tablet	Red blood cell production and oxygen transport
14	Vitamin B12	Capsule, tablet, liquid	Energy production
15	Probiotics	Capsule, powder	Gut health and immune support
16	Collagen peptides	Powder	Joint health and muscle repair
17	Electrolyte mix	Powder, drink	Hydration and electrolyte replenishment
18	Nitric oxide booster (e.g., Arginine)	Capsule, powder	Enhances blood flow and oxygen delivery
19	Coenzyme Q10 (CoQ10)	Capsule	Cellular energy production and antioxidant support
20	HMB (β-hydroxy β-methylbutyrate)	Powder, capsule	Reduces muscle breakdown and promotes growth
21	Taurine	Powder, capsule	Improves endurance and reduces muscle fatigue
22	Green tea extract	Capsule, liquid	Boosts metabolism and fat oxidation
23	Beetroot extract	Liquid, powder	Enhances endurance and nitric oxide production

(continued)

Table 1 (continued)

No	Supplement name	Formulation	Primary use
24	Spirulina	Powder, tablet	Natural protein source with antioxidants
25	Chlorella	Powder, tablet	Detoxification and energy boost
26	Tart cherry extract	Capsule, liquid	Reduces muscle soreness and inflammation
27	Ashwagandha	Capsule, powder	Stress reduction and improved recovery
28	Rhodiola Rosea	Capsule	Enhances endurance and reduces fatigue
29	Vitamin C	Tablet, capsule, powder	Antioxidant and immune support
30	Calcium	Tablet, capsule	Bone strength
31	Selenium	Capsule, tablet	Antioxidant support
32	CLA (conjugated linoleic acid)	Capsule	Supports fat metabolism
33	Medium-chain triglycerides (MCT oil)	Liquid	Energy source for endurance activities
34	Fenugreek extract	Capsule	Enhances strength and libido
35	Alpha-lipoic acid (ALA)	Capsule	Antioxidant and supports energy metabolism
36	Resveratrol	Capsule, tablet	Antioxidant support
37	Lycopene	Capsule, tablet	Cardiovascular health
38	Turmeric with curcumin	Capsule, tablet	Anti-inflammatory benefits
39	Caffeine	Tablet, capsule, powder	Enhances focus and energy
40	Maca root powder	Powder, capsule	Energy and stamina boost
41	Gingko biloba	Capsule	Supports cognitive performance
42	Ginseng	Capsule, powder	Energy boost and stress relief
43	D-Ribose	Powder	Supports energy production in cells
44	Chromium	Capsule	Supports glucose metabolism
45	MSM (methylsulfonylmethane)	Capsule, powder	Joint health
46	Pycnogenol	Capsule	Supports circulation and reduces oxidative stress
47	Acetyl-L-Carnitine	Capsule, powder	Cognitive and energy support
48	Phosphatidylserine	Capsule	Brain health and recovery aid
49	DIM (diindolylmethane)	Capsule	Hormonal balance
50	Echinacea	Capsule	Immune system support
51	Aloe vera juice	Liquid	Anti-inflammatory and digestive health

(continued)

Table 1 (continued)

No	Supplement name	Formulation	Primary use
52	Astragalus root	Capsule	Immune system booster
53	Bromelain	Capsule	Anti-inflammatory and digestive aid
54	Coconut water powder	Powder	Natural electrolyte source
55	Ginger root powder	Capsule, powder	Anti-inflammatory support
56	Papaya enzymes	Chewable tablet	Digestive support
57	Lecithin	Capsule, granules	Supports cell membrane and fat metabolism
58	Plant-based protein	Powder	Vegan protein source for recovery
59	Flaxseed oil	Capsule, liquid	Omega-3 source and inflammation control
60	Hemp protein powder	Powder	Plant-based protein with fiber

Carriers: Encapsulating and protecting nutrients during digestion.

Vehicles: Transporting nutrients to specific tissues and cells.

Release Modulators: Controlling the timing and rate of nutrient release.

Bioavailability Enhancers: Improving the absorption and utilization of nutrients.

Moving forward, the following areas require further investigation:

Long-term safety studies of various nanomaterials in humans.
Development of more biocompatible and biodegradable nanomaterials.
Advancements in nanosensor technology for improved accuracy and reliability.
Creation of user-friendly data platforms for personalized nutritional guidance.
Integration of nanotechnology with other personalized medicine approaches like genomics and proteomics [46].

3 Benefits of Nanotechnology in Personalized Sports Nutrition

Nanotechnology is a game changer in various fields including environmental science electronics and health. One of its most promising applications is in sports nutrition, since it offers new possibilities of performance optimization, recuperation enhancement and diet customization. Nanotechnology addresses long-standing issues associated with traditional supplements such as insufficient bioavailability, solubility issues and stomach pain which are more plausible for more effective and efficient delivery of nutrients. This paper looks at the implications of nanotechnology on sports nutrition to improve performance, recovery, nutrient time, digestion and bioavailability while decreasing side effects [47].

3.1 Precision in Nutrient Timing and Dosage

Athletics performance can be greatly impacted by the exact time and amount of nutrients in sports nutrition. In order to improve their training and recuperation, athletes frequently worry about the best times and methods to eat nutrients. However, the ability to precisely control nutrient release is limited by conventional supplementing techniques. Conversely, nanotechnology makes it possible to create specialised nutrient delivery systems that release nutrients at precise intervals to meet the body's needs at different phases of activity and recuperation [48].

3.1.1 Timed Nutrient Release via Nanoparticles

Nanoparticles like liposomes, micelles and polymeric carriers are also able to encapsulate the nutrients and are able to release them at definite timings depending on the requirements of the body. For example, Vitamin A and D are liposoluble vitamins and thus are carried by liposomal carriers. These carriers ensure that the nutrients are not degraded by stomach acids increasing the chances of absorption in the small intestine. This means, such nutrients will be released at the perfect timing when they are required such as before or during an exercise [49].

Nanocarriers of polymers can be designed in such a way that they release glucose, or amino acids depending on specific triggers like pH of the body or any enzymatic activity, in a way that enables the athletes to get the energy at the right time. For example, long distance athletes like marathon runners or cyclists often find it difficult to have the required supply of energy through all the activity done over a longer time period. The solution to these issues can be provided through nanotechnology in such a way that there is a continuous supply of glucose of any energy substrates throughout the physical activity [50].

3.1.2 Targeted Delivery for Optimal Dosage

Another important advantage of nanotechnology in sports nutrition is its ability to provide specific quantities of nutrients according to the requirements of the sportsman. Instead of relying on large and frequent doses of amino acid supplements, which may cause adverse effects or fail to be absorbed effectively, nanoparticles deliver precisely the dose quantitatively needed to the affected tissue. For example, a popular supplement used to boost muscle performance, creatine, can be delivered via a nanocarrier without the need for high doses which may cause bloating or discomfort, and still increase its absorption by muscle cells. By allowing communication throughout the muscle, nanotechnology ensures the muscles receive the nutrients in the best suitable time and in the most appropriate amounts for optimal performance [51].

3.2 Improved Digestion and Absorption of Macronutrients and Micronutrients

Optimal digestion and absorption of nutrients are essential for athletes to perform and recover efficiently. Nonetheless, there remain difficulties in solubility, stability, and/or absorption of specific nutrients in the digestive tract. Nanotechnology can improve the bioavailability of macro-and micronutrients [52].

3.2.1 Overcoming Solubility and Stability Challenges

In particular, the solubility and stability issues associated with fat-soluble vitamins A, D, E, and K, essential fatty acids such as omega-3s. Natural foods contain these nutrients in their primitive forms, which are not readily water-soluble and hence impossible for the body to absorb easily. The specifics of nanotechnology break these barriers by down-sizing the vitamin particles to the nanoscopic level, increasing surface area, and thereby bioavailability [53].

One of the delivery strategies for fat-soluble nutrient compounds to increase their solubility and absorption is nanoemulsions, stable dispersions of nanoparticles. For example, nanoemulsions of vitamin D allow for better absorption of this nutrient—as it is critical to athletes for its role in ensuring healthy bones, optimizing the performance of the immune system and enhancing muscle performance. Likewise, omega-3 fatty acids, which have well-established anti-inflammatory properties and protect cardiovascular health, are absorbed more efficiently when administered as a nanoemulsion.

Moreover, nanotechnology can be used to encapsulate and protect nutrients that are truly sensitive to degradation by heat, light, or exposure to stomach acids from these degrading agents. This barrier guarantees that by the time they arrive at the intestines, where absorption is most efficient, nutrients are preserved and bioavailable [54].

3.2.2 Enhancing Bioavailability of Key Nutrients

Nanotechnology can help address one of the biggest challenges faced by micronutrients: bioavailability. For instance, iron is an essential mineral for transporting oxygen and producing energy, but is poorly absorbed in its conventional state. And this poses a particular challenge for athletes, who require more iron because of the demands of physical activity. Iron nanoparticles coated with polymers are engineered to circumvent the acidity of the stomach, delivering iron directly to the small intestine, where it is absorbed more effectively.

Magnesium is crucial, for muscle function as it helps with both contraction and relaxation. Can sometimes be inefficiently absorbed by the body when taken in supplement form. This issue can be mitigated by utilizing nanocarriers to enhance the delivery of magnesium to muscle cells where it contributes to preventing cramps and aiding in recovery. The targeted approach of delivering magnesium in this manner also reduces the likelihood of experiencing problems associated with regular magnesium supplements, like diarrhea or bloating [55].

3.2.3 Improved Protein Absorption for Muscle Growth

Protein is essential for muscle growth and repair, particularly after intense physical activity. Athletes often rely on protein powders, amino acids, or branched-chain amino acids (BCAAs) to support muscle recovery. However, traditional protein supplements can be subject to digestive inefficiencies, limiting their effectiveness.

Nanotechnology addresses this issue by improving the absorption of proteins and peptides. Through nanoencapsulation, proteins are protected from degradation during digestion, ensuring that they reach the bloodstream intact. This allows for more efficient muscle protein synthesis, leading to better muscle recovery and growth. Furthermore, nanoparticles can be engineered to release amino acids directly into muscle tissue, promoting faster recovery after exercise and reducing soreness and fatigue [56].

3.3 Enhanced Athletic Performance and Recovery

Nanotechnology directly impacts athletic performance by enhancing energy availability, muscle recovery, and overall endurance. By improving the effectiveness of nutrient delivery, nanotechnology ensures that athletes can perform at their best for longer periods, while also accelerating recovery post-exercise [57].

3.3.1 Optimizing Endurance with Controlled Energy Release

Endurance athletes, such as long-distance runners, cyclists, and swimmers, require a consistent and sustained energy supply throughout their events. Traditional carbohydrate supplements often lead to energy spikes and crashes due to their rapid absorption, which can hinder performance. Nanotechnology addresses this issue by providing controlled-release energy supplements.

Glucose-loaded nanoparticles can gradually release glucose into the bloodstream, maintaining steady energy levels during prolonged physical activity. This ensures that athletes do not experience the rapid energy fluctuations commonly associated with traditional carbohydrate supplements. Similarly, nanoencapsulated creatine helps improve high-intensity performance by ensuring a steady supply of energy to muscles, allowing athletes to perform optimally for extended periods [58].

3.3.2 Speeding Up Post-exercise Recovery with Nanoencapsulated Amino Acids

Post-exercise recovery is a critical phase for athletes, as it involves repairing muscle fibers that have been damaged during intense physical activity. Protein and amino acids play a vital role in muscle recovery, but their absorption can be hindered by digestive inefficiencies. Nanotechnology enhances the bioavailability of amino acids, ensuring that they are delivered directly to muscle tissues for faster recovery.

Nanoencapsulation of amino acids allows them to bypass the stomach and reach the small intestine intact, where they are absorbed and transported to muscle tissues. This accelerated nutrient delivery helps reduce recovery time and muscle soreness, enabling athletes to train harder and more frequently. Additionally, amino acids such as leucine, which stimulate muscle protein synthesis, are more effectively delivered to muscle cells using nanocarriers [59].

3.3.3 Antioxidants for Recovery and Reducing Inflammation

Intense physical activity generates oxidative stress, which can damage cells and slow down recovery. Antioxidants, such as curcumin, resveratrol, and quercetin, play a key role in reducing oxidative stress and inflammation. However, these antioxidants often suffer from poor bioavailability, limiting their effectiveness.

Nanotechnology enhances the absorption and stability of antioxidants, making them more effective in reducing inflammation and promoting faster recovery. For example, nanoencapsulated curcumin has been shown to have superior anti-inflammatory properties compared to traditional curcumin supplements. This helps reduce muscle inflammation and accelerates the recovery process after intense training sessions or competitions [60].

3.4 Reduction in Gastrointestinal Issues Associated with Conventional Supplements

Many athletes experience gastrointestinal discomfort when using traditional supplements, particularly in high doses. These side effects can include bloating, nausea, and diarrhea, which may negatively impact performance and overall health. Nanotechnology offers solutions that enhance nutrient delivery while minimizing gastrointestinal issues [61].

3.4.1 Addressing Digestive Discomfort and Nausea

One of the main reasons athletes experience digestive discomfort when taking supplements is the need for high dosages of certain nutrients. By improving the bioavailability and absorption of nutrients, nanotechnology reduces the need for large doses, mitigating common side effects like bloating and nausea.

For example, zinc oxide nanoparticles provide sufficient zinc levels without causing gastrointestinal disturbances, such as nausea, which are often associated with conventional zinc supplements. Similarly, nanoencapsulation techniques allow nutrients to be delivered more gently to the digestive system, reducing the risk of irritation [62].

3.4.2 Nanoemulsions for Improved Digestive Compatibility

Nanoemulsions are particularly beneficial for improving the digestive compatibility of sports supplements. By increasing the surface area of fat-soluble nutrients, nanoemulsions enhance their solubility, making them easier for the body to absorb. This reduces the need for high doses of these nutrients and minimizes gastrointestinal discomfort.

For instance, omega-3 fatty acids are often associated with digestive issues when taken in large doses. However, when delivered through nanoemulsions, omega-3s are absorbed more efficiently, allowing athletes to receive the benefits without the discomfort [63].

3.5 Personalization of Nutritional Needs Using Nanotechnology

One of the most exciting aspects of nanotechnology in sports nutrition is its ability to personalize nutritional plans based on individual needs. Athletes have unique metabolic profiles, and their nutritional requirements can vary greatly depending on their sport, activity level, and genetics. Nanotechnology offers the potential to develop highly personalized supplements that meet these specific needs, improving overall effectiveness and efficiency [64].

3.5.1 Tailoring Supplements to Individual Metabolic Profiles

Nanotechnology enables the creation of customized nutritional solutions based on an athlete's metabolic profile. For example, athletes with higher energy expenditure or specific nutritional deficiencies can benefit from tailored supplement regimens that provide optimal doses of vitamins, minerals, and macronutrients.

By using nanotechnology, sports nutritionists can create supplements that are personalized not only to an athlete's metabolic rate but also to their unique responses to exercise. For instance, an athlete who struggles with muscle recovery may benefit from a protein-based nanoparticle supplement designed to improve amino acid delivery to muscles [65].

3.5.2 Precision Dosage for Different Types of Athletes

Nanotechnology also allows for the precise dosage of nutrients based on the type of sport or training regimen. Endurance athletes, powerlifters, and recreational athletes all have unique nutritional needs. Nanotechnology can deliver nutrients in the exact amounts and forms that each athlete requires, ensuring that they receive the most effective supplementation without wasting resources.

For example, powerlifters may require higher doses of creatine and protein to support muscle strength, while endurance athletes need sustained carbohydrate release for prolonged energy. Nanotechnology makes it possible to deliver these nutrients in precise amounts, ensuring that athletes can perform at their best [66].

3.6 Enhanced Safety and Reduced Risk of Side Effects

Another benefit of nanotechnology in personalized sports nutrition is its ability to reduce the risk of side effects associated with traditional supplements. By improving the bioavailability of nutrients, nanotechnology allows athletes to achieve the same or better results with lower doses, reducing the likelihood of adverse reactions.

3.6.1 Minimizing Toxicity and Overdose Risk

Nanotechnology's controlled-release mechanisms ensure that nutrients are delivered in optimal doses, minimizing the risk of toxicity. For instance, by using nanoparticles to release vitamin A slowly over time, the risk of vitamin A toxicity can be reduced, even when athletes require higher doses for performance benefits.

3.6.2 Improved Bioavailability with Reduced Side Effects

Nanoparticles also enhance the bioavailability of nutrients, ensuring that they are absorbed effectively by the body without causing digestive issues or other side effects. This is particularly important for athletes who may be sensitive to certain supplements. By improving nutrient absorption while minimizing side effects, nanotechnology can help athletes optimize their supplementation without negative consequences [67].

3.7 Novel Applications in Personalized Sports Nutrition

3.7.1 Smart Nutritional Systems

The integration of nanotechnology with wearable devices offers exciting new possibilities for personalized nutrition. Wearables that track an athlete's physiological data—such as glucose levels, hydration status, and heart rate—can be combined with nanotechnology to adjust nutrient delivery in real-time. For example, wearable devices could trigger the release of glucose or electrolytes from nanoparticles when an athlete's blood sugar or hydration levels drop [68].

3.7.2 Combining Nanotech with Other Cutting-Edge Technologies

In the future, nanotechnology may be combined with other advanced technologies, such as genetic testing and artificial intelligence, to create ultra-personalized nutrition plans. By analyzing an athlete's genetic makeup, activity levels, and performance data, AI-powered systems could recommend and adjust supplements in real time, ensuring that athletes always have the nutrients they need to perform at their best [69].

3.8 Environmental Impact and Sustainability of Nanotechnology in Sports Nutrition

3.8.1 Reducing Supplement Waste

Nanotechnology offers significant sustainability advantages in sports nutrition. By improving the efficiency and bioavailability of nutrients, nanotechnology reduces the need for large doses of supplements, thereby minimizing waste. This reduction in waste not only benefits athletes but also has positive implications for the environment, as it reduces the production of unnecessary packaging and materials [70].

Table 2 Comparison of conventional supplements vs. nanotechnology-enhanced nutrients

Nutrient	Challenges with conventional form	Nanotechnology solution
Fat-soluble vitamins	Poor solubility and absorption	Nanoemulsions for improved solubility and absorption
Proteins and peptides	Rapid digestion and incomplete absorption	Nanoencapsulation for targeted release
Iron	Low bioavailability and gastrointestinal side effects	Coated nanoparticles for increased absorption and minimal side effects
Curcumin (Antioxidant)	Low bioavailability and poor absorption	Nanoencapsulation to improve absorption and anti-inflammatory properties
Magnesium	Low solubility and poor bioavailability	Nano-formulation to improve absorption and reduce gastrointestinal irritation

3.8.2 Eco-friendly Nanomaterials in Sports Supplements

Researchers are also working on developing environmentally friendly nanomaterials, such as biodegradable nanoparticles, that can further reduce the environmental footprint of sports nutrition products. These innovations help ensure that the benefits of nanotechnology are balanced with sustainability, allowing athletes to benefit from personalized nutrition without contributing to environmental harm [71]. Table 2 shows Comparison of Conventional Supplements versus Nanotechnology-Enhanced Nutrients.

4 Safety and Toxicity of Nanomaterials in Human Consumption

Nanomaterials are defined as the materials with at least one dimension between 1 and 100 nm. These materials exhibit unique physical, chemical and biological properties but distinct from their bulk counterparts. These unique properties arise from their small size, high surface area to volume ratio and quantum effects. So that, nano materials are utilized to improve packaging such as moisture, gases, mechanical strength and antimicrobial activity. Generally, silver nanoparticles are commonly used for anti-microbial properties which can inhibit the growth of microorganism such as fungi, bacteria on food surfaces. These nanoparticles reduced spoilage and enhancing food safety. Titanium dioxide nanoparticles are good for UV protection and light absorption. It helps to maintain the colour and nutrient content of packaged foods but zinc oxide nanoparticles having more anti-microbial effects [72].

Table 3 Toxic and health effects of nanomaterials

Titanium dioxide	Effects on liver, spleen, lungs
Zinc oxide	Toxicity associated with oxidative stress
Gold	Gold nanoparticles decreased cell proliferation rate, adhesion and motility of human dermal fibroblasts
Cerium oxide	Cytotoxic and genotoxic in human dermal fibroblasts

Generally, nanomaterials are considered as good food packaging materials due to their potential to address critical challenges in the food industry such as reducing foodborne illness, extending shelf life and minimizing food waste. However, nanomaterials interact with food and surrounding environment which have emerged adverse effects on human health and ecosystem. The main drawback is migration of nano materials from packaging materials into food products. The migration of nano materials involves the transfer of nanoparticles from the packaging matrix to the food. Various factors are involved in migration of nano materials such as temperature, P^H, mechanical stress during storage and transportation. The toxicological effects of nanomaterials used in food packaging to understand their interactions with biological systems (Table 3) [73].

Nanomaterials can exhibit toxicity such as genotoxicity, cytotoxicity through various mechanisms such as oxidative stress, inflammation, genotoxicity, disruption of cellular function. NM's can induce oxidative damage to cells and DNA which alter cellular signalling pathways and compromise cellular integrity. *In-vivo* studies provided insights into the systemic distribution of NM's ingestion or inhalation which showed potential targets for toxicity such as liver, kidney and GIT tract. Moreover, the environmental factors impact of NM's released during production, use and disposal of food packaging materials. These nanomaterials can accumulate in soil, water bodies, air which effect on ecosystems and potentially entering in the food chain through agricultural products and aquatic organisms. Significantly, nanomaterials having various toxicity mechanisms such as oxidative stress, inflammation, DNA damage, disruption of cellular processes etc. For the safety assessment methods, included in-vitro testing method, in-vivo testing method, computational modelling etc. [74].

4.1 Regulatory Frameworks and Approval for Nanotech Based Products

In 2003, the USA enacted the twenty-first century Nanotechnology Research and Development act of 2003 to establish a National Program (NNP) for funding for nanotechnology R&D over 4 years. It provided authorisations under 6 of the act for a subset of the NNI. On the basis of the highest multi-agency R&D priorities namely as NSF, Department of Energy (DOE), NASA, National Institute of Standards and Technology (NIST) and the Environment Protection Agency (EPA). NNI is a collaborative, multi—agency, cross cut programme among 26 federal entities with a range of research, industry, trade, educational and regulatory roles and responsibilities. However, the Toxic Substances Control Act, Occupational Safety and Health Act, Food, Drug and Cosmetic Act but the major environmental laws such as clean Air Act and Resource Conservation and Recovery Act provide some legal basis for reviewing and regulating nano materials [75].

Regulatory frameworks involved in nano-technology based products:

- **FDA (US)**: regulates nano-technology based products in food, cosmetics, pharmaceuticals, medical devices etc.
- **EFSA (EU)**: evaluates and detects the safety of nano-technology based food products.
- **ISO**: develops standards for nanotechnology safety assessment and product development [76].

The organisation for Economic Co-operation and Development (OECD) refers to nanotechnology as a set of technologies which enables the manipulation, study or exploitation of very small structures and systems. Nanotechnology applications have been reported across a number of specific product areas including foods where enhanced protection by improved food packaging materials or improved delivery of a functional ingredient or a nutrient in food and medical products which enhanced bioavailability etc. In India, Department of Science and Technology constituted a task force for the regulatory framework for nanotechnology. Council of Scientific and Industrial Research (CSIR) initiated a major project "Nano-SHE" in 12th 5 year plan to evaluate and collect a database on various toxicological aspects of nanostructured materials. Whereas, guidelines are developed safe work practices in laboratories dealing with nanomaterials complied and circulated as guidelines by centre for knowledge management of nanoscience and technology (CKMNT) [77]. In India, regulatory matrix has been developed and categorised under the following tools:

- Environmental safety
- Information disclosure
- Occupational safety
- Research and development
- Hazardous chemicals
- Waste disposal
- Product specific

4.2 *Ethical Concerns in Data Collection for Personalized Nutrition*

Ethics is the systematic reflection on the moral aspects of life and health and also it's conflict. Food means much more than nutrition, food is enjoyment as well as cultural and personal identity. So that, food is an important aspect of human happiness and well-being but we cannot consider as instrument for health [78]. Researcher's are discussed about whether personalized nutrition contribute to a good life or not? (Table 4).

Generally, personalized nutrition is based on genetic information. SO, genetically modified food was the result of an effort to apply genetic knowledge and gene technology for improvement of food. Generally, personalized nutrition refers to tailored nutritional recommendations aimed at the promotion or maintenance of health and prevention against diseases. Personalized nutrition approaches that potential to spearhead the information processing representations such as digestion, absorption, metabolism etc. (Fig. 2).

Table 4 Personalized nutrition

Advantages	Limitations
Individualized approach	Higher costs
Potential for improved health outcomes	Complexity, limited availability
Enhanced adherence	Evolving science
Integration of advanced technologies	Potential privacy concerns

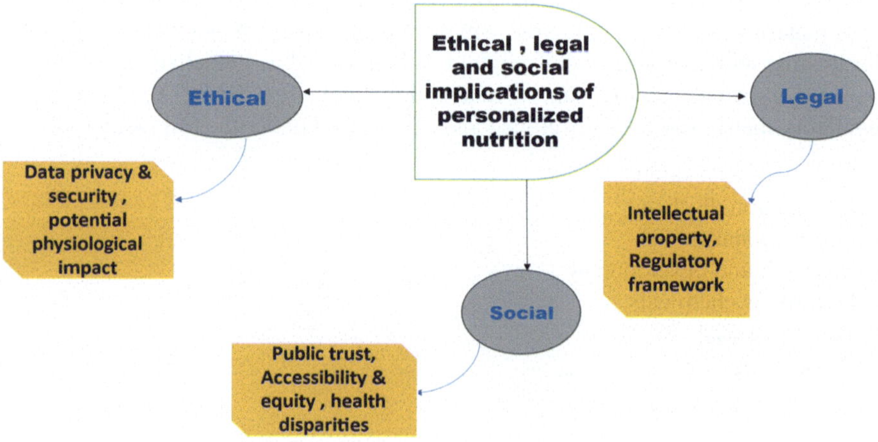

Fig. 2 Ethical, legal and social implication of personalized nutrition

Personalized nutrition approaches that potential to spearhead the information processing representations such as digestion, absorption, metabolism etc. Theseinteraction provide a linkage between molecular events and health outcomes through the integration of data, combination of multiscale models with health outcomes through advanced machine learning models, generation of non-intuitive hypotheses etc. Personalized nutrition and health research can be beneficial from building research infra structures. The development of personalized nutrition user friendly platforms that can be enable to interrogation of data from various resources, multiple studies, research groups at different level comprised of molecular, cellular, tissue, organ and population level etc. For data collection of personalized nutrition, it should provide clear explanation of how data is used to generate personalized recommendations, preventing the exploitation of personal data for commercial gain without individuals' knowledge or consent, ensuring transparent data and decisions etc.

Mitigating ethical concerns; establish clear guidelines and regulations for data collection and use implement robust data protection, ensure transparency and explainability in data collection and use practices etc. [79].

4.3 Public Perception and Acceptance of Nano Technology

Public perception and acceptance of nanotechnology are crucial for it's successful development and implementation. Various factors are involved in public perception, such as—

- Lack of understanding
- Media coverage
- Risk perception
- Benefits and applications

Over the course of the last decades, the number of consumer goods containing nano materials has been rapidly increasing. Whereas, silver, titanium, carbon are the most common materials used skin care, electronics, textiles and sports equipment are prominent consumer good areas in which nano technology has been successfully applied to improve modern lifestyle products. Recently, nano technology has also gained traction in the medical field with promising outlooks on improved drug delivery and new therapeutic approaches. Public perceptions of emergent technologies have become increasingly important tools to understand in part due to the worldwide backlash against genetically modified foods which effectively stalled a new industry.

Even with the widespread use of nanomaterials in everyday life, consumers knowledge about the functionality, benefits and possible dangers of nanotechnology is still modest [14]. As with any developing technology, it's public perception has direct implications on future policies and has to be taken into the academia and industry alike. Consequently, a comprehensive understanding of the interactions between society and technology is required. The value of gaining insight into opinions voiced by the general public and their understanding of technologies is reflected by the growing number of agencies engaged in opinion research. While, nano technology is no longer in it's infancy and various applications have been well established in different markets such as sport equipment's, textiles and cosmetics etc. [80].

Some strategies for improving public acceptance in nanotechnology:

- **Education and outreach**: clear, concise and proper communication about nanotechnology's benefits and risks.
- **Transparency and regulation**: Proper regulatory frameworks and open communication about safety assessments.
- **Public engagement**: Involvement of the public in decision making processes and encouraging feedbacks.
- **Responsible innovation**: Encouraging industry to adopt responsible innovation practices, prioritizing safety and sustainability.

To investigate the current state of nano technology related knowledge and public opinion, the Nan-O-Style project team launched on online survey. The survey promoted through a multi-channel approach which mainly media outlets and scientific events and a congress open to the general public [81].

5 Future Directions and Innovations

The combination of artificial intelligence (AI) with nanotechnology has become a frontier with revolutionary potential in the quickly changing world of modern technology and innovation [1]. Together, these two fields—each revolutionary in its own right—have the potential to expand human potential and make possible innovations that were previously only found in science fiction. The drive for scientific and technical innovation has undergone a tremendous development with the convergence of AI and nanotechnology. Understanding the history of both AI and nanotechnology and how each has influenced the modern world separately is essential to appreciating the significance of this convergence [82].

Information technology, a variety of sensors, nanotechnology, and the introduction of computers and smartphones have all been used to continuously monitor human nutrition and physical activity. The use of AI is promising, but it necessitates quick and accurate analysis of a large number of variables produced during monitoring. AI can learn both dependent and independent variables' linear and nonlinear correlations, creating a nonlinear input–output mapping that reveals and interprets previously hidden but incredibly helpful information for decision-making [83].

Even if artificial intelligence (AI) is not yet widely applied in the fields of fitness and nutrition, it is discovered that the technology currently in use is conducive to its application because these technologies gather a lot of data, which AI might be very helpful in mining. Answering the question of whether fitness and nutrition-related decision support and monitoring systems can be continuously trained and learned throughout their lifetime without losing their generality is one of the biggest challenges facing new technologies in this area. The inherent system's capacity to learn continuously throughout time by assimilating new information while holding onto previously acquired experiences serves as evidence for this concept [84].

Wearable technology holds great promise for non-invasive human health monitoring and tailored medicine. Devices that detect physical activity, like heart rate and body movement, were the main focus of early development. Real-time chemical and biological detection of bodily fluids, including sweat, interstitial fluids, saliva, breath, and blood, is now possible using next-generation sensors. These sensors are capable of identifying biomarkers linked to underlying medical disorders, such as the presence of glucose in different body fluids for the purpose of monitoring diabetes. Therefore, it is anticipated that wearable sensors will play a significant role in the transition to digital health and personalized care.

Together with the monitoring software, a wearable flexible headband device was created to track a person's health, particularly their level of drowsiness. The system distinguishes between alertness and drowsiness because of the way it was designed. The electronics module and the textile electrode continuously gather signals, which are then sent to the monitoring station via the Bluetooth module. In order to determine alertness and drowsiness, the monitoring station's software decodes the signal into useful information. To estimate the blink rates and AR-PSD of the EOG and EEG signals, respectively, a number of preliminary tests were conducted. In particular, a blink distinction was used to derive a correlation between EEG and EOG. The EOG signal is used to extract closed eyes, open eyes, normal blinks, and slow blinks. The decision and threshold criteria to measure drowsiness were then established based on these preliminary studies. The software's final display function used the derived decision criteria to show the discriminated factor. Thus, it is possible to distinguish between alertness and drowsiness with effectiveness. By comparison with other tests that measure drowsiness, the test results are validated and proven. Developing a wearable system, establishing a correlation between EEG and EOG, identifying sleepiness and wakefulness, validating and qualifying the system against current methods, and creating parametric decision criteria based on EEG/EOG are some of the main obstacles encountered when creating a real-time monitoring system to identify alertness and drowsiness [85].

Human nutrition is greatly impacted by nanotechnology, which presents a plethora of chances to enhance food safety and quality. Food scientists have been working hard in recent years to create novel food items with improved functional qualities. Making nanocolloids for use in food applications, for instance, has become a viable substitute for encapsulating nutrients and bioactive substances, enhancing their bioavailability and absorption in the digestive tract. Proteins, polysaccharides, or lipid compounds—which can be made up of a single molecular species or a mix of many molecular components—are commonly used to create these nanostructures [86].

Nonetheless, the use of altered nanoparticles in food products undoubtedly increases the exposure of humans to oral nanoparticles. One of the main ways that people are exposed to nanoparticles is by eating food that has been improved with them or by ingesting nanoparticles that have moved from packaging. Ingestion of nanoparticles has been associated with several biological effects, including DNA damage, oxidative stress response stimulation, and protein denaturation. Numerous studies have connected gastrointestinal and secondary organ damage to oral exposure to nanomaterials, especially solid nanoparticles. This highlights the importance of closely monitoring the use of nanoparticles in food. Therefore, before proper monitoring of nanoparticles in food can be implemented, it is imperative to ascertain the level of exposure and develop effective methods to examine the toxicity and food safety of nanoparticles [87].

Few studies have examined the potential of nanocompounds to enhance exercise performance. Resveratrol, iron, nitroxide radicals, and folic acid were the active ingredients in the nanocomposites used in these investigations [88]. Table 5 shows Summary of nutraceutical foods, including nanoencapsulated bioactive components

Nanotechnology holds promise for human civilizations since it can fulfill both basic needs and the desire for a comfortable life. Human existence requires a variety of essentials, including food, clean water, energy, clothing, shelter, health, and a healthy environment. The aspirations for a comfortable life include the implementation of automation in all professions, space exploration, longer lifespans, and other objectives. Over the past 30 years, there has been a significant advancement in the realization of nanotechnology-based components and products in a variety of

Table 5 Summary of nutraceutical foods, including nanoencapsulated bioactive components

Title of the item	Bioactive compounds	Heath benefits
Creatine nano	Purified creatine nanomonohydrate	Increases muscle strength
IOGO®	Vitamin D and calcium	Accomodating for developed bones
B-B-Red label tea	Ursolic acid, glycyrrhetinic acid, 6-gingerol, and withaferin A	Promote vitality
7XORIGINAL 4XVEGGI HOT MEALS IN A NANO™ COMBI	Vitamins, minerals, macro nutrients and micronutrients	Wellbeing promotion
Ingredient in NUTRASORB™ PROCES	Gingerols, polyphenols, and healthy alkaloids	Wellbeing promotion
NanOmega®	Folic acid, vitamins B6, B12, E, and D3	Crucial for heart, brain, and eye health
Nanomineral by NanoGreens	Each parcel contains twelve distinctive minerals	Helps in weight misfortune and boosts muscle growth
NutriNocheMagnessium supplement 30 PPM nano magnesium	Minerals and vitamins	Fundamental for the heart, muscles, and kidneys to work at their best
Tea pills from Nanolabs™ worldwide	Alkaloids, cannabinoids, terpenoids, and polyphenols (mostly flavonoids, stilbenes, and lignan-amides)	Increment digestion system and burns fat

fields, including agriculture, food product and packaging, water purification, automobiles, atmospheric cleaning, renewable energy, energy storage, consumer goods, sports equipment, cosmetics, cloth and fabrics, construction materials, shelter, health, information and communication technology, high speed manufacturing, space travel, military, sustainable environment, and computer [89].

The increasing integration of nanoscale science and engineering knowledge and nanosystems, nanotechnology will find extensive applications in sports, computing, manufacturing, and medicine in the next years, as well as in a better understanding and preservation of the natural world. The rapid development of nanotechnology on a global scale supports this assumption of its expected advancement in transforming society and the future of living beings. National governments and research institutes studying nanotechnology must, however, reconsider the sluggish pace of actual advancement in comparison to the anticipated roadmap and undertake more concerted efforts. In this study, we analyze the benefits, limitations, and disadvantages of nanotechnology, a predicted game-changing general-purpose technology of the twenty-first century [90].

The advantages of nanotechnology and individualized sports nutrition are listed in this paper, along with the problems and ethical issues that come with it. Nanoparticles are also used in supplement creation. According to assessments on the timeline of nanotechnology commercialization, there is a delay in commercializing nanotechnology research, which results in an increase in the duration. When examining the reason for the delay from a business perspective, it is critical to keep in mind that the successful commercialization of a specially targeted nanotechnology depends on a number of factors, including the status of the economy, market size, potential, and innovative products [91].

References

1. Stavitz J, Koc T (2024) Exploring the experiences and perspectives of division III athletes regarding personalized nutrition plans for improved performance-a qualitative investigation. Healthc (Basel, Switzerland) 12(9)
2. Nakajima M, Wang Z, Chaudhry Q, Park HJ, Juneja LR (2015) Nano-science-engineering-technology applications to food and nutrition. J Nutr Sci Vitaminol (Tokyo) 61(Suppl):S180–S182
3. Blue MN, Trexler ET, Hirsch KR, Smith-Ryan AE (2019) A profile of body composition, omega-3 and vitamin D in national football league players. J Sports Med Phys Fitness 59(1):87–93
4. Choi SJ, McClements DJ (2020) Nanoemulsions as delivery systems for lipophilic nutraceuticals: strategies for improving their formulation, stability, functionality and bioavailability. Food Sci Biotechnol 29(2):149–168
5. Lobo MA, Hall ML, Greenspan B, Rohloff P, Prosser LA, Smith BA (2019) Wearables for pediatric rehabilitation: how to optimally design and use products to meet the needs of users. Phys Ther 99(6):647–657
6. Surya Ulhas R, Ravindran R, Malaviya A, Priyadarshini A, Tiwari BK, Rajauria G (2023) A review of alternative proteins for vegan diets: sources, physico-chemical properties, nutritional equivalency, and consumer acceptance. Food Res Int 173(Pt 2):113479

7. Unnikrishnan Meenakshi D, Narde GK, Ahuja A, Al Balushi K, Francis AP, Khan SA (2024) Therapeutic applications of nanoformulated resveratrol and quercetin phytochemicals in colorectal cancer—an updated review. Pharmaceutics 16(6):761
8. Close GL, Sale C, Baar K, Bermon S (2019) Nutrition for the prevention and treatment of injuries in track and field athletes. Int J Sport Nutr Exerc Metab 29(2):189–197
9. Walshe A, Daly E, Ryan L (2024) A qualitative exploration of perceived challenges and opportunities in the implementation of injury prevention and management in amateur female sport. Front Sport Act Living 6:1430287
10. Evans R, McNamee M, Guy O (2017) Ethics, nanobiosensors and elite sport: the need for a new governance framework. Sci Eng Ethics 23(6):1487–1505
11. Lis DM, Kings D, Larson-Meyer DE (2019) Dietary practices adopted by track-and-field athletes: gluten-free, low FODMAP, vegetarian, and fasting. Int J Sport Nutr Exerc Metab 29(2):236–245
12. Khademi R, Mohammadi Z, Khademi R, Saghazadeh A, Rezaei N (2023) Nanotechnology-based diagnostics and therapeutics in acute lymphoblastic leukemia: a systematic review of preclinical studies. Nanoscale Adv 5(3):571–595
13. Amawi A, AlKasasbeh W, Jaradat M, Almasri A, Alobaidi S, Hammad AA et al (2023) Athletes' nutritional demands: a narrative review of nutritional requirements. Front Nutr 10:1331854
14. Jäger R, Mohr AE, Carpenter KC, Kerksick CM, Purpura M, Moussa A et al (2019) International society of sports nutrition position stand: probiotics. J Int Soc Sports Nutr 16(1):62
15. Siddiqui SA, Bahmid NA, Taha A, Abdel-Moneim AME, Shehata AM, Tan C et al (2022) Bioactive-loaded nanodelivery systems for the feed and drugs of livestock; purposes, techniques and applications. Adv Colloid Interface Sci 308:102772
16. Ngambenjawong C, Sylvestre M, Gustafson HH, Pineda JMB, Pun SH (2018) Reversibly switchable, pH-dependent peptide ligand binding via 3,5-diiodotyrosine substitutions. ACS Chem Biol 13(4):995–1002
17. Duhan JS, Kumar R, Kumar N, Kaur P, Nehra K, Duhan S (2017) Nanotechnology: the new perspective in precision agriculture. Biotechnol reports (Amsterdam, Netherlands) 15:11–23
18. Saini M, Trehan K, Thakur S, Modi A, Jain SK (2024) Advances in iron deficiency anaemia management: exploring novel drug delivery systems and future perspectives. Curr Drug Deliv
19. Liu Q, Huang H, Chen H, Lin J, Wang Q (2019) Food-grade nanoemulsions: preparation, stability and application in encapsulation of bioactive compounds. Molecules 24(23)
20. Liu Y, Qiao L, Zhang S, Wan G, Chen B, Zhou P et al (2018) Dual pH-responsive multifunctional nanoparticles for targeted treatment of breast cancer by combining immunotherapy and chemotherapy. Acta Biomater 66:310–324
21. Garland NT, Schmieder J, Johnson ZT, Hjort RG, Chen B, Andersen C et al (2023) Wearable flexible perspiration biosensors using laser-induced graphene and polymeric tape microfluidics. ACS Appl Mater Interfaces 15(32):38201–38213
22. Xie X, Zhang W, Abbaspourrad A, Ahn J, Bader A, Bose S et al (2017) Microfluidic fabrication of colloidal nanomaterials-encapsulated microcapsules for biomolecular sensing. Nano Lett 17(3):2015–2020
23. Ye G, Guo B (2019) Integration of micro sensors with mobile devices for monitoring vital signs of sleep apnea patients. Zhongguo Yi Liao Qi Xie Za Zhi 43(1):29–31
24. Rawson ES, Miles MP, Larson-Meyer DE (2018) Dietary supplements for health, adaptation, and recovery in athletes. Int J Sport Nutr Exerc Metab 28(2):188–199
25. Higgins MR, Izadi A, Kaviani M (2020) Antioxidants and exercise performance: with a focus on vitamin E and C supplementation. Int J Environ Res Public Health 17(22)
26. Seo BR, Payne CJ, McNamara SL, Freedman BR, Kwee BJ, Nam S et al (2021) Skeletal muscle regeneration with robotic actuation-mediated clearance of neutrophils. Sci Transl Med 13(614):eabe8868
27. Shen Y, Zhang C, Dai C, Zhang Y, Wang K, Gao Z et al (2024) Nutritional strategies for muscle atrophy: current evidence and underlying mechanisms. Mol Nutr Food Res 68(10):e2300347
28. Rashwan AK, Karim N, Xu Y, Hanafy NAN, Li B, Mehanni AHE et al (2023) An updated and comprehensive review on the potential health effects of curcumin-encapsulated micro/nanoparticles. Crit Rev Food Sci Nutr 63(29):9731–9751

29. Romanov V, McCullough J, Gale BK, Frost A (2019) A tunable microfluidic device enables cargo encapsulation by cell- or organelle-sized lipid vesicles comprising asymmetric lipid bilayers. Adv Biosyst 3(7)
30. Reifenberg P, Zimmer A (2024) Branched-chain amino acids: physico-chemical properties, industrial synthesis and role in signaling, metabolism and energy production. Amino Acids 56(1):51
31. Andrès E, Lorenzo-Villalba N, Terrade JE, Méndez-Bailon M (2024) Fat-soluble vitamins A, D, E, and K: review of the literature and points of interest for the clinician. J Clin Med 13(13)
32. Ma Y, Yu S, Mu D, Cheng J, Qiu L, Cheng X (2023) Liquid chromatography-tandem mass spectrometry in fat-soluble vitamin deficiency. Clin Chim Acta 548:117469
33. Pacheco-Torres J, Mukherjee N, Walko M, López-Larrubia P, Ballesteros P, Cerdan S et al (2015) Image guided drug release from pH-sensitive Ion channel-functionalized stealth liposomes into an in vivo glioblastoma model. Nanomedicine 11(6):1345–1354
34. Oryani MA, Nosrati S, Javid H, Mehri A, Hashemzadeh A, Karimi-Shahri M (2024) Targeted cancer treatment using folate-conjugated sponge-like ZIF-8 nanoparticles: a review. Naunyn Schmiedebergs Arch Pharmacol 397(3):1377–1404
35. Chen X, Wang L, Zhang J, Yan H, Wang S, Xiao J (2024) Controlled release of ceria and ferric oxide nanoparticles via collagen hydrogel for enhanced osteoarthritis therapy. Adv Healthc Mater e2401507
36. Sato T, Murakami Y (2022) Temperature-responsive polysaccharide microparticles containing nanoparticles: release of multiple cationic/anionic compounds. Mater (Basel, Switzerland) 15(13)
37. Weisany W, Yousefi S, Soufiani SP, Pashang D, McClements DJ, Ghasemlou M (2024) Mesoporous silica nanoparticles: a versatile platform for encapsulation and delivery of essential oils for food applications. Adv Colloid Interface Sci 325:103116
38. Li C, Zhang J, Zu YJ, Nie SF, Cao J, Wang Q et al (2015) Biocompatible and biodegradable nanoparticles for enhancement of anti-cancer activities of phytochemicals. Chin J Nat Med 13(9):641–652
39. Dini I (2022) Contribution of nanoscience research in antioxidants delivery used in nutricosmetic sector. Antioxidants (Basel, Switzerland) 11(3)
40. Altemimi AB, Farag HAM, Salih TH, Awlqadr FH, Al-Manhel AJA, Vieira IRS et al (2024) Application of nanoparticles in human nutrition: a review. Nutrients 16(5)
41. Gai Y, Wang E, Liu M, Xie L, Bai Y, Yang Y et al (2022) A self-powered wearable sensor for continuous wireless sweat monitoring. Small Methods 6(10):e2200653
42. Ok J, Park S, Jung YH, Kim TI (2024) Wearable and implantable cortisol-sensing electronics for stress monitoring. Adv Mater 36(1):e2211595
43. Liu Q, Wang C, Guo X, Du Q, Keshavarzi M (2024) Curcumin and its nano-formulations combined with exercise: from molecular mechanisms to clinic. Cell Biochem Funct 42(4):e4061
44. Domingues-Faria C, Vasson MP, Goncalves-Mendes N, Boirie Y, Walrand S (2016) Skeletal muscle regeneration and impact of aging and nutrition. Ageing Res Rev 26:22–36
45. Zhu J, Huang Q (2019) Nanoencapsulation of functional food ingredients. Adv Food Nutr Res 88:129–165
46. Sadr S, Lotfalizadeh N, Ghafouri SA, Delrobaei M, Komeili N, Hajjafari A (2023) Nanotechnology innovations for increasing the productivity of poultry and the prospective of nanobiosensors. Vet Med Sci 9(5):2118–2131
47. Li X, Zhang H, Mao X (2024) Liposomes delivery systems of functional substances for precision nutrition. Adv Food Nutr Res 112:257–300
48. Burke LM, Castell LM, Casa DJ, Close GL, Costa RJS, Desbrow B et al (2019) International association of athletics federations consensus statement 2019: nutrition for athletics. Int J Sport Nutr Exerc Metab 29(2):73–84
49. Mukhopadhyay D, Sano C, AlSawaftah N, El-Awady R, Husseini GA, Paul V (2021) Ultrasound-mediated cancer therapeutics delivery using micelles and liposomes: a review. Recent Pat Anticancer Drug Discov 16(4):498–520

50. Hoogkamer W, Snyder KL, Arellano CJ (2018) Modeling the benefits of cooperative drafting: is there an optimal strategy to facilitate a sub-2-hour marathon performance? Sports Med 48(12):2859–2867
51. Naderi A, de Oliveira EP, Ziegenfuss TN, Willems MT (2016) Timing, optimal dose and intake duration of dietary supplements with evidence-based use in sports nutrition. J Exerc Nutr Biochem 20(4):1–12
52. Yuan D, Qin L, Niu Z, Zhou F, Zhao M (2024) Maintained particulate integrity of soy protein nanoparticles during gastrointestinal digestion via genipin crosslinking enhancing stability and bioavailability of curcumin. Int J Biol Macromol 274(Pt 2):133213
53. McClements DJ, Öztürk B (2021) Utilization of nanotechnology to improve the handling, storage and biocompatibility of bioactive lipids in food applications. Foods (Basel, Switzerland) 10(2)
54. Chen F, Fan GQ, Zhang Z, Zhang R, Deng ZY, McClements DJ (2017) Encapsulation of omega-3 fatty acids in nanoemulsions and microgels: impact of delivery system type and protein addition on gastrointestinal fate. Food Res Int 100(Pt 1):387–395
55. Yamada M, Kuroda H, Shimoyama S, Ito R, Sugama Y, Sato K et al (2016) Efficacy of levo-carnitine for tyrosine kinase inhibitor-induced painful muscle cramps in patients with chronic myelogenous leukemia. Rinsho Ketsueki 57(4):489–491
56. Martinho D V, Nobari H, Faria A, Field A, Duarte D, Sarmento H (2022) Oral branched-chain amino acids supplementation in athletes: a systematic review. Nutrients 14(19)
57. Casazza GA, Tovar AP, Richardson CE, Cortez AN, Davis BA (2018) Energy availability, macronutrient intake, and nutritional supplementation for improving exercise performance in endurance athletes. Curr Sports Med Rep 17(6):215–223
58. Oosthuyse T, Florence GE, Correia A, Smyth C, Bosch AN (2021) Carbohydrate-restricted exercise with protein increases self-selected training intensity in female cyclists but not male runners and cyclists. J Strength Cond Res 35(6):1547–1558
59. Philpott J, Kern M, Hooshmand S, Carson I, Rayo V, North E et al (2023) Pistachios as a recovery food following downhill running exercise in recreational team-sport individuals. Eur J Sport Sci 23(12):2400–2410
60. Vaiserman A, Koliada A, Zayachkivska A, Lushchak O (2019) Nanodelivery of natural antioxidants: an anti-aging perspective. Front Bioeng Biotechnol 7:447
61. Pyne DB, West NP, Cox AJ, Cripps AW (2015) Probiotics supplementation for athletes—clinical and physiological effects. Eur J Sport Sci 15(1):63–72
62. Shaw KA, Zello GA, Rodgers CD, Warkentin TD, Baerwald AR, Chilibeck PD (2022) Benefits of a plant-based diet and considerations for the athlete. Eur J Appl Physiol 122(5):1163–1178
63. Cuenoud B, Rochat I, Gosoniu ML, Dupuis L, Berk E, Jaudszus A et al (2020) Monoacylglycerol form of omega-3s improves its bioavailability in humans compared to other forms. Nutrients 12(4)
64. Nikityuk DB, Kobelkova IV (2020) Sports nutrition as a model of maximum individualization and implementation of integrative medicine. Vopr Pitan 89(4):203–210
65. Edenfield KM (2020) Sports supplements: pearls and pitfalls. Prim Care 47(1):37–48
66. King A, Kwan K, Jukic I, Zinn C, Helms E (2024) Fueling for and recovering from resistance training: the periworkout nutrition practices of competitive powerlifters. Nutrition 122:112389
67. Hu S, Lin S, He X, Sun N (2023) Iron delivery systems for controlled release of iron and enhancement of iron absorption and bioavailability. Crit Rev Food Sci Nutr 63(29):10197–10216
68. Ghosh A, Nag S, Gomes A, Gosavi A, Ghule G, Kundu A et al (2022) Applications of smart material sensors and soft electronics in healthcare wearables for better user compliance. Micromachines 14(1)
69. Feretzakis G, Juliebø-Jones P, Tsaturyan A, Sener TE, Verykios VS, Karapiperis D et al (2024) Emerging trends in AI and radiomics for bladder, kidney, and prostate cancer: a critical review. Cancers (Basel) 16(4)
70. Abdelnour SA, Alagawany M, Hashem NM, Farag MR, Alghamdi ES, Hassan FU et al (2021) Nanominerals: fabrication methods, benefits and hazards, and their applications in ruminants with special reference to selenium and zinc nanoparticles. Anim Open Access J MDPI 11(7)

71. Close GL, Kasper AM, Walsh NP, Maughan RJ (2022) "Food first but not always food only": recommendations for using dietary supplements in sport. Int J Sport Nutr Exerc Metab 32(5):371–386
72. Suvarna V, Nair A, Mallya R, Khan T, Omri A (2022) Antimicrobial nanomaterials for food packaging. Antibiot (Basel, Switzerland) 11(6)
73. Baghi F, Gharsallaoui A, Dumas E, Ghnimi S (2022) Advancements in biodegradable active films for food packaging: effects of nano/microcapsule incorporation. Foods (Basel, Switzerland) 11(5)
74. Rolo D, Assunção R, Ventura C, Alvito P, Gonçalves L, Martins C et al (2022) Adverse outcome pathways associated with the ingestion of titanium dioxide nanoparticles-a systematic review. Nanomater (Basel, Switzerland) 12(19)
75. McPartland J, Shaffer RM, Fox MA, Nachman KE, Burke TA, Denison RA (2022) Charting a path forward: assessing the science of chemical risk evaluations under the toxic substances control act in the context of recent national academies recommendations. Environ Health Perspect 130(2):25003
76. Amenta V, Aschberger K, Arena M, Bouwmeester H, Botelho Moniz F, Brandhoff P et al (2015) Regulatory aspects of nanotechnology in the agri/feed/food sector in EU and non-EU countries. Regul Toxicol Pharmacol 73(1):463–476
77. Thiruvengadam M, Rajakumar G, Chung IM (2018) Nanotechnology: current uses and future applications in the food industry. 3 Biotech 8(1):74
78. Chaddha R, Agrawal G (2023) Ethics and morality. Indian J Orthop 57(11):1707–1713
79. Kolodziejczyk AA, Zheng D, Elinav E (2019) Diet-microbiota interactions and personalized nutrition. Nat Rev Microbiol 17(12):742–753
80. Parrella JA, Leggette HR, Lu P, Wingenbach G, Baker M, Murano E (2024) Nanofood insights: a survey of U.S. consumers' attitudes toward the use of nanotechnology in food processing. Appetite 201:107613
81. Hudson-Smith NV, Alvarez-Reyes W, Yao X, He J, Rodriguez RS, Mitchell S et al (2023) NanoAdventure: development of a text-based adventure game in English, Spanish, and Chinese for communicating about nanotechnology and the nanoscale. J Chem Educ 100(6):2269–2280
82. Lu L, Lu T, Tian C, Zhang X (2024) AI: bridging ancient wisdom and modern innovation in traditional Chinese medicine. JMIR Med Inform 12:e58491
83. Wassan S, Dongyan H, Suhail B, Jhanjhi NZ, Xiao G, Ahmed S et al (2024) Deep convolutional neural network and IoT technology for healthcare. Digit Health 10:20552076231220124
84. Adams RC, Rashidieh B (2020) Can computers conceive the complexity of cancer to cure it? Using artificial intelligence technology in cancer modelling and drug discovery. Math Biosci Eng 17(6):6515–6530
85. Chang WD, Cha HS, Kim K, Im CH (2016) Detection of eye blink artifacts from single prefrontal channel electroencephalogram. Comput Methods Programs Biomed 124:19–30
86. Saini R, Paserkar N, Varma A, Acharya DK, Upadhye V, Pathak D et al (2023) Emerging nanotechnological applications in preserving and improving the shelf life of food. Pharm Nanotechnol
87. Gaillet S, Rouanet JM (2015) Silver nanoparticles: their potential toxic effects after oral exposure and underlying mechanisms–a review. Food Chem Toxicol Int J Publ Br Ind Biol Res Assoc 77:58–63
88. Lima MR, Moreira BJ, Bertuzzi R, Lima-Silva AE (2024) Could nanotechnology improve exercise performance? Evidence from animal studies. Braz J Med Biol Res Rev Bras Pesqui medicas e Biol 57:e13360
89. Abou Elmaaty TM, Elsisi H, Elsayad G, Elhadad H, Plutino MR (2022) Recent advances in functionalization of cotton fabrics with nanotechnology. Polymers (Basel) 14(20)
90. Bhana S, Wang Y, Huang X (2015) Nanotechnology for enrichment and detection of circulating tumor cells. Nanomedicine (Lond) 10(12):1973–1990
91. Rambaran T, Schirhagl R (2022) Nanotechnology from lab to industry—a look at current trends. Nanoscale Adv 4(18):3664–3675

Nanoparticles in Fat Metabolism and Weight Management

Rajdip Goswami, Arunava Ghosh, Bhaskar Pal, Saikat Santra, Mahenderan Appukutty, and Ayan Chatterjee

Abstract Nanoparticles are an effective new strategy in the management of obesity and metabolic disorders by improving fat burning and controlling body weight. These nanosized particles allow the direct targeting of the therapeutic agents into the adipose tissues and other metabolic organs hence increasing the effectiveness of the therapeutic agents while minimizing the adverse effects on the systemic circulation. The thermogenesis induced by certain nanoparticles causes the activation of brown adipose tissue, leading to an increase in the amount of energy expended and the number of calories burned. Furthermore, the nanocarriers can be designed in a way that permits controlled release of the medication, thereby achieving prolonged therapeutic effects that enhance compliance in patients. They also enable modulating the action of enzymes, such as lipase, which could help reduce the uptake of fats, or promote lipolysis, which helps in breaking down fats. Additionally, the use of nanoparticles enhances the efficiency of insulin by introducing anti-inflammatory agents to the adipose tissue which is also a major problem in obesity and insulin resistance. Nano particle technology will also help in the control of food intake by enabling targeted delivery of hormones (for instance, leptin or GLP-1) to certain areas of the brain thus providing a more holistic approach to weight management.

R. Goswami
Department of Pharmaceutical Technology, School of Health and Medical Sciences, Adamas University, Barasat, Kolkata, West Bengal, India

A. Ghosh
Bengal School of Technology, Sugandha, Delhi Road, Hooghly, West Bengal, India

B. Pal
Department of Pharmacology, Charaktala College of Pharmacy, Charaktala, Mothabari, Malda, West Bengal, India

S. Santra
JRSET College of Pharmacy, Panchpota, Chakdah, Nadia, West Bengal, India

M. Appukutty
School of Sports Science and Recreation, Universiti Tecknologi MARA, Shah Alam, Malaysia

A. Chatterjee (✉)
School of Health Science and Technology, Medhavi Skills University, Topakhani, Sikkim, India
e-mail: ayan4189@yahoo.com

A. Chatterjee et al. (eds.), *Nanofuel: The Future of Sports Nutrition*, Smart Nanomaterials Technology, https://doi.org/10.1007/978-981-96-5471-0_17

The employment of nanotechnology in the amassing and burning down of fats proves worthy in treating medical conditions such as obesity. One of the reasons is the enhanced efficiency and safety of the treatment, all thanks to prevention or targeting the effects of the medication on certain organs. Further developments are needed in order to improve these applications addressing long-term effects as well as moral issues in real clinical practice.

Keywords Nanoparticles · Fat metabolism · Obesity management · Thermogenesis · Controlled drug release

1 Introduction

The increasing prevalence of obesity and its related metabolic diseases has become one of the public health greatest problems. Although the mechanisms of energy balance and adipose tissue have been greatly improved within the scientific literature, an application of such interventions as diet, physical activity, and pharmacotherapy to weight management proves ineffective in most cases. Hence, New solutions including nanotechnology, for instance, are being sought. Nanoparticles (NPs) have found many applications in medicine and nutrition science because of their small sizes, higher surface area to volume ratios and tunable surface properties among others [1]. Nanoparticles have a lot of potential to improve the targeted delivery and the overall outcomes of treatment in relation to achieving fat loss and controlling body-weight. The benefit is their capacity to focus on certain tissues or cells like fat or inflammatory cells residing in fat stores, which allows for the alteration of fat storage or utilization without too many off-target effects. Moreover, nanoparticles may be designed to release lipolytic agents, appetite control substances, anti-inflammatory agents and other bioactive materials in a predetermined and prolonged order of action, which goes beyond the scope of conventional drug delivery systems. Studies done recently have underscored the ability of nanoparticles to modulate processes such as adipogenesis, lipolysis, and thermogenesis. For example, lipid-core nanoparticles have been used to deliver polyunsaturated fatty acids and phytochemicals sourced from plants, which aid in the browning of white adipose tissues [2]. The latter is a process where weight gaining fats are converted to weight losing fats. Moreover, several doped metal, and polymeric nanoparticles are also explored in their ability to combat oxidative stress and inflammation, which are important in obesity-associated co-morbidities. This chapter delves into various roles nanoparticles can play in fat metabolism and weight control with a clear focus on mechanisms, efficacy and concerns in practice. It also describes the health problems and barriers to entry that must be overcome for nanology to be effectively used in medicine [3]. This new direction is emerging as a way to combine nanotechnology and metabolic biology which are two disciplines that go patiently hand in glove, providing solutions to metabolic diseases such as obesity which are becoming one of the critical health challenges of the current times and to which the world is seeking a heal [4].

2 Mechanism of Nanoparticles in Fat Metabolism

2.1 Targeted Delivery to Adipose Tissues

The adipose tissue most important in the energy storage system functions in metabolic regulation and endocrine elements. It is also the site where abnormalities related to adipose metabolism become evident in obesity and various metabolic diseases such as diabetes and heart diseases. The present advancements in nanotechnology have discovered unexplored routes for targeted delivery of therapeutic agents to the adipose tissues [5].

2.1.1 Nanoparticles: A Precision Tool for Adipose Tissue Targeting

Nanoparticles (NPs) are multifunctional nanocarriers that have been tailored to improve drug delivery through augmenting bioavailability, delaying circulation time, and enabling site-specific targeting. Specifically for their targeting of adipose tissue, nanoparticles can take advantage of some of the unique physiological and molecular characteristics of their fat deposits [6].

Enhanced Permeability and Retention (EPR) Effect

During inflammation and disease, adipose tissues develop increased vascular permeability, which allows nanoparticles to exploit this phenomenon for passive accumulation in adipose tissues. This action-mode would enable preferential localization of the delivered therapeutic agents in the adipose regions while sparing the organs, due to the reduced off-target effects. Nanoparticles can gain entry into adipose tissues due to the phenomenon, which ensures enhanced vascular permeability of the adipose tissues in inflammatory and diseased states. The EPR effect does ensure that therapeutic agents preferentially localize to the adipose regions, minimizing off-target effects [1, 7].

Surface Functionalization for Active Targeting

Yet another type of precise transfer would be achieved by targeting the nanoparticle from the surface with ligands, such as an antibody or peptide, or aptamer that is able to recognize specific markers on adipocyte, or even on adipose vasculature. An example of this is the peptides directed toward adipocyte-specific receptors such as prohibitin or integrins on endothelial cells, which could deliver the drug more precisely [8–12].

Lipophilic and Biocompatible Nanomaterials

Lipophilic nanoparticles are made from liposomes or polymeric micelles, which can target sites in adipose tissue efficiently due to the lipid-rich deposits constituted in fat deposits. Biodegradable polymers like PLGA (poly(lactic-co-glycolic acid)) thereby also ensure that a sustained release occurs yet with less toxicity [13].

2.1.2 Mechanisms in Fat Metabolism Modulation

Nanoparticles interact with adipose tissues to influence fat metabolism through various mechanisms:

Browning of White Adipose Tissue (WAT)

Some nanoparticles can actually deliver agents like uncoupling proteins 1 (UCP1) inducers to white fat tissue for browning. In other words, it promotes turning white adipose tissue into brown-like adipose tissue. “Browning,” or the actual process that turns white into brown fat, promotes thermogenesis, like elevating energy expenditure and combating obesity [14].

Lipolysis Stimulation

Nanoparticles have the capability to deliver lipolytics directly to adipocytes, where they initiate the cleavage of triglycerides into free fatty acids and glycerol. Such compounds can induce signaling pathways of cyclic adenosine monophosphate (cAMP), hormone-sensitive lipase (HSL), and fat reduction [15].

Anti-Inflammatory Effects

High levels of inflammation in the adipose tissue can impair metabolic function and often lead insulin resistance. Target-specific nanoparticle delivery of anti-inflammatory drugs or gene silencing agents (e.g., siRNA) would be able to block the action of pro-inflammatory cytokines and restore metabolic equilibrium [16].

Angiogenesis Modulation

The growth of adipose tissue is intimately associated with the development of blood vessels. Nanoparticles that cause inhibition of blood vessel growth can restrict the nutrients flowing to the adipose tissues in expansion and thus result in a decrease in mass of fat depots. Instead, possibly, an induction of angiogenesis might help in remodelling of adipose tissues to more efficient metabolic function [17].

2.2 Interaction with Metabolic Organs

The NPs are emerging as excellent tools for regulating metabolic function, particularly fat metabolism. Their special physical and chemical properties: small size, large surface area, and functional versatility; enable them to interact profitably with metabolic organs, including liver, adipose tissue, skeletal muscle, and pancreas. It could potentially be used in the treatment of metabolic disorders such as obesity, diabetes, and fatty liver disease [18].

2.2.1 Liver Interaction

Liver acts a major metabolic hub in lipid metabolism by coordinating lipogenesis, fatty acid oxidation and lipoprotein synthesis. Nanoparticles are directed toward hepatic cells to modify these pathways, such as lipid-based NPs delivery of therapeutic agents including siRNA or small molecules to hepatocytes, disrupting some genes involved in lipid synthesis (e.g., SREBP-1c), or pathways into fatty acid oxidation pathways. Additionally, several NPs, including gold or silver, act as an oxidative stress and inflammatory modulator in hepatic tissues and improve conditions, such as non-alcoholic fatty liver disease (NAFLD) [19].

2.2.2 Adipose Tissue Interaction

The role of adipose tissue is majorly, the site of fat deposition in the human body, and also exerts a significant influence on the homeostasis of energy throughout the system. Nanoparticles have a way of inducing alteration in the functions of adipocytes with respect to the storing and mobilization of fat. Polymers such as NPs loaded with bioactive compounds in the form of curcumin or resveratrol promote lipolysis and minimize adipogenic effects. NPs formulated into brown adipose tissue (BAT) activated agents could also elicit thermogenesis, thus promoting energy expenditure with a decrease in fat mass. Surface modifications of nanoparticles with ligands targeting adipose tissue receptors such as CD36 or integrins would increase the targeting efficiencies of the nanoparticles while causing minimal off-target effects [20].

2.2.3 Skeletal Muscle Interaction

Skeletal muscle is an important site for fatty acid oxidation and energy use. Therefore, augmenting the fatty acid oxidation ability of skeletal muscle might greatly enhance lipid metabolism. Moreover, incorporation of nanoparticles could describe any mitochondrial-targeted agent such as antioxidants or oxidative enhancers, directly into the tissues of skeletal muscle cells and improve metabolic function and fatty acid utilization. On the other hand, NPs containing drugs such as metformin or AMPK activators stimulate metabolic pathways in skeletal muscle leading to energy expenditure and reduced systemic lipid levels [4, 21–26].

2.2.4 Pancreatic Interaction

The pancreas then intervenes here via the regulation of glucose and lipid uptake through its actions of creating insulin and glucagon. Insufficient insulin signalling increases lipid accumulation and adipogenesis. The nanoparticles, which target pancreatic beta cells, either improve insulin secretion or improve insulin sensitivity. For example, cerium oxide NPs diminish oxidative stress in pancreatic beta cells, protecting their function and metabolic balance [5, 27–32].

2.2.5 Cross-Talk Among Organs

The effects of nanoparticles on one organ can regulate the metabolism of the rest of the organs. For example, if a nanoparticle improves hepatic lipid metabolism, it can simultaneously reduce the fat load on adipose tissue and skeletal muscle. Enhanced insulin secretion from the pancreas may also translate into improved lipid handling across other metabolic tissues. With an intrinsic ability to modulate fat metabolism with key metabolic organs, nanoparticles are now considered a platform for such modulating mechanisms. These NPs would, thus, target lipid biosynthesis, storage, and oxidation, as well as offer treatment avenues for some metabolic diseases. In the future, nanoparticle biocompatibility, biodistribution, and long-term safety must be optimized to maximize these nanoparticles' therapeutic potential within the application intended [33–37].

Figure 1 illustrates the mechanism of nanoparticles in fat metabolism, highlighting their role in enhancing lipid breakdown, regulating energy balance, and promoting overall metabolic efficiency [38–40].

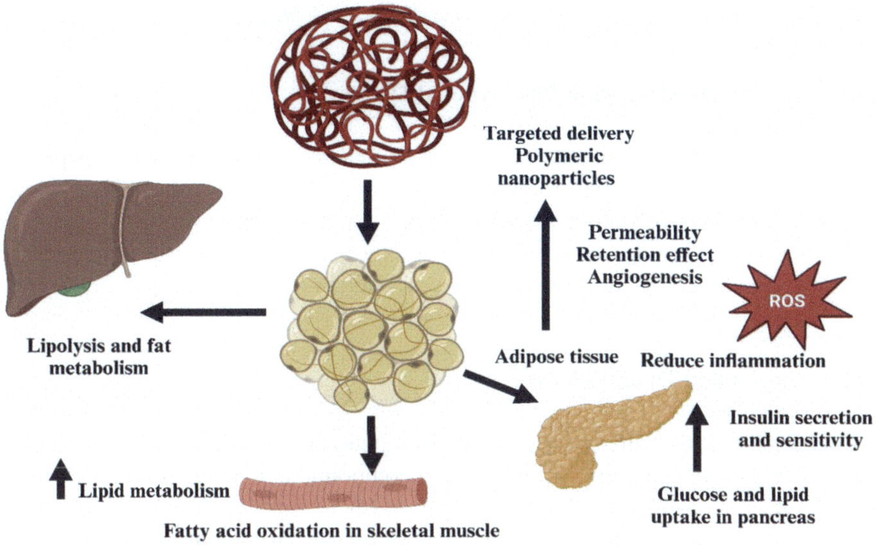

Fig. 1 Mechanism of nanoparticles in fat metabolism

2.3 *Minimizing Systemic Side Effects*

Nanoparticles have started being considered as possible agents or carriers for obesity and fat metabolism disorders treatments. However, the development of such nanoparticles without general systemic side effects is a very meticulous and careful work. This review addresses the optimization of nanocarriers for use in fat metabolism while alleviating adverse effects [41].

2.3.1 Mechanisms of Systemic Side Effects

The altered metabolic processing of fats in systemic tissues due to nanoparticulate side effects is also often associated with immune interaction patterns and biodistribution. Nanoparticles may, therefore, induce oxidative stress, inflammation, or cytotoxicity in healthy tissues. Accumulation in non-target organs: liver, kidneys, spleen, etc., also disrupts physiological functions. This long-term biodegradation and clearance produces chronic toxicity [42].

2.4 Strategies to Minimize Systemic Side Effects

2.4.1 Surface Modification for Targeted Delivery

Surface functionalisation of nanoparticles with ligands such as antibodies, peptides, or aptamers has the potential in improving their specificity towards adipose tissue, thereby decreasing off-target distribution and systemic exposure. Nanoparticles conjugated with peptides targeting the overexpressed receptors can deliver these payloads specifically to fat cells using these receptors [43].

2.4.2 Controlled Drug Release

With encapsulation of therapeutic agents into nanoparticles designed with mechanisms of release triggered by stimuli, active compound can be released only on the target tissue. An example includes use of pH-sensitive, enzyme-responsive nanoparticles for localized drug delivery to the acidic microenvironment of adipose tissue accompanied by preventing systemic leakage [44].

2.4.3 Biocompatible and Biodegradable Materials

By using biocompatible polymers, such as polylactic-co-glycolic acid (PLGA) or lipid-based carriers, toxicity and, respectively, promote safe clearance from the body. These materials degrade into non-threatening byproducts, greatly diminishing the long-term systemic risks [45].

2.4.4 Reducing Immune Activation

Now-a-days, coating nanoparticles with hydrophilic polymers such as polyethylene glycol (PEG) reduces immune system recognition, thus prolonging circulation time and decreasing inflammation. This stealth strategy ensures the effective delivery of the nanoparticles to a particular target tissue without inciting systemic immune activation [44].

2.4.5 Optimizing Dosage and Particle Size

Fine-tuning the size and dose of nanoparticles could lessen the potential for systemic accumulation. Smaller nanoparticles (< 100 nm) are more likely to stray from renal filtration during clearance, whereas larger particles may accumulate in the liver. Optimization of these parameters thus guarantees effective targeting while keeping low systemic retention. Nanoparticles hold tremendous hope of revolutionizing fat metabolism therapeutics; however, one of the most important points of advancement is to keep them from having systemic side effects. With adequate targeting strategies, biocompatible materials, and optimization of nanoparticle properties, the whole promise of therapeutic efficacy can effectively be utilized while ensuring patient safety [46].

3 Thermogenesis and Energy Expenditure

3.1 Activation of Brown Adipose Tissue (BAT)

The brown adipose tissue (BAT) is important in shivering-independent thermogenesis and energy expenditure, which are fundamental mechanisms for maintaining human metabolic homeostasis. While the white adipose tissue primarily serves the purpose of energy storage in the form of fat, in contrast, BAT consumes such stores actively and continually to burn calories and generate heat. Uncoupling protein 1 (UCP1) performs this activity: a protein found within the mitochondria, unique to BAT, that dissipates the heat generated by protons rather than the generation of ATP. Recent advancements within nanotechnologies have developed novel techniques for the manipulation of brown adipose tissues and enhancement of thermogenesis, signifying great promise in the treatment of obesity and metabolic disorders [47].

3.1.1 Mechanism of Thermogenesis in Brown Adipose Tissue

Thermogenesis in brown adipose tissue (BAT) is initiated by the activity of the sympathetic nervous system (SNS) following cold and other metabolic stimuli. Norepinephrine released from SNS terminals binds to β3-adrenergic receptors on BAT cells, resulting in a downstream cascade of intracellular signals, including cyclic AMP (cAMP) production and activation of protein kinase A (PKA). This cascade upregulates UCP1 expression and promotes the oxidation of fatty acids, producing heat. Besides the cold activation of brown fat, pharmacological and genetic interventions to stimulate brown fat thermogenesis have shown much interest because of their potential in managing an energy imbalance associated with obesity [29].

3.1.2 Nanoparticles in BAT Activation

Nanotechnology presents accurate and efficient delivery systems to the BAT appropriate for its thermogenic potential. Nanoparticles are designed to improve BAT activation, delivered bioactive compounds, genetic materials, or thermal sources directly into their adipose depots. These nanoparticles could be bio-disabling barriers, stable drugs, and controlled release, making them the best tool for BAT-targeted therapies [43].

3.1.3 Thermal Activation

Magnetic nanoparticles generate heat localization at an external magnetic field. This localized heating create conditions similar to those of cold exposure that can directly stimulate BAT thermogenesis. This strategy avoids systemic exposure to cold by providing controlled localized activation [48].

3.1.4 Drug Delivery

Thus, localized and sustained activation of BAT with minimal off-target effects could be achieved with NPs loaded with β3-adrenergic agonist or BAT-activators. This precision is accompanied by a reduction in systemic side effects often associated with pharmacological interventions [30, 35, 37, 48–50].

3.1.5 Gene Editing and RNA Delivery

Nanopapteral screaming genes like small interfering RNA or the CRISPR components have thereby delivered to rats exogenously to induce expression of UCP1 in the tissues, or modulate pathways functionally related to BAT thermogenesis. Hence, this is an upstream strategy for molecular enhancement of BAT function [25].

3.1.6 Nutraceuticals and Bioactive Compounds

Embraced into nanomedicines, polyphenols and other bioactives recognized for their thermogenic properties, are expected to enhance their bioavailability and target delivery to BAT. These compounds not only increase energy expenditure, but also help in improving metabolic outcomes [42].

3.1.7 Implications for Energy Expenditure and Metabolic Health

Nanoparticle-mediated strategies have consequently led to increased energy expenditure owing to enhancement in BAT activity, thus decreasing excess fat stores and improving glucose and lipid metabolism. Such strategies impart a significant effect in the management of obesity, type 2 diabetes, and associated metabolic disorders. Activation of BAT through nanotechnology also paves the way for personalized medicine as NPs can be designed to meet individual metabolic profiles. Nanotechnology can carry the concept of emerging techniques into the activation of BAT to improve thermogenesis and energy expenditure. Targeting and activating nanoparticles makes it possible to transfer the activation energy of brown adipocytes into the organization of thermogenesis by reaping the benefits of developing revolutionary treatments for the management of obesity and metabolic conditions. Further research and clinical translation are essential to realize fully the therapeutic potential from these technologies [11].

3.2 Thermogenic Effects of Nanoparticles

Thermogenesis, that is heat generation among organisms, plays a very crucial role in energy expenditure and subsequently maintaining metabolic balance. Nanotechnology over the last recent years has proved to be a breakthrough in unlocking the capabilities of nanoparticles (NPs) towards emerging applications in thermogenesis. This will especially prove to be useful in managing or potentially remedying the fatty metabolism. These nanomaterials are engineered to have unmatched physicochemical properties that actualize specific interactions with biological tissues, thus forming a new avenue for combating obesity or metabolic disorders [32].

3.2.1 Mechanisms of Nanoparticle-Induced Thermogenesis

Nanoparticles enhance thermogenesis through multiple mechanisms.

Thermal Conversion

Gold and iron oxide nanoparticles are quite unique in their ability to convert electromagnetic energy absorbed into heat on application of external stimuli: infrared light or magnetic fields. Their localized heating can, in turn, trigger thermogenic cells [8].

Drug Delivery

Nanoparticle delivery of thermogenic activators such as adrenergic agonists or mitochondrial uncouplers directly to brown adipose tissue or white adipose tissue undergoing browning minimizes systemic effects and maximizes efficacy [47].

Gene Modulation

Engineered nanoparticles can deliver unique nucleic acids, siRNA or components of CRISPR to modulate thermogenesis-related genes such as UCP1 in fat cells [51].

Inflammatory Pathway Modulation

A few nanoparticles, such as liposomes or polymer-based systems, can decrease chronic low-grade inflammation in adipose tissue and restore functioning to thermogenic optimum. Nanoparticles are promising agents for thermogenesis; however, biocompatibility, toxicity, and precise targeting hinder their development. Future studies should include the long-term effects of nanoparticle-based therapy on humans and their safety concerning clinical use. Developments in future nanoparticle engineering-particularly biodegradable systems and advanced targeting mechanisms-have the potential to revolutionize metabolic therapies. Nanotechnology, combined with other strategies such as diet and exercise, may yield new possibilities for treating obesity and improving metabolic health. In short, nanoparticles represent a state-of-the-art tool that can be used to take advantage of the body's natural thermogenic mechanisms-a new mechanism for regulating fat metabolism and energy expenditure. Continued innovation, should, however, lead to their redefinition towards the metabolic disease management landscape [43].

3.3 Implications for Weight Management

Growing incidence of obesity and metabolic conditions has resulted in evolving new strategies in the management of weight. One area that is catching up is nanotechnology for enabling thermogenesis and energy expenditure modulation in fat metabolism. This unique idea basically uses advanced material sciences to directly modulate the energy balance of the body by employing activation of brown adipose tissue (BAT), browning of white adipose tissue (WAT), and the metabolic pathways. Nanoparticle integration in fatty metabolism epitomizes a paradigm change in weight management. Targeting fat tissues specifically, such techniques may offer relatively safe and effective treatment options for obesity, minimizing the need for systemic drugs bearing extensive side effects. Modulation of energy expenditure at the cellular level creates yet more avenues for fighting obesity-associated metabolic disorders. Nanoparticle derived from great fence has challenges like biocompatibility issues, long-term toxicities and hurdles imposed through safety regulations. But as more research emerges, this should be part of the future building blocks in precision medicine against weight management, providing hope for more sustainable individual solutions to obesity treatment [29].

4 Controlled Drug Release

4.1 Design of Nanocarriers for Sustained Medication Release

Nanocarrier, which is new as a drug delivery system, offers an accurate means of designing a profile according to which therapeutic agents must be released. When it comes to fat metabolism, these systems are potentially used considerably for metabolic disorders like obesity, fatty liver disease, and dyslipidemia. Nanocarriers must be designed for sustained medication release in fat metabolism overcoming the challenges of drug stability, bioavailability, and targeting adipose tissues or associated metabolic pathways [52].

4.1.1 Nanoparticle Composition

Nanocarriers are to be constructed by biocompatible and biodegradable materials, viz. lipids, polymers, or hybrid systems. Those lipid-based nanoparticles such as liposomes and solid lipid nanoparticles (SLNs) provide natural lipid structures that facilitate compatibility with metabolic pathways of fat. For the sustained release, polymers like polylactic-co-glycolic acid (PLGA) have turned out to be the most suitable due to tunable degradation rates as well as their encapsulation of hydrophobic and hydrophilic drugs [52].

4.1.2 Targeted Delivery

Targeting of adipose tissues or specific metabolic pathways requires functionalizing nanoparticles with ligands like peptides, antibodies, or small molecules. These ligands interact with receptors that are overexpressed in adipose cells or related tissues for accurate targeting. Conjugate folate or transferrin will improve targeting efficacy about pathways associated with fat metabolism [44–46, 52].

4.1.3 Sustained Drug Release Mechanism

Continuous delivery of therapeutics is ensured through either a matrix-controlled design or a reservoir-controlled design. With matrix systems, the drug gets dispersed throughout the carrier material and, consequently, the release takes place through diffusion or degradation of the material. On the other hand, reservoir systems completely encapsulate the drug within a core membrane offering the precise release rate. The degradation rate can be modified via materials such as PLGA or lipid bilayers to attain the desired therapy timeline [29].

4.1.4 Mechanisms of Sustained Release

The sustained release of medication from nanocarriers occurs through the following mechanisms.

Diffusion-Controlled Release

Drugs slowly diffuse through the matrix or porous structure of the nanocarrier. This allows for slow, prolonged release of lipase inhibitors or metabolic stimulants for fat metabolism, producing consistently effective therapeutic levels in adipose tissues [21].

Erosion/Degradation-Controlled Release

The action of biodegradable nanocarriers, as in the case with PLGA, is to slow down the release of the drugs, which happens simultaneously with the degradation of these carriers over long periods of time. This appropriately and adequately meets the need of fat metabolism therapies for prolonged release of the compound in the body for continuous lipolysis or metabolic enhancement [42].

Stimuli-Responsive Release

Certain nanocarriers respond to specific stimulants such as pH, temperature, or enzymatic action. For example, functionalized liposomes that specifically target adipose tissue are likely to provide drug release in response to local enzymatic changes, diverting the effects of the drug to the fat cells [9].

Swelling-Controlled Release

Nanocarriers based on hydrogel swelling have the ability to absorb body fluids and deliver the IPMC encapsulated drug gradually with time. Such approaches ensure a sustained therapeutic exposure to the drug involved in pathways of fat metabolism [11].

4.1.5 Encapsulation Efficiency and Drug Stability

Efficiency of drug encapsulation has been associated with maximum payload and minimal wastage. Emulsification, nanoprecipitation, or solvent evaporation are some of these high drug loading methods which contribute to the stability of sensitive materials like hormonal secretion, or enzymes involved in fat metabolism.

Advanced nanocarrier systems synergizing stimuli-responsive mechanisms such as pH-sensitive, enzyme-responsive, or thermosensitive materials can refine drug release even further. The other element is the integration of imaging agents to nanocarriers to be used for theranostics-thus allowing treatment and monitoring of metabolic disorders. There have been great advances for developing nanocarriers for controlled release of drugs during fat metabolism. A combination of biocompatibility, specific targeting, and advanced release among other properties can offer very effective treatment in the management of metabolic disorders with as few side effects as possible [53].

4.2 Benefits of Prolonged Therapeutic Effects

Table 1 describes the benefits of Prolonged Therapeutic Effects in Controlled Drug Release Using Nanoparticles in Fat Metabolism.

This modality of control for drug release using nanoparticles is highly fruitful in creating bright prospects in the treatment of metabolic disease [55].

4.3 Enhancing Patient Compliance

Nanotechnology is something very fascinating to the medical world and has brought drastic changes in controlled drug release activities. Nanoparticles play a very important role in the potential treatment of diseases such as obesity, dyslipidemia, and related metabolic syndromes by the precise targeted delivery of drugs at the site of action. However, patient compliance is inevitable towards therapeutic success. Achieving a real and effective treatment demands optimisation in compliance strategies against nanoparticle-based drug delivery [23].

4.3.1 Improved Drug Delivery Systems

Sustained and reliable dispersal of drugs is assured by nanoparticle-based controlled drug release systems thereby bringing about the maximised avoidance of frequent dosing. Lipophilic drugs can be enclosed into these nanoparticles for facilitating long-acting dosing regimens in fat metabolism disorders and potentially reducing fluctuations in drug concentrations. Therefore, compliance is improved due to fewer doses required for patients since compliance is largely dependent on reduced burden to patients [57].

Table 1 Benefits of prolonged therapeutic effects in controlled drug release using nanoparticles in fat metabolism

Aspect	Description	Reference
I. Prolonged drug activity	Controlled drug release ensures sustained delivery of therapeutic agents, maintaining optimal drug levels over time	[16]
II. Enhanced patient compliance	Reduces the frequency of drug administration, which is especially beneficial for chronic conditions related to fat metabolism	[54]
III. Targeted delivery	Nanoparticles can be engineered for specific tissue or cellular targeting, such as adipocytes or liver cells	[42]
IV. Reduced side effects	By releasing drugs at controlled rates, off-target effects and toxicity are minimized, improving safety profiles	[55]
V. Improved bioavailability	Encapsulation of lipophilic drugs within nanoparticles enhances solubility and bioavailability	[55]
VI. Precision in fat metabolism	Controlled release enables precise modulation of metabolic pathways, improving efficacy in obesity or lipid disorders	[56]
VII. Protection of drug integrity	Nanoparticles shield sensitive drugs from enzymatic degradation, ensuring they remain effective until delivered	[30]
VIII. Minimized drug resistance	Sustained therapeutic levels can prevent peaks and troughs, reducing the likelihood of resistance development	[55]
IX. Customizable kinetics	Drug release profiles can be tailored to meet specific metabolic rates or therapeutic windows in fat metabolism	[56]
X. Economic and time efficiency	Prolonged effects mean fewer doses, leading to reduced healthcare costs and better resource utilization	[30]

4.3.2 Biocompatible and Non-toxic Nanoparticles

Patient Safety is one of the imperative attributes for compliance. Compliance with biocompatible and nontoxic nanoparticles such as PLGA (poly(lactic-co-glycolic acid))- or lipid-based nanocarriers would provide great mitigation to adverse effects. Lesser side effects, more trust gained by good confidence in therapy, also increase the compliance [39].

4.3.3 Targeted Delivery to Fat Metabolism Sites

Nanoparticles can be designed with the ability to target either adipose tissues or liver cells or enzymes involved in lipid metabolism through specific ligands or peptides. For instance, the surface-modified nanoparticles can be designed to bind specifically onto receptors on the adipocytes, rendering them direct targeting for the drug. Such strategy would not only improve the therapeutic performance of drugs used in adipose tissues but also reduce side effects possible for off-target uptake, hence enhancing acceptability of treatment [43].

4.3.4 Personalized Medicine Approaches

Individualizing drug treatment plans can greatly enhance compliance in patients. Specifically tailoring the formulations of nanoparticles to the metabolic rate, body composition, or genetic predisposition of each patient therefore renders the therapy vastly more effective and less intrusive. Patients are more willing to comply with personally prepared regimens that show result speed and visibility [41].

4.3.5 User-Friendly Administration Methods

It's possible to achieve high compliance rates using nanoparticle formulations through non-invasive or minimally invasive routes of delivery such as oral or transdermal systems. For instance, lipid-based nanoparticles for oral delivery will always be more convenient for the patient than injections [18].

4.3.6 Long-Term Therapeutic Efficacy

Sustained release formulations using nanoparticles can keep therapeutic drug concentrations for extended periods without repeated doses. For example, nanoparticles for once a month or twice-daily delivery have reduced treatment burden and improved compliance [53].

4.3.7 Patient Education and Engagement

It's necessary to educate patients regarding nanoparticle based treatment-it's merits and mechanisms. This outreach developed trust and compliance, by demonstrating how treatment targets fat metabolism-improvement in outcomes [25].

4.3.8 Monitoring and Feedback Systems

It is argued that integrated nanoparticles with smart drug delivery systems to enable real-time monitoring and feedback will improve adherence. As an example, nanoparticles emitting signals detectable by wearable devices could alert patients or doctors regarding treatment adherence and outcome, thus allowing the proactive management of the therapeutic regimen. Improving the patient's compliance for controlled drug release by nanoparticles in fat metabolism is an intricate affair. There are convenience-based, safe, and effective strategies, which must be coupled with patient involvement and education towards maximization of therapeutic outcome. Promise of nanotechnology in this area rests not just in changing the whole concept of drug delivery but potentially making it simpler and nicer for the patient as well [54].

5 Enzyme Modulation in Fat Metabolism

Table 2 describes the concepts related to lipase inhibition, lipolysis promotion, and their impact on fat absorption and utilization.

6 Enhancing Insulin Sensitivity

6.1 Delivery of Anti-inflammatory Agents to Adipose Tissue

Moreover, excess fat deposition indicates a particular role in energy homeostasis and metabolism, and in obesity, the excess storage of pro-inflammatory cytokines leads to paranoid conditions of chronic low-grade inflammation, inducing changes in insulin signaling pathways. This ultimately results in the development of insulin-resistance; a prime metabolic diseased condition resembling that of type 2 diabetes. The advent of a novel strategy of localized delivery of anti-inflammatory agents into our adipose tissues has given a good novel instrumentation to reduce the inflammation and restore the normal physiological sensitivity levels of insulin. Nanoparticle drug delivery systems are showing promise in their usage because they can be designed to target specific tissue, shield the drug from degradation, and improve therapeutic effectiveness [8].

6.1.1 The Role of Inflammation in Insulin Resistance

Chemokines are important mediators in the obesity-activated recruitment of inflammatory macrophages to fat. Adipocytes hypertrophied to several micrograms in weight release a chemokine (MCP-1), which pricks macrophages in the adipose

Table 2 Tabular breakdown of the concepts related to lipase inhibition, lipolysis promotion, and their impact on fat absorption and utilization

Concept	Description	Role of nanoparticles	Impact on fat absorption and utilization	Reference
Lipase inhibition	Involves blocking pancreatic or lipoprotein lipase enzymes responsible for breaking down triglycerides into absorbable free fatty acids and monoglycerides	Nanoparticles can deliver lipase inhibitors like orlistat in a controlled manner to the gastrointestinal tract	– Reduces fat absorption from the intestine – Promotes calorie deficit, aiding in weight management	[55]
Mechanism	Competitive or non-competitive binding to lipase active sites, preventing triglyceride hydrolysis	Nanocarriers can encapsulate hydrophobic inhibitors and target lipase-rich environments, enhancing bioavailability	– Limits dietary fat digestion – Alters fat utilization dynamics in the body	[52]
Lipolysis promotion	Enhances the breakdown of stored triglycerides in adipose tissue into free fatty acids and glycerol for energy production	Nanoparticles loaded with stimulators (e.g., caffeine, catecholamines) can target adipocytes for sustained stimulation of lipolysis	– Mobilizes stored fat, reducing fat mass – Increases energy expenditure	[8]
Mechanism	Activates enzymes like hormone-sensitive lipase (HSL) and adipose triglyceride lipase (ATGL) through signaling pathways (e.g., cAMP-PKA pathway)	Functionalized nanoparticles with ligands targeting adipocyte receptors (e.g., β-adrenergic receptors)	– Boosts metabolic rate – Reduces adipocyte size	[34]
Controlled drug release	Sustains drug concentration in the target area for prolonged efficacy	Nanoparticles enable slow release of inhibitors or stimulators, reducing dosing frequency	– Improves therapeutic outcomes with minimized side effects	[53]

(continued)

Table 2 (continued)

Concept	Description	Role of nanoparticles	Impact on fat absorption and utilization	Reference
Targeted delivery	Focuses on delivering drugs directly to tissues or enzymes involved in fat metabolism	Nanoparticles functionalized with antibodies or ligands for adipose tissue or GI tract targeting	– Increases drug efficiency – Reduces systemic side effects	[58]
Fat absorption	Process by which digested fats are absorbed through intestinal cells into the lymphatic system	Lipid-based nanoparticles can block absorption pathways or deliver agents that regulate fat uptake	– Decreases overall caloric intake from fats – Promotes leaner body composition	[59]
Fat utilization	Refers to the metabolic processes that convert stored or absorbed fats into usable energy	Nanoparticles can deliver metabolic boosters like L-carnitine to enhance mitochondrial fat oxidation	– Enhances energy production – Reduces reliance on carbohydrate metabolism	[25]
Biocompatibility	Ensures the nanoparticle does not elicit adverse immune reactions or toxicity	Use of biodegradable materials like PLGA or lipid nanoparticles ensures safety	– Ensures patient compliance and long-term usability	[41]
Therapeutic efficacy	The overall effectiveness of the treatment in achieving desired outcomes (e.g., weight loss, reduced lipid levels)	Enhanced by precise targeting and controlled release using nanoparticles	– Maximizes fat metabolism benefits – Improves lipid profile	[15]
Smart drug delivery systems	Systems capable of responding to biological signals to release drugs in a regulated manner	Responsive nanoparticles (pH-sensitive, enzyme-triggered) release drugs only under specific conditions	– Reduces unnecessary drug exposure – Improves dynamic fat metabolism management	[55]

(continued)

Table 2 (continued)

Concept	Description	Role of nanoparticles	Impact on fat absorption and utilization	Reference
Side effect mitigation	Reducing adverse effects like gastrointestinal discomfort, often associated with lipase inhibitors	Nanoparticles can localize drug activity to minimize systemic exposure	– Enhances patient adherence	[32]
Patient compliance	Ensuring consistent adherence to prescribed treatment protocols	Nanoparticle systems reduce dosing frequency and improve safety profiles	– Improves long-term treatment success	[24]

tissue. They are M1 macrophages and secrete a variety of pro-inflammatory cytokines like TNF-α and IL-6. The importance of these molecules relates to the down-regulation they result in protein down-streams of the insulin receptor signalling pathway, mainly due to their action towards the serine-phosphorylation of IRS-1. This redistribution of resources within the organism ultimately results in systemic insulin resistance through reduced glucose uptake and storage. So, if we want to restore metabolic health, we have to reduce the inflammation associated with adipose tissue [27].

6.1.2 Nanoparticles for Targeted Delivery

Nanoparticles, such as liposomes, polymeric nanoparticles, and lipid nanoparticles, provide an amazing platform for targeted delivery of anti-inflammatory agents to inflamed adipose tissue. Their small size, customizable surface properties, and their capacity to load a variety of therapeutic agents, make them excellent carriers [9].

6.1.3 Targeting Adipose Tissue

The nanoparticles can be functionalized with ligands or antibodies, targeting receptors overly expressed in either adipose or inflammatory cells, such as CD44, CD163, or scavenger receptors. Moreover, passive targeting via increased permeability and retention (EPR) in inflamed tissues can contribute to enhancement in adipose tissue accumulation [27].

6.1.4 Anti-inflammatory Agents

Nanoparticle-mediated delivery can involve the encapsulated corticosteroids or NSAIDs into nanoparticles as well as biologics like TNF-α inhibitors. Studies on curcumin-loaded nanoparticles reported significant improvements in adipose tissue inflammation and insulin sensitivity. Moreover, nanoparticles can also be used to deliver RNA-based therapeutics, such as siRNA or miRNA, which suppresses the expression of inflammatory genes specifically in adipocytes or macrophages [28].

6.1.5 Enhancing Insulin Sensitivity

Nanoparticle-bearing delivery mechanisms restore the integrity of insulin signaling pathways by attenuating inflammation. Research has demonstrated that delivering anti-inflammatory agents packaged in nanoparticles into the adipose tissue shifts the macrophage polarization from the pro-inflammatory M1 phenotype toward the anti-inflammatory M2 phenotype. This switch reduces the cytokine-induced insulin resistance and thereby promotes the uptake of glucose with increased metabolic flexibility. Yet every possibility comes with challenge. Biocompatibility, off-target effects, and tissue-specific delivery are among the critical issues. The use of innovative nanoparticle design technologies, such as hybrid nanoparticles and stimuli-responsive systems, will be the immediate solutions to these issues. Furthermore, combining nanoparticle-based therapies with lifestyles or diets may result in even further synergy of insulin sensitivity. Nanoparticle approaches are innovative and novel approaches to improving insulin sensitivity that deliver anti-inflammatory agents directly into adipose tissue. These systems could change the therapeutic landscape of insulin resistance and its associated metabolic disorders by targeting and destroying the inflammation in adipose tissue. Future research into optimizing delivery systems and translating preclinical success into the clinic will be of paramount importance to realize that promise [29].

6.2 Addressing Inflammation-Related Insulin Resistance

The greatly responsible factor for almost all metabolic disorders, that is type 2 diabetes, is insulin resistance (IR). It is closely associated with low-grade chronic inflammation, which eventually interferes with the normal process of insulin signaling so that one's body would not be able to regulate the normal blood sugar levels. One of the innovative ways involving nanoparticles for improving fatty acid metabolism, as well as restoring insulin sensitivity, can be used for handling inflammatory states leading to insulin resistance [9].

6.2.1 Inflammation and Insulin Resistance

Inflammation that persists for a long stretch of time is a very powerful cause of insulin unresponsiveness by being the big fat accumulators, oxidative stress generators, and unbalanced nutrient imbalances. Adipose tissues are a big part of this role by having secretion of pro-inflammatory cytokines such as TNF-α, IL-6, and also adipokines such as resistin and leptin, all of which impair the function of insulin receptors, thus decreasing its action. Hence, a reduced action of the insulin-body in turn indicates increased blood glucose levels.

The inflamed fat cells are becoming less effective to store the lipids but start to release fatty acids into the circulation, which contributes to insulin resistance. This further intensifies the metabolic derangement, into which blood sugar management becomes extremely difficult [60].

6.2.2 Nanoparticles and Fat Metabolism

Nowhere has Nanotechnology sought to transform methods of biomedical research in a unique way by delivering targeted approaches to the therapy of specific pathways in the disease. In insulin resistance, nanoparticles can be used for fat metabolism enhancements, reducing inflammation and improving insulin sensitivity [26].

6.2.3 Mechanisms of Action

Targeted Delivery

Nanoparticles engineered with surface modification to target inflamed adipose tissue effectively deliver therapeutic agents via receptor binding with specific ligands to the area of inflammation while minimizing off-target effects [34].

Reducing Inflammation

Such types of Nanoparticles can take anti-inflammatory molecules, curcumin, or resveratrol as drugs, directly to adipocytes. They suppress the production of inflammation cytokines and adipokines; thus, a negative effect is removed from insulin signaling pathways [55].

Improving Fat Metabolism

They can also bring drugs for promoting fatty acid oxidation or triggering thermogenesis in fat cells, otherwise known as adipocytes. Reductions in the amounts of fat stored in those cells facilitate insulin resistance, and again, improved metabolic health can be achieved through using fat fully as an energy source [56].

Restoring Insulin Sensitivity

Nanoparticles have the charm of helping restore the functioning of insulin receptors, so that insulin is able to perform its task of reducing blood sugar levels, mainly through targeting the inflammatory effects and improving the fat metabolism. Restoration of insulin sensitivity would be very therapeutically beneficial to those with obesity-induced resistance toward the hormone. Using nanoparticles for enhanced fat metabolism and reduced inflammation, a new strategy for insulin resistance has been born. It would perhaps impact improvement in insulin sensitivity by targeting the root causes of inflammation and metabolic dysfunction to manage diseases such as type 2 diabetes. With advances in nanomedicine research, nanoparticle-based therapy may evolve into a significant modality in the management of inflammation-related insulin resistance: an even more efficient, targeted, and personalized treatment strategy [61].

7 Appetite Control and Hormone Regulation

7.1 Targeted Hormone Delivery to the Brain

The appetite regulation and hormone management is vital for energy balance and metabolism-oriented health. Any hazardous disruption of these processes closely follows obesity, metabolic disorders, and the different ensuing complications. In this regard, recently, targeted hormone delivery in the brain, particularly using nanoparticles, has developed into a novel approach for modulating either appetite or fat metabolism. The goal of this paradigm is both to increase the efficiency and accuracy of hormone delivery and to have it act on specific targeted metabolic pathways. Hunger is regulated by the brain, more so with the hypothalamus, through signaling mechanisms of hormones such as leptin, ghrelin, insulin and glucagon-like peptide, ghrelin. These release signals to act in either way at the respective sites in the brain that mediate appetite. In fact, the signals get disturbed in metabolic disorders like obesity and hence lead into over-eating and poor fat metabolism. In order to address these complications, one possible idea is developing targeted hormone delivery systems, which enhance activation of hypothalamic sites in response to such hormones. Nanoparticles, owing to their diminutive size and surface characteristics, have recently emerged as possible vehicles for targeted hormone delivery across various biological barriers, including the blood–brain barrier (BBB) [62]. These nanoparticles may be further tailor-made to encapsulate appetite-regulating hormones, which will then, at targeted sites in the brain, be released to elicit local effects with a minimum of systemic side effects. For example, nanoparticle functionalization with ligands able to bind and internalize with receptors expressed in neurons involved in appetite control is anticipated to precisely deliver hormones such as leptin or GLP-1 into the hypothalamus. One of the most major benefits of using nanoparticles when delivering hormones in localized therapy is the prospects of overcoming the problems related to oral or systemic administration route of hormones. For example, leptin and GLP-1 have poor bioavailability when taken orally which is due to digestive tract degradation. Encapsulation within nanoparticles ensures hormone stability as well as optimizing pharmacokinetics for further delivery to the brain. Hormones might, thus, be delivered straight into the whole minutae of fat metabolism via nanoparticles, such that they can modulate the various important pathways regulated by these active hormones controlling lipid storage and the expenditure of energy, among other factors. For instance, leptin, the ob-gene product secreted by adipose tissue, signals the brain to depress appetite and increase energy expenditure. Leptin can thus be delivered directly to the neurons in the hypothalamus that are involved in the regulation of fat storage and energy balance. Similarly, it could also be GLP-1, a specific hormone involved in insulin secretion and satiation-induction, which can be targeted to those areas in the brain that regulate appetite and fat metabolism. Studies recently have proven that nanoparticles made delivery of hormones such as leptin and GLP-1 to the animal models of obesity cause diminished food intake, increased insulin sensitivity, and have dissimilar effects on weight reduction. Besides, these also have the potential of being therapeutic agents for metabolic disease patients who became resistant to these hormones. Finally, the switching of hormone delivery to the brain using nanoparticles appears to provide a novel avenue for the appetite regulation and hormonal modulation of fat metabolism. Such a strategy can offer excellent opportunities in designing therapeutic modalities for obesity and all the metabolic diseases associated with it, leading the path toward correction of normal metabolic function and improvement in an individual's energy balance. More studies into safety, effectiveness, and long-term impact of therapies involving nanoparticles will be instrumental in the process of putting such approaches into clinics [63].

7.2 *Effects of Leptin and GLP-1 on Food Intake*

In fact, the phenomenon of food intake is a multi-faceted mechanism partly revealed by judicious actions of hormones and neural signals where leptin and glucagon-like peptide-1 (GLP-1) are mainly needed for appetite regulation and fat metabolism. Leptin is actually a secretory product from adipose tissue that indicates the brain about energy reserves and decreases appetite when fat stores are larger. GLP-1 is secreted from the intestines as a morsel is ingested and decreases the appetite by stimulating the secretion of insulin. New avenues in nanotechnology have now opened windows which can amplify the effects of these hormones on appetite and fat metabolism [64–67].

7.2.1 Leptin and Appetite Control

Leptin is a hormone responsible for mediating the long-term control of body weight by sending signals to the hypothalamus to balance energy intake and expenditure, communicating that fat stores should be sufficient or low. Increased leptin levels in the body tell the brain to suppress hunger and promote energy expenditure when fat stores are well established. Sometimes, however, this can lead to leptin resistance in obese individuals because the brain becomes less responsive to leptin and instead overconsumes food to gain even more weight. Nanoparticles are now equipped with potential solutions for reversing that relationship with leptin. For example, nanoparticles may be designed to facilitate the delivery of leptin directly into the brain tissue or augment its receptor signaling, thereby restoring leptin sensitivity and appetite control in obese individuals [68–71].

7.2.2 GLP-1 and Appetite Regulation

GLP-1, a peptide hormone, inhibits food intake, insulin secretion induced, and evokes improved satiety. After meal ingestion, the peptide is released from the intestinal mucosa and subsequently acts on the brain to reduce hunger. In addition, it decreases gastric emptying and improves glucose metabolism. In therapeutic settings, these analogs of GLP-1, liraglutide being the most known one, have been beneficial with regard to type-2 diabetes and obesity; however, the minute half-life of GLP-1 in circulation restricts its effective time duration. Thus, systems based on nanoparticles for the delivery of GLP-1 may serve to stabilize and enhance the bioavailability of the hormone such that more prolonged and effective results on food intake and metabolism could be achieved. By designing nanoparticles capable of directing therapeutic agents specifically to tissues, including adipose tissue, one can cause site-specific action. Nanoparticles can be developed for phosphatase enhancement in fat metabolism, e.g., leptin and GLP-1. Lipid-based nanoparticles can be made to incorporate and deliver these hormones to the adipose tissue site for improving fat oxidation and reducing fat storage. Additionally, nanoparticles can serve as direct carriers of small molecules or genetic material to enhance modulation by leptin or GLP-1 signalling pathways. Both leptin and GLP-1 are two of the essential hormones responsible for the regulation of appetite and fat metabolism. Nanoparticle-based delivery systems have shown tremendous promise for improving efficacy in appetite regulation and fat metabolism. Parasitizing the hurdles of leptin resistance and short half-life of GLP-1, nanoparticles may open newer avenues in treating obesity and other associated metabolic disorders. Future research in developing targeted nanoparticle therapies may potentially provide new ways to control appetite and fat metabolism with better efficacy [72–75].

7.3 Nanotechnology in Appetite and Satiety Management

Control of appetite and regulation of satiation are the basic principles of a healthy body weight and the prevention of obesity-related diseases. Be it through fat metabolism, nanotechnology is now emerging in promising modifications in these phenomena. Nanoparticles, as per their size and surface area, may be employed to modulate specific physiological pathways in the appetite regulatory fat-storage exponent and be a new arena for obesity management [76–79].

7.3.1 Nanoparticles and Appetite Control

Regulation of appetite is highly complex, under the control of multiple interacting hormones, including ghrelin, leptin, and insulin. These hormones signal to the brain hunger or satiety, and nanoparticles can alter either the release or action of those hormones. For example, some nanoparticles could be manufactured to encapsulate and release specific compounds that would mimic the same satiety-inducing signals to induce fullness and thus reduce the urge to eat. Nanoparticles, by targeting the gut-brain axis, might exert appetite control by making signaling more effective, as with hormones like leptin, which informs the brain of satiation.

One of the approaches is to direct the active ingredients into the intestine using nanoparticles with appetite-suppressing drugs or compounds. An example could be as nanoparticles engineered to release active ingredients that control the activity of the ghrelin receptor, which controls transmission of hunger stimulation. The theory would then be that nanoparticles could control blood levels of ghrelin and thus decrease appetite and/or overeating. Nanotechnology is making remarkable advances in one of the most important areas of fat metabolism. Adipocyte (fat cells) metabolism, which consists of the storage and breakdown of fat, is crucial to energy balance and weight management. It is thus interesting to design nanoparticles that directly affect the processes of fat absorption and fat storage for the management of weight and satiety. Nanoparticles can improve bioavailability and stability for lipid metabolism regulatory compounds. For example, although nanoencapsulated antioxidants, polyphenols, or anti-obesity agents like resveratrol improved fat breakdown by enhancing lipolysis (fat breakdown) enzymes' activity and reducing lipogenesis (fat storage), they may especially target adipose tissue and locally increase action with potential improvements for fat metabolism with limited systemic outcome. That can be done by adding nanoformulations to dietary supplements or functional foods in order to increase the action of fat-burning agents facilitating their better absorption in the gut and thereby optimizing fat burning by the body. This can be especially beneficial in cases of obesity-related conditions where there is an excessive accumulation of fat, affecting the population's metabolic profile and inducing an increase in appetite. The possibility of nanoparticles in appetite and satiety control is glaring, but there are indeed challenges to be overcome. The biocompatibility and safety of nanoparticles should be established because indefinite use within the body leads to unintentional side effects. Additionally, targeting delivery systems should be developed to have the nanoparticles reach the desired tissues without affecting organs. In one word, nanotechnology can transform appetite control and satiety management by altering fat metabolism and hormone regulation through nanoparticles. Targeting specific biological pathways, nanoparticles constitute a promising arsenal in the battle against obesity and metabolic diseases for more effective and precise treatments of appetite and weight management. However, the long-term effects and safety of these technologies need further reviews [80–84].

8 Safety, Efficiency, and Ethical Considerations

8.1 Evaluating the Safety of Nanoparticles in Obesity Treatment

Obesity is an international health menace. Novel approaches are implored for therapy. One such area to venture into might be application of nanoparticles (NPs) on fat metabolism regulation. The efficacy of nanoparticles probably relies on their targeting effects on the specific cells and pathways, thus rendering the treatment more specific. However, in the advent of using nanoparticles in obesity therapy, a thorough assessment will have to be carried out, evaluating safety, efficacy, and ethical concerns [85–89].

8.2 Safety Considerations

Nanoparticles are constituent elements of obesity therapy, for instance lipid-based NPs, polymeric NPs and inorganic NPs, designed for interfacing biological systems at a molecular level. Thereby, this creates safety issues in respect to their biocompatibility, toxicity and long-term effects. It has been observed that NPs being liposome or micelle type are often regarded as biocompatible, but some NPs, such as metal-based NPs, can produce reactive oxygen species (ROS), causing oxidative stress and subsequent damage to tissues. Furthermore, their deposition in non-target tissues like the liver or kidneys could increase the chances of organ toxicity. There is a need for researchers to address the activation or suppression of the immune system, which could lead to adverse therapies. Moreover, any nanoparticles intended for human application for treatment of obesity will have to undergo extensive preclinical and clinical safety testing as per the regulations set forth by FDA and EMA [90–94].

8.3 Ethical Considerations

The employment of nanoparticles comes with many ethical dilemmas in the treatment of obesity. One is that high development and production costs of NP-based therapies would restrict access to them, which would aggravate inequities in health care access to therapies-especially since obesity is strongly associated with low-income across the population. Secondly, nanoparticle exposure poses potentially unknown threats to effects in the long run, raising a dilemma on informed consent as well as the principle of "do no harm". In fact, it is rather critical that patients have been quite informed about all possible ways of risks and benefits. Above all, the ethical implications of performing genetic interventions through NPs must be considered. Genetic modification of metabolic pathways might give rise to unanticipated side effects, such as changes in a person's natural energy balance, or increased vulnerability to other metabolic diseases. All forms of development for such therapies should be accompanied by transparent communication informed by ethical guidelines. Nanoparticles have great potential to transform obesity treatment-with their great promise toward precision and effectiveness. Equally, safety evaluations should be thoroughly undertaken to weed out the negative adverse effects, and ethical consideration must be borne in the development and application of nanoparticles for harm prevention. A balanced approach that encompasses safety, fairness, and ethical integrity is what will fulfil the promise of nanoparticles in obesity [95–98].

8.4 Ethical and Moral Considerations

Nanoparticle application has an enormous potential to revolutionize fat metabolism and help address global health problems like obesity, metabolic disorders, and even cardiovascular diseases. Along with promises of greater safety and efficacy, there are many ethical and moral issues associated with these applications, the vital ones being associated with health risks, social inequalities, informed consent, and environmental long-term impact [99].

8.5 Health Risks and Safety Concerns

Because of their small size, nanoparticles have been endowed with unique properties that allow them to interact with biological systems at a cellular and molecular level. This permits high precision targeting of fat cells and metabolic pathways while introducing possible unintended side effects such as accumulation in organs, toxicity, or an unexpected immune response. An ethical argument becomes essential to advocating convincing risk-benefit analysis that takes into consideration the most guarded interference patient safety. Clinical trials must also be rigorous in their factualness and inclusiveness so as to comprehend and weigh these risks [100–102].

8.6 Equity and Accessibility

The introduction of therapies based on nanoparticles may further exacerbate health inequalities. Advanced therapies are often expensive, meaning most of them would become limited to the wealthy and leave out the marginalized. This raises ethical questions of justice and fairness. To ensure equity, this will only be achieved when policymakers, researchers, and health providers focus on matters that concern cost barriers and develop models of accessibility for the underprivileged [103–106].

8.7 Informed Consent and Autonomy

Some patients may also find it difficult to grasp the significance of all this in relation to their own nanotechnology treatments. Informed consent is thus the kind of critical ethical requirement. Patients must receive complete, clear, and unbiased information regarding the benefits, risks, and uncertainties related to these nanoparticle interventions to allow them to take decisions according to their values and conditions and in consideration of their autonomy [107].

8.8 Long-Term Implications and Environmental Impact

The effects of nanoparticles on the human body and the environment in the long run are still unclear. The long-term durability of these particles in ecosystems and their potential for entering food chains through medical or industrial applications raise ethical concerns. Their life cycles must be taken into account when researchers then produce, use, or dispose of nanoparticles—a process that should strive to minimize environmental harm. Such responsibility extends to accounting for unintended consequences that would probably impact future generations [108–110].

8.9 Potential for Misuse

Ethanol nanoparticles form metabolic pathways to bring ethical manipulation for its malicious intents, such as using technology for a beauty condition or performance enhancement in ways contrary to health. Therefore, the latest regulations must be established in order to prevent such ethical error misuse with relevant technologies.

Nanoparticles certainly have a promise of change in fat metabolism, but their development and deployment is a serious ethical and moral issue. To achieve safety, equity, and sustainability, science and ethics must be synthesized through interdisciplinary approaches that prioritize transparency, inclusion, and accountability. This, then, would make up the final leverage in which the stakeholders can reap the luxuries of nanoparticle technologies while answering those to larger social and moral questions [111].

9 Future Directions and Research Needs

The crossroads of nanotechnology and metabolic research has abundant opportunities, especially in terms of understanding and acting upon fat metabolism. Nanoparticles (NPs) have potential usage in the concerned area in diagnosis, prognosis, and treatment of metabolic diseases like obesity, type 2 diabetes, and cardiovascular diseases. Numerous advancements have been made, yet there still exist some gaps in research that need to be addressed for the full exploitation of the potential benefits. It becomes imperative to have a more refined understanding of the molecular and cellular aspects of nanoparticles, particularly within the influences of fat metabolism. Future research should confirm the interaction between NPs and the ideal metabolic pathways, including lipogenesis and lipolysis, as well as the oxidation of fatty acids. For example, research could involve how such NPs would affect enzymes like hormone-sensitive lipases, as well as transcription factors such as PPARγ that mediate the effects under fat storage and mobilization. Further advanced experimental modalities would be needed in vitro as well as in vivo to understand these modalities measures for their safety and efficacy. Nanoparticles take a while to manufacture and put in practice, but they hold hope of delivering bioactive molecules preferentially to adipose tissues or other relevant metabolic organs such as liver and pancreas. At the same time, it remains a colossal challenge to develop systems that achieve high specificity while reducing any off-target effects. Future research should focus on functionalized NPs coupled with ligands targeting adipocyte specific receptors or other metabolic markers. Meanwhile, diversifying the research effort should target nanoparticle designs that can overcome barriers posed by dense extracellular matrix-like structures of adipose tissue. Nanoparticles can provide a prospect for real-time imaging and biosensing for the monitoring of metabolic processes. Research could propel exploration and development of nanoparticles as probes for in vivo imaging of lipid storage or degradation using fluorescence, MRI, or PET images. These tools may open doors to investigate the dynamics of metabolism and enable the early diagnosis of metabolic diseases and their therapeutic responses. Nanoparticles could serve to directly modulate fat metabolism, perhaps by delivering RNA interference (RNAi) molecules to knock down genes associated with obesity, or via nanoparticle-induced heating to promote thermogenic activity. But further work needs to be done in preclinical and clinical studies to answer questions about dosage, biodistribution, and long-term effects. This is a nanoparticle frontier that will revolutionize research on fat metabolism. A multidisciplinary approach with biologists, chemists, material scientists, and clinicians will answer the many existing challenges. Most importantly, new research could bridge these gaps, delivering novel solutions for metabolic health and diseases related to it in the future [112–114].

10 Conclusion

Nanotechnology has become one of the great revolutionary sciences of modernity, and the nanotechnology application in this field promises to have the greatest application in understanding and manipulating fat metabolism. Nanoparticles, or NPs, will prove to be great tools in the last decades for better understanding lipid metabolism narrowly and therapeutically delivering therapeutic agents, paving the way for very radical innovative treatments of metabolic disorders, especially obesity and dyslipidemia. Fat metabolism is an complex process that lipids undergo for absorption, transport, storage, and breakdown. Fat metabolism is critical for energy homeostasis. The disturbance in any of the above mentioned processes often leads to metabolic disorders, many of them affecting the global health. Most of the available treatment options for these disorders fail due to problems such as bioavailability of drugs, off-target effects, and insufficient tissue-specific delivery. Nanoparticles seem to settle nearly all of these issues by enabling precise delivery of drugs for the therapeutic purposes. An example of one of the major applications of nanoparticles in fat metabolism is utilizing direct targeted delivery of bioactive substances, such as lipase inhibitors, hormones, and genetic materials, to the relevant organs such as adipose or liver tissues. For example, lipid-derived nanoparticles such as liposomes and their solid lipid equivalents are used to encapsulate hydrophobic molecules-increasing the solubility and stability of the drugs. Similar applications additionally have polymeric nanoparticles and gold nanoparticles for delivering anti-obesity agents and their respective siRNA to regulate lipid metabolism at a molecular level. Nanoparticles are also very promising even in imaging and diagnosis related to lipid research. For example, both quantum dots and iron oxide nanoparticles have recently been successfully used for in situ observation of lipid uptake and trafficking into cells in real time. These improvements have greatly enhanced our understanding of metabolic disorders and opened up new avenues for exploring therapeutic targets for possible treatments. Even though there are so many hopes resting on their aptitude, nanoparticles for fat metabolism will also challenge the existing paradigm. Unresolved issues are lingering around long-term safety, possible toxicity, and environmental impact. According to some studies, certain nanoparticles induce oxidative stress or inflammatory responses, requiring thorough assessment for biocompatibility. Further, large-scale production and the regulatory approval process are serious barriers to penetrating this field's market for clinical use. Thus, it currently appears that the applications of nanoparticles depend on the further understanding of fat metabolism progress. Also, they can increase drug. The maximum use of nanoparticles in real time, measuring targeted valuable tissues, unlocks a region unprecedented in metabolic disorders treatment. Such wines, however, are not that easily sipped; just such challenges must be met by research setting its future goals at improving the designs of these entities to ensure that they are safe, effective, and environmentally friendly. Such integration could one day reshape metabolic medicine and help fight the growing burden of obesity and its comorbidities.

References

1. Hiwrale A, Bharati S, Pingale P, Rajput A (2023) Heliyon nanofibers : a current era in drug delivery system. Heliyon [Internet] 9(9):e18917. https://doi.org/10.1016/j.heliyon.2023.e18917
2. Abdel-Mageed HM, Radwan RA, AbuelEzz NZ, Nasser HA, El Shamy AA, Abdelnaby RM et al (2019) Bioconjugation as a smart immobilization approach for α-amylase enzyme using stimuli-responsive Eudragit-L100 polymer: a robust biocatalyst for applications in pharmaceutical industry. Artif Cells Nanomed Biotechnol [Internet]. 47(1):2361–2368. https://doi.org/10.1080/21691401.2019.1626414
3. Abdulhameed S, Pradeep NS, Sugathan S (2017) Bioresources and bioprocess in biotechnology. Bioresources Bioprocess Biotechnol 1:1–443
4. Aguilar-Pérez KM, Ruiz-Pulido G, Medina DI, Parra-Saldivar R, Iqbal HMN (2023) Insight of nanotechnological processing for nano-fortified functional foods and nutraceutical—opportunities, challenges, and future scope in food for better health. Crit Rev Food Sci Nutr 63(20):4618–4635
5. Wilczewska AZ, Niemirowicz K, Markiewicz KH, Car H (2012) Nanoparticles as drug delivery systems. Pharmacol Rep 64(5):1020–1037
6. Nathani K, Damani M, Mangrulkar S, Sawarkar S (2024) Polymeric nanofibers derived therapeutic approach for skeletal muscle regeneration: the promising future of muscle repair. J Drug Deliv Sci Technol [Internet] 94(1):105503. https://doi.org/10.1016/j.jddst.2024.105503
7. Williams GR, Raimi-Abraham BT, Luo CJ (2023) Nanofibres in drug delivery applications. Fibers 11(Figure 1):21
8. Hanson RL, Imperatore G, Bennett PH, Knowler WC (2002) Components of the "metabolic syndrome" and incidence of type 2 diabetes. Diabetes 51(10):3120–3127
9. Alkaladi A, Abdelazim AM, Afifi M (2014) Antidiabetic activity of zinc oxide and silver nanoparticles on streptozotocin-induced diabetic rats. Int J Mol Sci 15(2):2015–2023
10. Langwald SV, Ehrmann A, Sabantina L (2023) Measuring physical properties of electrospun nanofiber mats for different biomedical applications. Membranes (Basel) 13(5):20–24
11. Vaiserman A, Koliada A, Zayachkivska A, Lushchak O (2020) Nanodelivery of natural antioxidants: an anti-aging perspective. Front Bioeng Biotechnol 7
12. Liu J, Zhang L, Wang N, Zhao H, Li C (2022) Nano energy nanofiber-reinforced transparent, tough, and self-healing substrate for an electronic skin with damage detection and program-controlled autonomic repair. Nano Energy [Internet] 96(February):107108. https://doi.org/10.1016/j.nanoen.2022.107108
13. Jeong GJ, Castels H, Kang I, Aliya B (2022) Nanomaterial for skeletal muscle regeneration. Tissue Eng Regen Med [Internet] 19:253–261. https://doi.org/10.1007/s13770-022-00446-4
14. Anusiya G, Jaiganesh R (2022) A review on fabrication methods of nanofibers and a special focus on application of cellulose nanofibers. Carbohydr Polym Technol Appl [Internet] 4(November):100262. https://doi.org/10.1016/j.carpta.2022.100262
15. Stevenson R, Hueber AJ, Hutton A, McInnes IB, Graham D (2011) Nanoparticles and inflammation. Sci World J 11:1300–1312

16. Adetunji CO, Michael OS, Rathee S, Singh KR, Ajayi OO, Adetunji JB et al (2022) Potentialities of nanomaterials for the management and treatment of metabolic syndrome: a new insight. Mater Today Adv 13
17. Cedervall T, Hansson LA, Lard M, Frohm B, Linse S (2012) Food chain transport of nanoparticles affects behaviour and fat metabolism in fish. PLoS ONE 7(2):1–6
18. Bahadori E, Farjami Z, Rezayi M, Lngari H, Darroudi M, Avan A et al (2019) Recent advances in nanotechnology for the treatment of metabolic syndrome. Diabetes Metab Syndr Clin Res Rev 13(2):1561–1568
19. Hirst SM, Karakoti A, Singh S, Self W, Tyler R, Seal S et al (2013) Bio-distribution and in vivo antioxidant effects of cerium oxide nanoparticles in mice. Environ Toxicol 28(2):107–118
20. Hu H, Guo Q, Wang C, Ma X, He H, Oh Y et al (2015) Titanium dioxide nanoparticles increase plasma glucose via reactive oxygen species-induced insulin resistance in mice. J Appl Toxicol 35(10):1122–1132
21. Chalasani KB, Russell-Jones GJ, Yandrapu SK, Diwan PV, Jain SK (2007) A novel vitamin B12-nanosphere conjugate carrier system for peroral delivery of insulin. J Control Release 117(3):421–429
22. Thomas TP, Goonewardena SN, Majoros IJ, Kotlyar A, Cao Z, Leroueil PR et al (2011) Folate-targeted nanoparticles show efficacy in the treatment of inflammatory arthritis. Arthritis Rheum 63(9):2671–2680
23. Khalil LM, Abdallah OY, Elnaggar YSR, El-Refaie WM (2022) Novel dermal nanobilosomes with promising browning effect of adipose tissue for management of obesity. J Drug Deliv Sci Technol 74
24. Sarmento B, Ribeiro A, Veiga F, Sampaio P, Neufeld R, Ferreira D (2007) Alginate/chitosan nanoparticles are effective for oral insulin delivery. Pharm Res 24(12):2198–2206
25. Bae KH, Choi SH, Park SY, Lee Y, Park TG (2006) Thermosensitive pluronic micelles stabilized by shell cross-linking with gold nanoparticles. Langmuir 22(14):6380–6384
26. Chen H, Dorrigan A, Saad S, Hare DJ, Cortie MB, Valenzuela SM (2013) In vivo study of spherical gold nanoparticles: inflammatory effects and distribution in mice. PLoS One 8(2)
27. Jafari-Giv Z, Avan A, Hamidi F, Tayefi M, Ghazizadeh H, Ghasemi F et al (2019) Association of body mass index with serum calcium and phosphate levels. Diabetes Metab Syndr Clin Res Rev 13(2):975–980
28. Chhabra G, Chuttani K, Mishra AK, Pathak K (2011) Design and development of nanoemulsion drug delivery system of amlodipine besilate for improvement of oral bioavailability. Drug Dev Ind Pharm 37(8):907–916
29. Elkasabgy NA, Elsayed I, Elshafeey AH (2014) Design of lipotomes as a novel dual functioning nanocarrier for bioavailability enhancement of lacidipine: in-vitro and in-vivo characterization. Int J Pharm 472(1–2):369–379
30. Fujioka K, Takada Y, Sato S, Miyata T (1995) Novel delivery system for proteins using collagen as a carrier material: the minipellet. J Control Release 33(2):307–315
31. Yu T, Zhao S, Li Z, Wang Y, Xu B, Fang D et al (2016) Enhanced and extended anti-hypertensive effect of VP5 nanoparticles. Int J Mol Sci 17(12)
32. Ayala-Fuentes JC, Gallegos-Granados MZ, Villarreal-Gómez LJ, Antunes-Ricardo M, Grande D, Chavez-Santoscoy RA (2022) Optimization of the synthesis of natural polymeric nanoparticles of inulin loaded with quercetin: characterization and cytotoxicity effect. Pharmaceutics 14(5)
33. Pan Y, Neuss S, Leifert A, Fischler M, Wen F, Simon U et al (2007) Size-dependent cytotoxicity of gold nanoparticles. Small 3(11):1941–1949
34. Rodrigues MS, de Paula GC, Duarte MB, de Rezende VL, Possato JC, Farias HR et al (2021) Nanotechnology as a therapeutic strategy to prevent neuropsychomotor alterations associated with hypercholesterolemia. Colloids Surf B Biointerfaces 201
35. Sarmento B, Ribeiro A, Veiga F, Ferreira D (2006) Development and characterization of new insulin containing polysaccharide nanoparticles. Colloids Surf B Biointerfaces 53(2):193–202
36. Kuzuya T, Nakagawa S, Satoh J, Kanazawa Y, Iwamoto Y, Kobayashi M et al (2002) Report of the committee on the classification and diagnostic criteria of diabetes mellitus. Diabetes Res Clin Pract 55(1):65–85

37. Sahoo SK, Labhasetwar V (2003) Nanotech approaches to drug delivery and imaging. Drug Discov Today 8(24):1112–1120
38. Chen Y, Hazan V (2009) Developmental factors and the non-native speaker effect in auditory-visual speech perception. J Acoust Soc Am 126(2):858–865
39. Mei L, Zhang Z, Zhao L, Huang L, Yang XL, Tang J et al (2013) Pharmaceutical nanotechnology for oral delivery of anticancer drugs. Adv Drug Deliv Rev 65(6):880–890
40. Brigger I, Dubernet C, Couvreur P (2012) Nanoparticles in cancer therapy and diagnosis. Adv Drug Deliv Rev 64(SUPPL.):24–36
41. Xue Y, Xu X, Zhang XQ, Farokhzad OC, Langer R (2016) Preventing diet-induced obesity in mice by adipose tissue transformation and angiogenesis using targeted nanoparticles. Proc Natl Acad Sci U S A 113(20):5552–5557
42. Alam T, Khan S, Gaba B, Haider MF, Baboota S, Ali J (2017) Nanocarriers as treatment modalities for hypertension. Drug Deliv 24(1):358–369
43. Kim SY, Shin IG, Lee YM (1998) Preparation and characterization of biodegradable nanospheres composed of methoxy poly(ethylene glycol) and DL-lactide block copolymer as novel drug carriers. J Control Release 56(1–3):197–208
44. Timko M, Koneracká M, Tomašovičová N, Kopčanský P, Závišová V (2006) Magnetite polymer nanospheres loaded by Indomethacin for anti-inflammatory therapy. J Magn Magn Mater 300(1)
45. Antal I, Kubovcikova M, Zavisova V, Koneracka M, Pechanova O, Barta A et al (2015) Magnetic poly(D,L-lactide) nanoparticles loaded with aliskiren: a promising tool for hypertension treatment. J Magn Magn Mater 380:280–284
46. Fattal E, Couvreur P, Dubernet C (2004) "Smart" delivery of antisense oligonucleotides by anionic pH-sensitive liposomes. Adv Drug Deliv Rev 56(7):931–946
47. Kobyliak N, Virchenko O, Falalyeyeva T, Kondro M, Beregova T, Bodnar P et al (2017) Cerium dioxide nanoparticles possess anti-inflammatory properties in the conditions of the obesity-associated NAFLD in rats. Biomed Pharmacother 90:608–614
48. Kim SY, Ha JC, Lee YM (2000) Poly(ethylene oxide)-poly(propylene oxide)-poly(ethylene oxide)/poly(ε- caprolactone) (PCL) amphiphilic block copolymeric nanospheres: II. Thermo-responsive drug release behaviors. J Control Release 65(3):345–358
49. Alberti KGMM, Zimmet P, Shaw J (2005) The metabolic syndrome—a new worldwide definition. Lancet 366(9491):1059–1062
50. Danielsen PH, Risom L, Wallin H, Autrup H, Vogel U, Loft S et al (2008) DNA damage in rats after a single oral exposure to diesel exhaust particles. Mutat Res Fundam Mol Mech Mutagen 637(1–2):49–55
51. Recent advances in nanotechnology for the treatment of metabolic syndrome—ScienceDirect [Internet] [cited 2024 Dec 25]. https://www.sciencedirect.com/science/article/abs/pii/S1871402119300797
52. Lee JH, Yeo Y (2015) Controlled drug release from pharmaceutical nanocarriers. Chem Eng Sci 24(125):75–84
53. Assi RA, Abdulbaqi IM, Yee CS (2021) The evaluation of drug delivery nanocarrier development and pharmacological briefing for metabolic-associated fatty liver disease (Mafld): an update. Pharmaceuticals 14(3)
54. Karthikeyan E, Sivaneswari S (2024) Advancements in transdermal drug delivery systems: enhancing medicine with pain-free and controlled drug release. Intell Pharm [Internet] [cited 2024 Dec 25]. https://linkinghub.elsevier.com/retrieve/pii/S2949866X24001072
55. Ansari A, Hussain A, Wadekar R, Altamimi MA, Malik A, Mujtaba MA et al (2023) Nanovesicles based drug targeting to control tumor growth and metastasis. Adv Cancer Biol Metastasis 1(7):100083
56. Ambroselli D, Masciulli F, Romano E, Catanzaro G, Besharat ZM, Massari MC et al (2023) New advances in metabolic syndrome, from prevention to treatment: the role of diet and food. Nutrients 15(3)
57. Khutoryanskiy VV, Tirelli N (2008) Oxidation-responsiveness of nanomaterials for targeting inflammatory reactions. Pure Appl Chem 80(8):1703–1718

58. Hu W, Zhai C, Sun H, Gong XY, Cui L, Cai L et al (2023) The global burden of disease attributable to metabolic risks in 204 countries and territories from 1990 to 2019. Diabetes Res Clin Pract 196
59. Kolovou GD, Anagnostopoulou KK, Salpea KD, Mikhailidis DP (2007) The prevalence of metabolic syndrome in various populations. Am J Med Sci 333(6):362–371
60. Amin SA, Adhikari N, Kotagiri S, Jha T, Ghosh B (2019) Histone deacetylase 3 inhibitors in learning and memory processes with special emphasis on benzamides. Eur J Med Chem 15(166):369–380
61. Nomura DK, Casida JE (2016) Lipases and their inhibitors in health and disease. Chem Biol Interact 259:211–222
62. Kumar A, Chauhan S (2021) Pancreatic lipase inhibitors: the road voyaged and successes. Life Sci 271
63. Liu PK, Weng ZM, Ge GB, Li HL, Ding LL, Dai ZR et al (2018) Biflavones from Ginkgo biloba as novel pancreatic lipase inhibitors: inhibition potentials and mechanism. Int J Biol Macromol 118:2216–2223
64. Xiao S, Yu R, Ai N, Fan X (2015) Rapid screening natural-origin lipase inhibitors from hypolipidemic decoctions by ultrafiltration combined with liquid chromatography-mass spectrometry. J Pharm Biomed Anal 104:67–74
65. Belfeki H, Mejri M, Hassouna M (2016) Antioxidant and anti-lipases activities in vitro of Mentha viridis and Eucalyptus globulus extracts. Ind Crops Prod 89:514–521
66. Kim JH, Kim HJ, Kim C, Jung H, Kim YO, Ju JY et al (2007) Development of lipase inhibitors from various derivatives of monascus pigment produced by Monascus fermentation. Food Chem 101(1):357–364
67. Ahn JH, Liu Q, Lee C, Ahn MJ, Yoo HS, Hwang BY et al (2012) A new pancreatic lipase inhibitor from Broussonetia kanzinoki. Bioorganic Med Chem Lett 22(8):2760–2763
68. Patil SG, Patil MP, Maheshwari VL, Patil RH (2015) In vitro lipase inhibitory effect and kinetic properties of di-terpenoid fraction from Calotropis procera (Aiton). Biocatal Agric Biotechnol 4(4):579–585
69. Kato E, Yama M, Nakagomi R, Shibata T, Hosokawa K, Kawabata J (2012) Substrate-like water soluble lipase inhibitors from Filipendula kamtschatica. Bioorganic Med Chem Lett 22(20):6410–6412
70. Liang LF, Wang T, Cai YS, He WF, Sun P, Li YF et al (2014) Brominated polyunsaturated lipids from the Chinese sponge Xestospongia testudinaria as a new class of pancreatic lipase inhibitors. Eur J Med Chem 79:290–297
71. Kumar A, Chauhan GS (2010) Extraction and characterization of pectin from apple pomace and its evaluation as lipase (steapsin) inhibitor. Carbohydr Polym 82(2):454–459
72. Houghton D, Wilcox MD, Chater PI, Brownlee IA, Seal CJ, Pearson JP (2015) Biological activity of alginate and its effect on pancreatic lipase inhibition as a potential treatment for obesity. Food Hydrocoll 49:18–24
73. Ercan P, El SN (2016) Inhibitory effects of chickpea and Tribulus terrestris on lipase, α-amylase and α-glucosidase. Food Chem 205:163–169
74. Flores-Sanchez IJ, Gang DR (2013) Inhibition of hydroxycinnamoyl-CoA thioesterases in ginger (Zingiber officinale Rosc.) and turmeric (Curcuma longa L.) by lipase inhibitors. Plant Physiol Biochem 72:46–53
75. Luyen NT, Tram LH, Hanh TTH, Binh PT, Dang NH, Van Minh C et al (2013) Inhibitors of a-glucosidase, a-amylase and lipase from Chrysanthemum morifolium. Phytochem Lett 6(3):322–325
76. Wu X, He W, Zhang H, Li Y, Liu Z, He Z (2014) Acteoside: a lipase inhibitor from the Chinese tea Ligustrum purpurascens kudingcha. Food Chem 142:306–310
77. Chupak LS, Zheng X, Hu S, Huang Y, Ding M, Lewis MA et al (2016) Structure activity relationship studies on chemically non-reactive glycine sulfonamide inhibitors of diacylglycerol lipase. Bioorganic Med Chem 24(7):1455–1468
78. Delorme V, Raux B, Puppo R, Leclaire J, Cavalier JF, Marc S et al (2014) Supported inhibitor for fishing lipases in complex biological media and mass spectrometry identification. Biochimie 107(Part A):124–134

79. Mulvihill MM, Nomura DK (2013) Therapeutic potential of monoacylglycerol lipase inhibitors. Life Sci 92(8–9):492–497
80. Yu M, Lange S, Richter S, Tan T, Schmid RD (2007) High-level expression of extracellular lipase Lip2 from Yarrowia lipolytica in Pichia pastoris and its purification and characterization. Protein Expr Purif 53(2):255–263
81. Tiss A, Lengsfeld H, Carrière F, Verger R (2009) Inhibition of human pancreatic lipase by tetrahydrolipstatin: further kinetic studies showing its reversibility. J Mol Catal B Enzym 58(1–4):41–47
82. Hiol A, Jonzo MD, Rugani N, Druet D, Sarda L, Comeau LC (2000) Purification and characterization of an extracellular lipase from a thermophilic Rhizopus oryzae strain isolated from palm fruit. Enzyme Microb Technol 26(5–6):421–430
83. Afzal O, Kumar S, Kumar R, Firoz A, Jaggi M, Bawa S (2014) Docking based virtual screening and molecular dynamics study to identify potential monoacylglycerol lipase inhibitors. Bioorg Med Chem Lett 24(16):3986–3996
84. Chen H, Li J, Yao R, Yan S, Wang Q (2020) Mechanism of lipid metabolism regulation by soluble dietary fibre from micronized and non-micronized powders of lotus root nodes as revealed by their adsorption and activity inhibition of pancreatic lipase. Food Chem 305
85. Holtfrerich A, Hanekamp W, Lehr M (2013) 4-Phenoxyphenyl)tetrazolecarboxamides and related compounds as dual inhibitors of fatty acid amide hydrolase (FAAH) and monoacylglycerol lipase (MAGL. Eur J Med Chem 63:64–75
86. Birari RB, Bhutani KK (2007) Pancreatic lipase inhibitors from natural sources: unexplored potential. Drug Discov Today 12(19–20):879–889
87. Latino-Martel P, Cottet V, Druesne-Pecollo N, Pierre FHF, Touillaud M, Touvier M et al (2016) Alcoholic beverages, obesity, physical activity and other nutritional factors, and cancer risk: a review of the evidence. Crit Rev Oncol Hematol 99:308–323
88. Dollinger LM, Howell AR (1998) A 2-methyleneoxetane analog of orlistat demonstrating inhibition of porcine pancreatic lipase. Bioorg Med Chem Lett 8(8):977–978
89. Zou Y, Li J, Lu C, Wang J, Ge J, Huang Y et al (2006) High-fat emulsion-induced rat model of nonalcoholic steatohepatitis. Life Sci 79(11):1100–1107
90. Fan Y, Xu Y, Han QG, Shen LL, Xu H, Wu ZB et al (2018) Exploring inhibition mechanism and nature of lipase by Ligupurpuroside A extracted from Ku-Ding tea. Med Chem Res 27(7):1822–1833
91. Lowet ME (2002) The triglyceride lipases of the pancreas. J Lipid Res 43(12):2007–2016
92. Zhou Y, Li S, Ai M, Chen H, Liu Y, Li B et al (2020) 1,25(OH)2D3 mitigate cancer-related fatigue in tumor-bearing mice: integrating network pharmacological analysis. Biomed Pharmacother 128
93. Wang Y, Tao B, Wan Y, Sun Y, Wang L, Sun J et al (2020) Drug delivery based pharmacological enhancement and current insights of quercetin with therapeutic potential against oral diseases. Biomed Pharmacother 128
94. Alsamri H, Athamneh K, Pintus G, Eid AH, Iratni R (2021) Pharmacological and antioxidant activities of rhus coriaria l. (sumac). Antioxidants 10(1):1–28
95. Wen X, Zhang B, Wu B, Xiao H, Li Z, Li R et al (2022) Signaling pathways in obesity: mechanisms and therapeutic interventions. Sig Transduct Target Ther 7(1)
96. Oliveira RF, Gonçalves GA, Inácio FD, Koehnlein EA, de Souza CGM, Bracht A et al (2015) Inhibition of pancreatic lipase and triacylglycerol intestinal absorption by a Pinhão coat (Araucaria angustifolia) extract rich in condensed tannin. Nutrients 7(7):5601–5614
97. Pai SA, Martis EAF, Joshi SG, Munshi RP, Juvekar AR (2018) Plumbagin exerts antiobesity effects through inhibition of pancreatic lipase and adipocyte differentiation. Phyther Res 32(8):1631–1635
98. Gu Y, Hurst WJ, Stuart DA, Lambert JD (2011) Inhibition of key digestive enzymes by cocoa extracts and procyanidins. J Agric Food Chem 59(10):5305–5311
99. Quiroga PR, Grosso NR, Lante A, Lomolino G, Zygadlo JA, Nepote V (2013) Chemical composition, antioxidant activity and anti-lipase activity of Origanum vulgare and Lippia turbinata essential oils. Int J Food Sci Technol 48(3):642–649

100. Roh C, Jung U (2012) Nepeta japonica Maximowicz extract from natural products inhibits lipid accumulation. J Sci Food Agric 92(10):2195–2199
101. Kim J, Lee YS, Kim CS, Kim JS (2012) Betulinic acid has an inhibitory effect on pancreatic lipase and induces adipocyte lipolysis. Phyther Res 26(7):1103–1106
102. Koga H, Suzuki K, Okazaki T, Nishimura K, Lane GJ, Yamataka A (2013) Treatment of a densely fused fissure during thoracoscopic right middle lobectomy. Open J Pediatr 03(02):105–107
103. Buchholz T, Melzig MF (2016) Medicinal plants traditionally used for treatment of obesity and diabetes mellitus—screening for pancreatic lipase and α-amylase inhibition. Phyther Res 30(2):260–266
104. Chater PI, Wilcox M, Cherry P, Herford A, Mustar S, Wheater H et al (2016) Inhibitory activity of extracts of Hebridean brown seaweeds on lipase activity. J Appl Phycol 28(2):1303–1313
105. Miyata M, Koyama T, Yazawa K (2010) Water extract of Houttuynia cordata thunb. Leaves exerts anti-obesity effects by inhibiting fatty acid and glycerol absorption. J Nutr Sci Vitaminol (Tokyo) 56(2):150–156
106. Yang T, Chen J, Huang L, Wang T, Xiao C, Sun Z et al (2015) Extremely short-lived reaction resonances in $Cl + HD (v = 1) \rightarrow DCl + H$ due to chemical bond softening. Science (80-) 347(6217):60–63
107. Eom SH, Lee MS, Lee EW, Kim YM, Kim TH (2013) Pancreatic lipase inhibitory activity of phlorotannins isolated from Eisenia bicyclis. Phyther Res 27(1):148–151
108. Gu H, Kunze K, Takatani M, Minamizawa K (2015) Towards performance feedback through tactile displays to improve learning archery. In: UbiComp ISWC 2015—Adjunct proceedings of the 2015 ACM international joint conference on pervasive and ubiquitous computing and proceedings of the 2015 ACM international symposium on wearable computers, pp 141–144
109. Liu M, Liao DQ, Zhu QQ, Wang FH (2012) α-times integrated C-semigroups. Adv Pure Math 02(03):211–215
110. Sakulnarmrat K, Srzednicki G, Konczak I (2015) Antioxidant, enzyme inhibitory and antiproliferative activity of polyphenolic-rich fraction of commercial dry ginger powder. Int J Food Sci Technol 50(10):2229–2235
111. Nomura S, Maeda Y, Ishii K, Katayama Y, Yagi H, Fujishima N et al (2016) Relationship between HMGB1 and PAI-1 after allogeneic hematopoietic stem cell transplantation. J Blood Med 7
112. Cao ZH, Gu DH, Lin QY, Xu ZQ, Huang QC, Rao H et al (2011) Effect of pu-erh tea on body fat and lipid profiles in rats with diet-induced obesity. Phyther Res 25(2):234–238
113. Li Q, Liu Z, Huang J, Luo G, Liang Q, Wang D et al (2013) Anti-obesity and hypolipidemic effects of Fuzhuan brick tea water extract in high-fat diet-induced obese rats. J Sci Food Agric 93(6):1310–1316
114. Ho JN, Jang JY, Yoon HG, Kim Y, Kim S, Jun W et al (2012) Anti-obesity effect of a standardised ethanol extract from Curcuma longa L. fermented with Aspergillus oryzae in ob/ob mice and primary mouse adipocytes. J Sci Food Agric 92(9):1833–1840

Boosting Cognitive Function in Athletes with Nano-nutrients

Debajit Karmakar, Anindita Das, Karuppasamy Govindasamy, and Valentina Stefanica

Abstract This chapter will discuss how nanonutrients are increasingly used to help athletes by presenting recent developments in the use of nanotechnology in the sports nutrition industry which is quite developed at the moment. Athletes have factors that are barriers to their optimal performance for instance reaction time and decision making as well as mental strength thus there is need to improve brain performance. Nanonutrients are designed seventeen billion times smaller than the common nutrients which gives room for improving cognitive abilities include memory, attention as well as stress management and brain protection. This section explains how nanonutrients improve nutritional absorption by focusing on vitamins and minerals among other compounds. Such nutrients are known as nanonutrients go through nanonization and so can cross the blood brain barrier and thus improve neurons and synapses in the brain. Furthermore it looks into how these minuscule nutrients help reduce stress and inflammation which're factors associated with cognitive decline and mental fatigue in sports related situations. This section illustrates the power of technology in sports by emphasizing studies on nanoscale delivery systems such as liposomes and nanoemulsions which are able to target specific parts of the brain in a bid to enhance lasting and potent effects on brain activity. Nanoscale technology is able to shift the paradigm of sports abilities as covered in this chapter, highlighting the potential of

D. Karmakar (✉) · A. Das (✉)
Department of Physical Education Pedagogy, Lakshmibai National Institute of Physical Education, Gwalior, Madhya Pradesh, India
e-mail: debajitkarmakar2200@gmail.com

A. Das
e-mail: anindita416@gmail.com

K. Govindasamy
Department of Sports, Recreation and Wellnes, Symbiosis International (Deemed University), Hyderabad Campus, Modallaguda (V), Nandigama (M), Rangareddy, Telangana 509217, India
e-mail: gowthamadnivog@gmail.com

V. Stefanica
Department of Physical Education and Sport, Faculty of Sciences, Physical Education and Informatics, National University of Science and Technology Politehnica Bucharest, Pitesti University Center, Pitesti, Romania
e-mail: valentina.stefanica@upb.ro

A. Chatterjee et al. (eds.), *Nanofuel: The Future of Sports Nutrition*, Smart Nanomaterials Technology, https://doi.org/10.1007/978-981-96-5471-0_18

nanonutrients on improving the physical and mental capacities of athletes making them stronger competitors.

Keywords Neuroplasticity · Bioavailability · Neurotransmitters · Neuroimaging · Cognitive resilience · Phospholipids · Athletic performance

1 Introduction

Nanotechnology is revolutionizing a number of fields, and in nutrition, a new concept of nanonutrients has been introduced. These are nanoscale engineered compounds that have specific properties that set them apart from other bulk substances [1]. Nano-nutrients not only have high compositional specificity but also have enhanced bioavailability, targeted delivery and increased stability, which allows them to work better with the body. By acting at a size similar to biological structures, these nutrients cannot suffer from the usual barriers to absorption and thus reach their sites of utilization in cells and tissues [2]. This mechanism allows for nutrition supplementation timing that optimally benefits the body, depicting nanonutrients as a novel tool in improving athletes' physical and cognitive well-being [3].

The beginning stages of sports nutrition have dramatically changed from placing an emphasis on physical attributes to encompassing a broader view that equally prioritises mental strategy [4]. Athletes are not only appraised on their physical abilities but even their control of the head, reasoning, as well as concentration, have become essential factors for achievement. Recent studies also show nutrition and cognition are closely interlinked and fostering one's mental health can in return enhance one's athleticism [2]. This advancement has placed a lot of focus on nanonutrients which have quite sophisticated characteristics that fit in well with the determinants of precision nutrition. When blended into their diet, nanonutrients can allow athletes to stay focused, react in a timely manner, and deal with stress more efficiently all of which are crucial when competing in a highly competitive environment [3].

Nanotechnology and cognitive function work hand in hand as an innovative approach to athletic performance. While conventional nutritive measures can be helpful, they have some constraints in regard to bioavailability, as well as nanoparticles' targeted effects [4]. Nanotechnology meets these challenges by creating nutrient delivery systems that are compatible with the wiring pathways of the brain. This is not only a way to enhance the impact of conventional nutrients on cognitive performance; it also supplies entirely new compounds designed for their brains first and foremost [5]. For athletes of whom their cognitive requirements include decision-making in reaction time, strategies' formulation and mental stamina, these innovations provide one of the kind opportunity. What nanotechnology can do with aspect of sports nutrition is to reinvent the possibility of achieving enhanced human performance, thereby giving a new dimension to cognition in sports [1].

2 Neurophysiological Basis of Cognitive Performance in Athletes

2.1 Neuroplasticity and Cognitive Plasticity in High-Performance Athletes

Neuroplasticity is the process of change of the neural pathways in the brain as a result of new experience, learning, development or in the aftermath of an injury. This is even more evident in high performance athletes given the constant train and compete demeanour [5]. Coordination and training in exercise teaching foster architectural and functional modifications, neuritic changes, increased neural functional demands and neural plasticity [6]. It makes it easier to develop the need motor skills for the corresponding sport, quick thinking and cognitive skills. Therefore, we find that neuroplasticity is vitally helpful in enhancing the specific cerebral abilities needed to support the best in human performance: responsiveness, rationality, and focus [4].

Neuroplasticity's sister concept is cognitive plasticity, that refers to the brain's ability to modify its cognitive processes depending on the stimulus or other factors [7]. The cognitive plasticity of athletes is accomplished by practice, experience, use of cognitive input during training sessions, novelty, and various aptitudes necessary for athletic competition [8]. The above experiences foster improved cognitive plastics, which allow the athlete to make changes when performing depending on the circumstance that presented itself. Such thinking flexibility may affect an athlete's decision making, problem solving, and attention control, in a situation of high stress or pressure. In addition, a combination of neuroplasticity and cognitive plasticity is conducive to long-term enhancement of mental functioning, so athletes can reach their best mental state during their careers [5].

2.2 Neurometabolic Demands During Intense Physical and Mental Exertion

This Concept basically states that during periods of high energy production, sportspeople are subjected to tremendous neuro-metabolic load that is a test to their CNS. Physical activity especially vigorous WAP demands energy substrates like glucose and oxygen for both muscular activity and the working of brain [8]. These cells are the brain, which despite making up 2% of the body weight uses 20% of all energy in the body. At such moments, the brain requires more energy to perform such functions as decision making, concentration and coordination as the body exercises intensively [9]. With this we have an increased energy requirement together with an increased turnover rate of neurotransmitters which mediate neuronal communication. Therefore any rationing of these energy substrates for instance through glucose depletion deters mental performance conquering areas of attention, promptness and eradicating decision making potentialities [10].

Besides energy demands, extreme physical exercise results to other changes in the brain such as shifts in ion levels and changing neurotransmitter activity [11]. The raise in physical activity adds to the catecholamines, such as dopamine and norepinephrine that excite the arousal and attention related areas of the brain. But excess or prolonged increase in these neurotransmitters affects the brain and may cause mental sluggishness, rigidity and decreased efficiency [8]. There is a delicate scale of neuronal interaction between the stimuli of the excited state and the inhibitory post-synaptic neuron. Disruptions of this system, as caused by over training or by prolonged physical exertion can lead to impaired cerebral function and consequently a decline in performance [9].

3 Advanced Nano-encapsulation Technologies for Targeted Nutrient Delivery

It is possible to identify close relationship between cognitive efficiency of athletic performance and the neural substrate depending on brain functioning. Working memory, attention, decision making and other executive functions are mediated by a network of neurotransmitters and synapses as well as brain regions including the prefrontal cortex and hippocampus [12]. In athletes, these cognitive processes are crucial for the application of the mental challenge required in the specific sport as well as for the overall performance of a sport performer. Dopamine, serotonin and acetylcholine are the important neurotransmitters that regulate arousal, attention and memory all of which are essential for optimal cognitive performance [13]. Furthermore, since the brain is biological, it is resilient from repeated physical and mental strains while at the same time the brain being plastic allows an individual to fine tune performance to achieve peak efficiency [8].

Over the past few years, there has been a revelation of superior approaches to nano-encapsulation on targeted delivery to the brain. These technologies refer to the packaging of bioactive compounds and nutrients in nanoscale carriers where the desired release takes place within certain areas of the brain [10]. Feasibility of nano-encapsulation for delivery of compounds like Omega 3 Fatty Acids, antioxidants and B-vitamins distinctly to CNS increases the bio availability and effectiveness of these nutrients in enhancing cognitive function [12]. Increased particle surface area and small size as features of nanoparticles make it easier for them to penetrate the blood–brain barrier, an issue in the delivery of nutrients. Since the body is able to prioritize distribution of nutrients to the brains most demanding parts during competition through training and use of these technologies, then there lies great potential to enhance the general cognitive function of the athletes [13].

Pronounced superior efficiency of nutrient release and substantially decreased rates of side reaction because of reduction of compounds' systemic exposure at the targeted nutrient delivery through nano-encapsulation [14]. Such precision may lend itself in regulating neuronal communication mechanisms, growth, and protection, all of which are critical for optimising cognitive endeavour under stress. Specifically, athletes get benefits related to the improvement in mental state and quickness of the process of cognitive restoration after the physical load in the case of the individual approach to the supply of nutrients [9]. Moreover, enhancing the brain performance evidenced by these delivery systems could also benefit for any sport in terms of speed in decision making processes, coordination and stamina or mental well-being. The pairing of newly developed nanotechnology with the realm of orthomolecular nutrition science could also prove to be most beneficial to the cognitive performance of athletes at all levels of intensity [15].

4 Comparison of Traditional Versus Nano-enhanced Nutrient Absorption in Athletes

Neurophysiological performance in athletes has been established to be a product of neural organization, neurochemistry and brain activity. Neural feedback pathways, chemical mediators, and synaptic connectivity are the primary mechanisms that underlie cognitive function such as motor output, decision-making as well as reaction time [14]. They use their brain to reason very fast when it comes to any kind of vigorous exercise that they engage in. Chemicals in the brain such as dopamine, norepinephrine and serotonin affect the functions in the brain like attention, memory and moods other than affecting performance. These neurotransmitter systems are directly affected by metabolic and nutritional status and therefore determine an athlete; focus, learning and reaction to various stimuli during performance [13]. A clear understanding of neurophysiological processes is crucial for any athlete in order to have excellent working capacity and a fully developed capacity to think in constrained and highly pressurized environments [14].

Incorporation of nutrients is very crucial s in supporting the well being of the brain area. Originally, nutrient utilization goes through the digestive Tract where enzymes and nutrient transporters transmembrane proteins enable the uptake of macronutrients and micronutrients [16]. This process, although being very effective, can be affected by several aspects such as gut health, enzymatic activities and the ability of the certain nutrients to be absorbed in the system [7]. Because nutrients form part of energy substrates during and immediately after physical activity, the turnover rate is generally rapid and may become a limiting factor in nutrient solubility within the gastrointestinal tract in athletes [17]. Moreover, processes such as water depletion, intestinal upset, and nutrient interference can interfere with absorption causing temporary memory disturbances during training and competition.

Nano-enhanced nutrient absorption as a new strategy to delivering nutrients is among the promising methodologies to end the conventional nutrient delivery challenges. Nanotechnology is the process of handling materials at the nanoscale level hence improving on solubility, stability and bioavailability of nutrients [17, 18]. Nano encapsulation even has greater absorption over centres like the blood–brain barrier; for example, nano-encapsulation as a protected nanostructure containments within active ingredients. It allows for the straight delivery of such compounds such as omega 3 fatty acids, antioxidants and B vitamins to the brain which will improve neuro plasticity, decrease oxidative stress and also elevate natural memory and attention [13]. Nano-enhanced nutrients can help athletes enjoy improved and more prolonged mental acuity by making desirable brain nutrients always available whenever they are most required. This advancement has implications for the enhancement of physical and mental function in athletes, as it overcomes limitations of delivery of nutrients under conditions of stress, which are typical of athletic competition [4].

5 Key Nano-nutrients for Cognitive Enhancement for Athletes

5.1 Nano-formulated Phospholipids

In fact phospholipids are essential in the structure of cells and are more important in the cells that make up the brain since they are the cells that coordinate the body's functions or brain. In subsequent developments, phospholipids have been redesigned into nanoparticles to improve their uptake and delivery into particular regions of the brain [16]. Such nano-formulated phospholipids as phosphatidylserine and phosphatidylcholine are now being investigated for enhancing athletes' cognition [19]. It is very important that these phospholipids are shrunk down to a few nanometres allowing for large a surface area to come into contact with the blood brain barrier and be absorbed accordingly. This advanced delivery system assures that these important molecules get to neuronal membranes and where they offer their positive impact on the brain and its functions [13].

Phosphatidylserine and phosphatidylcholine are two essential acids that play a fundamental role in the preservation of synapse and neuronal transmission. When templated at the nanoscale they can promote the changes in the biophysical properties of synapse membranes that are required for neuronal plasticity and synaptic transmission [19]. In the process of identifying periods most optimal for athletic performance, this issue of synaptic membrane fluidity is relevant. It allows for quicker signal relay across neurons meaning that reaction time, decision making processes and overall brain performance when in sports will be enhanced [20]. Moreover, the having high permeability of these membranes affects the fluctuating distribution of the receptors and neurotransmitters; the neural substrate of the information processing in the brain under physically strenuous conditions is affected [18].

On the molecular level, the enhancement of cognitive performance using nano-formulated phospholipids is associated with the ability of these compounds to modulate major biochemical signals within the brain [20]. In its capacity as a signal transducing molecule, phosphatidylserine is involved in the regulation of enzymes for example protein kinase C this enzyme is important in processes of learning and remembering. In contrast, phosphatifidylcholine is a structural component which yields acetylcholine, a true neurotransmitter essential to attention, learning and memory processes [10]. Therefore, while increasing the bioavailability of these phospholipids to the brain, FHN improves the synthesis of acetylcholine in the body and hence enhances brain productivity. Moreover, nanoscale formulations help to maintain higher concentrations of these molecules in the brain, countering fatigue for longer loosen and providing cognitive improvements that athletes require for optimum performance for several hours [21]. The molecular mechanisms involved they therefore provide the background justifying why phospholipids formulated in the nano scale could enhance learning abilities in sportsmen and women and consequently physically and mentally strenuous high intensity sports [20].

5.2 Nano-engineered Antioxidants

The new wave of antioxidants developed using nanotechnology has the potential of improving brain performance particularly of athletes whose body undergoes significant oxidative stress resulting from activity [20]. It is unfortunate that conventional antioxidants hindered by low bioavailability, and they are not as precise in reaching areas where oxidation commonly occurs. However, by depositing nanotechnology, antioxidants can be designed at the molecular level with great efficiency to combat free radicals [22]. These molecules are man-made to have an increased surface area and high solubility, which enable them to diffuse into tissues far and reach areas where the drug's activities are desired such as the brain and muscles. Indeed, this programmatic approach serves a more effective pick-up of free radicals, thus decreasing the cellular injury and, at the same time, eliminating the negative impact of oxidative stress on cognitive functions such as memory, concentration, and clarity of mind [20].

This is because many athletes suffer from mitochondrial diseases that lead to less energy production in the brain of the athletes engaged in high intensity exercises. The mitochondria, often referred to as the powerhouse of the cell, is responsible for ATP synthesis that supports cognitive and physical activities [22]. The antioxidants nano-engineered on the other play a big role in averting this oxidative attack on the mitochondria by stabilizing mitochondrial membranes and improving their functions [18]. They also contain antioxidants, which can apparently influence mitochondria, maintaining their structure and enhancing the utilization of energy-bearing substrates. Therefore, athletes are able to train for longer; recover at a faster rate; and think more clearly, which are major assets in the development of mentally and physically high octane bodies needed during training and competition [23].

Definitely, the developments of the improving nano-engineered antioxidants include the nano-curcumin as well as the nano-resveratrol, because such product has positive impacts to the cognition. Curcumin, the bioactive compound found in turmeric, is an anti-inflammatory and antioxidant compound of remarkable potency [24]. However the bioavailability of this drug is very low thus it cannot be effectively administered through ingestion. The form of curcumin used in the study is curcumin nanoparticles wherein the compound gets modified to better penetrate the blood brain barrier and offer the neuroprotective effect [22]. Nano-curcumin has an ability to construe neuro inflammation, acting as neuron shield against oxidative stress, and boosting activeness of brain; all these property makes nano-curcumin as an appropriate supplement for athletes with a vision of improving their cerebral functions [24]. In the same way, another compound is nano-resveratrol, which is a polyphenol from red wine, which increases cognitive function by altering the activity of mitochondria and decreasing oxidative stress. Nano-resveratrol improves cerebral functions and at the same time helps to regulate circulation and oxygenation during dynamic physical activities. When used in compliance with these specific recommendations in the present to combat tissue oxidative stress, and to gain a significant increase in activity levels, these advanced antioxidant strategies can improve cognition and athletic performance [22].

5.3 Nano-mineral Complexes

Nano-mineral complexes prove to be an innovative concept of improving cognitive abilities in the sportsmen because they are endowed with better richness, and they require complexation to form complexes that will elicit better properties in athletes [25]. If incorporated into nano particles the mineral such as zinc, magnesium and selenium are capable of passing barriers in a cell and are efficiently transported to the different parts of the body. These minerals are essential for many physiological functions, especially in the brain where they have very important functions related to learning and memory [7]. The nano-formulation of these minerals increases its bioactivity; and makes sure that the mineral particles are present at the cellular level in sufficient quantity to have positive effects on neuronal functioning, signalling and cognition. Through nanotechnology, the essential minerals are transported to the areas of the brain that require them aiding in mental definition, focus and the time demand by athletes [23].

Zinc, magnesium and selenium are necessary for synthesis of neurotransmitter and neural signal transmission which are two key aspects of normal brain function. Zinc is involved in control of mechanisms of dopamine, glutamate, serotonin and other neurotransmitters which are important for mood, motivation and flexibility [25]. Magnesium, involved in synaptic plasticity and neuroprotection, regulates the flow of a signal between neurons and activates enzymes associated with the regulation of neurotransmitter release [26]. These minerals are complimented by Selenium, an antioxidant that guards against oxidative stress to neuronal tissues thus interfering with signal transduction. Combined these minerals help to maintain the homeostasis of neurotransmitters in the body so that signals between neurons in the brain are clear and unwavering [7]. These increased functions of neurotransmitters are especially advantageous for the athletes since they entail mental concentration, quick responses, and ability to make decisions during a high pressure performance, aspects which form part of the champions' performance.

The enhanced cognitive resilience and neuronal stress resistance by zinc, magnesium, and selenium nano-formulations are especially important to athlete experiencing the mental and physical pressures of training and competition. These minerals help the brain to cope with stressors and rehabilitate when complexity of cognitive loads is high [26]. The metal affects the hypothalamic–pituitary–adrenal axis of the brain so as to attain a balance in response to stress influences from the environment. Magnesium is said to be a natural relaxant for the central nervous system, which means it will fat down the stress pathways and increase tolerance to both short term and long term stress [7]. Selenium has antioxidants that prevent neurons from being damaged by reactive species which are produced during stressful moments as long as the brain cells are functioning optimally. Combined, these nano-mineral complexes support the fundamental structure of cognitive attitudes and allow athletes to think clearly and produce optimal results also under high stressful conditions resulting from trainings or competitions [26].

6 Neuroimaging and Biomarker Analysis

Neuroimaging techniques have evolved as rather elaborate devices in assess of cognitive performance, especially in the sporting context. These methods resolve the need to invade the brain and provide a detailed look into its activity, morphology and connectivity patterns and can provide a unique view of how different therapies including nano-nutrients can impact cognition. Scans like fMRI and PET scans: The two aforementioned methods are two advanced neuroimaging techniques that have changed the way the brains reaction to these external factors such as nano-nutrient supplementations are viewed fMRI gives real-time information on the blood oxygen levels of the brain and the PET scans showing the most active areas of the brain during a specific task [24]. This makes it possible to capture the variations in the neural activity as may be caused by the enhancement of cognition through nano-nutrients. In direct contrast, PET scans allow the visualisation of biochemistry by observing the distribution and uptake of radio labeled compounds in the brain. This gives a glimpse into the molecular activity of nano-nutrients and whether they are neuroprotective or have attendant cognitive boosting capabilities [27].

Actual findings from fMRI and PET scans have revealed the works of nano-nutrients in stimulating circuits in the neuronal realm. The brains of humans contain many neural networks with connections throughout them and disruptions to the networks may cause cognitive impairments that effects athletic performance [28]. Nano-nutrients can pass through the blood–brain-barrier and hence come into direct contact with neurons and glial cells. Neuroimaging studies have shown that the intake of some nano-nutrients result in changes of the brain's electrical wiring, especially those areas of the brain associated with memory, attention, and executive function [27]. These findings imply that nano-nutrients might help improve higher mental activities via optimizing neural activity in these pivotal parts of the human brain, Decision making and Motor coordinating are some crucial tasks for every athlete, fMRI has revealed stronger connections between the various networks that are crucial for such cognitive functions [28]. PET scans have also proved that the nano-nutrient supplementation affect the neurotransmitter system and this is a further suggestion that these nutrients can be of help in enhancing the cognition through affecting biochemical activity at this level [29].

Molecular biomarkers are central to the comprehension of the processes that underlie cognitive improvement. These biomarkers are founded on easy to measure characteristics of biological activities and can be used to determine the status of brain health and function. Since nano-nutrients are nanometre-scale particles, molecular biomarkers provide an understanding of how these substances affect cell changes that lead to enhanced cognition. The BDNF, neurotrophins and cytokines are some of the marker molecules that are used in related research to evaluate cognitive ability [24]. BDNF for example is one such protein that has a component in the brain's ability to change, grow new neurons and from synaptic formation also referred to as synaptic plasticity [30]. They found out that high levels of BDNF are related to increased memory and learning capacity which are critical in athletes. It also alter the density of BDNF and other neuotrophic factors: positive neural regeneration; improves cognitive function. Moreover, biomarkers prognosticate the efficacy of the nano-nutrient intercessions meaning that the specialists in this field will be able to present proof to the efficacy of these substances as enhancers of cognitive functions [29].

Another important process by which improvement in cognition might be achieved using nano-nutrients include attenuation of neuro inflammation and enhancement of synaptic density. It is well established that, long-standing neuro inflammation contributes to cognitive dysfunction and is implicated in a spectrum of neurodegenerative diseases [7]. In athletes, neuro inflammation is frequently caused by excessive physical exertion that triggers oxidative stress and produces inflammatory neurotoxins that damage the brain. Nano-nutrients seem to decrease neuro-inflammation since they possess minimal inflammation and reduced markers of inflammation have been demonstrated in animals and humans [31]. As anti-inflammatory and anti-hydroxicity agents, these nano-sized compounds can direct the alteration of molecular signals of inflammation and oxidative stress in the nervous system toward a more conducive physiology for cognition. Moreover, it will be shown that nano-nutrients can have neuro-protective and neuro-restorative effects through improved synaptic functioning and neurotransmission efficiency [7]. Increased integration of the prefrontal cortex and the hippocampus demonstrates better efforts to remember, faster decision making and better co-ordination all of which are components of the best of the physical performances. Therefore not only are nano-nutrients anti-neuro inflammatory but also help in rewiring or making the neural connections that facilitate proper cognition specific to athletic durability [31].

7 Personalized Nano-nutrient Cognitive Optimization

Designing special nano-nutrients for enhancing the cognitive ability is another very effective and Genius way to take the particular intervention plan of athlete based on their physiological and genetic background that exists within them. That genetic variance is at the centre of how the body processes and reacts to nutrients, nano-nutrient supplementation can then be regulated to match the genetic blue-print of the athletic individual [32]. Polymorphisms in genes encoding the factors for nutrient uptake, transport, and utilization, including vitamin, minerals, and bioactive substances can affect cognitive performance, stamina, and rate of recovery [31]. In this case, using single nucleotide polymorphisms or biomarkers linked to nutrient processing enzymes or neurotransmitter genes, appropriate nano-nutrient regimes can be recommended to augment human cognitive performance during high performance athletes thereby improving originality, attention, and choice making [33]. It also proceeds to receive the most cognitive profit while excluding the undesirable negligence that is bound to develop a nutrient density, density ratio, or an overall malady that is hazardous to the health and work productivity of the athlete, thus adopting the best of the best and leaving out the worst of the worst, and accomplishing a balanced and scientifically unparalleled technique towards the accomplishment of the extraordinary in the sporting industry [9].

AI and machine learning have therefore played a very crucial role in enhancing enhancements of personalized nano-nutrient cognitive enhancement protocols. By incorporating in big data genetic profile, biometric and performance data, machine learning models, can be used to forecast and fine-tune specific cognitive impact stoke points pertaining to nano-nutrient in-take from large database sets pertaining to the athlete [29]. Such AI systems can also look at an athlete's previous ability, state of health, and even cognitive values that correspond to the particular circumstances of an athletic event to guide him through a particular set of nano-nutrient mixes appropriate to his physically peculiar character. Real time solutions allow AI models to modify and enhance the suggestions on cognition permanently, thus providing a chance for the progressive strategy in cognition improvement [34]. For instance, it involves making adjustment to the nano-nutrient intake depending on the mental fatigue level of an athlete or depending on the result of a cognitive test, or depending on the level of intensity of the competition environment. Such a level of customization enables athletes to achieve nano-nutrient productivity that can be drawn on to create better training regimens, build stronger and more robust minds, and produce superior game performances [9].

Precision nutrition, which forms the basis of individual athletic training, in addition to genetics considers lifestyle, activity level, and specific physical exercise requirements to enhance cognitive ability through Nano-Nutrient solutions. Unlike the traditional approach of prescribing the macronutrient values for an athlete, precision nutrition provides tailored nutritional plans depending on the athlete's individual nature in terms of metabolism [33]. This include the integration of genetic data including genetic tests, microbiome data and blood markers in developing nutrient profiles for each person. However, in situation where cognitive performance is of myocardial concern precision nutrition for athletes entails more than just Vitamins and minerals balance. Beginners of high-intensity sports get better neuroprotective effects of certain antioxidants than those in low intensity sports, while those that involve quick thinking and decision making could benefit from certain amino acids required for neurotransmitter synthesis [35]. With this understanding of the problem, athletes are provided with the nuanced and evidence-based approach on how to sustain emotionally healthy and focused brain in high-pressure training and competition environments for maximizing athletic performance.

8 Potential Risks and Ethical Considerations

The use of nano-nutrient application-based strategies is an innovative, emerging field of research for improving athletic performance, but it has many risks that must be considered. Arguably one of the most important areas of research in this context is nano-toxicology or the scientific study of the toxicological effects of nanomaterials [28]. As nano-carbons, nano-nutrients inherently have different physical and chemical characteristics than their bulk counterparts. Because of their small size and high surface area, these particles are interacting with biological systems in ways that are still poorly understood [34]. Not enough time to determine the fate of a nanoparticle-enriched product, once inside the human body, so could potentially be stored through the organs or tissue, causing adverse effects like inflammation, oxidative stress, or tissue damage [28]. Owing to their size, long-term safety evaluations are needed to ascertain if such particles can indeed remain within the body harmlessly, or whether risks will develop with time, such as neurotoxicity or genotoxicity. With the growing applications of these nano-based strategies, it is crucial to provide rigorous studies to ensure their safety and minimize side effects, especially regarding chronic toxicity of these new materials [36].

At this point, the regulatory environment of nano-nutrient supplementation is in current flux. However, the body of evidence supporting the benefits of nano-nutrients in sports performance is growing. The problem has been that regulatory agencies have been playing catch-up with developments in nano-technology [34]. In most countries, for example, the use of nanomaterials in food products, supplements, or pharmaceuticals is not clearly defined within regulatory frameworks. Thus, you could have a situation where nano-nutrient-based products are introduced to the market without proper testing and surveillance, and they might be misused or used in ineffective or even damaging supplementation [37]. Such frameworks should therefore also consider the potential environmental impact that goes along with their use, as, besides human health, nanomaterials might pose risks to ecosystems [34].

Ethical considerations arise, too, when the tactics merge with nano-nutrient approaches in more customizable strategies for cognitive enhancement relating to athletic performance. It questions fairness and equity in the competitive environment, and more important, whether there is a potential of coercion embedded in the dynamic [36]. While emerging cognitive enhancement technologies may be advantageous, misuse may trigger an arms race where athletes feel compelled to use such technologies to remain competitive, irrespective of how they feel about the intervention on an individual basis. This may seriously distort sport as those who cannot afford or simply do not want to use such technologies are left at great competitive disadvantages [35]. The issue of whether or not the use of nanotechnology to achieve cognitive enhancement through nano-nutrients compromises the natural integrity of athletic performance raises ethical questions. Should mental capacities be allowed to develop artificially, or must everything depend on the natural endowments of athletes? The general implications of cognitive enhancement above and beyond the physical domain are issues that have to be thought through rather carefully in the name of a balanced drive toward fairness in competitive sports [36].

Transparency and informed consent are fundamental to ethical practices in implementing nano-nutrient interventions for athletic excellence. Being new and an extension of complex technologies, athletes should always be provided clear information, made as simple as possible, on risks, benefits, and possible side effects of such interventions [35]. Informed consent should, thus, not be just an administrative part of the whole process. One should allow dialogue between the athletes and healthcare professionals about the science of the intervention, its possible long-term effects, and possible health risks [10]. Transparency enables the athletes to make well-informed decisions about their role and participation in nano-nutrient-based strategies. Such consideration must include adequate consultation and review by medical practitioners, bioethics committees, and any other necessary reviews involving the regulatory principles to respect the autonomy of athletes [38]. The nature of nanotechnology is rapidly developing and advances on new findings must be carried forth in the information supplied to athletes, ensuring consent is consistently informed with the passage of the intervention [10]. Ethics should ensure that athletes remain safe and unharmed in the way interventions are done, always ensuring that they promote the best interests of individuals.

9 Future Directions and Emerging Technologies

CRISPR technology in conjunction with nano-nutrients may open up new avenues for the optimization of cognition in athletic capacities. Carbon-based nanomaterials, including nanotubes and nanofibers, have gained immense popularity in sports and exercise science primarily due to their versatile applications in sport [38]. They can be utilized as drug delivery systems, blood substitutes, and solid-state medicines, helping athletes from various fields partake in optimizing their cognitive processes.

When combined with CRISPR, nano-nutrients could help create personalized supplementation strategies that allow for precise genetic modifications that enhance absorption and utilization and increase the efficiency of cognitive-boosting nano-nutrients [39]. This synergy would not only optimize brain function and neuroplasticity but also contribute to cognitive resilience and recovery from cognitive fatigue. Thus, ability-based nutrient profiling at the level of genes with regard to genetic predispositions in athletes for concentration, decision-making, and overall performance improvement possesses a potential avenue toward elevating threshold boundaries in any sport [40].

Emerging nanodelivery platforms represent one of the key advances in sports cognitive enhancement. They allow, because of their enhanced bioavailability over classical delivery systems, a more effective absorption of nutrients into the bloodstream and through the blood–brain barrier [38]. They have been designed to deliver an accurate dose of nano-nutrients into target regions of the brain, namely, the hippocampus, prefrontal cortex, and amygdala-essential for memory, attention, and emotional regulation. New delivery systems, such as lipid-based nanoparticles and dendrimers provide enhanced stability and release profiles with control over efficient therapeutic activity while minimizing side effects [39]. Such benefits offer long-lasting cognitive performance with less need for frequent supplementation, which could result in enhanced concentration, decreased mental fatigue, and quicker recovery. These systems can also develop individualized cognitive enhancement regimens suited to the unique needs of the athletes, thereby benefitting them in short-term performances and long-term cognitive health [37].

Interdisciplinary science needs to arise between nano-nutrient research and athletic performance. For holistic applications to enhance cognitive function in athletes, nanotechnology, biochemistry, neurobiology, sports science, and genetics must come together [32]. Such collaboration is necessary between material scientists, bioengineers, nutritionists, and sports psychologists in developing cognitive enhancement approaches. For instance, a sports scientist could advise how to apply nano-nutrients in practical athletics, while neurobiologists could investigate the effects on brain structure and function [38]. Genomic researchers could demonstrate the genetic variance related to nutrient absorption and utilization ability and personalize the nano-nutrient supplementation [32]. With such interdisciplinary study, the optimizing mental performance may just lead to a zenith of yielding great discoveries for an athlete to, hence, outperform their opponents via cognitive superiority along with the help of training.

Nano-nutrients are now being modeled to predict performance enhancement using cognitive enhancers. Computerized models allow investigators to test how certain nano-nutrient composites might affect cognitive performance in one athletic setting versus another. These models can combine data from multiple sources (e.g. genetic profiles, nutritional status, brain activity, training regimens) to enable predictions on how a given nano-nutrient intervention will modulate cognitive outcomes [41]. If nano-nutrient doses had predictive validity, supplementing with the appropriate amounts of certain nano-nutrients could provide tailored cognitive enhancement, without the need for trial-and-error methods that inhibit performance gains and are financially inefficient [39]. We can use those models of prediction to observe the effect of cognitive supplementation on long-term neurodegeneration and decline in cognitive function in athletes, redetecting how nano-nutrients can be used over time to manipulate cognitive health towards better performance during the lifetime of an athlete [32]. Predictive modelling, as the computational tools and data mature, will without a doubt be the one of the most prominent drivers of the scientific advancement of cognitive enhancement in sport.

10 Conclusion

Research using neuroimaging techniques and biomarker analyses has provided substantial evidence supporting the effectiveness of nano-nutrient interventions in improving memory, focus, and decision-making abilities in athletes [41]. By combining personalized approaches, fueled by genetic profiling and artificial intelligence, tailored cognitive enhancement protocols have been created to cater to the specific requirements of individual athletes [36]. Nevertheless, this emerging field also raises significant concerns regarding safety, regulation, and ethics, particularly in relation to long-term effects and fairness in competition. As this field progresses, collaboration between different disciplines and the emergence of technologies like CRISPR and advanced delivery platforms present exciting prospects for future advancements [39]. The integration of nanotechnology with sports nutrition represents a ground breaking achievement in athletic performance enhancement, potentially transforming the way athletes approach cognitive optimization while underscoring the importance of ongoing research and thoughtful evaluation of both advantages and potential drawbacks.

References

1. Tran DMD, Westbrook RF (2017) A high-fat high-sugar diet-induced impairment in place-recognition memory is reversible and training-dependent. Appetite 110:61–71. https://doi.org/10.1016/j.appet.2016.12.010
2. Cooper SB, Bandelow S, Nevill ME (2011) Breakfast consumption and cognitive function in adolescent schoolchildren. Physiol Behav 103:431–439. https://doi.org/10.1016/j.physbeh.2011.03.018
3. Hassmén P, Blomstrand E, Ekblom B, Newsholme EA (1994) Branched-chain amino acid supplementation during 30-km competitive run: mood and cognitive performance. Nutrition 10:405–410
4. Winnick JJ, Davis JM, Welsh RS, Carmichael MD, Murphy EA, Blackmon JA (2005) Carbohydrate feedings during team sport exercise preserve physical and CNS function. Med Sci Sports Exerc 37:306–315. https://doi.org/10.1249/01.mss.0000152803.35130.a4
5. Roberts SP, Stokes KA, Trewartha G, Doyle J, Hogben P, Thompson D (2010) Effects of carbohydrate and caffeine ingestion on performance during a rugby union simulation protocol. J Sports Sci 28:833–842. https://doi.org/10.1080/02640414.2010.484069

6. Hoyland A, Dye L, Lawton CL (2009) A systematic review of the effect of breakfast on the cognitive performance of children and adolescents. Nutr Res Rev 22:220–243. https://doi.org/10.1017/S0954422409990175
7. Cheng Y-C, Huang Y-C, Huang W-L (2020) The effect of vitamin D supplement on negative emotions: a systematic review and meta-analysis. Depress Anxiety 37:549–564. https://doi.org/10.1002/da.23025
8. Nilsson A, Radeborg K, Björck I (2009) Effects of differences in postprandial glycaemia on cognitive functions in healthy middle-aged subjects. Eur J Clin Nutr 63:113–120. https://doi.org/10.1038/sj.ejcn.1602900
9. Meftahi GH, Hatef B, Pirzad Jahromi G (2023) Creatine activity as a neuromodulator in the central nervous system. Arch Razi Inst 78:1169–1175. https://doi.org/10.32592/ARI.2023.78.4.1169
10. Murray-Kolb LE, Beard JL (2007) Iron treatment normalizes cognitive functioning in young women. Am J Clin Nutr 85:778–787. https://doi.org/10.1093/ajcn/85.3.778
11. Przybelski RJ, Binkley NC (2007) Is vitamin D important for preserving cognition? A positive correlation of serum 25-hydroxyvitamin D concentration with cognitive function. Arch Biochem Biophys 460:202–205. https://doi.org/10.1016/j.abb.2006.12.018
12. Vellekkatt F, Menon V (2019) Efficacy of vitamin D supplementation in major depression: a meta-analysis of randomized controlled trials. J Postgrad Med 65:74–80. https://doi.org/10.4103/jpgm.JPGM_571_17
13. Peeling P, Binnie MJ, Goods PSR, Sim M, Burke LM (2018) Evidence-based supplements for the enhancement of athletic performance. Int J Sport Nutr Exerc Metab 28:178–187. https://doi.org/10.1123/ijsnem.2017-0343
14. Martin-Niedecken AL, Schättin A (2020) Let the body'n'brain games begin: toward innovative training approaches in esports athletes. Front Psychol 11:138. https://doi.org/10.3389/fpsyg.2020.00138
15. Gómez-Pinilla F (2008) Brain foods: the effects of nutrients on brain function. Nat Rev Neurosci 9:568–578. https://doi.org/10.1038/nrn2421
16. Dye L, Lluch A, Blundell JE (2000) Macronutrients and mental performance. Nutrition 16:1021–1034. https://doi.org/10.1016/s0899-9007(00)00450-0
17. Chambers ES, Bridge MW, Jones DA (2009) Carbohydrate sensing in the human mouth: effects on exercise performance and brain activity. J Physiol 587:1779–1794. https://doi.org/10.1113/jphysiol.2008.164285
18. Staśkiewicz-Bartecka W, Kardas M, Zydek G, Zając A, Chycki J (2024) Changes in body composition and nutritional periodization during the training macrocycle in football-a narrative review. Nutrients 16:1332. https://doi.org/10.3390/nu16091332
19. DiFrancisco-Donoghue J, Balentine J, Schmidt G, Zwibel H (2019) Managing the health of the eSport athlete: an integrated health management model. BMJ Open Sport Exerc Med 5:e000467. https://doi.org/10.1136/bmjsem-2018-000467
20. Ferrauti A, Weber K, Strüder HK (1997) Metabolic and ergogenic effects of carbohydrate and caffeine beverages in tennis. J Sports Med Phys Fitness 37:258–266
21. Aslam H, Green J, Jacka FN, Collier F, Berk M, Pasco J, Dawson SL (2020) Fermented foods, the gut and mental health: a mechanistic overview with implications for depression and anxiety. Nutr Neurosci 23:659–671. https://doi.org/10.1080/1028415X.2018.1544332
22. Stuart GR, Hopkins WG, Cook C, Cairns SP (2005) Multiple effects of caffeine on simulated high-intensity team-sport performance. Med Sci Sports Exerc 37:1998–2005. https://doi.org/10.1249/01.mss.0000177216.21847.8a
23. Scholey AB, French SJ, Morris PJ, Kennedy DO, Milne AL, Haskell CF (2010) Consumption of cocoa flavanols results in acute improvements in mood and cognitive performance during sustained mental effort. J Psychopharmacol 24:1505–1514. https://doi.org/10.1177/0269881109106923
24. Fila M, Chojnacki C, Chojnacki J, Blasiak J (2021) Nutrients to improve mitochondrial function to reduce brain energy deficit and oxidative stress in migraine. Nutrients 13:4433. https://doi.org/10.3390/nu13124433

25. Gao S, Jin Y, Hall KS, Liang C, Unverzagt FW, Ji R, Murrell JR, Cao J, Shen J, Ma F, Matesan J, Ying B, Cheng Y, Bian J, Li P, Hendrie HC (2007) Selenium level and cognitive function in rural elderly Chinese. Am J Epidemiol 165:955–965. https://doi.org/10.1093/aje/kwk073
26. Duncan MJ, Taylor S, Lyons M (2012) The effect of caffeine ingestion on field hockey skill performance following physical fatigue. Res Sports Med 20:25–36. https://doi.org/10.1080/15438627.2012.634686
27. Ari Y, Çakir E (2022) Correlation between participation in physical activity and healthy nutrition: an example of a sports science faculty. Baltic J Health Phys Activity 13. https://doi.org/10.29359/BJHPA.13.3.05
28. Jakobsen LH, Kondrup J, Zellner M, Tetens I, Roth E (2011) Effect of a high protein meat diet on muscle and cognitive functions: a randomised controlled dietary intervention trial in healthy men. Clin Nutr 30:303–311. https://doi.org/10.1016/j.clnu.2010.12.010
29. Pontifex KJ, Wallman KE, Dawson BT, Goodman C (2010) Effects of caffeine on repeated sprint ability, reactive agility time, sleep and next day performance. J Sports Med Phys Fitness 50:455–464
30. Lieberman HR (2003) Nutrition, brain function and cognitive performance. Appetite 40:245–254. https://doi.org/10.1016/s0195-6663(03)00010-2
31. Goulart JB, Aitken LS, Siddiqui S, Cuevas M, Cardenas J, Beathard KM, Riechman SE (2023) Nutrition, lifestyle, and cognitive performance in esport athletes. Front Nutr 10:1120303. https://doi.org/10.3389/fnut.2023.1120303
32. Vicente-Salar N, Fuster-Muñoz E, Martínez-Rodríguez A (2022) Nutritional ergogenic aids in combat sports: a systematic review and meta-analysis. Nutrients 14:2588. https://doi.org/10.3390/nu14132588
33. Philippou E, Constantinou M (2014) The influence of glycemic index on cognitive functioning: a systematic review of the evidence. Adv Nutr 5:119–130. https://doi.org/10.3945/an.113.004960
34. Cooke MB, Rybalka E, Williams AD, Cribb PJ, Hayes A (2009) Creatine supplementation enhances muscle force recovery after eccentrically-induced muscle damage in healthy individuals. J Int Soc Sports Nutr 6:13. https://doi.org/10.1186/1550-2783-6-13
35. Grzywacz E, Jaron A (2022) Well-being and mental health—diet, supplements, exercise or sleep? A review of reports from the last five years. Baltic J Health Phys Activity 12. https://doi.org/10.29359/BJHPA.12.2.08
36. Rawson ES, Miles MP, Larson-Meyer DE (2018) Dietary supplements for health, adaptation, and recovery in athletes. Int J Sport Nutr Exerc Metab 28:188–199. https://doi.org/10.1123/ijsnem.2017-0340
37. Wittbrodt MT, Millard-Stafford M (2018) Dehydration impairs cognitive performance: a meta-analysis. Med Sci Sports Exerc 50:2360–2368. https://doi.org/10.1249/MSS.0000000000001682
38. Meeusen R, Decroix L (2018) Nutritional supplements and the brain. https://doi.org/10.1123/ijsnem.2017-0314
39. Baker LB, Conroy DE, Kenney WL (2007) Dehydration impairs vigilance-related attention in male basketball players. Med Sci Sports Exerc 39:976–983. https://doi.org/10.1097/mss.0b013e3180471ff2
40. Du K, Markus E, Fecych M, Rhodes JS, Beverly JL (2018) Satiety and memory enhancing effects of a high-protein meal depend on the source of protein. Nutr Neurosci 21:257–267. https://doi.org/10.1080/1028415X.2016.1277055
41. Jacka FN, Cherbuin N, Anstey KJ, Butterworth P (2014) Dietary patterns and depressive symptoms over time: examining the relationships with socioeconomic position, health behaviours and cardiovascular risk. PLoS ONE 9:e87657. https://doi.org/10.1371/journal.pone.0087657

Safety and Efficacy of Nano-supplements in Sports

Pulen Das and Emmanuel Achumboro Apanga

Abstract This chapter explores the burgeoning field of nanotechnology in sports nutrition, examining its potential benefits and associated risks. This chapter delves into how nanoparticles enhance nutrient absorption, target delivery, and improve athletic performance parameters such as muscle strength, endurance, and recovery. Furthermore, chapter discusses the critical safety concerns, including potential toxicity, tissue accumulation, and the lack of robust regulatory frameworks. The ethical implications, including fairness in competition, informed consent, and anti-doping considerations, are also addressed. Finally, chapter highlight the need for further research, stricter regulations, and a multi-stakeholder approach to ensure the responsible and ethical development and utilization of nanotechnology in sports nutrition.

Keywords Nanotechnology · Sports nutrition · Nanoparticles · Athletic performance · Safety · Toxicity · Regulation · Ethics · Anti-doping · Informed consent

1 Introduction

Nanotechnology has enabled substantial advances in sports nutrition. Nano-supplements enhanced vitamin absorption and utilization when compared to other types of supplements. These cutting-edge advancements have the potential to increase productivity, minimize recuperation times, and improve overall health. Although there are benefits, we must not ignore the many serious risks. Because of their tiny size and strong reactivity, nanoparticles may pass through the intestinal wall as well as the blood–brain barrier. Although this helps the body absorb nutrients, it also

P. Das (✉)
Lakshmibai National Institute of Physical Education NERC Guwahati, Guwahati, India
e-mail: pulendas86@gmail.com

E. A. Apanga
Ho Polyclinic, Ghana Health Service, Ho, Volta Region, Ghana

A. Chatterjee et al. (eds.), *Nanofuel: The Future of Sports Nutrition*, Smart Nanomaterials Technology, https://doi.org/10.1007/978-981-96-5471-0_19

increases the likelihood of harmful substances entering the bloodstream. We cannot fully rule out unexpected biological consequences like oxidative stress or inflammation-induced damage. If nanoparticles disrupt the normal gut flora, an already weaker immune system or allergy tendency may worsen.

It may be difficult to determine the potential dangers linked with food and vitamin nanoparticles. Researchers discovered that nanoparticles may exacerbate oxidative stress in cells by releasing more reactive oxygen species (ROS). Sports' demanding schedules have already impacted a sizable segment of society. Silver and titanium dioxide nanoparticles may disturb the normal function of cells and tissues. Genotoxic nanoparticles damage DNA and increase cancer risk. Nanoparticles may disrupt metabolism by affecting ion channels and enzyme activity. Small quantities may not appear worrisome, but long-term interaction and buildup in the body may induce organ-specific poisons and instability. Nanosupplements has the potential to accumulate nanoparticles within tissues, which is a cause for concern. They dwell in the brain, kidneys, liver, and other vital organs and may survive long. Bioaccumulation may cause long-term inflammation, an indication of heart and immune system problems. Nanoparticles in the circulation may harm the brain, causing memory loss and neurological illnesses. Long-term exposure to carbon nanotubes and titanium dioxide increases cancer risk. The accumulation of nanoparticles in muscle cells may harm physical performance by making it difficult to repair and maintain endurance. Nano-supplements' inconsistent and poor regulatory structure raises safety concerns. Global discrepancies in nanoparticle nomenclature and categorization make monitoring tougher. Nanoparticles are distinct from other compounds; therefore, risk assessment approaches don't always work. Many nations lack adequate nano-supplement testing before sale, allowing potentially harmful items to enter the market. The FDA and EFSA have not yet created sports-specific regulations. This indicates coverage deficiencies. The supplement business is worldwide, and new items come forth frequently, making it impossible to enforce laws. The absence of sporting restrictions exacerbates the situation, as their metabolism and body demands differ significantly from those of typical individuals. Nano supplements are riskier due to unclear safety measures. Different manufacturing processes affect product quality. Some vitamins include nanoparticles of unclear safety or composition. Many products are marketed without testing, and labels seldom mention nanoparticles, leaving users in the dark. The long-term consequences of nano supplements and how nanoparticles interact and cumulatively affect us are yet unknown. Manufacturers aren't transparent, so authorities and buyers can't tell how safe something is. For gamers who consume foreign meals, other nations' safety standards might be confusing and harmful. Because nanoparticles affect biochemical and physiological processes, further difficulties arise. They make nutrients more accessible, but they may also alter nutrient absorption, causing anomalies. Nanoparticles may reduce mitochondrial activity, which runners require to create energy. Some nanoparticles mimic hormones or disrupt hormone function. This may disrupt the endocrine system and hinder growth, calorie burning, and healing. Nanoparticles may weaken or over activated the immune system, which is undesirable for athletes. Nanoparticles may delay recovery after strenuous exercise by increasing oxidative stress and inflammation, which is against their purpose. Sports nutrition can undergo a revolution with the careful use of nano supplements [1]. Chemical impacts, tissue development, and long-term health implications demonstrate the need for stringent safety standards and government regulation. Without these guidelines, it's tougher to ensure these medications are safe and effective. Athletes, instructors, and doctors must weigh the advantages and downsides. We need stronger standards, more manufacturer disclosure, and more scientific research to fix these issues. The sports community must understand nano-supplements' safety before they can employ them to enhance performance, health, and fair play.

2 Safety Concerns of Nano-supplements and Sports

People are using nanotechnology to produce food, particularly for gamers. Sports science now provides more options. Nano-supplements may improve nutrition absorption and distribution. However, safety concerns about these inventive solutions are growing. This article discusses the risks of nanomaterials in supplements, nanoparticle toxicity, how they build up in tissues, regulatory challenges, the absence of conventional safety measures, and how they alter metabolic and physiological systems.

2.1 Potential Risks Associated with Nanomaterials in Supplements

Nanoparticles are engineered materials under 100 nm. These substances improve supplement nutrient solubility, absorption, and bioavailability. Still, their small size, large surface area, and responsiveness pose risks: Nanoparticles may readily cross biological barriers, including stomach walls and the blood–brain barrier, due to their size. This could potentially introduce harmful substances into the body.

1. **Unpredictable Interactions**: Nanomaterials may cause oxidative stress, inflammation, or physiological malfunctions.
2. **Eating Risks**: Nanomaterials may interact with gut microorganisms, disrupting the delicate balance essential for health.
3. **Allergy reactions**: Certain individuals may experience intolerance to nanomaterials.

2.2 Toxicological Aspects of Nanoparticles in Foods

The toxicological profile of nanoparticles in food and supplements remains under-researched, raising significant concerns:

ROS from nanoparticles causes oxidative stress in cells: Athletes already under stress from intense training find this particularly problematic [2]. Nanoparticles: like TiO_2 and AgNPs injure cells and tissues. DNA damage by nanoparticles may induce alterations, cancer, and cellular dysfunction. Nanoparticles may disrupt biological systems, including ion receptors and enzymes, disrupting homeostasis. Nanoparticles may look harmless in tiny concentrations, but long-term interaction is harmful.

2.3 Accumulation in Tissues and Potential Long-Term Effects

Nanoparticles' capacity to accumulate in tissues is their deadliest feature. If consumed, nanoparticles may accumulate in organs such as the brain, liver, kidneys, and lungs, causing sustained effects.

1. **Chronic inflammation**: Research has linked chronic inflammation to problems with the heart and immune system.
2. **Neurotoxicity**: Bloodstream nanoparticles might cause memory loss and progressive disorders in the brain.
3. **Cancer risk**: Long-term exposure to nanoparticles like carbon nanotubes and titanium dioxide increases cancer risk. Nanoparticles may accumulate in muscles and tissues, reducing vitamin efficacy.

2.4 Regulatory Framework for Nano-supplements in Sports

Sports nano supplement rules change often. Building a strong government can be challenging for various reasons.

1. Different classifications of nanoparticles make regulation challenging.
2. Poor risk assessment protocols: Traditional safety evaluations seldom include nanoparticles' unique features and dangers.
3. Fewer tests before sale: Many nations don't test nanoparticle supplements before sale, allowing harmful items to be marketed [3].
4. The FDA and EFSA have not provided sports nutrition nanoparticle recommendations.
5. Law enforcement issues: The supplement industry is global, and the constant introduction of new ideas makes enforcement challenging.
6. Specific Athlete Concerns: The current standards ignore athletes' biological and physical demands, creating safety gaps.

2.5 Lack of Standardized Safety Protocols

Nanosupplements are riskier without safety regulations.

1. Uncontrolled production: There are several ways to create nanoparticles with unknown safety or composition, and with varying quality standards [4].
2. Poor product testing and labeling: Many individuals aren't aware certain items contain nanoparticles due to unclear labels.
3. Issues with Research: Nanoparticle effects and combinations aren't fully understood by long-term investigations.
4. Lack of transparency: Nanotechnology businesses don't have to disclose product safety, so regulators and purchasers can't assess them.
5. Different safety standards: Giving players unfamiliar meals might put them in danger since safety laws differ by nation and group.

2.6 Impact on Metabolic and Physiological Processes

(a) Nanoparticles alter metabolic and physiological systems, a major issue.
(b) Nanoparticles may alter nutrition absorption, causing imbalance.
(c) Nanoparticle metabolic stress may reduce mitochondrial and athletic energy production.
(d) Nanoparticles may disrupt the endocrine system, affecting metabolism, growth, and reproduction.
(e) Immune system changes: Nanoparticles may overstimulate or weaken the immune system, increasing or decreasing illness and inflammation risk [5]. Nanoparticles may slow recovery after intense exercise by altering inflammation and oxidative stress.

Nano-supplements are excellent and harmful for sports science. While they offer innovative ways to enhance games, we must not overlook their safety concerns. Be cautious due to chemical effects, tissue development, and long-term health hazards. Lack of safety measures and standards exacerbates the issue. Teachers, athletes, and health care professionals must consider the pros and cons and advocate for stronger restrictions and greater research. Athletes will only accept the promise of nano-supplements once they have established their safety.

The application of nanotechnology in sports nutrition has resulted in the creation of innovative nutritional supplement solutions for athletes and health-conscious individuals. In terms of distribution equity, bioavailability, and efficacy, contemporary techniques outperform prior ones. Table 1 shows nanotechnology's many sports nutrition applications and their pros and cons. Technological advancements also present information on new teaching options.

Nano-encapsulation is the process of encasing nutrients in nano-carriers like liposomes, micelles, or polymeric nanoparticles. In addition to making it easier for nutrients to reach various parts of the body, they also act as barriers, preventing outside

Table 1 Comparative table: methods in nanotechnology for sports nutrition

Method	Description	Advantages	Challenges
Nano-encapsulation [6]	Enclosing nutrients in nano-carriers for targeted delivery	Targeted release; enhanced stability	Expensive production; regulatory concerns
Nano-emulsions [7]	Lipid-based systems for fat-soluble nutrients	Improved solubility; faster absorption	Potential toxicity in large doses
Nanoparticles in proteins [8]	Nano-sized particles enhance protein absorption	Faster muscle recovery; increased efficiency	Unknown long-term effects on metabolism
Nano-coatings [9]	Protective nano-layer around supplements	Enhanced stability; improved shelf life	May introduce chemical residues
Nanoclusters [10]	Clusters improve the bioavailability of trace minerals	Improved absorption; higher efficiency	Limited research on health impacts
Nano-sensors [11]	Embedded in packaging to monitor quality	Real-time tracking of supplement integrity	High implementation costs
Nanofibers [12]	Fibers in capsules for steady nutrient release	Consistent nutrient supply; reduced side effects	Manufacturing complexity
Nano-powders [13]	Ultra-fine powders for easy dissolution	Faster absorption; easier ingestion	Potential inhalation risks during production
Nano-liposomes [14]	Encapsulate antioxidants for targeted delivery	Protects sensitive nutrients; better bioavailability	Expensive; potential allergenic reactions
Nano-hydroxyapatite [15]	Used in calcium supplements for improved bone health	Better integration with bone tissue	Limited studies on long-term effects

forces from damaging them. Unfortunately, nanoemulsions may be harmful in high enough concentrations to cause damage, despite their obvious benefits. The manufacturing of nanoemulsions often necessitates the use of specialized equipment and the expertise of professional individuals. Nanoparticle modification produces changed proteins that are more bioavailable and move to muscle tissues more efficiently. Using this approach, athletes may speed up muscle recovery after exercises and promote tissue regeneration, both of which may have considerable advantages. The body's increased rate of absorption of nanoparticles increases the potential for them to induce a faster and stronger response. Many people wonder whether nanoparticles are safe because they are concerned that extended exposure to them would affect their health. Currently, we have not thoroughly explored their interactions with metabolic pathways to determine potential long-term impacts.

A protective nano-layer may increase nutritional supplements' stability, shelf life, and palatability. This specific process is known as nanocoating. When stored in a protective container, vitamins, minerals, and probiotics create a barrier that shields them from external elements like moisture, light, and heat. The controlled release of nutrients enabled by nano-coating ensures a steady supply over an extended period of time. Customers may be concerned that the coatings would emit hazardous compounds into their bodies.Nanoclusters, which form when nanoparticles combine, may boost nutritional bioavailability, particularly for trace minerals.

Future research should focus on enhancing sports nutrition by limiting nanotechnology's negative effects while maximizing its positive implications in the following areas:

It is crucial to conduct a thorough study into the immediate and long-term effects of nanoparticles on human health, as well as to continue safety testing. Regulations govern nano-supplements to ensure their safety throughout the manufacturing, labeling, and consumption processes. When addressing nano-supplements, "sustainable manufacturing" refers to their creation in a way that is both economically and environmentally appropriate. The term "athlete-centric guidelines" refers to the standards that nano-supplement formulations must meet to suit athletes' specific physiological demands. Consumer education attempts to empower people to make educated choices by teaching them about the advantages and downsides of nano-supplements. The application of nanotechnology in sports nutrition enhances performance, recuperation, and overall health. With this, we are progressing quickly. However, we need to address certain issues before we can implement this strategy. The only way to achieve these objectives is to strictly follow the regulations, do extensive research, and maintain constant contact in both directions. Nanotechnology has the potential to dramatically revolutionize the fitness and health industries if it can learn to stay ahead of the competition while simultaneously protecting consumers.

3 Efficacy of Nano-supplements in Sports

Nanotechnology has changed sports nutrition. Nanoparticles are used in nano-supplements to improve sports performance, well-being, and recovery. This chapter examines nano-supplements in sports by reviewing existing research, examining their effects on performance criteria, and addressing restrictions and concerns.

3.1 Review of Existing Research on the Use of Nano-supplements in Sports

Sports researchers are increasingly studying nano-supplements. Nanoparticle-based nanosupplements increase nutrition absorption, stability, and targeting [6]. They may be better than supplements. The investigation yielded the following key findings:

1. **Better bioavailability**: Nanoformulations improve nutrient absorption, according to research. The plasma concentration of nano-encapsulated vitamins like vitamin D is greater than that of normal vitamins.
2. **Targeted Delivery**: Nanoparticles may supply nutrients to muscles, reducing waste and improving local benefits.Antioxidants such as nano-curcumin and nano-coenzyme Q10 may reduce oxidative stress during exercising. Animal studies reveal nano-supplements boost stamina, muscular strength, and healing. However, human research is scarce.
3. **Different Results**: Varying research methods, dosages, and formulations may lead to varying effectiveness claims.

3.2 Impact of Nano-supplements on Athletic Performance

(a) **Muscle Strength and Power**

1. Nano-supplements have the potential to enhance muscle strength and power in various ways.
2. More efficiently absorbed nano-amino acids promote muscle protein synthesis and growth.
3. Nano-formulated creatine improves muscle cell absorption, increasing ATP production and power output during intensive exercise [7].
4. Nano-magnesium supplements boost neuromuscular performance, retaining muscle strength and preventing fatigue.
5. After exercise, nano-protein powders rebuild muscle strength faster than regular protein supplements.

(b) **Endurance Performance**

Endurance athletes can benefit from nano-supplements in several ways:

1. Oxygen Delivery: Nano-iron pills increase hemoglobin, which increases cell oxygen delivery and physical activity.
2. Nano-carbohydrates slowly release glucose, allowing you to exercise longer [8].
3. Nano-ionic combinations maintain cell hydration and electrolyte balance for long-term activity.
4. Researchers have found that nano-encapsulated omega-3 fatty acids reduce inflammation and improve performance.

(c) Recovery and Regeneration

Nano-supplements may aid healing, which is crucial to running quickly.

1. Nano-curcumin and nano-resveratrol reduce inflammation and speed muscle recovery.
2. Improved Glycogen Replenishing: Nano-carbohydrate vitamins boost glycogen production and energy recovery [9].
3. Faster cell healing: nanoparticles of zinc and selenium speed up cell repair.
4. Early clinical trials demonstrate that nanosupplements reduce pain and speed recovery for athletes.

More individuals believe cognitive ability is crucial to sports success. Brain-healthy nanosupplements are:

1. Antioxidant nanoparticles protect neurons against exercise-induced oxidation.
2. Better concentration: Nano-caffeine mixtures provide long-lasting energy that doesn't crash, making it simpler to concentrate and respond fast.
3. Lowering stress: Nano-adaptogens like nano-ashwagandha may reduce cortisol, reducing stress and worry's impact on performance [10].
4. Better Memory: Nano-phosphatidylserine improves memory and decision-making, which is excellent for sports that need quick recall.

The use of nanotechnology in sports nutrition research has created previously unthinkable prospects. Table 2 indicates that some tactics provide both considerable challenges and unexpected opportunities. The development of nanogels is one area where nanotechnology has the potential to improve food delivery. These water-based carriers with controlled release may help athletes maintain their energy and hydration levels throughout lengthy, tough events such as triathlons and marathons. Throughout the competition, they give a consistent supply of energy, and their delayed release of nutrients conforms to the demands of endurance. Furthermore, by solubilizing hydrophobic compounds such as vitamins D and E, nano-micelles help the environment [16]. This is necessary because it allows the body to properly absorb these nutrients. The creation of these supplements addresses a fundamental problem with conventional ones: poor absorption. Quantum dots, a pioneering diagnostic tool, now allow for real-time monitoring of nutrient distribution and absorption. This approach provides information on tailored diet programs that no other method does. Beyond just monitoring their caloric intake, these gadgets may allow athletes to make regular, on-the-fly modifications to their calorie intake. Researchers discovered that combining plant-based compounds, such as curcumin, with nano-phytosomes improves their anti-inflammatory capabilities [17]. These natural supplements containing nano-phytosomes are more soluble, resulting in greater benefits such as reduced oxidative stress and improved workout recovery. Similarly, nano-bioactive peptides promote muscle regeneration by enhancing protein uptake. They are therefore an excellent tool for athletes looking to recover rapidly after rigorous exercise.

Table 2 Comparative table: emerging methods in nanotechnology for sports nutrition

Method	Description	Advantages	Challenges
Nano-gels	Water-based carriers for controlled nutrient release	Enhanced hydration and sustained energy release	High production costs; limited long-term research
Nano-micelles	Spherical structures solubilizing hydrophobic nutrients	Improved absorption and bioavailability	Expensive and complex production
Quantum dots	Track nutrient absorption and bio-distribution	Personalized nutrition insights	Potential toxicity and regulatory challenges
Nano-phytosomes	Lipid-based carriers for plant-based nutrients	Increased solubility and efficacy of herbal supplements	High production costs; limited formulations
Nano-bioactive peptides	Protein fragments at the nanoscale for muscle recovery	Faster healing and better nutrient absorption	Safety for long-term use requires further study
Nano-capsules	Controlled nutrient release over time	Consistent delivery; reduced risk of nutrient spikes	High cost and technical complexity
Nano-enzymes	Synthetic enzymes to enhance digestion and absorption	Improved efficiency for high-protein diets	Interaction with natural enzymes needs investigation
Nano-probiotics	Enhanced survival and colonization in the gut	Better digestive health and immune function	Stability during storage and transport is challenging
Nano-mineral complexes	Minerals combined with nanoscale carriers for better absorption	Reduced gastrointestinal side effects; enhanced efficacy	Limited long-term safety research
Smart nano-delivery	Nano-sensors and carriers for targeted nutrient delivery	Optimized nutrient utilization; reduced waste	High cost and technical complexity

Maintaining metabolic equilibrium requires providing nutrients over a longer period of time, which time-release nano-capsules allow for. Athletes regularly provide carbohydrates and amino acids before, during, and after competition to replenish their energy. Thanks to the discovery of nanoenzymes that can act like natural enzymes, there are now more ways to improve digestion and the breaking down of complex compounds, especially proteins. Protein is required for muscle development and maintenance; hence these foods supplement high-protein diets. Researchers are now working to produce nano-probiotics to address the issue of gut health and its impact on an athlete's overall performance [18]. These nanoscale probiotics improve meal absorption and prevent gastrointestinal irritation, so you won't have any more gas or bloating when exercising. They function by stabilizing the intestinal flora. Nanomineral complexes, which improve calcium and magnesium absorption, may help reduce muscle cramps and bone loss. These complexes are suitable for sports that need top performance even under pressure, such as those demanding long periods of standing or jumping. Advanced nano-delivery systems include sensors and nanocarriers. These systems ensure targeted delivery of nutrients, enhanced treatment effectiveness, and reduced waste. These tools have the potential to transform the sports nutrition industry by providing personalized, laser-focused treatments. The graphic depicts several advantages, including enhanced bioavailability, longer nutrient release, and quicker recovery. However, the graphic highlights some of the problems connected with each strategy. Among these difficulties are the enormous costs and legal obstacles connected with each technique, as well as the necessity for extensive long-term impact assessments. This effectively eliminates any potential long-term damage to human health. Finally, Table 2 summarizes the current status of nanotechnology in sports nutrition and demonstrates how these tactics may have a substantial impact on the firm [19]. Addressing these barriers and allocating funds to research and development could lead to unprecedented levels of innovation in the sports nutrition sector. This would significantly improve the health and performance of athletes globally. We will have met our aim if these tools can encourage people to adopt better eating habits. As a result, efficient and effective personalized sports nutrition will be accessible shortly.

3.3 Limitations and Challenges in Evaluating the Efficacy of Nano-supplements

Nanosupplements have great potential, but many issues prevent their complete evaluation and use.

1. Lack of long-term research: Most nanoparticle supplement studies focus on short-term effects, raising concerns regarding their safety and efficacy.
2. Nano-Formulations Variability: Variations in particle size, covering materials, and transport methods complicate the comparison of research findings.
3. Limited Human Trials: While preclinical studies are valuable, the evidence from large-scale human trials is not as robust [11].
4. Complex Interactions: We don't know how nanoparticles influence biological organisms, which might lead to unexpected effects.
5. Sham Effect: Advanced nano-supplements may alter research outcomes; hence controlled sham experiments must be developed.
6. Regulation issues: Without defined test and approval standards, efficacy is impossible to measure.

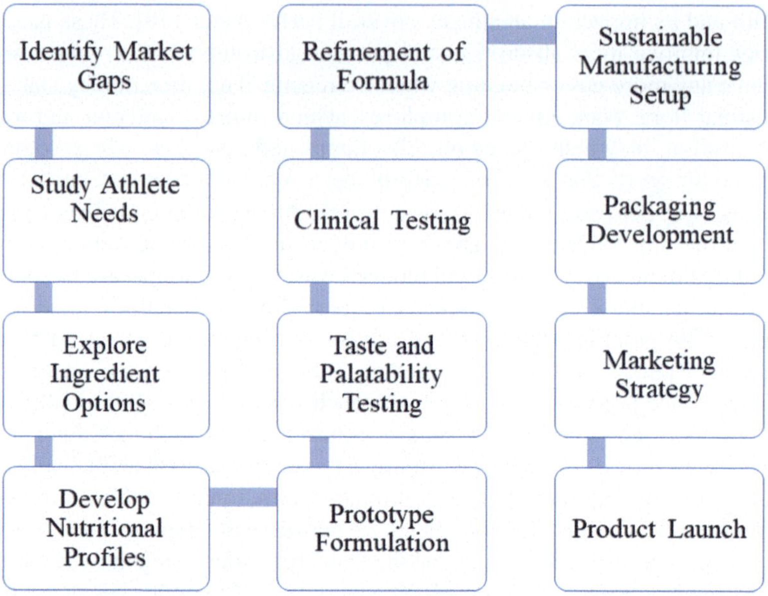

Fig. 1 Product development process

7. Cost and Accessibility: Nano-supplements cost more than ordinary supplements, making them difficult to consume.

Nano-supplements boost health, athletic performance, and rehabilitation. Studies have shown they may improve intelligence, strength, and injury recovery [12]. However, there are few long-term trials, formulations vary, and legal problems require additional investigation. Sports science must address these issues and develop success metrics as it studies nanotechnology. Nano-supplements may improve athletes and ensure fair and safe sports.

Figure 1 shows the main steps in the process of making a product, from finding gaps in the market to putting the finished product on the market [20]. The steps include finding out what athletes want, looking into other ingredients that can be used instead, making nutritional profiles, tweaking the recipe, trying it in clinical settings, designing packages, and coming up with a marketing plan.

4 Ethical and Regulatory Landscape

Nanotechnology in sports nutrition and supplements raises ethical and legal issues. Nano-supplements raise governance, anti-doping, sports justice, and safety issues. Nano-supplement ethics, anti-doping compliance, and global regulation are covered here.

4.1 Global Regulatory Frameworks Governing Nano-supplements

Every nation regulates nano-supplements uniquely due to their unique development, testing, and control methods. Global frames allow the FDA to regulate nano-supplements as dietary supplements and food additives without a category [21]. Despite manufacturers' predictions, nanoparticles' unique properties may prevent standard testing from proving their safety and efficacy.

1. The FDA's 2014 nanotechnology guideline recommends research but not nanoparticle testing.
2. Regulation (EU) 2015/2283 requires the European Food Safety Authority (EFSA) to approve modified nanoparticle-containing meals before they may be sold. The EFSA recommends a comprehensive risk evaluation that considers particle characterization, toxicity, and long-term safety.
3. Food Standards Australia New Zealand (FSANZ) carefully analyzes nano-supplements for safety. We emphasize clear labeling and nanotechnology use information.
4. India's Food Safety and Standards Authority (FSSAI) regulates nanotechnology in food and additives.
5. Global harmonization: The Codex Alimentarius Commission wants all foods, including nano-supplements, to be safe to ingest. But progress has been gradual, and no one agrees on a definition or set of criteria for nanomaterials in supplements.

4.2 Challenges in Establishing Ethical Boundaries in Sports

Nanosupplements in sports present ethical considerations beyond regulations. They question fairness, safety, and competitiveness. Some major issues are:

1. Fairness in Competition: Nano-supplements may help technologically advanced athletes perform better. People question whether these technologies violate fair play since it's tougher to distinguish legal medications from unfair advantages.
2. Informed Consent: Athletes may not grasp nano-supplements' hazards, particularly their long-term effects. Make the advantages, hazards, and unknowns of utilizing these items plain to acquire informed consent [22].
3. Safety and Health: Putting athletes at risk of nano-supplement health risks is unethical since their long-term consequences are unclear. Coaches, trainers, and sports organizations have ethical issues when they promote unsafe products.
4. Teen and amateur activities: Nanoparticles create safety, ease, and misuse problems.
5. Balancing Innovation and Fairness: Nanotechnology is a major scientific advance, but sports nutrition must combine innovation, fairness, and inclusivity.

4.3 Anti-doping Considerations and Compliance with WADA Guidelines

WADA monitors drugs that enhance sports performance. Nano-supplements complicate anti-doping: The Meaning of "Doping" Doping is when athletes utilize medicines or methods that improve performance or harm their health, according to WADA. Nano-supplements may satisfy this need if they outperform natural methods. Tracking nanoparticles is difficult since regular drug tests may not detect them. Detecting nanoparticles in live data requires mass spectrometry and nanoparticle tracking analysis. Uncontrolled chemicals: Nanosupplements may include novel chemicals or mixes not on WADA's list. Nanotechnology evolves swiftly, making exploration and explanation difficult for government bodies [23]. Therapeutic Use Exemptions (TUEs): Medically impaired athletes may get TUEs for vitamin or nanoparticle therapies. Clearance is difficult due to unclear rules. In one study, athletes tested positive for drugs due to nanoparticles. This requires improved quality control. Education and Awareness: To prevent accidental cheating, WADA advises players to learn about the hazards of vitamins, particularly nanotechnology-based ones. Legal and ethical issues surrounding sports nano supplements are complicated and evolving. Global regulating agencies monitor these drugs, yet there are still gaps in ensuring equitable, safe, and effective usage. Ethics emphasizes clear communication and strong norms for fairness and informed consent. Sports anti-doping operations validate themselves using WADA criteria. Nano-vitamins may make it impossible to discern whether supplements improve performance; therefore, they require additional testing and regulation. Politicians, scientists, sports teams, and players must collaborate to balance creativity, decency, and fairness to overcome these difficulties.

5 Case Studies and Current Research

With academics and practitioners looking into nanotechnology's potential to improve performance, recuperation, and general health, it has become very popular in sports nutrition and supplements recently. This section examines current research, practical uses, and lessons learnt from negative events with nano-supplements in sports.

5.1 Recent Studies on Nano-supplements in Sports Performance

1. Better bioavailability and performance: Sports Nutrition and Performance examined how nano-curcumin influenced endurance runners' oxidative stress

and repair. Researchers found that nano-curcumin reduced inflammation and accelerated recovery in athletes.

2. Faster recovery: Strength training and nano-zinc tablets improved muscle regeneration in a Journal of Athletic Performance research [24]. Protein production increased and healing times decreased in the research. This implies speedier muscle healing.
3. Nano-Caffeine and Cognitive Function: A Neuroscience Letters research study examined how nano-caffeine affects gaming players' reaction time and focus. Nano-caffeine helped individuals focus and recall longer without tiring them out like traditional caffeine tablets.
4. Omega-3 nanoparticles: A clinical investigation examined how omega-3 nanoparticles improved joint health and physical performance [25]. Joint flexibility and decreased pain helped athletes exercise more regularly and perform better.
5. Comparative Studies: New meta-analyses reveal that nano-supplements are more bioavailable and effective than normal supplements, although long-term safety is unknown.

5.2 Real-World Applications and Outcomes

1. Business patience Triathletes and ultramarathoners are increasingly adopting nano-electrolyte tablets to remain hydrated. Performance statistics and testimonials reveal higher energy and fewer cramps.
2. Strength training: Nano-creatine tablets improve weightlifting and bodybuilding. Better creatine phosphate absorption by cells boosts energy and lifting.
3. Nano-curcumin and nano-resveratrol assist professional football and basketball players recuperate [26]. Injury-related absences have decreased as players report less discomfort and quicker recovery.
4. Youth sports schools employ nano-protein powders to develop and rebuild muscle. Early studies indicate an increase in strength and endurance, but the long-term health implications remain uncertain.
5. E-Sports Performance: Nano-supplements like nano-caffeine and nano-l-theanine are becoming increasingly popular in e-sports, where rapid thinking and responses are crucial. Professional athletes report being more focused and less anxious during games.

5.3 Lessons Learned from Adverse Incidents

1. Nano-supplements contaminated with illegal chemicals led several athletes to test positive for drugs. This experience demonstrated the need for quality control in manufacturing.

2. Overdose and Toxicity: A clinical toxicology case study showing nano-supplement overdose is harmful. Endurance runners who took too many nano-iron supplements suffered stomach and liver enzyme issues.
3. Nanomaterials in jackets might cause allergic responses, particularly in sensitive athletes. This has prompted requests for stronger allergy labeling and pre-sale testing.
4. Unwanted Interactions: Nanoparticle vitamins and drugs may combine, according to research. Nano-curcumin affected blood thinner efficiency in players.
5. Regulation gaps: Inadequate supervision has led to the sale of defective items. We learned from adverse experiences that global regulations and testing techniques are essential.

Figure 2 links the advantages of a balanced health and well-being strategy. It discusses how getting enough energy, staying hydrated, controlling inflammation, eating protein after a workout, lowering stress, improving endurance, gut health, immune system support, metabolism speed, and avoiding artificial ingredients can improve health [27]. Nano-supplements may increase cognition, recovery, and athletic performance, according to current and case studies. Meanwhile, real-world applications have revealed safety flaws, legal gaps, and undesirable incidents. We must balance inventiveness, extensive testing, and ethical processes to ensure the safe and effective use of nano-supplements in sports.

Table 3 shows how revolutionary approaches in nano-supplements have enabled substantial advances in sports nutrition. These therapies may have various impacts on athletes, recovery times, and overall health. One example is the use of nano-hydrogels, water-based transporters that can transport nutrients and keep their recipients hydrated for extended periods. This method may aid endurance athletes who need water for lengthy periods of time while training [28]. These meal supplements

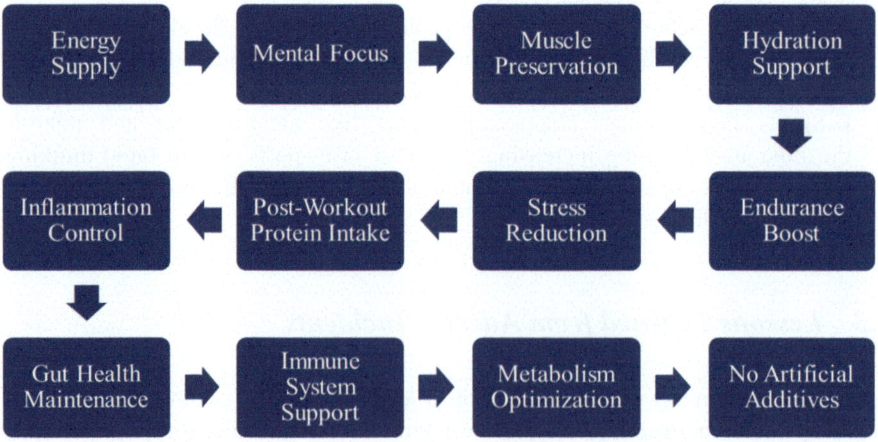

Fig. 2 Holistic health and wellness benefits

include hydrogel shells to ensure people stay hydrated and nourished throughout the day. The high production costs and complex manufacturing processes hinder people from adopting such goods. Using nanolipid carriers, it is possible to improve the bioavailability of fat-soluble chemicals and antioxidants, such as vitamins D and E. Lipid-based systems increase the solubility of these carriers, leading to an improvement in nutrient availability. Compared to conventional supplements, this method achieves increased bioavailability. While nano-lipid carriers offer certain benefits, their primary obstacles remain unresolved. This article discusses scalability, and the risks associated with lipid oxidation. Problems make it difficult for these airlines to maintain their economic viability. Scaffolds made of nanocomposite materials are one of the many innovative approaches to repairing tissue and improving joint function. These structures facilitate the healing and regeneration processes by quickly delivering nutrients to wounded areas. It is possible that athletes who struggle with connective tissue or joints might benefit from this individualized approach [29]. Only research organizations and specialist pharmaceutical companies use these scaffolds due to their high production costs and lack of clinical validation. Because nanosurfactants increase the solubility and dispersion of hydrophobic nutrients, they have the potential to enhance their bioavailability. This strategy may overcome molecular insolubility, potentially increasing the body's intake of essential nutrients. However, despite their obvious usefulness, their adoption remains delayed. Several issues arise, such as the presence of chemical residues and the strict regulations surrounding quality management. Nanosensors based on quantum mechanics allow for more individualized nutrition. These sensors enable real-time monitoring of the body's absorbance and distribution of nutrients, thereby enabling the creation of individualized diet plans. By increasing their intake and utilizing real-time data collection, athletes can enhance their performance. This rationale is accurate. Not only are these sensors prohibitively costly to rapidly scale up and deploy, but they also raise ethical concerns. Nanostructured emulsions are suitable for extended use due to their consistent provision of nutrients and energy. These emulsions cater to athletes who compete for extended periods and require consistent access to nutrients. Upon consumption, they provide a steady supply of energy. Because there has not been sufficient study to demonstrate the long-term safety of certain goods, we are required to consider the potential risks that they may pose. Researchers often use nanocrystals to stabilize and improve the bioavailability of minerals and vitamins, which are considered delicate substances [30]. The fact that these components are stable at the nanoscale guarantees that nutrients will continue to be beneficial throughout the course of time. The widespread use of nanocrystals is challenging, even though they increase both stability and efficiency. The costs involved and the production processes represent a barrier to entry. Branching nanodendrimers have the potential to transport intricate combinations of macronutrients. Due to the efficiency of dendrimers, athletes who are following particular diets might benefit from using them. Despite their tremendous potential, nano-dendrimers are costly to create and generally not available in the commercial market. The enhanced barrier qualities of nanoparticles included inside nanobiofilms enable innovative packaging solutions. This film not only extends the shelf life of supplements without compromising their quality but also ensures athletes use only the highest quality and most recently harvested components. The primary factors that restrict the use of nano-biofilms are worries about the environment and the high costs of manufacture. Intelligent nanopatches can achieve customized transdermal feeding. This stop degraded nutrients from entering the digestive tract. The use of transdermal patches, as opposed to oral medicines, enables athletes to immediately absorb nutrients into their vascular circulation. The effectiveness of patches remains a challenge due to concerns about the compatibility of adhesives and the nutritional composition. Limitations like this preclude widespread use.

Finally, Table 3's methodologies show how sophisticated nanotechnology has the potential to transform the sports nutrition industry. We need to address issues such as manufacturing costs, scalability, and long-term safety testing. Nonetheless, each strategy offers distinct advantages depending on the athlete's needs. These cutting-edge strategies are herald in a new era of tailored and successful sports nutrition, transforming performance, recovery, and overall health. The advent of a new age is approaching. As research and technological breakthroughs increase, we expect these tactics to have a significant influence.

6 Integration in Sports Science Methodologies

Medical professionals, coaches, and sports nutritionists must collaborate to include nanosupplements in sports science programs. We explore below these participants' roles in planning and monitoring nano-supplement consumption and their effects on training and competitiveness.

6.1 Role of Sports Nutritionists and Coaches in Nano-supplement Planning

Nutritionists help athletes create supplement programs based on their health, nutritional requirements, and performance objectives.

1. Nano-supplements are more soluble; thus, physicians must be cautious not to overdose patients.
2. Educating and increasing awareness: Coaches and physicians discuss nano-supplements' merits and downsides with athletes. Knowledge of a product's safety, side effects, and restrictions helps consumers make smarter judgments.
3. Collaboration with medical professionals: Working with sports physicians ensures nano supplements don't interfere with medical treatments or harm individuals. Regular medical visits may detect hazards and side effects early.
4. Moral Issues: Coaches and chefs must discuss nano-supplements to maintain fair play and anti-doping. Informing athletes that nano-supplements are constantly tested boosts confidence and duty.

Table 3 Comparative analysis of emerging methods in nano-supplements for sports nutrition

Method	Description	Advantages	Challenges
Nano-hydrogels	Water-based carriers designed for sustained nutrient and hydration delivery	Provides prolonged hydration and nutrient availability	High production costs; requires advanced manufacturing
Nano-lipid carriers	Lipid-based systems improving the bioavailability of fat-soluble nutrients	Enhanced absorption of fat-soluble vitamins and antioxidants	Scalability issues; potential for lipid oxidation
Nano-composite scaffolds	Structures promoting tissue repair and joint health through targeted delivery	Supports joint recovery and tissue regeneration	Limited clinical validation; expensive
Nano-surfactants	Compounds that increase solubility and dispersion of hydrophobic nutrients	Improved solubility; enhances nutrient bioavailability	May introduce chemical residues; requires stringent quality control
Quantum nano-sensors	Devices for real-time tracking of nutrient bioavailability and distribution	Enables personalized nutrition strategies; precise monitoring	High cost; raises ethical concerns over data privacy
Nano-structured emulsions	Nano-sized droplets delivering slow-release energy and nutrients	Provides sustained energy release during activities	Requires more safety testing for long-term use
Nano-crystals	Stabilized nano-sized particles for improved bioavailability of sensitive compounds	Enhances stability and efficacy of vitamins and minerals	Requires precise production techniques; potential high costs
Nano-dendrimers	Branched nanoscale structures delivering complex macronutrient formulations	Facilitates targeted and efficient nutrient delivery	Complex to synthesize; limited commercial availability
Nano-biofilms	Packaging films embedded with nanoparticles to improve barrier properties	Extends shelf life; maintains supplement quality	High production cost; environmental concerns
Smart nano-patches	Adhesive patches delivering nutrients transdermally	Convenient; avoids digestive system degradation of nutrients	Adhesive and nutrient compatibility issues

6.2 Monitoring Athletes' Response to Nano-supplements

Before introducing nano-supplements, health and performance measurements might determine their efficacy. Nano-supplements influence health, healing, and sports performance, say coaches. Wearable devices and sophisticated analytics may provide real-time bodily data. Regular blood testing and biomarketer investigations may detect nanoparticle toxicity and nutritional deficiencies. Players benefit from monitoring oxidative stress, inflammation, and system function. The emotional experiences of frequent gamers with nano-supplements enhance dose and recipe.

6.3 Training and Competition Considerations

1. When training and competing, you must take nano-supplements at the right time and volume. Nano-carbohydrates provide long-lasting energy before exercise, while nano-proteins aid recovery.
2. Changing your exercise plans: Nano-supplements may slow healing; therefore, you may need to reduce workout intensity and frequency to prevent overtraining. Nanosupplements affect training regimens; thus, coaches should adjust.
3. Competition-ready: Nano-supplements must fulfill anti-doping guidelines to remain competitive. Extra testing may be necessary to ensure that athletes' vitamin programs do not include illicit medicines.
4. Travel and Logistics: Nano-supplements must be simple to procure and maintain for trips and competitions. Local and sport regulations must be considered by coaches.
5. Psychological Factors: Nano-supplements may enhance athletes' confidence and mental toughness, so set reasonable objectives. Everyone must work together to design, monitor, and execute nano-supplements. Coaches and sports physicians assist athletes in using supplements and respecting all regulations and ethics. Monitoring athletes' behavior and tailoring their routines to incorporate nano-supplements can optimize their benefits while safeguarding their health and performance. By addressing these concerns, sportspeople may employ nano-supplements to enhance nutrition and performance.

7 Limitations and Future Plans

Sports nutrition holds great potential for nanotechnology, but its full use requires correction. This section examines research gaps, the necessity for long-term studies with larger sample numbers, and emerging advancements that potentially impact sports nutrition nanotechnology.

Figure 3 activities are crucial to a company's long-term strategy. These practises emphasise social and environmental responsibility via ethical farming, sustainable food procurement, energy-efficient manufacturing, zero-waste production, eco-friendly packaging, and carbon reduction.

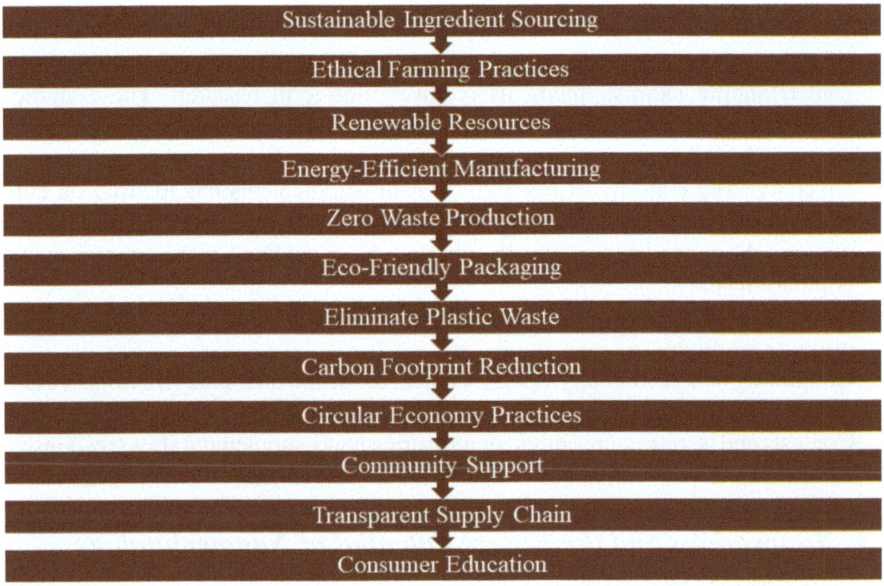

Fig. 3 Pillars of sustainable business practices

7.1 Current Gaps in Research and Understanding

1. There are insufficient mechanistic studies: nanoparticles offer numerous advantages, but their effects on biological systems remain unknown. Some of the most significant concerns involving nanoparticle entry, breakdown, and exit require further research.
2. Safety: This research doesn't examine bioaccumulation or long-term effects. Nano-supplements are challenging to employ without safety ratings. Few studies have examined how nanoparticles influence sensitive populations like young athletes and sick patients.
3. Nano-Formulations Variability: Particle size, surface charge, and covering materials make nano-supplement outcomes and standards inconsistent.
4. Regulation uncertainty: Nano-supplements are difficult to produce and evaluate without clear guidelines. There are no universal standards or testing methodologies; thus, there is uncertainty.
5. Issues with Real-World Applications: Controlled studies may not adequately capture the complexity of real-world sporting scenarios. People seldom consider how environment, diet, and metabolism change throughout time.

7.2 Need for Long-Term Studies and Broader Sample Sizes

1. Understanding nano-supplements requires long-term research. These studies should weigh the merits and downsides of metabolic pathway alterations or organ-specific toxins.
2. Diversity: The majority of research on nano-supplements has focused on elite athletes, leaving little understanding of their impact on recreational and novice players. Age, gender, and health must all be considered in a comprehensive population study.
3. We should investigate nano-supplements for brain health, healing, fitness, and injury prevention.
4. Combining nanotechnology, sports science, food, and medicine will improve and ensure proper usage of research projects.
5. Contests and testing camps might demonstrate nano-supplement stress responses.

7.3 Innovations and the Future Potential of Nanotechnology in Sports Nutrition

1. Customized Nano-Supplementation: Genetics and metabolomics research may allow nano-supplements personalized to an athlete's genes, activity regimen, and nutritional demands. Nano-supplementation is achievable because biosensors and personal electronics monitor bodily data in real time.
2. Smart Delivery Systems: Nanosupplements may release nutrients dependent on pH or enzyme activity. This concentrated approach may boost performance but reduce impact.
3. Sustainability and Green Nanotechnology: Safe nanoparticles may safeguard the environment and extend nanosupplement life. Biodegradable nanoparticles may reduce bioaccumulation.
4. AI and Big Data: Based on performance data, AI and big data analytics may predict supplement use, changing sports nutrition. Nanoparticle supplement performance may be improved by machine learning algorithms that identify player tendencies.
5. Track brain function: Nanotechnology may assist mentally demanding athletes make choices, manage stress, and think coherently. Because they make your brain wiser and healthier, nootropics may inspire new thoughts.
6. Therapeutic Uses: Nano-supplements may prevent injuries, speed recovery, and improve performance. Nanoparticles may deliver medications to damaged cells, speeding up healing. International bodies might simplify nano-supplement testing and approval. Clear labeling and safety assessments help athletes and customers trust a product. Nanotechnology might improve sports nutrition, but it needs further investigation. We need to study many individuals and real-life scenarios to understand long-term safety and recipe changes. As AI, eco-friendly technology, and smart transportation improve, diet-specific nano-supplements

will aid. Nanotechnology might revolutionize everything. Sports may assist by bringing individuals from diverse areas together to work ethically and legally. This ensures secure and informative replies worldwide.

8 Future Directions and Conclusion

Sports nutrition's incorporation of nanotechnology is a developing area with great potential but also calls much attention. This part lists the important topics for further studies, investigates possible uses, investigates ethical issues, and ends with important lessons for responsible advancement of the discipline.

8.1 Future Research Needs to Address Safety and Efficacy Concerns

1. We must study nanoparticle safety profiles over time to assess bioaccumulation, toxicity, and health risks.
2. Research should focus on nano-supplement-physiological system interactions in high-performance conditions.
3. Standardizing nano-supplement formulations may help ensure product quality, efficacy, and safety using established standards.
4. To substantiate the performance-enhancing claims of nano-supplements in a variety of situations, solid, thorough research is required.
5. To fully understand nano-supplements, future research should include women, teens, older athletes, and those with pre-existing medical conditions.
6. Studies mimicking real-world conditions, including different meals, environmental impacts, and competitive pressures, can help translate lab findings into practical applications.

8.2 Potential Applications of Nano-supplements in Personalized Sports Nutrition

1. Customized Nutrition Plans: Nanotechnology and genetic and biochemical profiling may provide players customized nourishment.
2. Future nano-supplements may release nutrients depending on pH, weariness, or water level using adaptive technology like smart nanoparticles.
3. Preventing damage and mending: Nanoparticle-based solutions that target inflammation or give antioxidants to tissues may speed up healing and reduce damage.

4. Better cognitive performance: Nanotechnology-based nootropics may improve mental attention, rapid reactions, and decision-making for physical and strategic sports.
5. Wearable Integration: Scientists can monitor athletes' reactions in real time and adjust their supplements using nano-supplements.

8.3 Ethical Considerations and Responsible Use of Nano-supplements in Sports

1. Nano-supplements are expensive; thus, athletes in amateur or resource-limited contexts may vary from those at the top.
2. Fair competition requires policymakers and interests parties to address these inequities. Doctors, coaches, and athletes need clear labeling of nano-supplement compounds and their risks to make informed decisions.
3. Compliance with anti-doping standards, especially WADA guidelines, is essential to sports integrity.
4. Athletes should understand nano-supplements' risks and benefits to make ethical and medically sound judgments.
5. Long-term effects: Scholars and professionals must consider the repercussions of widespread nano-supplement use on society, the environment, sustainability, and public health.

8.4 Concluding Remarks and Key Takeaways

1. Finding the Right Balance Between New Ideas and Safety: Nano-supplements might revolutionize athlete nutrition, but scientific data, societal concerns, and regulatory control should restrict their usage.
2. Collaboration: Scientists, sports organizations, politicians, and athletes must collaborate across disciplines to address challenges and use nanotechnology.
3. Education and Awareness: Comprehensive education initiatives for athletes, instructors, and others may make nano-supplements safer.
4. The future: Nanotechnology might improve sports nutrition for athletes worldwide in performance, healing, and health if research, innovation, and ethics continue.
5. Finally, sports nutrition Nano-supplements combine science, technology, and physical achievement. This review reiterates nano-supplement advantages and hazards, fairly portrays their role in sports, and emphasizes the need for greater study across many sectors to safely progress their usage. Nano-supplements improve sports nutrition by customizing dispersion and absorption. They boost endurance, healing, and efficacy. Sportspeople have better nutritional and recovery control thanks to flexible meal delivery systems and smart nanoparticles.

It has the potential to enhance more than just your physical performance. You may improve brain clarity, stress tolerance, and reaction time. Nano-supplements may be toxic, bioaccumulative, and harmful despite their benefits.Formula changes and lack of regulation create product quality and safety concerns. Adoption and application vary due to fair access and informed consent difficulties. The present research does not examine demographics, long-term consequences, or real occurrences, creating a knowledge gap. Nano-supplement safety and efficacy are difficult to assess without standardized testing techniques. Nano-supplements revolutionize sports nutrition by offering customized therapies. Nano-supplements may provide elite athletes a competitive edge for high-stakes performance. They may also be useful in leisure and amateur sports. High manufacturing and development costs limit availability, causing sports inequities. The ethical implications of performance enhancement and anti-doping should be considered. Sports nutritionists, coaches, and policymakers must work together to add nano-supplements. Awareness and education programs help athletes make smart choices. To ensure equitable and safe usage of contemporary supplements, regulators must set clear norms and structures.

1. Nanotechnology, sports science, nutrition, and regulation must collaborate to improve the game. Working together on research initiatives may help us comprehend nanoparticles' chemical characteristics and sporting usage.
2. Longitudinal research: To understand how nanosupplements affect health and function, we need long-term studies. We should study the long-term effects, how vitamins and medications affect different age groups and genders, and how they differ.
3. More real-world applications: Researchers should employ real-world examples to assess training strategies, competitive pressures, and outside influences. Wearable electronics and AI analytics might transform data collection and analysis in these instances.
4. Ethics and society: Nano-supplements influence justice, accessibility, and information sharing.

Scientists should study the environmental impact of nanotechnology-made nutrients'. Players may eat differently if nano-supplements improve performance and healing. Research, societal concerns, and government oversight must support it prior to its use. To attain their full potential, they must cover research gaps, foster cross-disciplinary collaboration, and ensure safety.

For sports nutrition to progress, new ideas and responsibilities must work together. Nanotechnology may provide everyone new chances if utilized fairly and safely. Micro- and nano-supplements may improve health and sports if studied and used collaboratively.

References

1. Nile SH, Baskar V, Selvaraj D, Nile A, Xiao J, Kai G (2020) Nanotechnologies in food science: applications, recent trends, and future perspectives. Nano-Micro Lett 12:45
2. Vieira IRS, Tessaro L, Lima AKO, Velloso IPS, Conte-Junior CA (2023) Recent progress in nanotechnology improving the therapeutic potential of polyphenols for cancer. Nutrients 15:3136
3. Sekhon BS (2010) Food nanotechnology—an overview. Nanotechnol Sci Appl 3:1–15
4. Ndlovu N, Mayaya T, Muitire C, Munyengwa N (2020) Nanotechnology applications in crop production and food systems. Int J Plant Breed 7:624–634
5. Brandelli A, Pinilla CMB, Lopes NA (2019) Nanotechnology in food preservation. Adv Process Technol Bio-Based Nanosyst Food 9:277–311
6. Taouzinet L, Djaoudene O, Fatmi S, Bouiche C, Amrane-Abider M, Bougherra H, Rezgui F, Madani K (2023) Trends of nanoencapsulation strategy for natural compounds in the food industry. Processes 11(5):1459. [Online]. Available: https://doi.org/10.3390/pr11051459
7. Mushtaq A et al (2023) Recent insights into nanoemulsions: their preparation, properties and applications. Food Chem: X 18:100684. https://doi.org/10.1016/j.fochx.2023.100684
8. Joseph TM, Kar Mahapatra D, Esmaeili A, Piszczyk Ł, Hasanin MS, Kattali M, Haponiuk J, Thomas S (2023) Nanoparticles: taking a unique position in medicine. Nanomaterials 13:574. https://doi.org/10.3390/nano13030574
9. Thakur D et al (2024) Nanoemulsion edible coating for shelf-life improvement and quality control in perishable products. Plant Nano Biol 10:100114. https://doi.org/10.1016/j.plana.2024.100114
10. Singh A et al (2024) In-depth exploration of nanoparticles for enhanced nutrient use efficiency and abiotic stresses management: present insights and future horizons. Plant Stress 14:100576. https://doi.org/10.1016/j.stress.2024.100576
11. Matindoust S, Baghaei-Nejad M, Shahrokh Abadi MH, Zou Z, Zheng L-R (2016) Food quality and safety monitoring using gas sensor array in intelligent packaging. Sens Rev 36(2):169–183. https://doi.org/10.1108/sr-07-2015-0115
12. Dharmaraj D, Chavan N, Likhitha U, Nayak UY (2024) Electrospun nanofibers for dermatological delivery. J Drug Delivery Sci Technol 99:105981. https://doi.org/10.1016/j.jddst.2024.105981
13. Naureen F et al (2024) Inhalable dry powder nano-formulations: advancing lung disease therapy-a review. Front Nanotechnol 6. https://doi.org/10.3389/fnano.2024.1403313
14. Rudzińska M, Grygier A, Knight G, Kmiecik D (2024) Liposomes as carriers of bioactive compounds in human nutrition. Foods 13(12):1814. https://doi.org/10.3390/foods13121814
15. Izzetti R, Gennai S, Nisi M, Gulia F, Miceli M, Giuca MR (2022) Clinical applications of nano-hydroxyapatite in dentistry. Appl Sci 12:10762. https://doi.org/10.3390/app122110762
16. Jain KK (2020) Role of nanobiotechnology in drug delivery. Methods Mol Biol 2059:55–73
17. Sajid A et al (2020) Characterization and antibacterial properties of *Eriobotrya japonica* extract loaded silver-nanoparticles. Curr Bioact Compd 17:58–63
18. Yao M, McClements DJ, Xiao H (2015) Improving oral bioavailability of nutraceuticals by engineered nanoparticle-based delivery systems. Curr Opin Food Sci 2:14–19
19. Jafari SM, McClements DJ (2017) Nanotechnology approaches for increasing nutrient bioavailability. Adv Food Nutr Res 81:1–30
20. McClements DJ (2020) Recent advances in the production and application of nano-enabled bioactive food ingredients. Curr Opin Food Sci 33:85–90
21. Mirbagheri R, Elhamifar D, Shaker M (2021) Yolk–shell structured magnetic mesoporous silica: a novel and highly efficient adsorbent for removal of methylene blue. Sci Rep 11:23259
22. Yetisgin AA, Cetinel S, Zuvin M, Kosar A, Kutlu O (2020) Therapeutic nanoparticles and their targeted delivery applications. Molecules 25:2193
23. Abolmaali SS, Tamaddon AM, Salmanpour M, Mohammadi S, Dinarvand R (2017) Block ionomer micellar nanoparticles from double hydrophilic copolymers, classifications and promises for delivery of cancer chemotherapeutics. Eur J Pharm Sci 104:393–405

24. Ognik K et al (2016) Effect of silver nanoparticles on the immune, redox, and lipid status of chicken blood. Czech J Anim Sci 61:450–461
25. Kothamasu P et al (2012) Nanocapsules: the weapons for novel drug delivery systems. Bioimpacts 2:71–81
26. McClements DJ (2023) Modeling the rheological properties of plant-based foods: soft matter physics principles. Sustain Food Proteins 1:101–132
27. Zhang L et al (2019) Detection of choline and hydrogen peroxide in infant formula milk powder with near infrared upconverting luminescent nanoparticles. Food Chem 270:415–419
28. Jäger A, Stefani V, Guterres SS, Pohlmann AR (2007) Physico-chemical characterization of nanocapsule polymeric wall using fluorescent benzazole probes. Int J Pharm 338:297–305
29. Vieira IRS, Conte-Junior CA (2024) Nano-delivery systems for food bioactive compounds in cancer: prevention, therapy, and clinical applications. Crit Rev Food Sci Nutr 64:381–406
30. Lima AL, Gratieri T, Cunha-Filho M, Gelfuso GM (2022) Polymeric nanocapsules: a review on design and production methods for pharmaceutical purpose. Methods 199:54–66

Optimizing Sleep and Recovery with Nanotech Interventions

Kirti Chhabra, Neha Kashyap, and Bajrang Dhundura

Abstract Nanotechnology has transformed sports by enhancing athletes' skills and ensuring their safety. With the use of nanotechnology, sports equipment might be made to precisely match each athlete's physiological and performance needs. Sleep is the most vital recovery strategy used by both elite and sub-elite athletes. Athletes' schedules could have a detrimental effect on their sleep and recovery and the rigorous and repeated nature of a competitive season can also put their bodies and minds to the test, which emphasizes the importance of getting enough good sleep. Emerging technologies like nanoscale therapies and nano devices help speed up the recovery process for injured athletes. Wearable sleep sensors monitor movement, heart rate, and heart rate variability (HRV) as indicators of sleep–wake cycles using actigraphy. These gadgets can help athletes who participate in high-intensity aerobic sports, like swimming and cycling, balance their resistance training regimens with good sleep hygiene, enabling enough time for rest and recovery. Novel approaches like targeted melatonin delivery systems and therapies based on nanomaterials optimize recovery by targeting hormonal and inflammatory pathways. Nanotechnology also create wearable exoskeletons or suits which assists athletes in returning to their pre-injury states, lowers the injury risk, thus improving the overall performance.

Keywords Nanotechnology · Sleep · Recovery

K. Chhabra (✉)
Department of Physiotherapy, RPIIMS, Karnal, Haryana, India
e-mail: kirtichhabra999@gmail.com

N. Kashyap
SAHS, Galgotias University, Greater Noida, Uttar Pradesh, India

B. Dhundura
Healthcare Australia, Osborne Park, Australia

A. Chatterjee et al. (eds.), *Nanofuel: The Future of Sports Nutrition*, Smart Nanomaterials Technology, https://doi.org/10.1007/978-981-96-5471-0_20

1 Introduction

Sleep and post exercise recovery is essential for athletes, that requires a careful balance between training stress and physical recovery to optimize adaptation and performance for upcoming sessions of training or competitions [1, 2]. Sleep is a crucial psychophysiological function for maintaining an athlete's health, performance, and general well-being. An athlete's psychophysiological recovery, which involves several restoring physiological functions and processes, relies primarily on their sleep cycle. These include immunological, metabolic, and neuroendocrine processes that are influenced during numerous stages of the sleep cycle [3]

Sleep is essential and generally regarded as being necessary for the best possible performance and recovery results [3]. It is a frequently neglected component of athlete health that can significantly affect sports performance [4]. Athletic performance is negatively impacted by inadequate sleep length and quality. Poor sleep can raise the risk of disease and injury, reduce athletic performance, jeopardize overall health, and impede the growth of athletes. Long training sessions and intense exercise can cause poor sleep quality, which can lead to daytime weariness, non-restorative sleep, increased fragmentation, and worsened sleep latencies. Additionally, the elite teams across a range of age groups are likely to experience impaired sleep quality due to competitive periods, travel, and intense training [5]. The homeostatic and circadian control of sleep is impacted by irregular sleep–wake rhythms, which lowers the quantity and quality of sleep. Lack of sleep is linked to lower anabolic and higher catabolic hormones, which affect muscle protein synthesis and may hinder recovery and training responses. The possibility that insufficient sleep affects both sports performance and wellbeing raise concerns about the high percentage of athletes who report having trouble sleeping. As a result, it is necessary to find tried-and-true methods for enhancing athletes' sleep [1, 2]. Thus, having techniques to assist reduce sleep disturbances is essential for improved sleep quality [4].

While optimizing sleep through sleep hygiene practices might help with some common sleep issues, but an athlete who has an undiagnosed sleep condition may not be able to receive the recuperation sleep they require to compete at their best. Sleep questionnaires (Athlete Sleep Screening Questionnaire), wearable technologies, clinical sleep interviews, and polysomnography—a sleep study used to detect sleep disorders—are some of the methods used to test athletes for sleep issues [4]. Strategies like a cave-like setting, a good pre-sleep routine, limiting nighttime light exposure, and strategically consuming caffeine and carefully controlling the macronutrient timing, type and dietary supplementation can all help athletes get better sleep and maximize performance [1, 2, 4, 5].

Athletes' physical and mental abilities may be put to the test during a competitive season due to its repetitive and demanding nature. In order to manage fatigue and improve their performance in upcoming training and competition, athletes must adopt recovery techniques that strike a balance between stress and recuperation [1]. In recent years, nanotechnology is changing the face of sports, improving athletes' abilities and guaranteeing their safety. Sports have always welcomed technology advancements and nanotechnology is one of the most recent of these developments [6]. It is the process of utilizing atoms and molecules to create new materials and is a technology used to investigate the constituent of materials that are smaller than 100 nano meters. At this level, material technology, its processing and research is referred to as nanotechnology [7].

Nanotechnology has the potential to create sports gear that is specifically suited to the physiological and performance requirements of each individual athlete. Racquets, clubs, and bicycles that are tailored to a player's strength, style, and technique can help them perform better and avoid injuries. For athletes who have suffered injuries, nanoscale treatments and medical equipment can hasten their recovery. These technologies can decrease downtime and increase an athlete's total career lifespan by encouraging tissue regeneration and healing at the cellular level [6].

The incorporation of nanotechnology has resulted in "smart textiles" that can communicate data in real-time and monitor athletes' vital signs. Better performance analysis and injury prevention are made possible as a result. These clothes are able to identify problems like dehydration or hyperthermia and notify athletes, coaches, or medical personnel. For prompt treatment and recovery, early injury detection is essential. Highly sensitive diagnostic instruments made possible by nanotechnology help identify injuries even before obvious symptoms show up. An unprecedented degree of control and performance may be given to athletes by incorporating nano sensors and actuators straight into their bodies, which also makes it feasible to optimize muscle contractions and track physiological parameters in real time [6]. Hence, this chapter explores how nanotechnology presents intriguing strategies to improve athletes' sleep and recovery, which could result in better performance outcomes and a lower chance of injury.

2 Nanotech Intervention for Sleep Optimization

1. **Wearable sleep monitors/trackers:**

There are numerous types of wearable sleep monitors, such as smartwatches, rings, armbands, headbands or sensor clips. The majority of them use sensors like skin temperature, pulse oximetry, actigraphy, and HRV monitoring. The measurement of blood oxygen levels, or pulse oximetry, improves sleep efficiency monitoring and sleep stage estimation. Electroencephalographic, pulse oximetry, and actigraphy sensors are combined in the most precise wearable sleep monitors [8].

Accelerometer-based sleep measures can be categorized as either variables that are "directly" measured by the device or that must be manually entered (such as wake after sleep onset, time at wake, time at lights out, and time at sleep onset) or variables that are derived from these measures (such as sleep efficiency, duration, and latency). Furthermore, photoplethysmography (PPG), which projects light onto the skin and measures the variation in light absorption to determine heart rate and heart rate variability, has become more widely used in more recent actigraphy devices.

When PPG was added to wearables, the readings' accuracy rose in comparison to those of devices without. The sensor technology advancements make it possible to forecast sleep stages and may even highlight warning signs of possible sleep disorders [3].

(a) Wrist/finger- worn wearables are frequently used wearable for monitoring or tracking sleep behaviour in sports settings. These permits longitudinal sleep measurements (over a season) and diagnose circadian rhythm disorders like jet lag or phase delay. Also, facilitate communication and interaction between athletes and performance professionals and encourages further testing or PSG for sleep problems [3].

Many devices have automated scoring algorithms that may be instantaneously sent to a smartphone, while others require a scientist to grade the data. These can be utilized at any time, whether it's the off-season, preseason, or playing season and generally advised for a minimum of two weeks in order to establish sleeping patterns. Examples include Readiband, Fitbit, Acti-Graph, Whoop, Acti-Watch, Oura, Garmin, Apple Watch. The Oura ring measures sleep and extracts sleep staging data using a variety of sensors, such as mobility, body temperature, heart rate, heart rate variability, and pulse wave variability amplitude [3, 9].

(b) Smart Pyjamas: Particularly, "Phyjamas," so called because of their suggested physiological measurement, are a tracker that is incorporated into fabric and has a network of fabric-based sensors that track and report the important sleep metrics. Heart rate, respiration rate, and ballistic movements can all be measured with sensor-enhanced loose-fitting sleepwear. Although the sleeping positions of these intelligent pyjamas have been verified, they have not yet been verified to measure crucial sleep parameters [3].
(c) Head worn sensors that focused on improving sleep monitoring take into account EEG, EOG, and EMG signals in addition to other metrics utilized by other wearables. In order to detect sleep and sleep staging, the Somfit (Compumedics, Australia) is a device that is applied to the forehead using a single gel adhesive strip. It measures oxygen saturation, heart rate, heart rate variability, EEG, and EOG/EMG (derived from EEG signal). The Dreem headband is another popular head-mounted sleep aid that uses a pulse oximeter to track heart rate, a 3D accelerometer to track position, movement, and breathing frequency, and five EEG electrodes (O1, O2, FpZ, F7, and F8) to assess brain cortex activity. These tools could be used to estimate two-stage sleep and could be an alternative to PSG for field-based sleep assessment [3].
(d) Nearables, as opposed to wearables, are typically placed on a bedside table or beneath a mattress and can detect sleep or sleep issues from a distance. These devices typically contain motion sensors to measure sleep, light, sound, humidity, temperature, and ballistocardiography vibration (for respiratory and heart rate, stroke volume). Devices like the SleepScore Max and ResMed S+ are placed close to the bed (i.e., on your bedside table) and use a combination of light and sound sensors to estimate sleep, whereas under-the-mattress products

(like EarlySense, EmFit, and Withings Aura) or on-the-mattress products (like Beddit, Eight mattress cover, and RestOn) use sensors like ballistocardiogram, respiration, and motion sensors [3, 9].

Numerous factors relating to sleep and wakefulness can be obtained via sleep-related trackers. When examining sleep/wake patterns, the following sleep metrics are advised to be used: Sleep Onset Latency, Sleep Onset Time, Total Sleep Time, Wake After Sleep Onset, Wake Time, Sleep Efficiency, Time in Bed, Sleep Onset Variance and wake variance but exhibit significant inter-individual variance [3].

Also, the data provided by these sleep -related trackers should be used as a reference for optimizing sleep in athletes. Based on the data, these ratings frequently indicate an athlete's level of recovery and may even recommend certain training regimens. A bad night's sleep, for instance, could lead to a low readiness score, which would then advise the athlete to rest or take today off. Furthermore, seeing a low readiness score the morning of a significant competition—for instance, an Olympic final—may have a detrimental effect on the athlete's motivation and performance, depending on their personality and level of trust in the gadget. As a result, care should be taken when utilizing these results to guide performance and training [3].

2. **The Vinci Power Nap and Space Power Nap neuro architecture system**

It is a revolution in rapid stress reduction and optimal regeneration, provided by a system of 20-min power naps, while swinging in a specially designed cocoon, in an effectively developed surrounding free from pollution, with silence from outside noise, and disconnection from phones—all designed to calm, relax, and harmonize all six senses simultaneously. This innovative tool aids in lowering stress, jet lag, controlling the circadian rhythm, enhancing concentration, productivity, and reaction times, halting bone loss, enhancing sleep quality, and promoting human health and welfare [10].

Sleep is essential for immune system strengthening, reset, and recovery. It also helps prevent mistakes brought on by jet lag and exhaustion, which is particularly crucial for athletes. The VPN technology's pendulum movement frequency revolutionizes the synchronization of breathing, heartbeat, and brain delta waves, which promotes the finest possible body and mind regeneration. Additionally, the pulsating pressure on the skin raises hormone levels, including serotonin, oxytocin, and dopamine. All of this helps to restore energy and sleep, enhancing people's longevity, psychological well-being, and sleep quality [10].

3. **Targeted delivery of Melatonin:**

An endogenous hormone called melatonin (N-acetyl-5-methoxytryptamine) is generated by the pineal gland and delivered into circulation. Daily and seasonal variations in the ambient light–dark cycle regulate the production of melatonin, which is governed by endogenous oscillators in the suprachiasmatic nucleus (SCN). Gland that transforms and sends light signals from the sympathetic nervous system into signals that are related to hormones [11]. Hormones and endocrine factors are chemicals made in organs or specific cells that function in locations other than their secretion sites, enabling communication between organs (for example, between peripheral tissues and the central nervous system). Maintaining homeostasis and adapting changes in the environment depend on this chemical communication [10].

Melatonin works by activating the high-affinity G protein-coupled receptors MT1 and MT2, which has several positive impacts on sleep, circadian irregularities, and mood disorders. Melatonin enhances the latency, length, and quality of sleep by coordinating with circadian rhythms. Furthermore, its benefits on neuronal survival and sleep regulation have been demonstrated. Since exogenous melatonin binds to MT1 and MT2 receptors and activates the same downstream pathways as endogenous melatonin, it can be useful in treating melatonin release deficiencies. A melatonin receptor agonist called Ramelteon acts on MT1 to enhance sleep by reducing wake-promoting impulses. It may also affect how long people sleep by acting on MT2 [11].

4. **Other special sleep- inducing substances:**

(a) Polyphenols Phlorotannins are considered as natural sleeping aids, which found to have sedative-hypnotic effects via the gamma-aminobutyric acid type A-benzodiazepine receptor. Additionally, the Ministry of Food and Drug Safety approves Ecklonia cava supplement containing phlorotannins as a component with health benefits that enhances the quality of sleep [12].

(b) Maydis stigma, an innovative medicinal substance encouraging sleep and minimizing sleep disturbances activating the melatonin receptors 1& 2 and genes involved in the circadian clock [11].

3 Nanotech Interventions for Recovery Enhancement

1. **Near-infrared spectroscopy-equipped wearable sensors**

Near-Infrared spectroscopy is a technique for measuring the oxygenation of living muscles. The wearable sensors equipped with near-infrared spectroscopy can gauge an athlete's muscle oxygen saturation levels and give them feedback on how well their muscles are recovering. In order to evaluate a player's ability to sustain performance during exercises without experiencing tiredness, these sensors may also continually record changes in the oxygen saturation of muscle tissue [8].

2. **Wearable exoskeleton and exosuits**

The control mechanism of exoskeletons is based on wearable sensors that measure the force generated by muscles or the angular displacement or velocity of the user's body joints. Through the use of several sensors, exoskeletons may identify the motion intentions of patients with restricted motor capabilities, thereby enhancing the dexterity of their crippled limbs and gently directing their movements toward a goal.

Exoskeletons may also help athletes increase their strength, agility, and endurance while avoiding injury-prone movement patterns [8].

Utilizing a number of cutting-edge wearable sensors, exoskeleton applications in athletic training should incorporate a comprehensive and detailed assessment of athletes' performance. Athletic performance will be measured by the wearable sensors, enabling the exoskeleton's training methods to be modified for better training results. Therefore, it is possible to think of rehabilitation exoskeletons as both ambitious trainers and mobility facilitators. The exoskeleton and a collection of wearable sensors serve crucial roles in assisting athletes in returning to their pre-injury level of performance, mirroring the data gathered from wearable sensors [8].

3. **Nanomaterials in clothing for recovery:**

In order to keep athlete dry and comfortable during vigorous physical exercise, textiles with moisture-wicking and breathable properties are made using nanofibers and nanoparticles. "Smart textiles" that can track athletes' vital signs and send data in real-time have been made possible by the incorporation of nanotechnology. Better performance analysis and injury prevention are made possible by this. These clothes can warn athletes, coaches, or medical personnel of problems like dehydration or hyperthermia. Shock-absorbing materials used in protective gear such as padding, shin guards, and helmets have been created using nanotechnology. In high-impact sports like football and hockey, these materials can more effectively absorb and disperse impact forces, lowering the chance of damage [6].

4. **Therapies based on nanoparticles:**

(a) Nanoparticles specifically target and regulate the expression of pro- and anti-inflammatory substances by phagocytosing inflammatory sensors or macrophages. Focusing on effector cells—particularly those that present antigens—may also improve immunological tolerance or cellular responsiveness. Because they are modulable, nanocarriers can target these cells passively or aggressively [13].

One form of Nanoparticles that has been developed to counteract the adverse effects of other traditional medicines is chitosan-poly (γ-glutamic acid) nanoparticles loaded with clofenac, which showed negligible cytotoxicity for human macrophages. In order to promote wound healing, Au NPs combined with antioxidative small molecules (α-lipoic acid and epigallocatechine gallate) increase angiogenesis and decrease inflammation. Furthermore, Au nanoparticles can improve the skin absorption of loaded medications [13].

(b) Silver nanoparticles are found to have antimicrobial properties because they prevent bacteria from synthesizing ATP and damaging their DNA. Apart from this, due to their antiflogistic property, burn wound healing get accelerated. Additionally, silver nanoparticles hasten tendon healing by promoting cell growth and increasing production of collagen and proteoglycans [13].

References

1. Doherty R, Madigan SM, Nevill A, Warrington G, Ellis JG (2021) The sleep and recovery practices of athletes. Nutrients 13(4):1330
2. Barnard J, Roberts S, Lastella M, Aisbett B, Condo D (2022) The impact of dietary factors on the sleep of athletically trained populations: a systematic review. Nutrients 14(16):3271
3. Driller MW, Dunican IC, Omond SE, Boukhris O, Stevenson S, Lambing K, Bender AM (2023) Pyjamas, polysomnography and professional athletes: the role of sleep tracking technology in sport. Sports 11(1):14
4. Bender A, Lambing K (2023) A practical guide to improve sleep and performance in athletes. Int J Sports Sci Coaching 19. https://doi.org/10.1177/17479541231201105
5. Vitale KC, Owens R, Hopkins SR, Malhotra A (2019) Sleep hygiene for optimizing recovery in athletes: review and recommendations. Int J Sports Med 40(08):535–543
6. Hassan S (2023) Nanotechnology in sports: from performance enhancement to injury prevention. J Nanosci Curr Res 8:200
7. Li J (2022) Application of nanotubes combined with ethnic sports rehabilitation therapy in the treatment of patients with knee arthritis. J Nanomater 2022(1):8679892
8. Park JH, Banarjee C, Fu J, White-Williams C, Coel R, Zaslow T et al (2024) Youth athletes and wearable technology. F1000Res 13:1381
9. De Zambotti M, Cellini N, Goldstone A, Colrain IM, Baker FC (2019) Wearable sleep technology in clinical and research settings. Med Sci Sports Exerc 51(7):1538
10. Filcek M (2023) Vinci Power Nap® synchronization technology—neuroarchitecture for better sleep, adding energy for life, reducing stress and jet lag, improving wellbeing and body-mind regeneration. helpful for leaders, doctors, soldiers, children, drivers pilots, astronauts, people traveling on earth and beyond. J Biotechnol Biomed 6:105–128
11. Kim RE, Mabunga DF, Kim HJ, Han SH, Kim HY, Shin CY, Kwon KJ (2022) Novel therapeutics for treating sleep disorders: new perspectives on maydis stigma. Int J Mol Sci 23(23):14612
12. Kim S, Kim D, Um MY, Yoon M, Choi JS, Choi YH, Cho S (2022) Marine polyphenol phlorotannins as a natural sleep aid for treatment of insomnia: a review of sedative-hypnotic effects and mechanism of action. Mar Drugs 20(12):774
13. Cardoza JV, Ali Z, Simon S, Thakkar D, George SS, Isaac SP (2024) The role of nanoparticles in accelerating tissue recovery and inflammation control in physiotherapy practices. Cureus 16(11):e73540

Nanotechnology in Anti-inflammatory Sports Supplements

Dipjyoti Sharma, Pollobi Porasar, Abhilash Bharadwaj, Arjit Prosad Roy, Tarun Kumar Dua, Aadesh Upadhyay, Darshana Hazarika, and Koushik Nandan Dutta

Abstract This chapter critiques the development of anti-inflammatory supplements and equipment for athletes and sports performance-focused nano recovery. It emphasizes on the impact nanotechnology has had on sports equipment and nutrition—improving the equipment made alongside the supplements provided to maximize athletic performance and recovery. Developments such as nanoparticles in composite equipment and bioavailable nano supplements formulated for efficient delivery of anti-inflammatory agents are the main highlights of the chapter. The chapter goes over nano delivery systems of anti-inflammatory phytoconstituents, including terpenoids, phyto-cannabinoids, phytosterols, carbohydrates, and polyphenols. These systems show promising results in improving the therapeutic efficacy of traditional supplements while potentially causing discomfort. A few recent studies are examined in detail such as the effects of omega-3 nanocapsules, nano-curcumin and glutamine supplements on athletic performance and recovery. Focus is also given to the safety and mechanisms regarding the said supplements and systems, especially with respect to nanomachines. The use of nanotechnology within sports has the potential to help improve sporting performance as well as reduce soreness and enhance recovery. As this chapter will show, the current applications of, and future possibilities for,

D. Sharma · P. Porasar · A. Bharadwaj · K. N. Dutta (✉)
Department of Pharmacognosy, NETES Institute of Pharmaceutical Science, NEMCARE Group of Institution, Shantipur, Assam, India
e-mail: koushik5dutta@gmail.com

A. P. Roy
DmbH Institute of Medical Science, Dadpur, Puinan, Hooghly, West Bengal, India

T. K. Dua
Department of Pharmaceutical Technology, University of North Bengal, Raja Rammohunpur, Darjeeling, West Bengal, India

A. Upadhyay
Project Lead-Oncology, One Clinical Research, Perth, WA, Australia

D. Hazarika
Department of Community Health Nursing, NEMCARE Institute of Nursing Sciences, NEMACRE Group of Institutions, Santipur, Assam, India

A. Chatterjee et al. (eds.), *Nanofuel: The Future of Sports Nutrition*, Smart Nanomaterials Technology, https://doi.org/10.1007/978-981-96-5471-0_21

nanotechnology in sports as well as the broader aspects of safety and regulatory concerns are extensively explored.

Keywords Nanotechnology · Anti-inflammatory supplements · Nano-delivery systems · Athletic recovery · Phytoconstituents

1 Introduction

Inflammation is the most common disease, which occurs when the body emits chemicals to fight away infection or repair damaged tissue. When the wound or infection heals, the inflammation stops [1]. Recent studies reported that inflammation is both an adaptive reaction to the body and a protective mechanism of the body[2]. Inflammation plays a crucial role in the body's defence mechanism. Our immune system uses inflammation as a protective mechanism to keep the body safe and aid in healing. It may be acute or chronic. Each has its unique pathways. Acute inflammation affects a wounded site's cells, the immune system and the local vasculature. On the other hand chronic inflammation is defined by tissue damage brought on by inflammatory processes as well as a gradual shift in the kind of immune cells that develop in the afflicted inflammatory tissues [3]. In sports, inflammation is an important factor for performance. Sports performance can be impacted by inflammation in both positive and negative ways. After exercise, the body naturally experiences inflammation as a part of its healing process. It facilitates muscle healing, recuperation and workout adaption. Inflammation from athletics has a detrimental impact on their general health as well as how long they can perform. Managing inflammation effectively is essential for maximizing sports performance and lowering injury risk [4]. In recent years, the field of nanotechnology has been gaining traction in the widespread use of its concepts and end products. Applications of nanotechnology research range from the prevention and treatment of diseases to newer and better technological devices. In sports, nanotechnology has shown great promise in reducing inflammation and has the ability to enhance a wide range of sectors of the sports industry [5].

In recent years, advancement in the biomedical field has undergone vast transitions in overcoming the shortcomings of the traditional system of medicine with the newer approaches. One such major and extensive approach is nanotechnology. This branch takes us to the nano-level dimension of materials with uncountable positive outcomes [6]. Nanotechnology in healthcare mainly provides for diagnosis and treatment of complex diseases using 'nano tools' like nanoparticles, liposomes, dendrimers, nanorods and hydrogels are some of them which can impact the health of a person at a nano level, ensuring much more accuracy. These nano tools or nanomaterials aid in the treatment, diagnosis, prevention and control of disease conditions of individuals. They are safer, target-specific and deliver the therapeutic to exactly the site of action where required. Such tools can easily cross biological barriers which are difficult to penetrate. In a nutshell, nanotechnology has a crucial role in biomedical therapeutics. These nanomaterials can have multiple modes of action like targeting various tissues and specific cellular organelles and enhancing the anti-inflammatory responses, prolonging the release of active compounds to minimize dosing frequency and extend the period of therapeutic effects, increasing the absorption of therapeutics with consequently increasing efficacy and bioavailability and finally they facilitate the cellular internalization of therapeutics which increases efficiency to treat inflammation [7].

Nanopharmacology is an emerging branch in medical sciences which deals with the utilization of new nanotechnological approaches based on novel pharmacological principles in order to increase the therapeutic efficacy of the medication and fix the problems related to disabled pharmacokinetics and unpredictable dose dumping issues. With advancements in the field of nanoscience, nano pharmacology has come up as a breakthrough approach to compile pharmacology with biomedicine. Some approaches like surface-engineered nanotubes, dendrimers and nanoparticles aim nontoxic, efficacious and biocompatible formulations [8]. The use of nano pharmacology coupled with nanotechnology can be a promising alternative to provide better outcomes for sportspersons which can be a connecting link between nano pharmacology and sports.

Nanotechnology has potentiated enormous applications in the fields of health and sports. The performance of athletes in sports is greatly impacted by the use of various nano approaches in sports. The use of various nano phytochemicals has a profound impact on recovery of the athletes from stress and inflammation. Moreover, the use of nanotechnologies like carbon nanoparticles, fullerenes, nanoparticles of silica derivatives and many others in designing structured sports equipment for better performance [5]. Research brings the use of various anti-inflammatory supplements based on nanotechnological approaches which have excelled in improving performance in sports. These supplements may improve muscle recovery, maintain good joint health by minimizing inflammation and joint pain in athletes, improve the immune system during the intense training period and mitigate exercise-induced oxidative stress. By using nano approaches like the use of phospholipid vesicles-based liposomes, biodegradable polymeric nanoparticles, dendrimers and nano emulsions which may enhance therapeutic delivery, increase drug encapsulation, increase solubility and solubility and finally stabilization of lipophilic compounds with the increase in absorption can be achieved easily [9, 10].

A recent study reported that during aerobic training (5 days at 18–20 m/min) of male Swiss albino mice co-administered with nano curcumin formulation (80 mg/kg/day) through oral gavage showed down-regulation of inflammatory MAPK/ERK signalling pathways showing anti-inflammatory response [11]. Another study revealed that nano-bubbled curcumin when administered for 6 weeks to male ICR mice showed increased glycogen content. The nanoformulation altered the gut microbiota retained energy and prevented fatigue during aerobic exercise [12]. Oral curcumin-loaded nanomicelles promoted muscle healing, lessened the percentage of collagen fibres and increased blood circulation in male adult rats as compared to normal curcumin at a higher dose. Thus, promising its potential anti-inflammatory action [13]. Cardiovascular defects become a great issue in sports. Nanoformulations of omega-3 fatty acids containing α-linolenic acid rich flaxseed oil-based oil-in-water nanoemulsion have been shown to improve 17β-estradiol and/or C6-ceramide internalization in vascular cells in human endothelial cells and vascular smooth muscle cells. The nanoformulation Increased C6-ceramide and 17β-estradiol uptake by cultured cells; increased endothelial and decreased vascular smooth muscle cells proliferation (p38 MAPK inhibition) [14]. Omega 3 fatty acid based atorvastation loaded lipid-based nanocarrier prepared by melt emulsification and ultrasonication technique have shown to increase the bioavailability of omega 3 fatty acids and exert strong antihyperlipidemic action; a condition essential in sports activity [15]. A 20% w/v flax seed oil nanoemulsion prepared by microfluidization technique was reported to prevent vascular damage and reduce arterial stenosis to 30% [16]. Apart from them, glutamine is a very good supplement for the recovery of muscular damage and reducing muscle fatigue. Nanoformulation of glutamine may improve the overall performance of sportspersons [17, 18].

2 Mechanism of Inflammation

Inflammation is a cellular homeostasis response to a damaging condition in the body (Fig. 1). Inflammation is an adaptive response to harmful effects that threatens the integrity of cellular homeostasis. Inflammation is initiated within a very short time in any host that has a functional immune system as it is a response to a proper stimulant, such as microbes or any other foreign invaders [19]. Immune cells such as macrophages, mast cells, neutrophils and lymphocytes play a crucial role in inflammatory response. Moreover, non-immune cells like epithelial cells, endothelial cells and fibroblast also plays a role in inflammatory response. These inflammatory molecules activates the pain receptors that increases the inflow of calcium and sodium ions into the neurons that induces action potential firing [20, 21]. Inflammatory pathways can vary significantly depending on the nature of inducer. In case of microorganisms' infection, the immune cells immediately sense the pathogens and activates pathogens-specific receptors that induces the production of mediators such as cytokines and chemokines. Through the modification of vascular endothelial permeability and recruitment of neutrophils and extra plasma into the site of infection, quickly accelerates the inflammatory process and the invading pathogens are simultaneously targeted and destroyed by the immune cells. The time period of inflammatory response depends on the level of damage caused by the infection [19]. Phospholipids that are released by the damage cells are converted to prostaglandins H2 (PGH2) via COXs. PGH2 is converted by prostaglandin synthases to either PGE2, prostacyclin (PGI2) and PGF2, that mediates fever, inflammation and increased pain or to thromboxane A2(TXA2), that regulates platelet formation. When injured tissues have recovered, inflammation usually subsides, but it can become chronic inflammation that continues for months or even years [22].

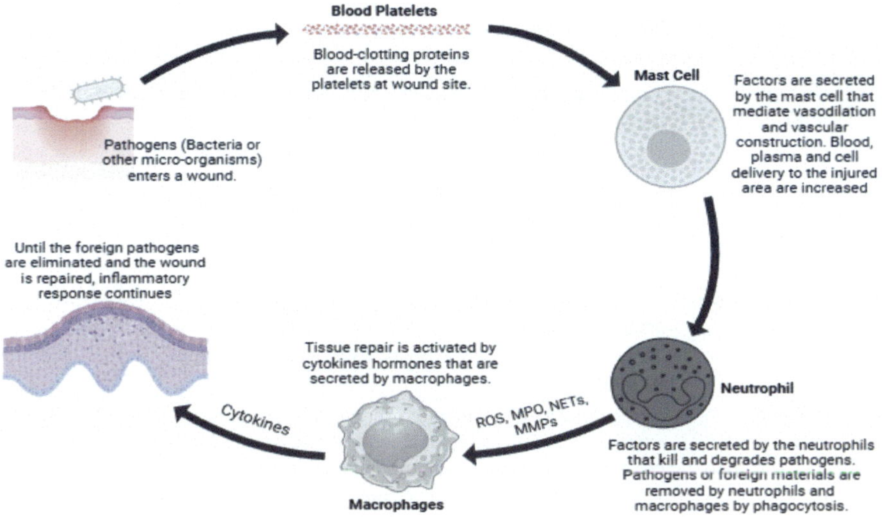

Fig. 1 Mechanism of action for Inflammation

Phospholipase A2 is released and then converted to arachidonic acid, following the inflammatory response. COX-1 and COX-2 of COX enzymatic pathway, is responsible for changing arachidonic acid into prostaglandins (PGs). Thromboxane and PGs are normally produced by COX-1 in platelets, renal tubule cells and gastrointestinal mucosal cells. At the site of inflammation COX-2 is upregulated and produces PGs causing inflammation and pain. When COX-2 is inhibited, it reduces production of PGs, that results in anti-inflammatory and analgesic effects [22].

3 Application of Nanotechnology

There are various applications for nanoparticles (Table 1). It is mostly used in bio medicine, biochemical analysis, engineering and magnetism, identification of gene, gene expression and processing of proteins encoded by genes [23]. Nanoparticles, a sophisticated biological system, when used in medical practises, it would change the fundamentals of detection of diseases, disease therapy and disease avoidance in the future years [24]. In recent years, the application of nanotechnology for regenerative medicine has exponentially increased as the interaction between the cells and the biomaterials at nanoscale level has increased [25]. Some application for nanoparticles in medical practice are listed below.

Table 1 Application of nanotechnology

S. No	Sector	Effect	References
1	Medical diagnosis	To tract disease development, nanoparticles can be used to bind several biomolecules, both inside and outside the cells, at the same time. Identification of components that causes diseases	[24]
2	Drug distribution	To release drug to specific location, nanocrystals are used. The adverse effects are greatly reduced as the required dosage of the drug is released within the affected area	[24]
3	Orthopaedics	It is easier for the calcium containing deposits to accumulate on prostheses by using tiny components such as mini plastics, graphite microfibers, porcelain, and nanotubes	[24]
4	Cancer	Tiny particles due to its microscopic size might be extremely helpful in the treatment of carcinoma. Using radiation therapy and quantum entanglement, detailed representation of tumour sites may be produced. With the help of nanocircuits detector diagnostic cards can be created that would require only a few jobs of patient's blood to discover and diagnose cancer in its early stages	[24, 26]
5	Cardiac therapy	Nano materials have the potential for use in concurrent heart physiology and enabling the biological discovery of fresh opportunities and improvements in treatment of serious heart issues	[24]
6	Dermatology	There are various types of engineered nanoparticles used for dermal and transdermal drug delivery. Transdermal drug delivery system ensures drug stability, precise drug deposition and controlled release. It also helps to avoid first-past metabolism and the gastrointestinal tract. Moreover, the administered drug amount can be changed, which permits the use of relatively potent drug without the risk of toxicity in the system	[27]
7	Dentistry	Nanostructures containing cisplatin shows high efficacy against oral carcinoma of squamous cells when distributed on polyethylene glycol poly (Glutamic acid) and produce noticeably less kidney damage than iron alone. Solid shape nanoparticles offers the advantages of increased drug stability, defence against proteolytic enzyme destruction and continuous release of absorbed components as they are effective transporters of hydrophobic or water liable radiation	[24]

4 Impact of Nanotechnology on Sports Industry

Innovations gives not only various advantages in the field of sports but also at level of sports quality, reducing the occurrence of injury [5]. In order to conduct scientific study and social practice, many scholars require measuring equipment, with advanced and sophisticated material, due to which nanotechnology has become a part in the field of material preparation, electronic information, aerospace and competitive sports [28]. In the past few years, the use of nutritional supplement has shifted focus from macronutrient like lipid, carbohydrate, fats, etc. to micronutrients obtained from plants or nature in form of fruits or extract. Fruits from plants helps in supporting health and exercise due to its rich content of macronutrients, micronutrients, minerals, vitamins, fiber, and various phytoconstituents like phenolic compounds. This also helps in great adaptations to the training stimulus [29]. The sports industry has undergone heavy transformation and development due to modern nanotechnology. These two complement, promote and influence each other and it has become the new norm [23]. There are several benefits in using nanotechnology in sports. It has tremendous potential to enhance sporting equipments. Advantages such as strength, equipment, athletes' comfort, safety and endurance, enhanced agility and decrease the risk of injury with comparison to the earlier equipments used [5].

Nanotechnology is used to skilfully combine the strength of humans and machines. For example, in pole vaults, sticks or bamboo pole were used but with the addition of nanomaterial elements, pole vault results were frequently updated and from that it has increased nearly two times. Due to the evolution of the material, the mass of the equipment became lighter and the air-resistance in the run-up phase was reduced. Additionally, nowadays high-tech fiber materials are used in sailboats for competition, in archery bowstring, in bicycle tire, golf stick, javelin, hockey stick and various other equipments [28]. Nanotechnology helps in developing nanoparticles made from natural fruits to be used as sports supplement to aid athletes in training and even in sports competitions. Introduction of nanotechnology helps in overcoming the drawbacks of herbal drugs such as low solubility and low bioavailability and enhance herbal dosage forms by increasing solubility, stability and bioavailability of the active substances [29]. Fatigue is the difficulty in sustaining or initiating voluntary activities. Deterioration in physical performance accompanies physical fatigue and it leads to physical endurance capacity reduction in athletes. Resveratrol-loaded solid lipid nanoparticle can be used as a novel nutritional supplement in sports against fatigue [30].

The equipments used in golf is enhanced by utilizing nanotechnology, has changed the way the golf is played. Nanomaterials used to design the golf stick has reduced its weight and increased the strength that ensures the speed of the golf ball head [5]. Nowadays swimming has become a popular sport as it is fun and convenient. Swimming industry has great importance and potential for development. Nanotechnology has great impact on swimming as it advances athlete performance, enhances equipment and develops the industry. By integrating nanotechnology in training and equipments, swimmers achieve better hydrodynamics and recovery. Nanomaterials in swimsuit reduces the drag and improves buoyancy, and nanomedicines promotes fast stamina recovery as it reduces fatigue and boost oxygen transport. In infrastructures, like pool coating and water purification treatment, nanotechnology usage has improved safety and accessibility for swimmers [23].

Nanomaterial, used in functional fabrics as the core technique for shell fabrics, that can be smaller than the diameter of hair, plays a important role for the portability and comfort of the clothing. The nanometer fiber fabric is used for making sportswear as it is light and has strong waterproof and permeability activity. Sport shoes made from nanosilver are used in various countries. Nanosilver are nontoxic, non-metamorphic, smelless and has no stimulation effect on human skin [28]. Other nanomaterials used in sports includes, fullerenes for its super conductive property, carbon nanotubes for use in nano-enhanced sports equipments, silicon dioxide nanoparticles utilized for producing glass characterized as a poor heat conductor, and carbon nanoparticles for use in road racing [5].

As this technology continues to evolve, its utilization in sports is guaranteed to grow further that connects human potential and technological innovation and ultimately developing and transforming the gameplay and experience of sports.

5 Nano Materials and Supplements

In recent years, the progression of pharmaceutical delivery methods has received more focus. Solid lipid nanoparticles (SLNs) have potential for the oral delivery of lipophilic drugs owing to several beneficial characteristics: Solid lipid nanoparticles (SLNs) are nontoxic, biocompatible, have a substantial payload capacity, and improve the solubility and stability of encapsulated pharmaceuticals [30]. In recent years, many nanomaterials have emerged in competitive sports, significantly enhancing athletic performance and advancing scientific and humanistic approaches to sports. Currently, nanomaterials are extensively used in the domain of sports engineering, including stadiums, athletic equipment, and sports apparel. Nonetheless, the size effect, surface impact, quantum size effect, and macroscopic effects of quantum tunnelling in nanomaterials may induce particular biological responses, potentially resulting in adverse consequences for the environment, biology, and human health [31]. Nutritional supplements have been shown to significantly improve exercise performance. Nutritional supplementation is primarily used to augment physical performance, expedite recuperation between workouts, promote antioxidant activity, reduce oxidative stress and inflammation, increase substrate availability, and improve mitochondrial function and quality. Nanotechnology permits the alteration of the chemical and physical properties of the nano compound surface, enabling targeted action, regulated release, prolonged half-life, and reduced toxicity. Notwithstanding the many advantages of nutritional nano compounds over conventional nutritional supplements, their effectiveness as ergogenic aids has been markedly under-explored. A thorough examination of the benefits of nutritional nanocompounds in relation to exercise performance may provide insights into novel methods for improving the effectiveness of nutritional supplements and promoting exercise performance [10].

Antioxidants are essential for safeguarding tissues from significant damage during exercise by neutralising reactive oxygen species (ROS). The intake of diverse sports supplements containing antioxidants may enhance both acute hard exercise and persistent training; hence, it is plausible that dietary supplementation with certain antioxidants might be advantageous. These sports supplements are functional meals that provide health advantages to athletes and enhance sports performance. This may be accomplished by reducing the influence of elements that induce weariness and hinder the execution of skilled activities. Certain natural items meet all the criteria for rigorous training and are a superior and more refined alternative to the chemical supplements now promoted in global marketplaces. Black ginger root, pomegranate peel, and moringa leaves are often used natural goods that enhance energy levels in the body, therefore demonstrating their efficacy as physical fitness supplements [32].

Nanomaterials may be classified as either metallic or non-metallic from a materials perspective. From the perspective of structural arrangement, it may be amorphous or crystalline, and may exist as either single-phase or multiphase. Nanostructure is a novel system designed or fabricated according to certain principles using tiny material components. It comprises a nanoarray system, a mesoporous assembly system, and a thin-film mosaic system. Due to the particle size of nanomaterials being smaller than 100 nm, they exhibit unique features in optics, chemistry, and magnetism that are not found in conventional materials, mechanics, and thermodynamics [33, 34].

6 Anti-inflammatory Phytoconstituents and Their Nanoparticle-Mediated Delivery

- **Polyphenols:**

Polyphenols (PPH) are a diverse and extensive group of ubiquitous compounds found as secondary metabolites in many vascular plants. These natural compounds vary from basic molecules to intricate structures, all characterised by the presence of benzene rings with one or more hydroxyl groups. In the plant, these active compounds are crucial for growth, reproduction, and resistance to pathogens, predators, and diseases [35]. The robust free radical scavenging capability is perhaps the most extensively researched property and contributes to the anti-inflammatory effects of these compounds. Quercetin (QC), resveratrol (RE) and tannins are specifically acknowledged as analgesics [36]. Consequently, the clinical uses of PPH are limited. To mitigate these disadvantages, nanodelivery methods capable of preserving the structural integrity of these bioactive compounds have been devised. In recent decades, there has been significant interest in the active molecules of turmeric known as curcuminoids, with curcumin (90%) being the most prevalent, which has been investigated for its antioxidant, anti-inflammatory, anticancer, antiviral and antifungal activities [37]. Recent in vitro studies indicate that curcumin's anti-inflammatory properties may be associated with its capacity to reduce the synthesis of inflammatory mediators, inhibit extracellular matrix degradation, prevent chondrocyte apoptosis and function as an antioxidant by diminishing the excessive production of reactive oxygen species [38].

- **Terpenoids:**

Terpenoids are helpful in the prevention and treatment of anti-inflammatory diseases. These chemicals are generated as secondary metabolites by many species, including plants, mosses, algae, lichens, insects, bacteria, and marine creatures [39]. Recent years have seen substantial study focused on the molecular mechanisms behind the anti-inflammatory effects of plant-derived terpenoids. Numerous studies have shown that terpenoid compounds may inhibit nuclear factor-κB (NF-κB) signalling, a primary regulator in the development of inflammatory disorders and cancer. Terpenoids are a very diversified class of compounds, characterised by remarkable variations in chemistry, structure, and function. These compounds are categorised based on the quantity of isoprene units and carbon atoms [40]. Terpenoids are a very diversified class of chemicals, exhibiting remarkable variation in chemistry, structure, and function. These compounds are categorised based on the quantity of isoprene units and carbon atoms. Squalene also functioned as both the active agent and the solvent carrier for the hydrophobic medication. In addition to its anti-inflammatory properties, squalene has been shown to possess in vitro cytoprotective action [41]. Lycopene is a carotenoid exhibiting significant pharmacological qualities, including antioxidant and anti-inflammatory effects, and is widely found in tomatoes, red carrots, watermelons, and other red fruits and vegetables [42]. Lycopene has restricted systemic absorption owing to its very poor water solubility. Lycopene nanosized delivery devices have recently been developed to address this limitation. Butnariu et al., in their study has created nanoemulsions using aqueous propolis and lycopene, which were evaluated for their preventive efficacy against acute UV-A-induced inflammation in rat paws. An improved therapeutic impact was seen in comparison to a conventional suspension, together with a prolonged duration for the evaluated parameters [43].

- **Phyto-cannabinoids:**

Phyto-cannabinoids are characterised as lipophilic compounds originating from the hemp plant, *Cannabis sativa* L. (Cannabaceae). The cannabis extract has about 460 chemicals, approximately 70 of which are Phyto-cannabinoids. Phyto-cannabinoids function by regulating cannabinoid receptors, which are integral to essential physiological processes within the central and autonomic nervous systems, as well as immunological, endocrine, reproductive, and cardiovascular functions [44]. Additionally, Phyto-cannabinoids interact with several receptors, including the potential vanilloid type-1 receptor, the deorphanized G protein-coupled receptor, the peroxisome proliferator-activated receptors, and the T-type Ca^{2+} channel. All these targets modulate the anti-nociceptive and anti-inflammatory responses while also participating in the regulation of sleep, obesity, epilepsy, and breast cancer cell migration [35].

- **Phytosterols: (plant sterols and stanols)**

These are naturally occurring chemicals that are widely present in vegetable oils, cereal grains, nuts, legumes, fruits, and vegetables. The name phytosterol specifically denotes about 200 distinct chemicals, with sitosterol, campesterol, sitostanol, and campestanol being the most prevalent. Phytosterols has several bioactive features, including the ability to decrease blood cholesterol by partial suppression of intestinal cholesterol absorption, anti-atherogenic actions, and, specifically for β-sitosterol, immune-stimulating and anti-inflammatory activity [45]. The use of nanotechnology has the potential to significantly contribute to the improvement of the pharmacokinetics of phytosterols. For the purpose of achieving this objective, Leong et al., has synthesised and characterised nano dispersions of water-soluble phytosterols. In specifically, he investigated the impacts of four distinct organic phases, namely hexane, isopropyl alcohol, ethanol, and acetone, as well as their proportions to the aqueous phase, and the use of conventional homogenisation as opposed to high-pressure homogenisation on the vectors that were generated. In the end, phytosterol nano dispersions were manufactured using the process of emulsification-evaporation with hexane [46].

- **Carbohydrates:**

The healthy growth and development of vegetables is dependent on carbohydrates, which are one of the most important components of plants and are crucial to their growth. As a matter of fact, carbohydrate components constitute ninety percent of the main cell wall and are essential to the functioning of the wall. Pectin, hemicelluloses, and cellulose are the three forms of polysaccharides that are primarily responsible for building the structure of plants. Polysaccharides are high-molecular-weight carbohydrates [47]. It was discovered that a pectin that was extracted from the aerial sections of the comaruman, *Comarum palustre* (Rosaceae), has immunomodulating and anti-inflammatory action. To be more specific, the galacturonan fragments that were produced from the partial hydrolysis of pectin (with a molecular weight of up to 10 kDa) were able to reduce the adherence of neutrophils to fibronectin in vitro to a larger degree than the original pectin [48] (Table 2).

Also, Jiang et al., in their study has reported about Rapid screening of many potential target natural compounds is made possible by the technique, which has revolutionised the screening model for natural nutritional candidates. Their systematic model significantly improves research efficiency. By mainly controlling NK cell numbers, echinacoside successfully controls the inflammatory immune response after exercise. This is the first report to suggest that echinacoside, a natural compound with promising therapeutic potential, may be taken orally to mitigate exercise-induced inflammation caused by an overabundance of natural killer (NK) cells by bringing down peripheral blood levels of NK cell-associated cytotoxic proteins like GZMB, PRF1, FASLG, and CCL4 [50].

Table 2 Nano sized delivery systems for anti-inflammatory carbohydrates

Bioactive principle	Nano-vector	Type of delivery system	Experimental model	Reference
Mannose-6-phosphate	Liposomes	Liposomes containing *Aloe vera* gel	In vitro	[48]
	Nanocapsules	Polyamide NC	In vitro	[49]

Nanotechnology is a strategy that has already given considerable leaps forward in the direction of bringing anti-inflammatory phytochemicals considerably closer to therapeutic applications. This is a promising approach. According to this interpretation, the rise in the number of nanotoxicological studies that have been published in the scientific literature is connected to an increasing knowledge and sensitivity about the ethics of nanotechnology. For the purpose of establishing a stronger correlation between intelligent nanoparticle design and rigorous safety assessment for future applications in biomedicine, additional research is required to improve the cost-effectiveness and long-term safety of nanosized drug delivery systems. This can be accomplished through the utilisation of biocompatible materials and release processes that do not cause any harm [35]. Given the enormous surface area of nanosized additives, it is possible that a very tiny quantity of these additives might be sufficient to offer a significantly enhanced experience of a flavour or flavour. It is possible that this will make it possible to reduce the amount of salt, fat, and other additives that are present in food, such as artificial colours, flavouring agents, and preservatives, while still providing greater aesthetic, nutritional, and health advantages. Other areas of use include the utilisation of nanoparticles for the purpose of antibacterial qualities, as well as the utilisation of nano-sized supplements that are said to have better absorption and bioavailability inside the body [51]. In spite of the fact that nutritional nano compounds provide a number of benefits in comparison to traditional nutritional supplements, the potential of these compounds as ergogenic aids has been investigated not nearly as thoroughly. Once a greater grasp of the advantages of nutritional nano compounds in relation to exercise performance will get, this may be able to gain insights into novel tactics that may optimise the effect of nutritional supplements and increase exercise performance. There have only been a few of research that have really investigated whether or not nanocompounds are useful in improving exercise performance. Nanocomposites containing the following active chemicals were used in the study done by Lima et al., are resveratrol, iron, nitroxide radicals, and folic acid [10].

In the context of nutrition, the term "nutritional supplements" (NS) refers to sources of nutrients that are concentrated and supplemented with additional substances that have been shown to have physiological or nutritional benefits. Proteins and amino acids, carbohydrates, vitamins, isotonic liquids, minerals and trace elements, creatine, herbal medicines, enzymes, and other substances are included in the category of NS. As a consequence of the fact that NS are not supplied as prescription medicine but rather as off-the-shelf goods, which can be purchased in fitness stores or over the internet, their use is common among both professional and amateur athletes, members of the military, members of the general population, adolescents, and other relevant groups [52, 53].

7 Sport Supplement for Inflammation

- **Dietary Carbohydrates**: A decrease in blood glucose levels has been associated with activation of the hypothalamic–pituitary–adrenal axis and heightened secretion of adrenocorticotropic hormone and cortisol. Sustaining plasma glucose levels can mitigate elevations in stress hormones and diminish immune alterations linked to inflammation. The use of glucose-containing beverages has been shown to elevate blood glucose levels, enhance endurance running ability, and reduce plasma cortisol levels and circulating leukocyte subsets. Carbohydrate consumption diminishes the inflammatory response to acute exercise by lowering levels of IL-6, total anti-inflammatory IL-1RA, and cortisol [54, 55].
- **Micronutrients**: The antioxidant potential of certain dietary micronutrients is noteworthy for strengthening the endogenous antioxidant defence mechanisms of athletes, mitigating the adverse effects of oxidative damage caused by free radicals. Zinc, antioxidant chemicals (such as vitamins C and E, beta-carotene, and N-acetylcysteine), vitamin B12, and folic acid (vitamin B9) are some of the specific supplements that have been investigated in athletes up to this point. In addition, a number of newly introduced supplements, including as b-glucan, curcumin, and quercetin, have lately been receiving a lot of attention [56, 57]. The concentrations of certain minerals are crucial for the immunological function of athletes. Low zinc levels can impair the immune system; dietary zinc is crucial for infection resistance. However, research indicates that zinc-deficient athletes should not take mega-doses of zinc (> 25 mg/d) to reduce superoxide free radical levels produced by activated neutrophils, as it may lead to excessive suppression of T-lymphocyte proliferation [58, 59].
- **Dietary Fat**: The amount and type of lipids that are consumed on a daily basis have the ability to influence the biochemistry of the cellular immune system, which includes the generation and expression of cytokines. In general, omega-6 dietary lipids are responsible for a rise in the levels of pro-inflammatory cytokines and inflammatory PGs, while omega-3 lipids have the potential to reduce the levels of these cytokines and inflammatory PGs. It has been seen in human research that an increase in the amount of fat consumed in the diet leads to an increase in endurance capacity [31]. After exercise, concentrations of the anti-inflammatory cytokine IL-2 were shown to be considerably elevated in individuals with elevated amounts of dietary fat. A reduction in the proliferative response to mitogen by peripheral blood mononuclear cells was also seen when there was an increase in the amount of dietary fat and physical activity. Both physical activity and a diet rich in fat have the ability to raise levels of interleukin-2 (IL-2) and lower levels of interleukin-6 (IL-6). However, the quantity of proinflammatory cytokines can be controlled by increasing the amount of fat consumed in the diet in conjunction with the anti-inflammatory effects of physical activity [59].

- **Omega-3 EFAs (Fish Oil)**: The active components in fish oil, EPA and DHA, facilitate the transformation of COX into prostaglandin E3. Prostaglandin E3, a natural anti-inflammatory drug, competitively inhibits the conversion of arachidonic acid to prostaglandin E2, a potent inflammatory compound. Omega-3 essential fatty acids, included in fish oil, may directly diminish the degenerative enzymes aggrecanase and matrix metalloproteinase, along with IL-1, TNF-alpha, and COX-2, so alleviating inflammation in synovial cartilage [60].

8 Benefits of Nano-technology

Nanomedicine focusses on utilising nanotechnology to enhance the efficacy and safety of medications. One method is to encase unprotected medications in biocompatible nanocarriers, such as dendrimers, micelles, nanoparticles, or liposomes. Size, shape, surface charge, and dosage are some of the design characteristics of nanoparticulate drug delivery systems that may be adjusted to improve drug circulation and to target particular tissues or subcellular organelles with great precision. To help treatments cross the blood–brain barrier and enter the central nervous system, NDDS surfaces may be modified with cell-penetrating peptides or ligands. By controlling geographical localisation and reducing dose and adverse effects, NDDSs may increase the efficacy of therapeutics. To enhance the efficacy of treatment, tiny molecules targeted targeting pain signalling receptors or a nanocarrier containing several analgesics may be used. These approaches may be able to deal with the duplication that exists in basic functions, such as the transmission of pain. To alleviate systemic, neuropathic, localised, and disease-related pain with minimal side effects and addiction risk, researchers are seeking novel drug delivery methods [61, 62]. The therapeutic index of various treatments has been the subject of much study into drug delivery methods based on nanotechnology. Nano-delivery technologies have often increased hazardous drug levels rather than decreased effective dosages to accomplish this. Using nanotechnology devices to either reroute the medicine away from healthy tissues and towards the sick location, or to replace the harmful solubilising chemicals employed in traditional drug formulations, tends to boost the safety of the medicinal product. Furthermore, nanodelivery systems have been used to either improve medication absorption across various biological membranes or to provide a way for sustained drug delivery via local or systemic administration, depending on the drug and its intended route of administration [63]. The family of medications known as non-steroidal anti-inflammatory medicines (NSAIDs) is nevertheless widely used in clinical practice today to treat inflammatory disorders including arthritis. A number of approaches exist in which nano-drug delivery may improve the therapeutic efficacy of nonsteroidal anti-inflammatory medications (NSAIDs), such as those used to treat inflammation, fever, and pain. For example, a nonsteroidal anti-inflammatory medication (NSAID) delivered via a nanodelivery system may have a lower toxicity, be more soluble in water and thus easier to dissolve and have a quicker beginning of action due to increased permeability across biological membranes [64]. Nanodelivery methods may improve the therapeutic efficacy of NSAIDs in several ways. Despite having relatively good bioavailability upon oral administration, most NSAIDs are weak acids and are categorised as class II medicines in the Biopharmaceutical Classification System (BCS) due to their limited water solubility at acidic pH. The dissolution of the medication from the dosage form strongly influences the rate of adsorption of these agents, which may significantly delay the start of analgesic effects [65]. Nanodelivery systems have the potential to improve the solubility of nonsteroidal anti-inflammatory drugs (NSAIDs), leading to a fast increase in plasma drug concentrations and consequently analgesia. This might be useful in the treatment of acute episodes of pain. Nanodelivery systems may also improve the effectiveness and safety of these agents by decreasing fluctuations in the reported drug levels and extending their circulation in the systemic circulation [66]. Biomaterials or polymers make up the majority of carriers, while organic NP (Nano particles) have superior biocompatibility. You may classify these polymers as synthetic or natural. The synthetic polymer poly (lactic co-glycolic acid, or PLGA) has recently gained a lot of attention as a possible drug delivery carrier because to its low hydrophobicity and high potential for application in targeting and treatment [67]. In a study that included the construction of an implantation model in mice, PLGA and stromal cell derived factor-1 (SDF-1) were used. Transplanting this solid scaffold into the subcutaneous region reduced the inflammatory reaction. Subsequent studies used an amino terminal peptide consisting of amino acids 2–26 (Ac2-26), which may resemble annexin A1, and encapsulated Ac2-26 with PLGA to target collagen IV (Col IV). Significant therapeutic efficacy of Col IV-Ac2-26 PLGA NPs was shown in an atherosclerosis animal model. The therapy of inflammatory bone disorders, including osteoarthritis, osteosarcomas, and bone cancer metastases, is one area where PLGA has shown potential [68]. Researchers have looked at the anti-inflammatory effects of liposomal nanoparticles using phosphatidylserine (PS) (PSL)-containing liposomes as a biomaterial carrier. Their Interleukin-10-conjugated PSL (PSL-IL 10) demonstrated strong anti-inflammatory and anti-obesity benefits in a mouse model of obesity. In addition, PSL may be recognised particularly by macrophages, transforms inflammatory M1 macrophages to anti-inflammatory M2 macrophages, and mimics apoptotic cells [69]. To alleviate inflammation, inorganic NPs are often combined with anti-inflammatory medications and inorganic carriers. Metal oxide NPs show great promise as a drug delivery mechanism in nanomedicine due to their increased stability. Zinc oxide nanoparticles, or ZnO NPs, are very beneficial in the medical field [70]. Numerous investigations on the potential theragnostic uses of the Quantum dot (QD), a nanoparticle made of fluorescent semiconductor, have been conducted. Moreover, recent research has shown that QD may have a role in medication distribution. Researchers found a correlation between celecoxib and QD. The inflamed paw of the mice was precisely targeted by these QD-Celecoxib conjugates, which not only had an accurate anti-inflammatory impact but also a bio-imaging effect [71]. Carbon nanotubes (CNTs) have attracted a lot of interest due to their promising future as nanocarriers. Whether as the main drug carrier or an adjuvant to improve or modify the release of pharmaceuticals from another parent delivery mechanism, carbon nanotubes (CNTs) may be used in combination with anti-inflammatory drugs. When treating arthritic pain, the two most common methods of administering nonsteroidal anti-inflammatory medications (NSAIDs) are subcutaneous injections and oral administration. In contrast to transdermal distribution, which typically results in low concentration and lacks precise control, oral therapy is associated with systemic side effects. CNTs have been used to regulate and enhance the release of nonsteroidal anti-inflammatory drugs (NSAIDs) at a low voltage, and they do not irritate the skin [72].

9 Safety of Nano-technology

A significant drawback to the long-term use of many NSAIDs is the variety of adverse effects associated with these medications, which encompass gastrointestinal and renal issues, along with an elevated risk of cardiovascular complications for patients. The risks and adverse events are more prevalent in individuals who have concurrent cardiovascular, cerebrovascular, gastrointestinal or renal conditions. By extending the systemic circulation of these agents and maintaining their release, a nanodelivery system can improve the efficacy and safety of these agents, minimising fluctuations in steady-state drug levels for long-term use [73]. There is considerable amount of evidence indicating the mucosal epithelium is a potential source of the gastrointestinal adverse effects of acidic NSAIDs. Injuries to the gastroduodenal mucosa, brought on by this local contact, may have toxic consequences and compromise the mucosa's protective abilities. A significant decrease in gastrointestinal side effects may be achieved by enclosing NSAIDs in nano delivery systems. This reduces their interaction with the mucus layer after oral administration, which in turn reduces their local irritating effects on the epithelium [74, 75]. Improving safety and reducing systemic toxicity are greatly linked to the deployment of nano-particulate delivery systems. These systems are highly effective in altering the biodistribution of encapsulated NSAIDs, efficiently directing them away from areas where drug toxicity may occur, such as the kidneys and heart. Consequently, although the encapsulation of NSAIDs does not affect their selectivity for COX-1 or COX-2, it modifies the in vivo bio-distribution of the encapsulated drug, inhibiting its buildup in tissues prone to the side effects linked with NSAIDs. Nanodelivery systems that possess the appropriate composition and particle size have demonstrated a propensity to accumulate in tissues impacted by inflammatory conditions, such as rheumatoid arthritis, owing to the enhanced permeability and retention effect (EPR). This approach to targeting, combined with a range of active strategies, has been explored to improve the therapeutic index of NSAIDs. A strategy of interest focusses on targeting activated macrophages that have folate receptors, which are abundant in the inflamed synovial membrane, by employing folate-linked nanodelivery systems [76–79]. This approach to targeting, along with various active strategies like focussing on activated macrophages that express folate receptors prevalent in the inflamed synovial membrane, has been investigated to enhance the therapeutic effectiveness of NSAIDs

[80]. By offering practical ways for the localised administration of NSAIDs, nanodelivery technologies may increase their safety by reducing their systemic exposure. Even when a local anti-inflammatory effect is required, NSAIDs have traditionally been delivered orally or parenterally. Because they are less likely to induce adverse effects, the topical NSAID formulations that are now available are preferred over the oral NSAIDs. Nevertheless, there is little proof that these topical preparations are equivalent to oral formulations in terms of efficacy. With the use of penetration-enhancing components and procedures, nano-delivery systems have the potential to enhance the dermal delivery of NSAIDs to deeper layers of the skin while minimising systemic exposure. Another option is for the formulations to provide improved transdermal administration of NSAIDs; this would allow the medicine to enter the systemic circulation and be released gradually over time. Possible reduction in NSAID gastrointestinal adverse effects and/or frequency of usage may be achieved with such formulations [81, 82]. Nano-delivery systems can be crafted to improve the targeted delivery of NSAIDs via parenteral administration, including intraarticular or subcutaneous injections close to the inflamed joints. These formulations are typically crafted to extend the presence of the encapsulated NSAIDs at the site of inflammation and regulate the release of the drug, ensuring a continuous therapeutic effect that minimises the necessity for frequent injections [83].

10 Conclusive on benefits and Safety of Nano Delivery for inflammation

In conclusion, nanomedicine has substantial advantages in terms of improving the efficiency and safety of drug delivery systems, especially with regard to non-steroidal anti-inflammatory medications (NSAIDs). It is possible to accurately target pharmaceuticals to particular tissues by the use of biocompatible nanocarriers, such as dendrimers, liposomes and nanoparticles. This allows for improved therapeutic results while simultaneously minimizing the risk of adverse effects. The danger of gastrointestinal and systemic toxicity that is often associated with nonsteroidal anti-inflammatory drugs (NSAIDs) may be reduced by using these nanodelivery systems, which have the ability to improve drug solubility, boost absorption and maintain constant drug levels. Furthermore, the capability of modifying the surfaces of these carriers enables enhanced penetration despite the presence of biological barriers, such as the blood–brain barrier and specific delivery to regions that are inflamed. By using this targeted approach, not only is the therapeutic index of nonsteroidal anti-inflammatory drugs (NSAIDs) increased, but the chance of unwanted effects is also greatly reduced. This makes therapies safer for patients, particularly those who already have illnesses that they are treating. In general, the application of nanotechnology to the process of medication distribution might be considered a potential innovation in the field of medical treatment. This could pave the way for therapeutic choices that are both more effective and safer.

References

1. Oronsky B, Caroen S, Reid T (2022) What exactly is inflammation (and what is it not?). Int J Mol Sci 23(23)
2. Xu Z, Chenli Z (2024) Inflammation BT—textbook of pathologic anatomy: for medical students. In: Chen K, Liang L, Li M, Pan Y (eds) Springer Nature Singapore, Singapore, pp 75–105. https://doi.org/10.1007/978-981-99-8445-9_4
3. Metsios GS, Moe RH, Kitas GD (2020) Exercise and inflammation. Best Pract Res Clin Rheumatol 34(2)
4. Dinetz E, Bocharova N (2024) Inflammation in elite athletes: a review of novel factors, the role of microbiome, and treatments for performance longevity. 16(10)
5. Ćibo M, Šator A, Kazlagić A, Omanović-Mikličanin E (2020) Application and impact of nanotechnology in sport. In: In: Brka M, Omanović-Mikličanin E, Karić L, Falan V, Toroman A (eds) BT—30th scientific-experts conference of agriculture and food industry. Springer International Publishing, Cham, pp 349–362
6. Gonzalez L, Loza RJ, Han KY, Sunoqrot S, Cunningham C, Purta P et al (2013) Nanotechnology in corneal neovascularization therapy—a review. J Ocul Pharmacol Ther 29(2):124–134
7. Haleem A, Javaid M, Singh RP, Rab S, Suman R (2023) Applications of nanotechnology in medical field: a brief review. Glob Heal J 7(2):70–77
8. Jain K, Mehra NK, Jain NK (2014) Potentials and emerging trends in nanopharmacology. Curr Opin Pharmacol [Internet] 15(1):97–106. Available from: https://doi.org/10.1016/j.coph.2014.01.006
9. Knapik JJ, Steelman RA, Hoedebecke SS, Austin KG, Farina EK, Lieberman HR (2016) Prevalence of dietary supplement use by athletes: systematic review and meta-analysis. Sport Med 46(1):103–123
10. Lima MR, Moreira BJ, Bertuzzi R, Lima-Silva AE (2024) Could nanotechnology improve exercise performance? Evidence from animal studies. Brazilian J Med Biol Res 57:1–11
11. (2024) Exercise and Nano-curcumin supplementation mediates cross-talk between MAPK/ERK signaling in the regulation of inflammatory disease. Mater Methods Anim 4(March):22–30
12. Chen YM, Chiu WC, Chiu YS, Li T, Sung HC, Hsiao CY (2020) Supplementation of nano-bubble curcumin extract improves gut microbiota composition and exercise performance in mice. Food Funct 11(4):3574–3584
13. Kazemi-Darabadi S, Nayebzadeh R, Shahbazfar AA, Kazemi-Darabadi F, Fathi E (2019) Curcumin and nanocurcumin oral supplementation improve muscle healing in a rat model of surgical muscle laceration. Bull Emerg Trauma 7(3):292–299
14. Deshpande D, Janero DR, Amiji M (2013) Engineering of an ω-3 polyunsaturated fatty acid-containing nanoemulsion system for combination C6-ceramide and 17β-estradiol delivery and bioactivity in human vascular endothelial and smooth muscle cells. Nanomed Nanotechnol Biol Med [Internet] 9(7):885–894. https://doi.org/10.1016/j.nano.2013.02.007
15. Sreedhar R, Kumar VS, Bhaskaran Pillai AK, Mangalathillam S (2019) Omega-3 fatty acid based nanolipid formulation of atorvastatin for treating hyperlipidemia. Adv Pharm Bull 9(2):271–280
16. Deshpande D, Kethireddy S, Janero DR, Amiji MM (2016) Therapeutic efficacy of an ω-3-fatty acid-containing 17-β estradiol nano-delivery system against experimental atherosclerosis. PLoS ONE 11(2):1–23
17. Coqueiro AY, Rogero MM, Tirapegui J (2019) Glutamine as an anti-fatigue amino acid in sports nutrition. Nutrients 11(4):863
18. Córdova-Martínez A, Caballero-García A, Bello HJ, Pérez-Valdecantos D, Roche E (2021) Effect of glutamine supplementation on muscular damage biomarkers in professional basketball players. Nutrients 13(6):1–9
19. Ahmed AU (2011) An overview of inflammation: mechanism and consequences. Front Biol China 6(4):274–281

20. Wu J, Ma T, Zhu M, Huang T, Zhang B, Gao J et al (2022) Nanotechnology reinforced neutrophil-based therapeutic strategies for inflammatory diseases therapy. Nano Today [Internet] 46:101577. https://doi.org/10.1016/j.nantod.2022.101577
21. Wang H, Zhou Y, Sun Q, Zhou C, Hu S, Lenahan C et al (2021) Update on nanoparticle-based drug delivery system for anti-inflammatory treatment. Front Bioeng Biotechnol 9(February):1–9
22. Bhansali D, Teng SL, Lee CS, Schmidt BL, Bunnett NW, Leong KW (2021) Nanotechnology for pain management: current and future therapeutic interventions. Nano Today [Internet] 39:101223. https://doi.org/10.1016/j.nantod.2021.101223
23. Liu S, Guo B (2022) The impact of the development of modern biotechnology and nanotechnology on the swimming sports industry. Adv Mater Sci Eng 2022
24. Rathi S, Reche A, Dhamdhere N, Bolenwar A (2023) Perspectives on the application of nanomaterials in medical and dental practices. Cureus 15(8)
25. Hajiali H, Ouyang L, Llopis-Hernandez V, Dobre O, Rose FRAJ (2021) Review of emerging nanotechnology in bone regeneration: progress, challenges, and perspectives. Nanoscale 13(23):10266–10280
26. Brannon-Peppas L, Blanchette JO (2004) Nanoparticle and targeted systems for cancer therapy. Adv Drug Deliv Rev 56(11):1649–1659
27. Raszewska-Famielec M, Flieger J (2022) Nanoparticles for topical application in the treatment of skin dysfunctions—an overview of dermo-cosmetic and dermatological products. Int J Mol Sci 23(24)
28. Wang P, Wang JY (2014) Development and application of nanotechnology in sports. Adv Mater Res 918:54–58
29. Chabib L, Trianloka AMB, Hidayat AUMJ, Awaluddin R, Firmansyah F (2020) Potential tropical fruits to aid sports performance and its prospect to be developed into nano supplement. IOP Conf Ser Earth Environ Sci 448(1):2–10
30. Qin L, Lu T, Qin Y, He Y, Cui N, Du A et al (2020) In vivo effect of resveratrol-loaded solid lipid nanoparticles to relieve physical fatigue for sports nutrition supplements. Molecules 25(22)
31. Li P (2022) Nanomaterials in sports training and its biological safety. Sci Program 2022
32. Gopi S, Jacob J, Varma K, Amalraj A, Sreeraj TR, Kunnumakkara AB et al (2017) Natural sports supplement formulation for physical endurance: a randomized, double-blind, placebo-controlled study. Sport Sci Health 13(1):183–194
33. Zhang G, Khan AA, Wu H, Chen L, Gu Y, Gu N (2017) The application of nanomaterials in stem cell therapy for some neurological diseases. Curr Drug Targets 19(3):279–298
34. Lv Y, Jin S, Wang Y, Lun Z, Xia C (2016) Recent advances in the application of nanomaterials in enzymatic glucose sensors. J Iran Chem Soc 13(10):1767–1776
35. Conte R, Marturano V, Peluso G, Calarco A, Cerruti P (2017) Recent advances in nanoparticle-mediated delivery of anti-inflammatory phytocompounds. Int J Mol Sci 18(4):1–23
36. González R, Ballester I, López-Posadas R, Suárez MD, Zarzuelo A, Martínez-Augustin O et al (2011) Effects of flavonoids and other polyphenols on inflammation. Crit Rev Food Sci Nutr 51(4):331–362
37. Chainani-Wu N (2003) Safety and anti-inflammatory activity of curcumin: a component of tumeric (Curcuma longa). J Altern Complement Med 9(1):161–168
38. Mathy-Hartert M, Jacquemond-Collet I, Priem F, Sanchez C, Lambert C, Henrotin Y (2009) Curcumin inhibits pro-inflammatory mediators and metalloproteinase-3 production by chondrocytes. Inflamm Res 58(12):899–908
39. Salminen A, Lehtonen M, Suuronen T, Kaarniranta K, Huuskonen J (2008) Terpenoids: natural inhibitors of NF-κB signaling with anti-inflammatory and anticancer potential. Cell Mol Life Sci 65(19):2979–2999
40. Reddy LH, Couvreur P (2009) Squalene: a natural triterpene for use in disease management and therapy. Adv Drug Deliv Rev [Internet] 61(15):1412–1426. https://doi.org/10.1016/j.addr.2009.09.005
41. Das B, Yeger H, Baruchel H, Freedman MH, Koren G, Baruchel S (2003) In vitro cytoprotective activity of squalene on a bone marrow versus neuroblastoma model of cisplatin-induced toxicity: implications in cancer chemotherapy. Eur J Cancer 39(17):2556–2565

42. Kırkıl G, Muz MH, Sancaktar E, Kaman D (2012) The effect of LYcopene SUpplementation. Int J Adv Eng Technol 8:98–104
43. Butnariu M, Butu A (2019) Plant nanobionics: application of nanobiosensors in plant biology. In: Prasad R (ed) Plant nanobionics: volume 2, approaches in nanoparticles, biosynthesis, and toxicity [Internet]. Springer International Publishing, Cham, pp 337–376. https://doi.org/10.1007/978-3-030-16379-2_12
44. ElSohly MA, Slade D (2005) Chemical constituents of marijuana: the complex mixture of natural cannabinoids. Life Sci 78(5):539–548
45. Trautwein EA, Demonty I (2007) Phytosterols: natural compounds with established and emerging health benefits. OCL 14(5):259–266
46. Leong WF, Lai OM, Long K, Che Man YB, Misran M, Tan CP (2011) Preparation and characterisation of water-soluble phytosterol nanodispersions. Food Chem 129(1):77–83
47. Caffall KH, Mohnen D (2009) The structure, function, and biosynthesis of plant cell wall pectic polysaccharides. Carbohydr Res [Internet] 344(14):1879–1900. https://doi.org/10.1016/j.carres.2009.05.021
48. Popov SV, Ovodova RG, Popova GY, Nikitina IR, Ovodov YS (2005) Adhesion of human neutrophils to fibronectin is inhibited by comaruman, pectin of marsh cinquefoil *Comarum palustre* L., and by its fragments. Biokhimiya 70(1):133–138
49. Takahashi M, Kitamoto D, Asikin Y, Takara K, Wada K (2009) Liposomes encapsulating aloe vera leaf gel extract significantly enhance proliferation and collagen synthesis in human skin cell lines. J Oleo Sci 58(12):643–650
50. Jiang F, Yang R, Xue D, Li R, Tan M, Zeng Z et al (2022) Effects of a natural nutritional supplement on immune cell infiltration and immune gene expression in exercise-induced injury. Front Nutr 9(September):1–13
51. Chaudhry Q, Groves K (2010) Nanotechnology applications for food ingredients, additives and supplements. In: Nanotechnologies in food. The Royal Society of Chemistry
52. Dietz P, Ulrich R, Niess A, Best R, Simon P, Striegel H (2014) Prediction profiles for nutritional supplement use among young German elite athletes. Int J Sport Nutr Exerc Metab 24(6):623–631
53. Kioukia-Fougia N, Georgiadis N, Tsarouhas K, Vasilaki F, Fragiadaki P, Meimeti E et al (2016) Synthetic and natural nutritional supplements: health "allies" or risks to public health? Recent Pat Inflamm Allergy Drug Discov 10(2):72–85
54. Walsh NP, Gleeson M, Pyne DB, Nieman DC, Dhabhar S, Shephard RJ et al (2011) Part two: maintaining immune health. Most 17:64–103
55. Riera J, Pons V, Martinez-Puig D, Chetrit C, Tur JA, Pons A et al (2013) Dietary nucleotide improves markers of immune response to strenuous exercise under a cold environment. J Int Soc Sports Nutr [Internet] 10(1):1
56. Nieman DC, Henson DA, McMahon M, Wrieden JL, Davis JM, Murphy EA et al (2008) β-Glucan, immune function, and upper respiratory tract infections in athletes. Med Sci Sports Exerc 40(8):1463–1471
57. Nieman DC, Stear SJ, Castell LM, Burke LM (2010) A-Z of nutritional supplements: dietary supplements, sports nutrition foods and ergogenic aids for health and performance: part 15. Br J Sports Med 44(16):1202–1205
58. Peters EM (1997) Exercise, immunology and upper respiratory tract infections. Int J Sport Med 18(S1):S69–S77
59. Buonocore D, Negro M, Arcelli E, Marzatico F (2015) Anti-inflammatory dietary interventions and supplements to improve performance during athletic training. J Am Coll Nutr 34(September):62–67
60. Maroon JC, Bost JW, Borden MK, Lorenz KM, Ross NA (2006) Natural antiinflammatory agents for pain relief in athletes. Neurosurg Focus 21(4)
61. Almeida JPM, Chen AL, Foster A, Drezek R (2011) In vivo biodistribution of nanoparticles. Nanomedicine 6(5):815–835
62. Blanco E, Shen H, Ferrari M (2015) Principles of nanoparticle design for overcoming biological barriers to drug delivery. Nat Biotechnol 33(9):941–951

63. Weissig V, Pettinger TK, Murdock N (2014) Nanopharmaceuticals (part 1): products on the market. Int J Nanomed 9:4357–4373
64. Kermanizadeh A, Powell LG, Stone V, Møller P (2018) Nanodelivery systems and stabilized solid-drug nanoparticles for orally administered medicine: current landscape. Int J Nanomed 13:7575–7605
65. Al Lawati H, Jamali F (2016) Onset of action and efficacy of ibuprofen liquigel as compared to solid tablets: a systematic review and meta-analysis. J Pharm Pharm Sci 19(3):301–311
66. Mennini N, Furlanetto S, Bragagni M, Ghelardini C, Di Cesare ML, Mura P (2014) Development of a chitosan-derivative micellar formulation to improve celecoxib solubility and bioavailability. Drug Dev Ind Pharm 40(11):1494–1502
67. Bala I, Hariharan S, Kumar MNVR (2004) PLGA nanoparticles in drug delivery: the state of the art. Crit Rev Ther Drug Carrier Syst 21:387–422
68. Gu W, Wu C, Chen J, Xiao Y (2013) Nanotechnology in the targeted drug delivery for bone diseases and bone regeneration. Int J Nanomed 8:2305–2317
69. Nagata S, Hanayama R, Kawane K (2010) Autoimmunity and the clearance of dead cells. Cell 140(5):619–630
70. Agarwal H, Shanmugam VK (2020) A review on anti-inflammatory activity of green synthesized zinc oxide nanoparticle: mechanism-based approach [Internet]. Bioorg Chem 94:103423.https://doi.org/10.1016/j.bioorg.2019.103423
71. Kalangi SK, Swarnakar NK, Sathyavathi R, Narayana Rao D, Jain S, Reddanna P (2018) Synthesis, characterization, and biodistribution of quantum dot-celecoxib conjugate in mouse paw edema model. Oxid Med Cell Longev 2018
72. Spizzirri UG, Hampel S, Cirillo G, Nicoletta FP, Hassan A, Vittorio O et al (2013) Spherical gelatin/CNTs hybrid microgels as electro-responsive drug delivery systems. Int J Pharm [Internet] 448(1):115–122. https://doi.org/10.1016/j.ijpharm.2013.03.013
73. Hatt KM, Vijapura A, Maitin IB, Cruz E (2018) Safety considerations in prescription of NSAIDs for musculoskeletal pain: a narrative review. PM R [Internet] 10(12):1404–1411. https://doi.org/10.1016/j.pmrj.2018.06.011
74. Wolfe MM, Lichtenstein DR, Singh G (1999) Gastrointestinal toxicity of nonsteroidal antiinflammatory drugs. N Engl J Med 340(24):1888–1899
75. Soehngen EC, Godin-Ostro E, Fielder FG, Ginsberg RS, Slusher MA, Weiner AL (1988) Encapsulation of indomethacin in liposomes provides protection against both gastric and intestinal ulceration when orally administered to rats. Arthritis Rheum 31(3):414–422
76. Arias JL (2011) Nanoparticle therapy in arthritis. In: Nanomedicine in health and disease. CRC Press Enfield, USA, p 293
77. Prasad LK, O'Mary H, Cui Z (2015) Nanomedicine delivers promising treatments for rheumatoid arthritis. Nanomedicine 10(13):2063–2074
78. Roy K, Kanwar RK, Kanwar JR (2015) Molecular targets in arthritis and recent trends in nanotherapy. Int J Nanomed 10:5407–5420
79. Pham CTN (2011) Nanotherapeutic approaches for the treatment of rheumatoid arthritis. Wiley Interdiscip Rev Nanomed Nanobiotechnol 3(6):607–619
80. Li M, Mondrinos MJ, Chen X, Gandhi MR, Ko FK, Lelkes PI (2006) Elastin blends for tissue engineering scaffolds. J Biomed Mater Res Part A [Internet] 79(4):963–973. http://www.ncbi.nlm.nih.gov/pubmed/16948146
81. Derry S, Conaghan P, Jap DS, Pj W, Ra M (2016) Topical NSAIDs for chronic musculoskeletal pain in adults. Rev Summary Findings Main Comparison 4
82. Vega E, Egea MA, Garduño-Ramírez ML, García ML, Sánchez E, Espina M et al (2013) Flurbiprofen PLGA-PEG nanospheres: role of hydroxy-β-cyclodextrin on ex vivo human skin permeation and in vivo topical anti-inflammatory efficacy. Colloids Surfaces B Biointerfaces 110:339–346
83. Zhang Z, Huang G (2012) Intra-articular lornoxicam loaded PLGA microspheres: enhanced therapeutic efficiency and decreased systemic toxicity in the treatment of osteoarthritis. Drug Deliv 19(5):255–263

Future Directions and Emerging Trends in Nano-fuels for Athletes

Pulen Das and Emmanuel Achumboro Apanga

Abstract This chapter investigates the emerging role of nanotechnology in sports nutrition, exploring its potential to revolutionize athlete performance and recovery. Chapter examines the rise of nano-supplements, analyzing their impact on key performance parameters like muscle strength, endurance, and cognitive function. Furthermore, Chapter delves into the advancements in sports nutrition technology, including skin-interfaced microfluidic systems for real-time sweat analysis and hydrogel-encapsulated carbohydrates for improved gut comfort. The chapter also explores the growing significance of nutraceuticals in sports nutrition, examining their potential benefits and challenges. Chapter analyze current trends in sports nutrition, including the increasing popularity of personalized nutrition and the growing awareness of the impact of the gut microbiome on athletic performance. Finally, Chapter discuss the future direction of nanotechnology in sports nutrition, emphasizing the need for further research, stricter regulations, and a multidisciplinary approach to ensure the safe and ethical development and utilization of these innovative technologies.

Keywords Nanotechnology · Sports nutrition · Nano-supplements · Nutraceuticals · Personalized nutrition · Microfluidics · Hydrogel · Gut microbiome · Performance enhancement · Safety · Regulation

Learning Outcomes

1. To understand the rise of sports naono supplements in current scenario
2. To understand the new innovation in sports nutritiontechnology
3. To understand the rise of nutraceuticalsin sports nanosupplements
4. To understand challenging trends in sport nutrigenomics in sports nutrition
5. To understand trends of uses most uses nano supplements by Athletes

---

P. Das (✉)
Lakshmibai National Institute of Physical Education NERC Guwahati, Guwahati, India
e-mail: Pulendas86@gmail.com

E. A. Apanga
Ho Polyclinic, Ghana Health Service, Ho, Volta Region, Ghana

A. Chatterjee et al. (eds.), *Nanofuel: The Future of Sports Nutrition*, Smart Nanomaterials Technology, https://doi.org/10.1007/978-981-96-5471-0_22

6. To understand nanocompound in sports nutrition
7. To understand Future direction and recommendation of nanotenchnology.

1 Introduction

The future of nano-supplements hinges on continued innovations. The field of sports nutrition has a rich history of advancements, driven by exercise physiologists, nutritionists, and dieticians. These experts have developed cutting-edge techniques and technologies inside their laboratories, utilizing tools including treadmills, ergometers to study elite athletes such as distance runners and cyclists. This research has not only advanced sports science but has also paved the way for the emergence of sports nutrition as a distinct discipline. Figure 1 provides a comprehensive overview of key innovations and their corresponding milestones in sports nutrition research, demonstrating the synergistic relationship between sports science and nutrition (Table 1).

The omics technologies: These refer to a suite of large-scale, high-throughput techniques used to study various biological molecules [3]

- Genomics: Studies the entire genome of an organism.
- Transcriptomics: Analyzes gene expression (RNA levels).
- Metabolomics: Investigates the complete set of metabolites in a biological system.
- Proteomics: Studies the entire set of proteins produced by an organism.
- Phenomics: Characterizes the observable traits of an organism.

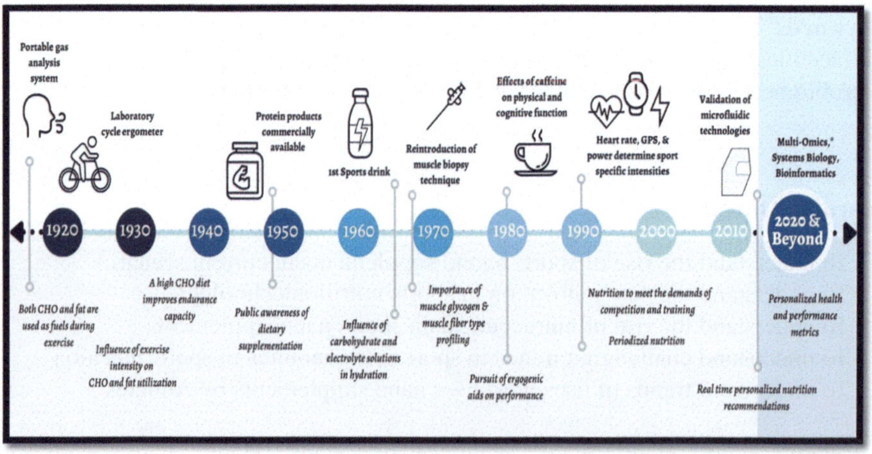

Fig. 1 Presents a time line showcasing the profound influence of nutritional breakthroughs in athletics on sporting domains. This timeline likely includes the incorporation of the array of '-omics' disciplines, encompassing genomics, transcriptomics, metabolomics, proteomics, phenomics, and conceivably epigenomic [1]

Table 1 Timeline of advancements in sports nutrition and performance research [2]

Year	Event/Discovery	Description
1920	Portable gas analysis system	The system and its applications
1920	Laboratory cycle ergometer	The device and its use in research
1940	Protein products commercially available	The development and impact of protein products
1950	1st sports drink	The first sports drink and its significance
1960	Reintroduction of muscle biopsy technique	The technique and its importance in research
1970	Effects of caffeine on physical and cognitive function	Research findings on caffeine's impact
1980	Heart rate, GPS, and power determine sport specific intensities	How these technologies are used to measure performance
1990	Validation of microfluidic technologies	The validation process and applications of microfluidic technologies
2000	Multi-omics, systems biology, bioinformatics	These fields and their contributions to sports research
2010	Personalized health and performance metrics	Personalized metrics and their use in training and nutrition
2020 and beyond	Real-time personalized nutrition recommendations	How personalized nutrition recommendations are evolving

- Epigenomics: Examines the study of epigenetic modifications (changes in gene expression not involving DNA sequence alterations).

By incorporating these omics technologies, sports nutrition research is gaining a deeper understanding of individual athlete responses to training, nutrition, and other factors, leading to more personalized and effective nutritional strategies.

In 2003, the Committee of International Olympic (IOC) recognized the critical role of nutrition in athletic success. Their special committee on sports nutrition stated that the quantity, types of food, and meal timing significantly impact an athlete's performance. Proper nutrition supports intense training, facilitates rapid recovery, enhances adaptation to training demands, and minimizes the risk of health issues and injuries.

Sports performance can be significantly influenced by the quantity, type, and schedule of nutritional intake [4].

1.1 The Human Body Prioritizes Survival Over Peak Performance

The human body is built for survival, not necessarily for pushing its limits. High-level physical or mental performance comes at a biological cost. We require to supply our bodies with nutrition required to satisfy these demands that promote healing if we want to succeed. Weightlifters and athletes were the primary focus of sports nutrition in the past, that concentrated on employing supplements including protein powders and creatine for improved muscle mass, strength, and recovery speed. However, a growing segment of health-conscious individuals, ranging from casual fitness enthusiasts to weekend warriors, is emerging. This shift marks the rise of Nano supplements, representing a new era of holistic health. These individuals seek a balanced approach to wellness, going beyond muscle building and time-trial improvements. Their focus extends to various aspects of health, including athletic performance, weight management, mood enhancement, metabolic function, and bone and joint health.

Weekend warriors and regular exercise enthusiasts belong to a growing generation of health-conscious individuals [5].

1.2 Convenience and Multi-benefit Supplements Are Gaining Popularity Among Athletes

Athletes frequently seek guidance from a range of professionals, including trainers, physiologists, coaches, doctors, and dietitians, regarding their nutritional needs, specifically what to eat and which supplements to incorporate into their routines. A recent survey conducted by FMCG Gurus [6] among global consumers of sports nutrition supplements revealed a significant shift in focus. Over three-quarters (79%) of respondents indicated a primary goal of improving their overall well-being rather than addressing specific health concerns. Furthermore, nearly half (45%) of these individuals consume only one supplement, highlighting a preference for convenient and streamlined solutions.

Recognizing this evolving consumer demand, Balchem has introduced a ‘Vital Trio’ of ingredients designed to offer multiple health benefits in a single, convenient dose. This innovative combination features Albion Magnesium bis-glycinate chelate, vitamin D3, and the world’s only patented double micro-encapsulated vitamin K2, K2VITAL™ DELTA [7]. The market for magnesium supplements is experiencing exceptional growth, with a remarkable 25.5% expansion rate in 2023, significantly surpassing the anticipated 2–3% growth. Similarly, the vitamin D market is on a strong trajectory, expected to boost at a 12.1% yearlyescalation rate from 2023 to 2029 [4]. The vitamin K2 market also presents significant opportunities, with a projected yearly growth rate of 14.9% from 2024 to 2030. These robust growth

Table 2 Key market trends and balchem's approach [8]

Aspect	Description
Consumer trend	Convenience and multi-benefit supplements are gaining popularity among athletes
Consumer focus	Improving overall well-being is prioritized over addressing specific health concerns
Supplement consumption	45% of consumers use only one supplement, indicating a preference for simplicity
Balchem's approach	Introduced "Vital Trio"—a combination of magnesium bisglycinate chelate, vitamin D3, and K2VITAL™ DELTA
Magnesium market	Strong growth, with a 25.5% expansion rate in 2023
Vitamin D market	Significant growth, with a projected 12.1% yearly escalation rate from 2023 to 2029
Vitamin K2 market	Projected yearly growth rate of 14.9% from 2024 to 2030
Consumer awareness	Increasing awareness and demand for essential nutrients like magnesium, vitamin D, and vitamin K2

projections underscore the increasing consumer awareness and demand for these essential nutrients (Table 2).

1.3 Crucial Considerations for Nano Supplements by Athlete

In the realm of combat sports, possessing accurate nutritional knowledge is paramount for athletes to achieve optimal performance and maintain overall health. Misconceptions surrounding dietary supplements, particularly the perceived necessity of steroids, can lead to detrimental health consequences. Therefore, a comprehensive understanding of nutritional requirements and a meticulous assessment of individual needs are crucial [9].

The daily dietary intake of combat athletes is increasingly supplemented with a diverse range of sports nutrition products, including nano-sized supplements. These supplements often incorporate vitamins, minerals, herbs (botanicals), amino acids, and antioxidants to enhance muscular performance when combined with strength training. Supplementing is mostly for performance improvement, however, it's additionally crucial to consider lean muscle mass, physical health, mental health, and body composition. [10]. It's crucial to remember that not all performance-enhancing substances are safe for health. Supplements are meticulously investigated by the World Anti-Doping Agency (WADA), that permits their application in sports if there are no health risks. For example, it is widely recognized that in healthy individuals, anabolic–androgenic steroids (AAS) [11] result in systemic inflammatory responses with multiple organ dysfunction. Therefore, vigilant approach to supplement use is necessary to ensure the safety and well-being of athletes.

The Athletes' are advanced with

1. **Importance of Nutrition Knowledge**: Emphasizes the need for accurate information to avoid misconceptions about supplements and steroids.
2. **Nutritional Needs of Combat Athletes**: Highlights the unique dietary requirements and the increasing role of sports nutrition supplements.
3. **Performance Enhancement and Health**: Acknowledges the dual goals of improving performance while maintaining overall health.
4. **WADA's Role**: Recognizes the importance of regulatory bodies like WADA in ensuring the safety and ethical use of supplements in sports.
5. **Potential Risks**: Warns about the dangers of substances like anabolic–androgenic steroids, which can have detrimental health effects.

2 Shifting Focus

2.1 The Present Scenario

Over the past five years, the focus within sports nutrition research has shifted from original investigations to a greater emphasis on review articles. A substantial portion of published articles, ranging from 17.6 to 20.2%, fall into the review category, with 4.4 to 6.0% specifically classified as meta-analyses. Traditionally, laboratory-based experimentation has been the cornerstone of most paradigms in sports nutrition. Recent advances in wearable technology and real-time monitoring have revolutionized field-based research by improving ecological validity. All sports science disciplines, including sports nutrition, are currently having a rare chance to take advantage of these technical advancements while applying them for advancement in field [12].

2.2 Potential Applications of Nano-supplements in Personalized Sports Nutrition

1. **Personalized Nutrition:** Nanotechnology, combined with genetic and biochemical profiling, has the potential to create highly personalized nutrition plans for athletes.
2. **Adaptive Nutrient Delivery:** Future nano-supplements may incorporate adaptive technology, such as smart nanoparticles, to release nutrients based on factors like pH levels, fatigue levels, or hydration status.
3. **Enhanced Recovery and Tissue Repair:** Nanoparticle-based solutions can accelerate healing and minimize tissue damage by targeting inflammation and delivering antioxidants directly to affected areas.

4. **Cognitive Enhancement**: Nanotechnology-based nootropics have the potential to enhance cognitive function, including improved mental focus, faster reaction times, and enhanced decision-making abilities, which are crucial for both physical and strategic aspects of sports performance.
5. **Real-time Monitoring and Adjustment:** Wearable technology can be integrated with nano-supplements to enable real-time monitoring of an athlete's physiological responses. This data can then be used to dynamically adjust supplement delivery and optimize performance.

2.3 Netnographic Analysis of Supplements and Gaining Popularity

Netnography is a qualitative research method that delves into the online behaviors and interactions of communities. In a specific study, [13] researchers analyzed over four million social media profiles, including those of athletes, examining 18,595 posts related to supplement use. Data demonstrated that Whey Protein, Branched Chain Amino Acids (BCAAs), Creatine, Multivitamin supplements, and Nitric Oxide boosters were the most often mentioned supplements. Weight loss (8%), higher energy levels (17%), and muscle building (23%), had been the primary motivations for supplement use. Despite the generally favorable online discourse surrounding supplements, a significant percentage of users (19%) reported experiencing a range of negative side effects. Acne (9%), water retention (9%), stomach pain (9%), rashes (7%), sexual dysfunction (7%), and weight gain (5%), were among the most prevalent complaints.

Furthermore, the study highlighted concerns regarding the quality and safety of supplements. Concerns over possible contamination (47%), counterfeit products (17%), and the existence of unidentified or hidden ingredients (11%) were highlighted by a substantial percentage of users. This ethnographic study provides valuable insights into the perceptions, experiences, and concerns of individuals, particularly athletes, regarding supplement use within the context of online social media communities (Table 3).

Table 3 The table provides a summary of automated online monitoring results for keywords related to sports supplements [13]

Category	Number of mentions	Percentage
Web	7452	40.1
Blogs	6568	35.3
Social	3564	19.2
Forum	723	3.9
News	158	0.8
Other	130	0.7
Total	18,595	100

The overall frequency of mentions for a group of generic keywords (for instance, 'most used sports supplement,' 'most sold sports supplement,' etc.) across different internet resources over the designated time period is displayed in Table 1.

Table 4 represents number of mentions of the most referenced supplements by athletes and other individuals in Social Media Reach.

Table 5 summarizes the primary motivations for supplement use, as identified through the analysis of online forum discussions. 'Muscle gain,' 'energy,' and 'weight loss' were the most frequently cited reasons.

Netnographic methods have enabled the collection and analysis of extensive real-world data on nano-supplement usage. The current research, that has been conducted across social media sites and online forums that are usually thought of as safe spaces for candid discussion, discovered a variety of issues and self-reported negative effects linked to these supplements. We keenly observed products including multivitamins, whey protein, creatine, BCAAs, and nitric oxide boosters.

Millions of individuals had been exposed to online comments, advice, and other content about supplements, according to the study. It's concerning to note that most supplement users got their knowledge from internet sources, with very few seeking advice from medical experts including doctors or dietitians. Given the existing absence of site-specific or legislative rules controlling the distribution of supplement information, this conclusion is alarming. This absence of oversight may allow for the promotion of uncertified and potentially inaccurate information about these products. The researcher's conclusions represent merely the preliminary insights, necessitating additional investigation. Big data and AI may improve pharmacovigilance and toxicovigilance operations and assist with predicting trends.

3 Innovation in Sports Nutrition

Example

Skin-interfaced microfluidic patches provide real-time sweat analysis for hydration guidance.

3.1 Skin-Interfaced Microfluidic System

Investigator Mr. Lindsay B. Baker and his research team (2020) proposed a paradigm for microfluidic technologies incorporated into wearable patches delivering athletes immediate input on sweat rate and composition [14].

Traditional sweat analysis methods involve using absorbent patches that require manual collection and subsequent laboratory analysis. This process is cumbersome and impractical for real-time monitoring during physical activity. In overcoming this concern, researchers have developed an innovative wearable microfluidic device

Table 4 Trends of discussed nano supplements products in online database [13]

Product	Mentions	Mentions per category	Social media reach	Trending area
Creatine	9375	• *Social: 2944(31.4%)* • *Forum: 394(4.2%)* • *Blogs: 804(8.6%)* • *News: 11(0.1%)* • *Web: 5222(55.7%)*	3.5 M	• Fitness • Workout • Bodybuilding
Branched chain amino acid/BCAA	8290	• *Social: 4076 (49.3%)* • *Forum: 141 (1.7%)* • *Blogs: 850 (10.2%)* • *News: 28 (0.3%)* • *Web: 3,195 (38.5%)*	3.2 M	• Fitness • Protein • Workout
Whey protein	17,438	• *Social: 6616 (37.9%)* • *Forum: 245 (1.5%)* • *Blogs: 1907 (10.9%)* • *News: 33 (0.2%)* • *Web: 8637 (49.5%)*	7.1 M	• Fitness • Supplements • Protein
Nitric oxide/nitric oxide booster	10,392	• *Social: 712 (6.8%)* • *Forum: 282 (2.7%)* • *Blogs: 1100 (10.6%)* • *News: 28 (0.3%)* • *Web: 8270 (79.6%)*	472,000	• Health • Nutrition • Healthy lifestyle
Multivitamin	9063	• *Social: 2396 (26.4%)* • *Forum: 114 (1.3%)* • *Blogs: 1815 (20.0%)* • *News: 54 (0.6%)* • *Web: 4684 (51.7%)*	2.7 M	• Health • Vegan supplement • Wellness

Table 5 Trends of motivation for supplement use [13]

Motivation	Percentages
• Muscle gain	23
• Energy	17
• Weight loss	8
• Weight gain	7
• Strength	5
• Injury recovery	4
• Fat loss	3
• Endurance	1
• Stress reduction	1
• Therapeutic effects	1
• Low blood sugar	1
• Stay youthful	1
• Organ health	1
• Cardiovascular health	1
• Bone health	1
• Testosterone production	1
• Others	24

and a smartphone-based image processing platform. This advanced method facilitates ongoing and instantaneous assessment of localized perspiration rate and sweat chloride concentration ([Cl^-]) directly on the epidermis.

The primary focus of this research lies in the realm of human performance and athletics. During physical activity and heat stress, significant fluid and electrolyte losses occur through sweat, leading to fluid imbalances. These imbalances can increase the strain on the cardiovascular system, potentially impairing both physical and cognitive performance. By providing real-time insights into sweat composition and fluid loss, this wearable technology can help athletes and individuals engaged in strenuous physical activity optimize their hydration strategies and mitigate the risks associated with dehydration (Fig. 2).

Protocols

(a) **Application**: The user begins by cleaning and drying the left ventral forearm. They then apply the sweat patch to the prepared skin surface.
(b) **Sweat Collection**: The patch passively collects sweat while the athlete completes their workout. During this time, colorimetric reactions occur within the patch.
(c) **Image Capture**: Subsequent to workout session, user captures an image of sweat patch utilizing corresponding smartphone application.
(d) **Data Processing**: The patch's image data is processed by the application. The user's body mass, sex, sport, length of exercise, weather, and energy expenditure

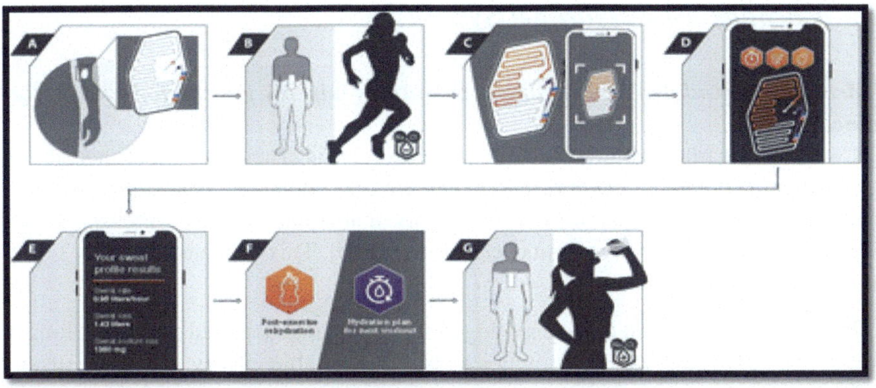

Fig. 2 Amodel of micro fluidic technologies integrated into wearable patches [14]

are among the other relevant inputs that are then integrated with this data. These typically originate via the user's profile within the application.

(e) **Sweat Profile Generation**: Application provides a thorough sweat profile based on the processed data, including estimations of total sweat loss, sweat sodium loss, and whole-body sweating rate. Following that, a comprehensive summary of the user's sweat characteristics during the workout is displayed on the screen.

(f) **Personalized Hydration Recommendations**: Utilizing the generated sweat profile, the application provides personalized fluid intake recommendations tailored to the individual's specific needs.

(g) **Hydration Implementation**: The user follows these personalized recommendations to ensure proper rehydration immediately after the workout. They also apply these realizations to their hydration plans for subsequent exercises with comparable levels of intensity, length, and environmental factors.

Microfluidic patch successfully permits real-time evaluation of sweat chloride concentration ([Cl^-]) and sweating rate under practical circumstances, as demonstrated by this work. This novel method is noteworthy since it doesn't demand specific knowledge or lab equipment, making it easily accessible and easy to use. Through the application of this inexpensive wearable sensor device, academics, practitioners, and athletes may acquire important physiological insights by gathering data on sweating rate and electrolyte loss. Hydration strategies in practical contexts might be improved and informed by these discoveries.

The applications of this technology extend beyond athletic performance and fitness, with potential benefits for military readiness and clinical medicine. This technology has the potential to revolutionize how we monitor and manage fluid and electrolyte balance in various contexts, promoting improved health and performance outcomes.

3.2 Carbohydrate Hydrogel System

Hydrogel-encapsulated carbohydrates potentially improve gut comfort and performance.

Consuming carbohydrates (CHO) during exercise is crucial for endurance activities, as demonstrated by recent research on a 2-h marathon [15]. An exogenous CHO supply greatly enhances performance at events of adequate time and intensity where availability of carbohydrates limits performance [16]. This exogenous supply serves multiple crucial functions:

- **Fuel Provision**: When muscle glycogen stores become depleted during prolonged exercise, exogenous CHO provides a readily available energy source to sustain performance.
- **Muscle Glycogen Sparing**: By providing an alternative fuel source, exogenous CHO helps to spare muscle glycogen stores, allowing them to be utilized more efficiently later in the event.
- **Liver Glycogen Preservation**: Exogenous CHO intake can also help to preserve liver glycogen stores, which play a vital role in maintaining blood glucose levels and supporting overall energy metabolism.

These mechanisms demonstrate the crucial role of carbohydrate intake during endurance events, particularly those of significant duration and intensity, in optimizing performance and mitigating the negative effects of glycogen depletion [17].

More recently, specialty sports foods which claim to solve these issues by employing 'hydrogel technology' have become commercially accessible. Alginate, a polymer derived from seaweed, and pectin, a soluble fiber, are combined with traditional CHO sources to produce supplements that encapsulate the CHO when exposed to low pH fluids which include stomach acid. This 'Hydrogel' may be transported to the small intestine by increased stomach emptying rates, then dissolves in a higher pH environment for absorption, reducing gastrointestinal discomfort, improving muscle CHO delivery, and improving performance (Fig. 3).

Hydrogel-encapsulated carbohydrates (CHO) form a gel-like structure in the stomach due to interactions between the CHO source, pectin, and alginate. This gelation process is triggered by the acidic environment of the stomach. Intestinal absorption of monosaccharides is rate-limiting step in application of exogenous CHO for energy production, although hydrogel formation may improve gastric emptying. This means that even with faster gastric emptying, the rate at which carbohydrates are broken down and absorbed in the small intestine ultimately determines how effectively they can be used as fuel [18].

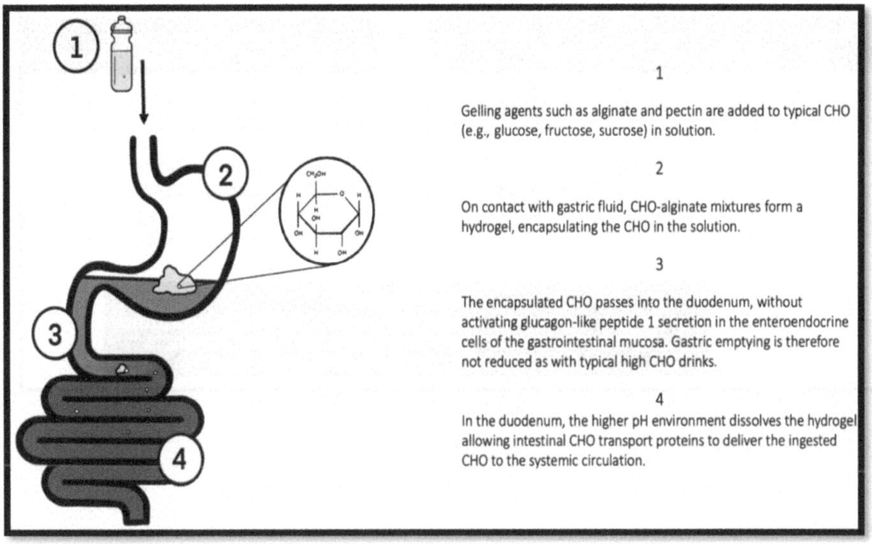

Fig. 3 Mechanisms of CHO hydrogel production and transport to the small intestine. CHO = carbohydrate [17]

4 The Rise of Nutraceuticals

Food-derived compounds known as nutraceuticals may provide additional health advantages over simple nourishment. Dr. Stephen DeFelice first used the term 'nutraceutical' in 1989 to refer to these chemicals, combining terms 'pharmaceutical' and 'nutrition' [19].

As the Greek physician Hippocrates, regarded as the father of medicine, once said, "Let food be your medicine," the principle underlying nutraceuticals is to concentrate on prevention [19].

4.1 Nutraceutical Categories

See Fig. 4, Tables 6 and 7.

4.2 Attitudes Towards Using Nutraceutical Supplements in Different Sports [27]

The nutrient intake variances are frequently recommended in various nutrition assessments associated with various sports (Fig. 5; Table 8).

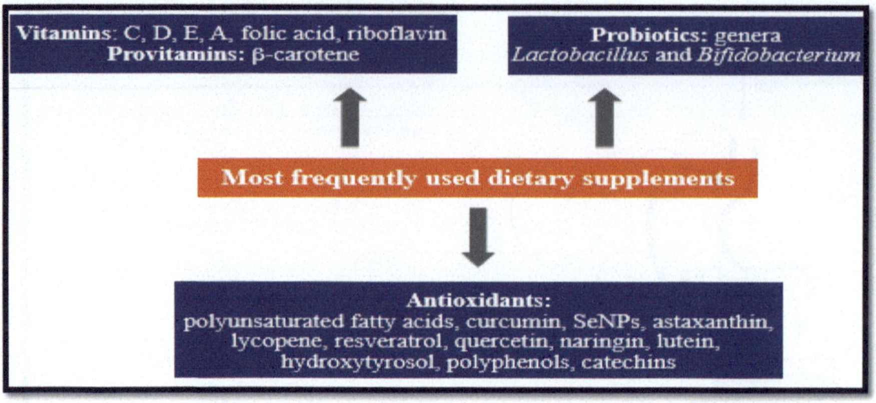

Fig. 4 Most frequently used human dietary nanosupplements [20]

Interpretation

(a) The most frequently consumed supplement has been vitamin D from the medical supplements sub-category (47%).
(b) The most frequently consumed supplement has been vitamin C from the antioxidants sub-category (51.8%).
(c) The most frequently consumed supplement has been vitamin E from the named product sub-category (32.3%).
(d) The most frequently consumed supplement has been testosterone from the prohormones and hormone boosters' sub-category (12.4%)

4.3 Recent Protocols for Nano Supplements

(a) **Protocols for Nutrient Timing of Combative Individual Sports**

Guidelines for timing nutrition in combat sports are provided by the International Society of Sports Nutrition (ISSN). Key recommendations include:

- **Carbohydrate**: Consume high-carbohydrate meals 4 h before competition to ensure adequate glycogen stores for high-intensity demands.
- **Protein**: Distribute protein intake throughout the recovery phase. Consider slow-absorbing proteins like casein before sleep to sustain amino acid availability.
- **Caloric Distribution**: The majority of calories should be consumed in several short meals throughout the day, as this has been proven in certain research studies to have beneficial impacts on weight loss [35].

Table 6 Classification of nutraceutical products [21]

Category	Examples	Description
Dietary supplements including botanicals:	• Vitamins, minerals, co-enzyme Q, carnitine • Gingsing, Gingko Biloba, Saint John's Wort, Saw Palmetto	Substances taken to supplement the diet, often for health or performance benefits
Functional foods	• Stanols (Benecol) in reduction of cholesterol adsorption • Prebiotics—oligofructose for control of intestinal flora • Omega-3 milk in prevention of heart disease • Oats, bran, psyllium, and lignin's for heart disease and colon cancer • Foods/cereals/snacks enriched with soluble fibers, vitamins, and minerals • Canola oil with lowered triglycerides for cholesterol reduction • A food product that is part of usual diet but has beneficial effects that go beyond the traditional nutritional effects • Examples: Yogurts—probiotics for intestinal health	Foods that provide additional health benefits beyond basic nutrition
Medicinal foods	• Transgenic plants for oral vaccination against infectious diseases • Transgenic cows and lactoferrin for immune enhancement • Health bars with added medications	Medication-containing health bars Transgenic cows for lactoferrin, plants for oral vaccinations

(b) **Protocol for Glycogen Restoration**

- **Pre-competition**: Consume high-glycemic-index (GI) carbohydrates 4 h before competition to fuel energy demands.
- **Post-exercise**: Ingest high-GI carbohydrates immediately after exercise for rapid recovery.
- **Daily Intake**: ISSN recommends 8–12 g/kg/day carbohydrates for athletes.
- **Rapid Glycogen Restoration**:
  - Protocol 1:1.2 g/kg/h high-GI CHO.
  - Protocol 2:0.8 g/kg/hCHO+0.2–0.4 g/kg/h protein+3–8 mg/kg caffeine [36].

Table 7 Recent trends of nano supplements in sports [20]

Category by use	Formulations/ Examples	Comments/Risk assessments	Disadvantages and risks	References
Sports foods: Specialized products used to provide a practical source of nutrients when it is impractical to consume them in everyday foods	Isotonic sports drink High-energy sports drink Electrolyte supplement Sports gel Vegetal protein Meat protein Whey protein Casein Sports bar Sports confectionary Liquid meals	Easy to use; provides an exact specified amount of specific nutrients for sports performance; does not contain large amounts of fiber, fat, or other components that could impair physical performance	Does not contain all nutrients necessary for everyday life and should not replace normal meals; reduces meal variety; often marketed with non-scientific information	[22]
Medical supplements: Used to treat clinical issues, including diagnosed nutrient deficiencies	Mineral complex Vitamin complex Iron Zinc Prebiotics/probiotics Alpha lipoic acid Transient receptor potential channel agonists Quinine N N-Acetylcysteine	Manufacturers observe strict regulations; impacts are extensively examined	Can lead to toxicity if administered in excess	[23]
Ergogenic supplements: Supplements that increase energy production rate	Glycerol Caffeine β-Alanine Bicarbonate Nitrate Sodium bicarbonate	WADA allows secure application in sports nutrition; evidence supports performance enhancement	Overdose can induce ill effects; may be purposely contaminated with banned substances	[24, 25]
Functional foods: nutrient-rich products of natural or fortified origin	Fruits Vegetables Nuts Legumes Whole grains Seafood Fermented foods Herbs and spices	Some products have clinical trials and can be used alone or as adjuvant therapy; natural products have high patient compliance	Even if heavily promoted, scientific data are often inconsistent; unregulated by authorities; biologically active substance inconsistencies	[26]

Sources Appl. Sci. 2022, 12, 8611. https://doi.org/10.3390/app12178611

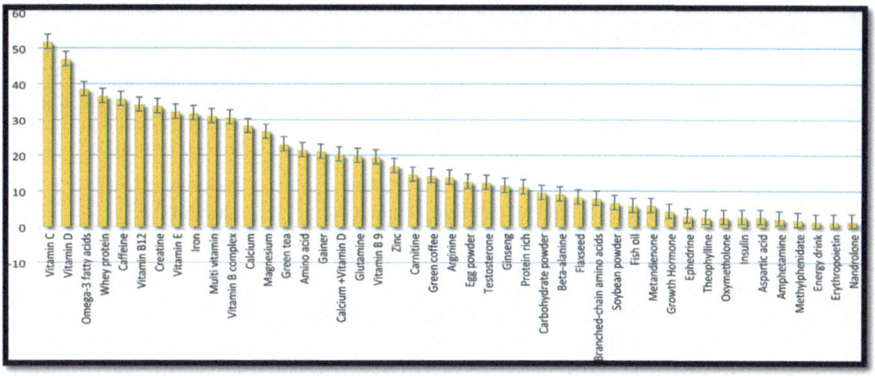

Fig. 5 Prevalence of supplement usage [28]. *Sources* Moradi et al. BMC Sports Science, Rehabilitation (2024) 16:150 https://doi.org/10.1186/s13102-024-00940-3

Protein Protocols:

- **Daily Intake:**
  - Athletes require higher protein intake than the general population (0.8–0.9 g/kg/day).
  - Recommended range: 1.2–2.1 g/kg/day.
  - May increase to 3.1 g/kg/day in calorie-restricted diets.
- **Benefits:**
  - Anticatabolic effects.
  - Faster recovery.
  - Injury prevention.
  - Muscle gain and anabolic effects.
  - Supports weight loss strategies [37].

(c) **Protocols for Creatine Loading**

Loading Phase (Optional):

- Duration: 5–7 days
- Dosage: 0.3 g/kg body weight per day, typically divided into 4 equal doses.
  - Example: For a 70 kg athlete, the daily dose would be 21 g (divided into 4 doses of 5.25 g each).
- Goal: Rapidly increase intramuscular creatine stores.

Table 8 Trends of nanofuels intake in different types of sports activities

Main findings		Sport type/Number and gender of subjects	Ref.
Insufficient CHO and protein intake compared to recommendations		Soccer/19 female	[29]
Of the male participants, 23.9% consumed insufficient amounts of protein, while 48.3% consumed insufficient amounts of CHO	Participants, with 23.9% of male participants consuming inadequate amounts of protein. Of the female participants, 21.8% consumed insufficient amounts of protein and 54.4% consumed insufficient amounts of CHO	Endurance/157 males and 83 female	
Of the male participants, 19.9% consumed insufficient amounts of protein, while 50.5% consumed insufficient amounts of CHO	Insufficient amounts of CHO and protein had been consumed by 80.5% and 41.8% of the female participants, respectively	Team/138 males and 104 females	[30]
24.2% of female participants and 69.2% of male participants consumed insufficient amounts of protein and CHO, respectively	73.0% of female participants consumed insufficient amounts of CHO, while 29.1% of female participants consumed insufficient amounts of protein	Strength/32 males and 39 female	[31]
Lower CHO intake compared with recommendations		Soccer/6 males	[32]
22.3% lower energy intake than recommended		Swimming/9 females	[33]
All athletes consumed insufficient amounts of calories and CHO; 46% consumed less protein and 9% consumed less fat than is recommended		Australian football/46 professional males	[34]
On training days, up to 77% failed to fulfill CHO requirements; 62% consumed less fat and 13% consumed less protein. Football/26 men		Football/26 males	[34]

**Sources* Appl. Sci. 2022, 12, 8611. https://doi.org/10.3390/app12178611

Maintenance Phase:

- Duration: Ongoing (throughout the training period)
- Dosage: 3–5 g per day, typically taken once or twice daily.

Key Considerations:

- Individualization: Individual factors including body weight, workout intensity, and goals could impact the optimal dosage.
- Cycling: Some athletes may cycle their creatine intake, using a loading phase periodically to replenish stores.
- Consumption: Creatine can be consumed with carbohydrates and/or protein to enhance absorption.
- Hydration: Adequate hydration is crucial when supplementing with creatine [38].

Benefits:

- Increased power and strength: Creatine supplementation can significantly enhance power output and muscular strength.
- Improved exercise performance: Increased performance in high-intensity exercises like weightlifting, running, and jumping can result from it.
- Enhanced recovery: Creatine can help accelerate recovery from intense training sessions and reduce muscle damage.

4.4 General Fluid Intake Protocol

As Per American College of Sports Medicine (ACSM), National Athletic Trainers' Association (NATA), Sports Dietitians Australia (SDA), International Olympic Committee (IOC) [39]

- **Before Exercise:**
  - **ACSM**: 2–4 h prior to exercise, consume 5–10 mL of fluid/kg body weight. Urine that is pale yellow is ideal.
  - **NATA**: Customize your plans. Euhydration or, with medical supervision, hyperhydration < 2%.
  - **SDA**: Adjust to anticipated requirements and fluid equilibrium. Hyperhydration pre-event could be beneficial.
  - **IOC**: Achieve plasma osmolality < 290 mmol/kg, urine specific gravity < 1.020, and BML < 1–2%.
- **During Exercise:**
  - **ACSM**: Consume about 0.4–0.8 L/h to replenish fluid i.e. lost as sweat and keep BML below 2%.
  - **NATA**: Limit BML < 2% while consuming sufficient liquid to compensate for sweat loss.
  - **SDA**: Individualized plans based on real-time assessment, thirst, tolerance, and experience.
  - **IOC**: Maintain body weightby minimizing loss refrain from gain. Practice plans in training conditions.
- **After Exercise:**
  - **ACSM**: To reduce urine loss, consume $\approx$ 1.5 L/kg BML at a moderate pace.
  - **NATA**: Within four hours, consume as much liquids as 150% of estimated fluid loss.
  - **SDA**: Consume up to 150% of BML to rapidly reverse moderate-severe fluid deficit.
  - **IOC**: Rehydrate to restore fluid balance slightly above BML (100–120%).

Fluid Intake in Hot Environments:

- **SDA**: Pre-event hyperhydration may be beneficial when fluid intake opportunities are limited.
- **IOC**: Emphasizes the importance of achieving proper fluid balance before exercise in hot conditions.

5 Challenges and Trends

5.1 *Sport Nutrigenomics:* "Eat According to Your Genes Takes Personalized Nutrition"

The field of nutrigenomics, which is expanding, examines the complex relationship among an individual's genetic composition and response to food. Nutrigenomics, which investigates how genetic variations influence how the body processes and uses nutrients, provides a significant tool for boosting sports performance through tailored dietary strategies [40]. Utilizing genomic technologies and genetic information, nutrigenomics addresses issues concerning body composition and performance metrics, nutrition and health, and other outcomes. Gene-diet interactions can alter metabolic pathways that are critical to athletes' health and performance, and genetic variants already alter our capacity to absorb, digest, ingest, and transport nutrients [41]. Genetic testing for customized sports is becoming more and more necessary, and coaches, dietitians, nutritionists, and other sports medicine professional's requirements for comprehending data that is now available in this emerging domain.

The Impact of Nutrigenomics on Athletic Performance

- **Individualized Approach**: Nutrigenomics recognizes that athletes are not created equal. Genetic variations can significantly impact factors such as:
  - **Nutrient absorption and metabolism**: How efficiently the body absorbs and utilizes key nutrients like carbohydrates, proteins, and fats.
  - **Muscle fiber type distribution**: Determining an athlete's predisposition towards endurance, strength, or power sports.
  - **Response to training**: How effectively the body adapts to exercise stress, including factors like muscle protein synthesis and recovery.
- **Tailored Nutrition Plans**: By analyzing an athlete's genetic profile, nutrigenomics can inform the creation of highly personalized nutrition plans. This may involve:
  - **Optimizing macronutrient intake**: Adjusting carbohydrate, protein, and fat intake based on individual genetic needs.
  - **Identifying nutrient deficiencies**: Pinpointing potential nutrient deficiencies based on genetic predispositions.

- **Selecting the most effective supplements**: Choosing supplements that are most likely to benefit the athlete based on their genetic makeup.

- **Enhanced Performance**: By addressing individual nutritional needs and optimizing nutrient utilization, nutrigenomics can potentially enhance athletic performance in several ways, including:
  - **Improved training adaptations**: Faster muscle growth, increased endurance, and quicker recovery.
  - **Reduced risk of injury**: By optimizing nutrient intake, athletes may reduce their risk of injuries related to nutritional deficiencies.
  - **Enhanced overall health and well-being**: A well-balanced diet, tailored to individual genetic needs, can support overall health and well-being, leading to improved athletic performance.

The genetic architecture of an athlete, encompassing variations in genes like ACTN3 and EPOR, plays a pivotal role in determining athletic prowess by influencing muscle fiber composition, oxygen utilization, and thereby overall performance. Variants in these genes can significantly affect an athlete's strength, endurance, and susceptibility to injuries, highlighting the importance of customizing training and dietary regimens to their genetic profiles [40] (Fig. 6).

Genetic variations (SNPs) influence individual responses to dietary factors, such as caffeine, impacting performance. For example, CYP1A2 rs726551 genotypes (AA, AC, CC) can differentially affect caffeine's impact on performance.

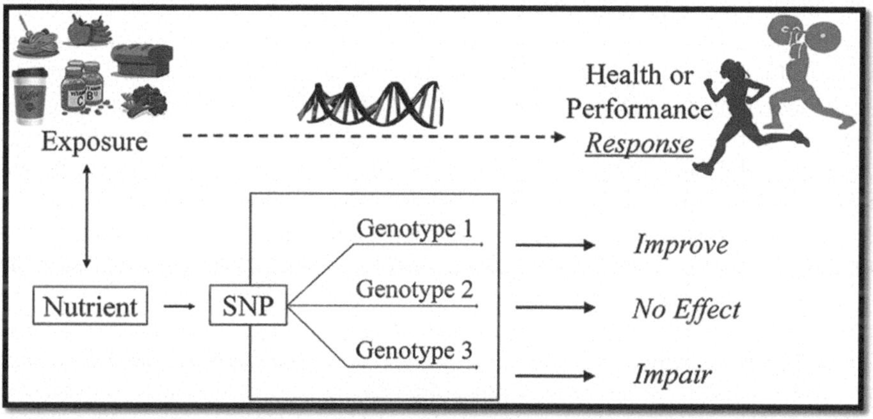

Fig. 6 The nutrigenomics approach to sport nutrition [42]

5.2 *Trends of Bioactive Supplements*

Caffeine as a Bioactive Nano supplements

Caffeine is a widely studied compound in the field of nutrigenomics, with numerous randomized controlled trials investigating how genetic variations influence its impact on athletic performance. Although many people have found that taking a caffeine supplement improves their workout performance, the strength of these effects varies greatly. Some individuals experience significant performance improvements, while others show little or no response to caffeine compared to a placebo. Genetic differences in genes related to caffeine metabolism, sensitivity, and the body's reaction to caffeine, such as CYP1A2 perhaps ADORA2, may be contributing to the inter-individual variability in caffeine response.

Several investigations have evaluated effect of caffeine and CYP1A2 genotype on cycling performance of male athletes. While moderate caffeine improved performance across all participants, the greatest benefit was observed in 'fast metabolizers' (AA genotype). These individuals experienced a significant performance improvement with both low and moderate caffeine doses, with the most pronounced effect at 4 mg/kg. However, in those with the CC genotype, 4 mg/kg caffeine reduced performance, whereas for individuals who had the AC genotype, it had no effect. These results imply that the reaction to caffeine during exercise is greatly influenced by the CYP1A2 genotype. Research on resistance training and basketball has demonstrated that caffeine enhances performance in activities requiring muscular endurance, notably resistance exercise sets and repeated steps, but mostly in individuals with AA genotype for CYP1A2 gene. According to these results, the CYP1A2 genotype has a major impact on ergogenic effects of caffeine during endurance-style exercise. In a tiny pilot study, 12 female cyclists' ergogenic effects were investigated concerning the ADORA2A gene (rs5751876). All subjects with the TT genotype demonstrated more accurate cycling time trial performance when caffeine was consumed, while just one subject with the C allele accomplished this. These first results point to a possible relationship between the ADORA2A genotype and the ergogenic effects of caffeine; however, bigger sample numbers are required for additional research to validate these results and explore possible variations between the CT and CC genotypes.

Another potential genetic factor impacting that caffeine effects performance is ADORA2A gene, which codes for adenosine A2A receptor. Adenosine A2A receptors play crucial roles in various physiological processes. In the cardiovascular system, they regulate myocardial oxygen demand and enhance coronary blood flow. In the brain, they modulate neurotransmitter release, impacting functions such as sleep and pain perception.

Adequate sleep is crucial for athlete recovery and performance. The ADORA2A gene, which encodes adenosine receptors in the brain, plays a vital role in regulating sleep–wake cycles. Caffeine blocks adenosine receptors, promoting wakefulness. Genetic variations within ADORA2A, specifically the rs5751876 SNP, influence individual responses to caffeine and sleep. Individuals with the TT genotype for this SNP may exhibit enhanced responses to caffeine. Studies have shown they may experience faster reaction times, particularly after sleep deprivation, and potentially less sleep disturbance following caffeine intake compared to those with the C allele. These findings suggest that the ADORA2A genotype may influence an individual's sensitivity to caffeine, impacting sleep quality and potentially athletic performance. It is essential to recognize that caffeine can interfere with sleep architecture, delaying sleep onset and diminishing deep sleep. Additional study is required to comprehensively elucidate the intricate relationship among ADORA2A genotype, caffeine intake, sleep quality, and athletic performance [43–45].

Key Points:

- Genetic Variability in Caffeine Response:
  - Individual responses to caffeine vary significantly.
  - Genes like CYP1A2 and ADORA2A play a crucial role in how individuals metabolize and respond to caffeine.
- CYP1A2 and Caffeine's Effects on Exercise:
  - CYP1A2 genotype influences caffeine's ergogenic effects.
  - 'Fast metabolizers' (AA genotype) experience greater performance benefits from caffeine, particularly in endurance-type exercise.
- ADORA2A and Caffeine's Effects on Performance and Sleep:
  - ADORA2A gene regulates sleep–wake cycles and influences caffeine's effects on the nervous system.
  - ADORA2A rs5751876 SNP may influence caffeine's impact on sleep quality and psychomotor performance.
  - TT genotype may experience enhanced responses to caffeine with less sleep disturbance.
- Importance of Sleep for Athletes:
  - Adequate sleep is essential for athlete recovery and performance.
  - Caffeine can disrupt sleep architecture.
- Need for Further Research:
  - Additional study is required to comprehensively elucidate the intricate relationship among ADORA2A genotype, caffeine intake, sleep quality, along with athletic performance.

5.3 Genetic Variations Unique Responses to Nutrients (Foods–Diets–Supplements)

Genetic variation influences the impact of micro- and macronutrients, together with bioactives that include caffeine, on performance. This micronutrient serves various vital functions that are linked to athletes' good health, immunity, and performance. Elite athletes exhibit enhanced visual capabilities, evidenced by their proficiency in specific visual skills, including contrast sensitivity, dynamic acuity, stereoacuity, and ocular judgment, essential for executing interceptive actions (e.g., hand–eye coordination) and discerning fine spatial details in various sports [46, 47]. Moreover, inadequate visuomotor reaction time (VMRT) has been associated with a heightened risk of musculoskeletal injury in sports necessitating enhanced visual stimulus detection and motor response execution [48]. Visuomotor abilities are essential for improved sports performance and, thus, necessitate optimal eye health.

(a) **Vitamin A**

- Insufficiencies in micronutrients such as vitamin A might impair the immune system, keeping athletes more susceptible to infections. Provitamin carotenoids are rapidly converted to vitamin A post-absorption by the BCMO1 enzyme, located in intestinal mucosal enterocytes [49]. Beta-carotene, the predominant provitamin A carotenoid in the diet, requires conversion to retinal or retinoic acid for the correct functioning of vitamin A. The BCMO1 gene variant rs11645428 modulates the small intestine's conversion of provitamin A carotenoids to active vitamin A, affecting plasma carotenoid levels [50].
- Vitamin A insufficiency may result from this conversion inefficiency in GG genotypes [51]. Low β-carotene responders may benefit from pre-formed vitamin A or vegan supplements for adequate vitamin A levels for eyesight, immunity, and normal development.

(b) **Anemia-Related Micronutrients: Iron, Folate, and Vitamin B12**

- Numerous studies have shown that poor iron stores and anemia might negatively impact sports performance [52–54]. Elite athletes are more likely to have anemia and low iron, folate, and vitamin B12 levels compared to the general population, which might negatively effect performance. Weariness, weakness, and, in extreme cases, shortness of breath or palpitations are the main symptoms [55–58].

Iron

Iron can be detected in oxygen-carrying protein hemoglobin and myoglobin. Menstruation and reduced food and energy intake indicate that female athletes at risk of iron deficiency [59]. In athletes, footstrike hemolysis, gastrointestinal bleeding, exercise-induced inflammation, NSAID use, and environmental factors including altitude may affect iron metabolism. Insufficient or improper vitamin B12, folate, and iron reserves induce megaloblastic anemia.

Iron Overload

Genetic variations, particularly in the HFE gene, influence serum iron levels. While some variants increase iron absorption and may benefit athletic performance by enhancing iron availability, they also raise the risk of hemochromatosis (iron overload). Excess iron can damage tissues through oxidative stress. Individuals with HFE gene variants associated with higher iron levels may have an athletic advantage, but careful monitoring is crucial to prevent iron overload. Iron supplementation should be avoided in these individuals to minimize health risks and maintain optimal performance.

Focuses on the key points:

- Genetic influence on iron levels: HFE gene plays a significant role.
- Potential athletic advantage: Increased iron can benefit performance.
- Risk of hemochromatosis: Excess iron is harmful.
- Importance of monitoring: Prevent iron overload and its consequences.
- Avoidance of supplementation: In individuals with HFE gene variants.

Low Iron Status

Genetic variations in three key genes—TMPRSS6, TFR2, and TF—play a significant role in iron status regulation. These genes influence hepcidin, a hormone that controls iron absorption. While iron deficiency anemia is a well-known performance limiter, even subclinical iron deficiency can negatively impact athletic performance; particularly in women. Certain gene genotypes increase the risk of low iron. For example, the GG genotype in TMPRSS6 is linked to lower transferrin saturation and hemoglobin levels. In contrast, the AA genotype in the TF gene is associated with low ferritin and elevated transferrin, indicating potential iron deficiency. Variations in TFR2 can impact red blood cell parameters.

These genetic indicators may diagnose iron deficiency risk and guide dietary iron consumption. However, excessive iron supplementation can be detrimental. Iron overload is damaging tissues through oxidative stress that impedes athletic performance, although being recommended. Therefore, a balanced approach is crucial. Genetic screening may monitor iron intake to ensure maximum performance and reduce the danger of iron overload and its health effects.

Genetic factors influencing iron status and their implications for athletes:

- Genetic Influence on Iron Regulation:
  - Three genes (TMPRSS6, TFR2, and TF) play a crucial role in iron regulation by influencing hepcidin, a hormone that controls iron absorption.
  - Certain gene genotypes increase the risk of low iron status.

    GG genotype in TMPRSS6: Linked to lower transferrin saturation and hemoglobin.

AA genotype in TF: Associated with low ferritin and elevated transferrin.
CC genotype in TFR2: Can impact red blood cell parameters.

- Impact of Iron Status on Athletic Performance:
  - Iron deficiency, even without anemia, can negatively impact athletic performance, particularly in women.
  - Inadequate iron can impair oxygen delivery, affecting exercise capacity and performance.
- Importance of Personalized Approach:
  - Genetic testing can help assess individual risk for iron deficiency.
  - This information can be used to guide personalized dietary iron intake, ensuring adequate levels for optimal performance.
- Risks of Iron Overload:
  - Excessive iron supplementation might cause medical issues.
  - Iron overload can cause oxidative stress, damage tissues, and impair athletic performance.
- Need for Balanced Approach:
  - A balanced approach is crucial to optimize iron status.
  - The goal is to achieve and maintain adequate iron levels for optimal performance while minimizing the risk of iron overload.

Vitamin B12

Red blood cell synthesis and aerobic capacity depend on vitamin B12. Megaloblastic anemia, which causes exhaustion, weakness, and diminished cell oxygen delivery due to blood oxygen-carrying capability, can result from B12 deficiency. Interestingly, genetic variations in the FUT2 gene (rs602662) influence B12 levels. Low B12 is more likely to occur among individuals with GG or GA genotypes, but only if their diet is deficient in easily absorbed sources. Studies have shown that people with the AA genotype tend to have higher B12 levels compared to those carrying the G allele.

5.4 Vitamin D, Vitamin C, and Protein Supplements Are Popular Choices

Vitamin D

While no direct link has been established between genetic variations and athletic performance outcomes specifically related to vitamin D, this vitamin plays a crucial role in several aspects important for athletes. Vitamin D improves bone health, immunological function, and training recovery, which may improve athletic performance.

Genetics might impact 25(OH) D levels, the primary type of vitamin D. Strong bones require calcium absorption, which this vitamin helps with, especially for athletes at risk of stress fractures. Studies indicate that persons with adequate vitamin D levels may receive advantages including diminished injury risk, enhanced muscle fiber growth, reduced inflammation, and a lower likelihood of respiratory infections. These advantages can facilitate recuperation and enable an athlete to acclimate to rigorous training.

Two genes, GC and CYP2R1, have been identified as significantly impacting vitamin D status. Genetic variations in these genes correlate with diminished 25(OH) D levels heightened risk. An investigation indicated that genetic variables accounted for a greater fraction of the diversity in vitamin D levels than supplementation, implying that genetics may be more influential than supplementation in predicting individual vitamin D status. The particular genetic variants occur in CYP2R1 (rs10741657) and GC (rs2282679) were significantly correlated with reduced vitamin D levels. Individuals possessing specific genotypes in these genes exhibited a markedly higher likelihood of having inadequate vitamin D levels. These findings correspond with extensive genome-wide investigations that consistently identified these genes as key determinants of vitamin D status. The quantity of risk variations an individual has is closely related to their chances of vitamin D deficiency.

These results imply that genetic diversity, rather than supplements or other lifestyle factors, might be a more powerful predictor of vitamin D deficiency, highlighting importance of considering individual genetic predispositions when assessing and optimizing vitamin D status in athletes.

Focuses on the key points:

- Vitamin D's importance for athletic performance: Impacts bone health, immunity, and recovery.
- Genetic influence on vitamin D levels: GC and CYP2R1 gene variations are crucial.
- Genetics versus supplementation: Vitamin D status may be more significantly determined by genetics.
- Specific genotypes associated with low vitamin D: GG or GA in CYP2R1, GG in GC.
- Implications for athletes: Understanding genetic predispositions can help optimize vitamin D status and potentially improve athletic performance.

Vitamin C

As a potent antioxidant, vitamin C is crucial for lowering oxidative damage brought on by exercise in athletes. However, excessive supplementation with vitamin C can have detrimental effects, potentially blunting beneficial training adaptations and even diminishing performance. While vitamin C supplementation was once believed to be beneficial for athletes, studies have shown that excessive intake can negatively impact training adaptations. Dietary intake of vitamin C, typically from fruits and vegetables, up to 250 mg daily, is generally considered sufficient to reduce oxidative stress without adverse effects.

In addition to its antioxidant properties, vitamin C is necessary for collagen-manufacturing, crucial building block of connective tissues including ligaments and tendons. Recent research suggests that supplementing with vitamin C, along with gelatin, can enhance collagen synthesis in athletes, potentially improving injury prevention and accelerating tissue repair. Individual responses to vitamin C intake can be influenced by genetics. The GSTT1 gene plays a significant role in vitamin C metabolism. Individuals with the GSTT1Del/Del genotype are more susceptible to vitamin C deficiency than those with the Ins allele, even when meeting the Recommended Dietary Allowance (RDA). Genetic testing could help identify athletes at risk of low vitamin C levels, which can increase oxidative stress and impair muscle and connective tissue repair.

While genes involved in vitamin C transport (SLC23A1 and SLC23A2) have been identified, their impact on individual responses to vitamin C intake remains unclear. Therefore, personalized dietary recommendations based on these genetic variations are not currently supported by scientific evidence.

Focuses on the key points

Vitamin C and Exercise:

- Vitamin C is an antioxidant that helps combat exercise-induced oxidative stress.
- High-intensity exercise markedly elevates the generation of reactive oxygen species (ROS) in athletes.
- While initially thought beneficial, excessive vitamin C supplementation can negatively impact training adaptations and potentially diminish performance.
- The consumption of vitamin C from fruits and vegetables, up to 250 mg daily, is typically adequate for reducing oxidative stress without negative consequences.

Vitamin C and Collagen Synthesis:

- Vitamin C is essential for collagen synthesis, a fundamental element of connective tissues.
- Supplementation with vitamin C, along with gelatin, can enhance collagen synthesis, potentially improving injury prevention and tissue repair in athletes.

Genetic Influence on Vitamin C Metabolism:

- The GSTT1 gene plays a significant role in vitamin C metabolism.
- Individuals with the GSTT1Del/Del genotype are more susceptible to vitamin C deficiency than those with the Ins allele, even with adequate dietary intake.
- Genetic testing could help identify athletes at risk of low vitamin C levels.

Limited Role of Vitamin C Transporters:

- While genes involved in vitamin C transport (SLC23A1 and SLC23A2) have been identified, their impact on individual responses to vitamin C intake remains unclear.
- There is currently insufficient evidence to support personalized dietary recommendations based on these genetic variations.
- There's a growing interest in testosterone boosters, highlighting the need for stricter regulations.

Protein

The FTO gene significantly influences body weight and composition. Individuals with specific genetic variations in the FTO gene (like the AA genotype at rs1558902) may be more susceptible to weight gain. However, dietary interventions can mitigate these genetic predispositions. Research indicates that a high-protein diet may be especially advantageous for individuals possessing AA genotype of FTO gene. Studies have demonstrated greater fat loss and improved body composition in these individuals when following a high-protein diet compared to those with other FTO genotypes. While a high-protein diet can be beneficial for some, excessive protein intake may not be advantageous for everyone. Individuals without the AA genotype may not experience the same degree of benefit from a high-protein diet and may even find it counterproductive. In these cases, adjusting macronutrient intake by substituting excess protein with carbohydrates, fiber, and healthy fats may be more beneficial for achieving and maintaining optimal body composition and athletic performance.

Key points:

- FTO Gene and Body Composition:
  - FTO gene is strongly associated with body weight and composition.
  - Genetic variations in FTO, particularly the rs1558902 (or rs9939609) polymorphism, influence an individual's susceptibility to weight gain.
- Dietary Protein and FTO Gene:
  - A high-protein diet can mitigate the effects of FTO gene variants on body composition.
  - Individuals with the AA genotype in the FTO gene may experience greater fat loss and improved body composition on a high-protein diet compared to those with other genotypes.
- Impact of FTO Genotype on Weight Loss:
  - Studies have shown that individuals with the A allele (particularly AA genotype) in FTO gene may experience greater weight loss on a hypocaloric diet, especially when combined with higher protein intake.

- Importance of Personalized Nutrition:
  - Dietary recommendations should be personalized based on individual genetic variations, including the FTO genotype.
  - Athletes with the AA genotype may benefit from a moderate-to-high protein diet (at least 25% of energy from protein) to optimize body composition.
  - Athletes without the AA genotype may benefit from a moderate protein intake (15–20% of energy) to avoid excess protein intake and ensure adequate intake of other essential nutrients.
- Benefits of Lean Mass:
  - Greater lean mass is associated with improved performance in strength and power sports, as well as some endurance events.
  - Lean mass also contributes to injury prevention.

Whey protein

Whey protein supplementation can enhance muscle building following resistance training. Research indicates that the consumption of whey protein prior to and following exercise might improve muscle build and strength development. Whey protein could also improve recovery by allowing athletes to perform more repetitions and lift heavier weights in subsequent training sessions.

While whey protein is rapidly absorbed, offering potential benefits immediately after exercise, there's no conclusive evidence that it surpasses other high-quality protein sources like casein or soy in terms of overall muscle growth within 24 h. Beyond muscle protein synthesis, whey protein may also offer immune system benefits. Research indicates that whey supplementation assists in maintaining glutathione levels, essential for immunological function, after rigorous exercise. Current guidelines advise athletes to ingest 1.3–1.8 g of protein/kg of body weight per day to facilitate muscle protein synthesis and recovery. Strength and power athletes maybe benefit from 1.6–1.8 g/kg protein consumption. In conclusion, while whey protein can be a valuable addition to an athlete's diet, particularly in the immediate post-exercise window, it's crucial to consider overall protein intake throughout the day. Muscle growth and health require a balanced diet with a variety of high-quality protein sources.

- Whey protein enhances muscle growth:
  - Whey protein before and after resistance training improves muscle mass and strength.
  - Whey protein can improve recovery by allowing for greater training volume in subsequent sessions.
- Whey protein may offer immune benefits:
  - Whey protein supplementation can help maintain glutathione levels, which are important for immune function, particularly after intense exercise.

- Importance of overall protein intake:
  - While whey protein can be beneficial, it's crucial to consider overall protein intake throughout the day.
  - Current recommendations for athletes are 1.3–1.8 g of protein/kg of body weight daily.
  - Protein consumption in this range may improve strength and power athletes.
- Whey protein versus other protein sources:
  - Whey protein is rapidly absorbed, providing potential benefits immediately after exercise.
  - However, there's no conclusive evidence that whey protein surpasses other high-quality protein sources (like casein or soy) in terms of overall muscle growth within 24 h.
- Personalized approach:
  - Athletes should consider their individual needs and training goals when incorporating whey protein into their diet.
  - Muscle growth and health require a balanced diet with a variety of high-quality protein sources.

6 Trends of Uses Most Five Nano Supplements by Athletes

There are several safe and advantageous sports supplements. BA, creatine, caffeine, nitrates, and protein are great performance supplements for athletes.

6.1 Beta-Alanine (BA)

The liver synthesizes BA, a non-proteogenic amino acid present in poultry, fish, and meat. It is a precursor of carnosine, a dipeptide that buffers hydrogen ions in skeletal muscles to reduce fatigue during intense activity. BA promotes anaerobic performance by increasing muscle carnosine, which maximizes intracellular buffering and delays muscle tiredness during high-intensity activities (Table 9).

Key Points

- **Anaerobic Performance**: BA significantly enhances anaerobic performance in activities lasting 1–4 min by increasing muscle carnosine, improving buffering capacity, and delaying fatigue.
- **Aerobic Performance**: BA offers modest benefits in aerobic activities up to 25 min, such as improving time to exhaustion in cycling and rowing. Effects on longer durations are less clear.

Table 9 Beta-alanine randomized controlled trials (Rcts) [57]

Reference	Supplement dose and duration	Results	Conclusion
• Guo et al. [60]	• 4.8 g daily for 8 weeks	• Improved vertical and • Horizontal jumps	• The impact of β-alanine on muscle power remained greater
• Ojeda et al. [61]	• 30 mg kg^{-1} and 45 mg kg^{-1} for acute trials	• Enhanced maximal aerobic speed in endurance athletes	• High β-alanine doses enhance performance in aerobic-anaerobic transition zones
• Sas-Nowosielski et al. [62]	• 4 g/day for 4 weeks	• Campus board performance improved with fewer 'slaps' and easier traverses	• In continuous climbing and repetitive upper body movements, β-alanine improves performance, but not in shorter climbs
• Shbib et al. [63]	• 4times/day for 4 weeks	• Higher serum carnosin, average and peak power, and lower tiredness index in placebo group	• Combining β-alanine supplementation with plyometric training enhances handball players' anaerobic power and carnosine levels
• Jaques et al. [64]	• 3.2 g/day for 4 weeks	• While body composition and 2 km rowing performance increased with time, there had been insignificant differences between placebo and β-alanine groups	• Supplementing with β-alanine enhanced rowing performance, but more research is required
• Smith et al. [65]	• 6.4 g/day for 4 weeks	• Insignificant effects had been observed on body composition, strength, endurance, or intermittent sprinting	College rugby athletes' body composition and performance are unaffected; more research is"required
• Glenn et al. [66]	• 800 mg + 8gdex-trose,4times/day for 28 days	• Significantly higher peak torque and total work in lower-body isokinetic strength without affecting body composition or handgrip strength	• Female master athletes who apply β-alanine supplements perform better on lower-body exercises without compromising their handgrip strength or body composition

(continued)

Table 9 (continued)

Reference	Supplement dose and duration	Results	Conclusion
• Bellinger et al. [67]	• 6.4 g/day for 4 weeks	• Enhanced TTE during supramaximal cycling, without affecting TT performance over 1 or 10 km	• Supramaximal cycling TTE is increased effectively, with minimal impact on 1 and 10 km TT performance
• Gross et al. [68]	• 4.8 g/day for 5 weeks	• Enhanced aerobic energy contribution, decreased oxygen deficit, and improved maximal and average (CMJ) power	• Supplementing with β-alanine improved the explosive and repeated jump performance of elite alpine skiers
• Kern et al. [69]	• 4 g/day for 8 weeks	• On average, the β-alanine group gained 2.1 lb of lean mass, while the placebo group gained 1.1 lb	• Gains in lean mass appear to be enhanced by β-alanine
• Baguet et al. [70]	• 5 g/day for 7 weeks	• Carnosine content rose 45.3% in soleus and 28.2% in gastrocnemius; performance in 2000-m rowing improved by 4.3 s	• Supplementing with β-alanine improves performance in trained rowers
• Derave et al. [71]	• 4.8 g/day for 4 weeks	• Although it did not affect 400 mt sprint performance, increased carnosine concentration in calf muscles improved knee extension torque	• In well-trained track and field athletes, β-alanine does not enhance 400 m sprint performance, but it increases muscle carnosine and decreases fatigue during dynamic contractions
• Van Thienen et al. [72]	• 2–4 g/day for 8 weeks	• During the latter sprints of a simulated bicycle race, β-alanine increased mean power output by 5.0% and peak power generation by 11.4%	• Consuming β-alanine supplements improves sprint performance during endurance cycling, particularly when high-intensity efforts are performed after extended exercise

Legend: *TT* time trial, *TTE* time to exhaustion, *CMJ* countermovement jump

- **Body Composition**: Body composition, which includes lean mass, fat mass, and body mass, is typically not greatly impacted by BA supplementation.

Summary: BA is a useful supplement for athletes primarily to improve anaerobic performance, especially during short-duration, high-intensity sports. It has insignificant impacts on body composition and aerobic performance, and these effects may alter based on training regimens and individual characteristics.

6.2 Caffeine

The most widely employed and well-researched ergogenic aid is caffeine. It is frequently consumed by athletes to enhance their perceived energy, mental clarity, and performance. Early research on caffeine examined its impact on affected muscle fatigue. Caffeine is most widely consumed in sports as a pre-workout supplement and in energy drinks. Adenosine receptor antagonism, phosphodiesterase inhibition, and intracellular calcium mobilization are some of the methods by which caffeine functions. It operates as an antagonist of the A1 and A2A adenosine receptors. This causes greater attention and concentration (Table 10).

Research consistently demonstrates that caffeine enhances endurance performance in sports like running and cycling. Studies have shown:

- **Increased Time to Exhaustion**: Dittrich et al. research reported that, participants using caffeinated chewing gum significantly extended their time to exhaustion on a treadmill compared to those using a placebo.
- **Improved Cycling Performance**: Hodgson et al. found that both caffeinated coffee and pure caffeine improved cycling performance in a time trial compared to decaffeinated coffee and placebo.
- **Genotype Influence**: Guest et al. observed that caffeine's effect on cycling performance varied based on an individual's CYP1A2 genotype, highlighting the importance of personalized caffeine strategies.

Recommendations:

- ISSN recommends a caffeine dosage of 3–6 mg/kg body mass for performance improvement.
- A smaller dosage of 1–3 mg/kg might be adequate for cognitive advantages.
- For best results, take your caffeine 30–60 min before working out.
- Caffeine can have a wide range of effects on different people.

6.3 Creatine

Among athletes, creatine is one of the most frequently employed ergogenic supplements in sports nutrition. The primary method that creatine is utilized by athletes is

Table 10 Caffeine RCTs [58, 59, 73]

Reference	Supplement dose and duration	Results	Conclusion
• Wu and Jiang [74]	• For 8 weeks, take 3 mg/kg1hour before exercise instead of 6 mg	• Dose of 6 mg/kg worked best	• The greater caffeine dose increased aerobic power, strength, and vertical leap
• Tamilio et al. [75]	• 45 min prior to working out, 3 mg/kg was consumed	• Increased repetitions to failure, drop jumps (DJ), and countermovement jumps (CMJ)	• Consuming caffeine improves muscle strength, power, and endurance, however, the results aren't always constant
• Berjisian et al. [76]	• 6 mg/kg caffeine consumed 60 min before workout	• Weight lifted during 1RM, repetitions to failure, and CMJ all increased; 1RM velocity remained constant	• Supplementing with caffeine can increase muscle strength, stamina, and vertical jump
• Dittrich et al. [77]	• 300 mg of caffeine consumed soon before workout	• Increase in time to exhaustion	• Caffeine improves time to exhaustion in running
• Guest et al. [43]	• Caffeine intake: 0, 2, or 4 mg/kg 25 min before warmup	• Comparing 4 mg/kg to a placebo, riding time reduced by 4%	• Cycling time is improved by caffeine
• Hodgson et al. [73]	• Eat 5 mg/kg an hour before working out. 4.3% improvement in the time trial for cycling A strong ergogenic aid for cycling is caffeine	• 1RM and repetitions to failure are identical	• Caffeine doesn't increase muscle strength or endurance

to increase intracellular creatine in the muscles, which increases the availability of creatine phosphate (PCr) in the muscles and increases an individual's capacity for acute activity to improve training adaptations [78]. Power or strength athletes have mostly used it to maximize training adaptations and recovery from physically taxing exercises (Table 11).

Creatine supplementation offers benefits beyond performance enhancement.

- **Recovery**: Combining creatine with carbohydrates and protein enhances its uptake and storage in muscles, aiding in post-exercise recovery. According to experiments, creatine may boost recovery from rigorous training and minimize muscle damage.
- **Reduced Injury Risk**: Athletes who consume creatine supplements have reported fewer instances of heat illness, cramping in their muscles, dehydration, and general ailments, which may result in fewer days missed from training.

Table 11 Creatine Rcts [79]

Reference	Supplement dose and duration	Results	Conclusion
• Desai et al. [80]	• Acute consumption of real creatine as opposed to false creatine, dextrose as a placebo, and nothing as a control	• For both the squat and the bench press, all treatments led to higher repetitions than the control	• The impact of the belief in taking creatine had been identical to that of the actual creatine
• Yamaguchi et al. [81]	• For 28 days, either 3 g creatine or a placebo	• The creatine group had less muscle soreness	• Creatine has been shown to improve recovery from muscle discomfort
• Mills et al. [82]	• Participants received 0.0055 g/kg of creatine or a placebo for 6 weeks duration	• Creatine group exhibited significant enhancements in chest press, leg press, overall body strength, and leg press endurance, whereas the placebo group demonstrated statistically insignificant changes	• Creatine enhanced muscle strength and endurance in those engaged in resistance training
• Candow et al. [78]	• 6-week investigation with two placebo groups, one receiving 0.15 g/kg of body weight and the other receiving 0.10 g/kg of body weight	• Muscle thickness surrounding the elbow significantly increased in both creatine groups relative to the placebo group. Men utilizing creatine exhibited greater enhancements in leg press performance compared to women employing creatine	• Creatine supplementation during resistance training promotes muscle growth around the elbow in comparison to a placebo
• Chilibeck et al. [83]	• Eight-week intervention with 5 g of creatine or a placebo	• Creatine group exhibited increased repetitions for both the bench press and leg press exercises	• Creatine enhances muscular endurance in rugby athletes
• Hoffman et al. [84]	• 10-week study evaluating the effects of 5 g creatine, 5 g creatine combined with beta-alanine, or a placebo	• Creatine combined with beta-alanine increased lean mass and body fat percentage, while both the beta-alanine and creatine groups enhanced strength; the creatine-only group elevated testosterone levels	• The combination of creatine and beta-alanine appears to exert the most significant influence on lean body mass and body fat percentage

(continued)

Table 11 (continued)

Reference	Supplement dose and duration	Results	Conclusion
• Bemben et al. [85]	• 9-week trial contained placebo group, control group, and group that consumed 20 g creatine for five days, then 5 g daily	• Creatine increased cell hydration compared to placebo and control	• Cell hydration may improve with creatine
• Stone et al. [86]	• Compared to a calcium pyruvate group only and a placebo group throughout 5-weeks, one group received 5 g creatine, or 60% calcium pyruvate, and 40% creatine for the other creatine group	• Body mass, lean body mass, one repetition maximum (RM) bench press, mixed 1RM squat and bench press, and static vertical jump (SVJ) increased significantly in the creatine and creatine + calcium pyruvate groups	• Body mass/ composition, maximum strength, and SVJ improved with creatine alone and creatine plus calcium pyruvate

- **Enhanced Recovery**: Research by Yamaguchi et al. and Greenwood et al. has demonstrated that supplementing with creatine assists in reducing the damage that exercise causes to muscles.
- **Strength and Muscle Growth**: Regional muscle thickness, strength, and power have all been demonstrated to increase with creatine treatment. According to Mills et al., athletes who used creatine supplements demonstrated notable increases in their muscle strength, leg press, and chest press.

According to research, creatine may be a useful supplement for athletes trying to maximize their training and recovery, that could enhance performance and lower their risk of injury.

6.4 *Nitrates*

The body produces nitric oxide (NO) from nitrates, which are bioactive substances. Green leafy vegetables, pomegranate extract, and beetroot juice are sources of dietary nitrates. Nitrates were initially identified as a signaling molecule and their function in cardiovascular health was investigated. Because NO dilates blood vessels, its impacts on exercise performance have been investigated (Table 12).

By lowering the body's energy expenditure, nitrates are thought to improve workout efficiency.

Table 12 Nitrate RCTs [87]

Reference	Supplement dose and duration	Results	Conclusion
• Yuschen et al. [88]	• Acute ingestion of beetroot juice before submax cycling for 30 min	• Beetroot juice improved mean exercise load	• Beetroot increases aerobic exercise capacity and cardiovascular health
• Giv et al. [89]	• 8 weeks of football practice and juice from beetroot supplementation	• The beetroot juice supplement had a substantial effect on aerobic power, respiratory exchange ratio, anaerobic threshold, anaerobic power, field performance, and fatigue index	• Soccer players might benefit from consuming beetroot juice
• Mosher et al. [90]	• mmol every day for 6 days in form of beetroot juice	• Increase in repetitions till failure and total weight lifted	• Nitrates enhance resistance training performance
• Vanhatalo et al. [91]	• For 15 days, consume 5.2 mmol of beetroot juice every day	• Reduced oxygen consumption and blood pressure. Enhanced peak power output and work rate at the gas exchange threshold	• Both acute and chronic nitrate supplementation reduces blood pressure while increasing oxygen consumption and exercise performance
• Bailey et al. [92]	• 5.5 mmol/day for 6 days in the form of beetroot juice	• Decrease in oxygen consumption increase in time to exhaustion	• Nitrate supplementation reduces the cost of oxygen and increases exercise tolerance
• Larsen et al. [93]	• Three days of 0.1 mmol/kg/day nitrates or placebo	• Reduced demand for O2At submaximal workloads, HR, lactate, VE, VE/VO2, and RER don't alter	• Decreased oxygen demands in submaximal work

Improved Aerobic Efficiency:

- Nitrate supplementation reduced oxygen consumption and increased gross efficiency during submaximal cycling.
- Reduction in oxygen uptake and an increase in time to exhaustion in both moderate and severe exercise following nitrate supplementation from beetroot juice.
- Beetroot juice supplementation during soccer training improved both aerobic and anaerobic performance.
- Nitrate supplementation enhanced aerobic exercise capacity and cardiovascular function during submaximal cycling.
- The effects of long-term nitrate supplementation on exercise performance include notable increases in work rate and peak power at gas exchange threshold.

6.5 Protein

Because protein supplements have been demonstrated to have numerous advantages for muscle growth, repair, and recovery, they have become increasingly popular among bodybuilders, athletes, and fitness enthusiasts. This section examines the principles, dosages, benefits, and possible disadvantages of protein supplementation, emphasizing research results from reliable sources and peer-reviewed studies. The process by which body regenerates and repairs muscle tissue, identified as muscle protein synthesis (MPS), depends on protein (Table 13).

Whey Protein: A Valuable Tool for Athletes

The advantages of whey protein supplementation for athletes are often shown by research.

- **Muscle Growth and Body Composition**: whey protein supplementation, when combined with appropriate training programs, can effectively increase muscle mass and reduce body fat percentage (Mhamed et al. and Kim et al.).
- **Enhanced Strength and Power**: Kim et al. discovered that the whey protein group significantly outperformed the placebo group in terms of muscle strength and power.
- **Superior to Collagen**: After resistance training, Jacinto et al. discovered that whey protein increased muscle growth more effectively than collagen peptides.
- **Mixed Protein Sources**: mixed protein supplement containing whey, casein, and pea protein can positively impact golf performance and muscle function.

Comparison to Soy Protein:

Whey protein might be slightly superior to soy protein in terms of its effects on VO2max and hormone responses, but soy protein can provide comparable advantages in terms of muscle adaption and antioxidant status.

Dosage and Considerations:

- Protein intake for athletes should be between 1.2 and 2.2 g/kg of body weight per day.
- Muscle protein synthesis is encouraged by consuming protein in equal amounts throughout the day.
- Every meal should contain 0.4–0.55 g of protein/kg body weight for optimum potential muscle protein synthesis.
- Whey protein supplementation, when utilized properly, can be an effective tool for athletes looking to increase muscle mass, strength, and recovery. For personalized advice, see a healthcare practitioner or certified dietitian.

Summary

Most effective sports nutritional supplements are protein, nitrates, creatine, caffeine, and BA; each offers unique benefits to athletes.

Table 13 Protein RCTs [94]

Reference	Supplement dose and duration	Results	Conclusion
• Mhamed et al. [95]	30 g of whey daily for 2 months	Whey > control for body fat reduction and leg muscle volume	Whey protein supplementation combined with 2-month endurance training leads to a reduction in body fat and increased leg muscle volume. ASAT, ALAT, and CK were also reduced
• Seo et al. [96]	8 weeks of a combined animal and plant-based protein (60 g/d)	Golfers improved driver distance, driver ball speed, and grip strength	The intake of a mixed protein containing both animal and plant proteins positively affected golf performance and muscle function
• Kim et al. [97]	• 4 weeks of supplementing with whey versus placebo	• The whey protein group had significantly larger increases in muscle mass and strength in multiple muscle groups, including the dominant knee and shoulder extensors	• Whey protein supplements improve muscular mass, strength, and endurance
• Jacinto et al. [98]	• 10 weeks of whey protein supplementation versus leu-matched collagen peptides whey protein had a larger rise in biceps brachii and vastus lateralis muscle thickness	• Whey protein had a larger rise in biceps brachii and vastus lateralis muscle thickness	• Whey protein supplement was more effective than a leucine-matched collagen peptide product in improving muscle growth, but not strength and power, following a 10-week resistance exercise program in young people
• Naclerio and Seijo [99]	• Varies, typically 20–40 g post-exercise	• Significant growth in muscle and strength increases	• Protein supplementation, especially after exercise, promotes muscular development and strength in people who do regular strength training
• Morgan and Breen [100]	• Varies protein hydrolysates used	• Significantly improves exercise-induced muscle recovery and adaption	• Protein hydrolysates promote post-exercise recovery and adaptability by rapid absorption and efficient amino acid delivery

(continued)

Table 13 (continued)

Reference	Supplement dose and duration	Results	Conclusion
• Mazzulla et al. [101]	• 20–40 g post-exercise; dispersed consumption (0.4–0.55 g/kg/meal)	• Improved muscle protein production throughout the day	• A uniform distribution of protein consumption across meals promotes muscle protein synthesis and minimizes amino acid deficits
• Antonio et al. [102]	• 3.4 vs. 2.3 g/kg/d	• No difference in LBM increase; higher dose led to greater fat mass reduction	• High protein intakes promote fat loss
• Devries and Phillips [103]	• Take 20–40 g of fast-digesting protein, such as whey, after exercise	• Faster healing, better muscle repair and growth	• Whey protein is excellent for post-exercise recovery due to its quick absorption and high leucine concentration

- **Protein**: is necessary for muscle growth, repair, and recuperation. A daily intake of 1.5–2.2 g/kg of body weight is advised. A decrease in fat mass could be attributed to increased consumption.
- **Nitrates**: Nitrates, which increase oxygen transfer to muscles and enhance endurance, are found in beetroot juice. Typical dose: 300–500 mg nitrate, taken 2–3 h before activity.
- **Creatine**: increases strength, power, and lean mass by recharging energy stores (PCr and ATP). Usual dosage is 3–5 g/day.
- **Caffeine**: enhances focus, alertness, and pain tolerance, leading to improved performance during exercise. Recommended dosage is from 3 to 6 mg/kg body weight, thirty to sixty minutes prior to an exercise.
- **BA**: boosts carnosine levels, which improves capacity to exercise by neutralizing muscle acidity and decreasing fatigue. Suggested dosage is 2–5 g each day.

These are trends of supplements strategy athletes can potentially achieve their specific performance and/or body composition goals.

7 Nanocompound in Sports Nutrition (Future Potential)

7.1 "Could Nanotechnology Improve Exercise Performance? Evidence from Animal Studies"

Nanoparticles, defined as colloidal structures between 1 and 100 nm in size, exhibit diverse shapes (rod, globular, conical, etc.) [104]. These cutting-edge materials find increasing applications in sports, encompassing equipment, clothing, and nutrition. Their integration aims to enhance athletic performance and drive the long-term evolution of sports. While conventional sports supplements offer benefits, they often suffer from limitations. Rapid degradation within the gastrointestinal tract and significant metabolism by enterocytes can significantly reduce their bioavailability, thereby limiting their ergogenic potential.

In contrast, nutritional nano-compounds offer several advantages:

- **Enhanced Solubility and Dissipation**: Nanoparticles can improve the solubility and distribution of active compounds within the body.
- **Increased Bioavailability**: Improved bioavailability translates to more efficient absorption and utilization of nutrients by the body.

Nanoparticles in Exercise Performance: Emerging Evidence

Limited research has explored the specific impact of nanocompounds on exercise performance. However, studies have investigated the effects of nanocomposites containing: [105, 106].

- **Resveratrol**: This polyphenol exhibits anti-inflammatory, antitumorigenic, and antioxidant properties. While promising for enhancing exercise performance, its low bioavailability has hindered its clinical application. Lipid-based nanosystems have shown the potential to overcome this limitation.
- **Iron**: Exercise training can induce iron deficiency, leading to fatigue and impaired performance. Iron-loaded liposomes may offer a more efficient way to deliver iron to the body.
- **Nitroxide Radicals**: These stable free radicals possess antioxidant properties. Subcutaneous infusion of redox-active nanoparticles containing nitroxide radicals enhanced time-to-exhaustion during treadmill jogging, according to an investigation conducted on rats.

7.2 Nanotechnology in Sports Nutrition: Overcoming Limitations of Oral Delivery

Most sports supplements are administered orally, subjecting them to the body's initial metabolic processes, which can significantly reduce their bioavailability and efficacy. This limitation is particularly pronounced for compounds like proteins and peptides.

Nanotechnology offers a promising solution by enabling targeted drug delivery, bypassing the limitations of oral administration. For example, Coenzyme Q10 (CoQ10), a potent antioxidant, can combat exercise-induced oxidative stress by neutralizing free radicals. However, oral CoQ10 often exhibits poor bioavailability.

By encapsulating CoQ10 within nanoliposomes, researchers have demonstrated enhanced delivery to the liver. This targeted delivery can improve the body's ability to utilize CoQ10, leading to:

- **Reduced lactic acid accumulation**: Minimizing muscle fatigue.
- **Inhibition of protein breakdown**: Protecting muscle tissue.
- **Maintenance of liver glucose levels**: Providing sustained energy during exercise.

These findings suggest that sports drinks incorporating CoQ10-loaded nanoliposomes may offer superior anti-fatigue effects compared to conventional CoQ10 supplements. Professionals in sports nutrition and science are urged to utilize advanced technologies and be informed of rapid developments to identify optimal solutions. Applications in precision medicine and biomedical research commonly utilize these technologies (Tables 14, 15 and 16).

7.3 Liposomal Technology and Vitamin C in Sports Nutrition

Vitamin C is a well-known antioxidant with potential to enhance athletic performance by:

- **Reducing the risk of upper respiratory tract infections (URTIs)**: Elite athletes are particularly susceptible to URTIs due to the high volume of training stress. Vitamin C has been shown to reduce the incidence, duration, and severity of colds in physically stressed individuals, minimizing training disruptions.
- **Improving respiratory function**: Vitamin C can help ameliorate exercise-induced bronchoconstriction.
- **Enhancing energy metabolism**: It plays a role in carnitine synthesis and the regulation of fatty acid utilization during exercise.

While oral vitamin C supplementation (e.g., 500 mg capsules) offers some benefits, its bioavailability may be limited.

Liposomal technology presents a promising avenue for enhancing the efficacy of vitamin C and other sports nutrition supplements. By encapsulating nutrients within liposomes, it is possible to:

- **Improve bioavailability**: Enhance the absorption and delivery of nutrients to target tissues.
- **Protect nutrients from degradation**: Shield them from the harsh environment of the gastrointestinal tract.
- **Enable targeted delivery**: Potentially enhance the delivery of nutrients to specific tissues, such as muscles [106].

Table 14 Illustrations of currently available or possible 'in-field' non-invasive technologies or approaches that might influence current or future nutrition research, interventions, and/or suggestions [03]

Theme	Innovation	Potential nutritional impact/suggestion
Methodologies	• Wearable patches that integrate microfluidic technologies	• Instant feedback on sweating rate and composition, that affects fluid composition and water intake
	• (DXA) Dual-energy X-ray absorptiometry	• Benchmark standard for measuring BMD
	• Double-labeled water (bolus of 2H218O water)	• Estimated energetic dietary requirements may change depending on the gold standard methodology for assessing free-living daily energy consumption
	• Glycine (the tracer bolus and collection of urine)	• Whole-body protein turnover (synthesis, breakdown, and net protein balance) could be determined through the excretion rates of 15 N in urine urea and ammonia
	• Continuous glucose monitoring assessment	• Assess the fluctuation of blood glucose concentrations
	• Biomarkers tracking health	• Performance
	• Isotopic methods. The investigation of metabolic flow utilizing stable isotope-tagged substrates	• Isotopic methods. The investigation of metabolic flow utilizing stable isotope-labeled substrates
	• DNA & RNA sequence of DNA and RNA	• The process of determining genes and gene expression (e.g., responders vs. non-responders) by recognizing the nucleic acid sequence
Food science	• Sports beverage formulations include glucose and fructose	• Enhance the quantity of transportable and oxidizable CHO to improve endurance ability and reduce digestive difficulties
	• CHO hydrogel sports beverage formulas	• Increase stomach emptying, which results in more transportable and oxidizable CHO, to improve endurance performance and reduce GI concerns
	• Beta-alanine with slow release (cellulose-based excipient)	• Reduce the quantity of urine losses of beta-alanine and the paraesthesia side effects of beta-alanine as a nutrition ergogenic supplement

(continued)

Table 14 (continued)

Theme	Innovation	Potential nutritional impact/suggestion
Sport science/ equipment integration	• Ergometer power meters	• Capability to more correctly measure exercise energy consumption to better project energetic nutrition requirements
	• Basic integrated tracking activity (heart rate, GPS, accelerometer)	• Ability to more correctly assess exercise energy expenditure and intensity to better project energetic nutrition requirements
	• Advanced integrated activity monitors (heart rate, GPS, accelerometer, core, and skin temperature)	• Ability to more correctly estimate in real-time exercise energy expenditure, activity intensity, and core body temperature to better project energetic nutrition, hydration, and cooling requirements coupled with pacing decisions

• Some indicated technologies/methodologies continue to have construct validity issues and require additional research validation. BMD is bone mineral density; CHO is carbohydrate; GI is gastro-intestinal; GPS is global positioning system; and HR is heart rate

Research Gap:

Despite the potential benefits, the application of liposomal technology in sports nutrition remains largely unexplored. While research on liposomal vitamin C and exercise performance is lacking, this technology holds promise for improving the delivery of other essential nutrients for athletes, such as antioxidants, amino acids, and vitamins.

8 Conclusion

8.1 Innovations and the Future Potential of Nanotechnology in Sports Nutrition

(a) Customized Nano-Supplementation: Genetics and metabolomics research may allow nano-supplements personalized to an athlete's genes, activity regimen, and nutritional demands. Nano-supplementation is achievable because biosensors and personal electronics monitor bodily data in real-time.
(b) Smart Delivery Systems: Nanosupplements may release nutrients dependent on pH or enzyme activity.This concentrated approach may boost performance but reduce impact.
(c) Sustainability and Green Nanotechnology: Safe nanoparticles may safeguard the environment and extend nanosupplement life. Biodegradable nanoparticles may reduce bioaccumulation.

Table 15 Comparative table: emerging methods in nanotechnology for sports nutrition [107]

Method	Description	Advantages	Challenges
• Nano-gels	• Water-based carriers for controlled nutrient release	• Enhanced hydration and sustained energy release	• High production costs; limited long-term research
• Nano-micelles	• Spherical structures solubilizing hydrophobic nutrients	• Improved absorption and bioavailability	• Expensive and complex production
• Quantum dots	• Track nutrient absorption and bio-distribution	• Personalized nutrition insights	• Potential toxicity and regulatory challenges
• Nano-phytosomes	• Lipid-based carriers for plant-based nutrients	• Increased solubility and efficacy of herbal supplements	• High production costs; limited formulations
• Nano-bioactive peptides	• Protein fragments at the nanoscale for muscle recovery	• Faster healing and better nutrient absorption	• Safety for long-term use requires further study
• Nano-capsules	• Controlled nutrient release over time	• Consistent delivery; reduced risk of nutrient spikes	• High cost and technical complexity
• Nano-enzymes	• Synthetic enzymes to enhance digestion and absorption	• Improved efficiency for high-protein diets	• Interaction with natural enzymes needs investigation
• Nano-probiotics	• Enhanced survival and colonization in the gut	• Better digestive health and immune function	• Stability during storage and transport is challenging
• Nano-mineral Complexes	• Minerals combined with nanoscale carriers for better absorption	• Reduced gastrointestinal side effects; enhanced efficacy	• Limited long-term safety research
• Smart nano-delivery	• Nano-sensors and carriers for targeted nutrient delivery	• Optimized nutrient utilization; reduced waste	• High cost and technical complexity

(d) AI and Big Data: Based on performance data, AI and big data analytics may predict supplement use, changing sports nutrition. Nanoparticle supplement performance may be improved by machine learning algorithms that identify player tendencies.

Table 16 Summary of studies investigating the effect of nano-compounds on exercise performance [104]

Study	Animal model	Nanocarrier	Protocol	Performance test	Effects (compared to non-nano encapsulated form)	Effects (compared to control/placebo)
Resveratrol						
• Sun et al. [108]	• C57BL/6 J mice	• SLN	• Running training, 120 min per day for 8 weeks, plus 25 mg/kg given 1 h before exercise, 6 days per week for 8 weeks	• Timeto exhaustion running test	• No impact. However, only nano-compounds showed differences from the control	• Running distance; Respiratory exchange ratio at low to medium intensity workout; Mitochondrial biogenesis is the removal of defective mitochondria; mitochondrial function
Iron						
• Xu et al. [109]	• Wistar rats	• LIP	• High-intensity running training for 4 weeks to induce anemia. After anemia was confirmed, 2 more weeks of training receiving ferric ammonium citrate liposomes or heme iron liposomes at 5 mg iron/100 g body weight	• No measurement of performance	• m Serum iron; m Red blood cells, hematocrit, and hemoglobin; m Liver iron content and superoxide dismutase in heme iron liposomes vs free heme iron; m Serum and liver superoxide dismutase in ferric ammonium citrate liposomes vs free ferric ammonium citrate; k Serum and liver malondialdehyde superoxide dismutase in heme iron liposomes vs free heme iron	

(continued)

Table 16 (continued)

Study	Animal model	Nanocarrier	Protocol	Performance test	Effects (compared to non-nano encapsulated form)	Effects (compared to control/placebo)
Nitroxide						
• Radicals Toriumi et al. [110]	• Fischer 344 rats	• RNP	• A subcutaneous injection of redoxactive nanoparticles (from 200 to 400 mg/kg) before exercising	• Time to exhaustion running test at maximal aerobic velocity	• K time to exhaustion in low molecular weight free antioxidant supplementation; prevalence of damage to mitochondria in low molecular weight free antioxidant supplementation	• Time till exhaustion in a dose-dependent manner; k hemolysis; exercise-induced" loss in red blood cells and rise in plasma free iron; oxidative stress in skeletal muscle. Blood markers that indicate muscle injury
Folic acid						
• Qin et al. [111]	• Kunming mice	• LDH	• 5 mg/kg for 28 days	• Time to exhaustion swimming test (load of 10% of body weight)	• A propensity to extend the time to exhaustion, improve muscle and hepatic glycogen levels, and decrease blood urea nitrogen and lactic acid levels (not significant)	• Time to exhaustion; muscle and hepatic glycogen levels blood urea nitrogen and blood lactate levels

(SLN): Solid lipid nanoparticles; self-nanoemulsifying drug delivery system (SNEDDS); liposomes (LIP); redox-active nanoparticles (RNP); and layered double hydroxide nanoparticles (LDH)

(e) Track brain function: Nanotechnology may assist mentally demanding athletes make choices, manage stress, and think coherently. Because they make your brain wiser and healthier, nootropics may inspire new thoughts.
(f) Therapeutic Uses: Nano-supplements may prevent injuries, speed recovery, and improve performance. Nanoparticles may deliver medications to damaged cells, speeding up healing. International bodies might simplify nano-supplement testing and approval.
(g) Clear labeling and safety assessments help athletes and customers trust a product. Nanotechnology might improve sports nutrition, but it needs further investigation.

We need to study many individuals and real-life scenarios to understand long-term safety and recipe changes. As AI, eco-friendly technology, and smart transportation improve, diet-specific nano-supplements will aid. Nanotechnology might revolutionize everything. Sports may assist by bringing individuals from diverse areas together to work ethically and legally. This ensures secure and informative replies worldwide.

Future Directions

Sports nutrition's incorporation of nanotechnology is a developing area with great potential but also calls much attention. This part lists the important topics for further studies, investigates possible uses, investigates ethical issues, and ends with important lessons for responsible advancement of the discipline.

8.2 Future Research Needs to Address Safety and Efficacy Concerns

This list outlines crucial considerations for the responsible development and use of nano-supplements. Here's a breakdown:

1. **Long-term Safety Assessment:**
    - **Bioaccumulation**: Understanding how nanoparticles accumulate in the body over time is critical.
    - **Toxicity**: Evaluating potential harmful effects, both short-term and long-term, is essential.
    - **Health Risks**: Assessing the overall impact on human health, including potential for chronic diseases.
2. **Physiological Interactions in High-Performance Conditions:**
    - **Nano-supplement-Physiological System Interactions**: Research should delve into how nanoparticles interact with the body's systems during intense physical activity.

- **High-Performance Conditions**: This includes examining effects under conditions like extreme endurance, strength training, and competitive stress.

3. **Standardization of Formulations:**
   - **Product Quality**: Ensuring consistent and reliable product quality across different brands and batches.
   - **Efficacy**: Verifying that the supplements deliver the claimed benefits.
   - **Safety**: Establishing clear safety standards to minimize potential risks to consumers.
4. **Robust Research to Substantiate Claims:**
   - **Performance-Enhancing Claims**: Rigorous scientific studies are needed to validate claims about improved athletic performance.
   - **Variety of Situations**: Research should cover a wide range of athletic activities and competitive scenarios.
5. **Inclusivity in Research:**
   - **Diverse Populations**: Studies should include a diverse range of participants, including women, adolescents, older adults, and individuals with pre-existing medical conditions.
   - **Understanding Individual Differences**: This will help tailor recommendations and ensure safety for different populations.
6. **Real-World Conditions:**
   - **Mimicking Real-World Scenarios**: Research should simulate real-life conditions, such as dietary habits, environmental factors, and the pressures of competition.
   - **Translational Research**: This will help translate lab findings into practical applications for athletes and improve the reliability of research findings.

References

1. IOC Consensus Statement on Sports Nutrition (2004) Available online at: https://stillmed.olympic.org/documents/commissions_pdffiles/medical_commission/ioc_consensus_statement_on_sports_nutrition_2003.pdf. Accessed 14 Dec 2021
2. Hawley JA, Maughan RJ, Hargreaves M (2015) Exercise metabolism: historical perspective. Cell Metab 22:12–17. https://doi.org/10.1016/j.cmet.2015.06.016
3. Jonvik KL, King M. Rollo I, Stellingwerff T, Pitsiladis Y (2022) New opportunities to advance the field of sports nutrition. Front Sports Act Living 4. https://doi.org/10.3389/fspor.2022.852230
4. Halder M et al (2019) Vitamin K: double bonds beyond coagulation insights into differences between vitamin K1 and K2 in health and disease. Int J Mol Sci 20(4):896. https://doi.org/10.3390/ijms20040896.9
5. FMCG Gurus (2023) Nutritional supplement trends in 2023

6. Schwalfenberg GK, Genuis SJ (2017) The importanceof magnesium in clinical healthcare. Scientifica (Cairo) 2017:4179326. https://doi.org/10.1155/2017/4179326
7. Trang HM, Cole DEC, Rubin LA, Pierratos A, Siu S, Vieth R (1998) Evidence that vitamin D3 increases serum 25-hydroxy vitamin D more efficiently than does vitamin D2. Am J Clin Nutr 68(4):854–858. https://doi.org/10.1093/ajcn/68.4.854
8. Beulens JW et al (2013) The role of Menaquinones (vitamin K2) in human health. Br J Nutr 110(8):1357–1368. https://doi.org/10.1017/S0007114513001013
9. Capling L, Beck K, Gifford J, Slater G, Flood V, O'Connor H (2017) Validity of dietary assessment in athletes: a systematic review. Nutrients 9:1313
10. Knapik JJ, Steelman RA, Hoedebecke SS, Austin KG, Farina EK, Lieberman HR (2016) Prevalence of dietary supplement use by athletes: systematic review and meta-analysis. Sports Med 46:103–123
11. Tirla A, Vesa CM, Cavalu S (2021) Severe cardiac and metabolic pathology induced by steroid abuse in a young individual. Diagnostics 11:1313
12. Muniz-Pardos B, Angeloudis K, Guppy FM, Keramitsoglou I, Sutehall S, Bosch A et al (2021) Wearable and telemedicine innovations for olympic events and elite sport. J Sports Med Phys Fitness 61:1061–1072. https://doi.org/10.23736/S0022-4707.21.12752-5
13. Catalani V, Negri A, Townshend H, Simonato P, Prilutskaya M, Tippett A, Corazza O (2021) The market of sport supplement in the digital era: a netnographic analysis of perceived risks, side-effects and other safety issues. Em Trends Drugs Addictions Health 100014
14. Baker LB, Model JB, Barnes KA, Anderson ML, Lee SP, Lee KA et al (2020) Skin-interfaced micro fluidic system with personalized sweating rate and sweat chloride analytics for sports science applications. Sci Adv 6:eabe3929. https://doi.org/10.1126/sciadv.abe3929
15. Caesar E (2019) Inside the race to break the two-hour marathon. Retrieved from https://www.newyorker.com/sports/sporting-scene/inside-the-race-to-break-the-two-hour-marathon-eliud-kipchoge
16. Stellingwerff T, Cox GR (2014) Systematic review: carbohydrate supplementation on exercise performance or capacity of varying durations. Appl Physiol Nutr Metab 39(9):998–1011. PubMed ID: 24951297 https://doi.org/10.1139/apnm-2014-0027
17. Coyle EF, Coggan AR, Hemmert MK, Ivy JL (1986) Muscle glycogen utilization during prolonged strenuous exercise when fed carbohydrate. J Appl Physiol 61(1):165–172. PubMed ID: 3525502 https://doi.org/10.1152/jappl.1986.61.1.165
18. King AJ, Rowe JT, Burke LM (2020) Carbohydrate hydrogel products do not improve performance or gastrointestinal distress during moderate-intensity endurance exercise. Int J Sport Nutr Exerc Metab 2020(30):305–314. https://doi.org/10.1123/ijsnem.2020-0102
19. Puri V, Nagpal M, Singh I, Singh M, Dhingra GA, Huanbutta K, Dheer D, Sharma A, Sangnim T (2022) A comprehensive review on nutraceuticals: therapy support and formulation challenges. Nutrients 14:4637. https://doi.org/10.3390/nu14214637
20. Peeling P, Castell LM, Derave W, de Hon O, Burke LM (2019) Sports foods and dietary supplements for optimal function and performance enhancement in track-and-field athletes. Int J Sport Nutr Exerc Metab 29:198–209
21. Garthe I, Maughan RJ (2018) Athletes and supplements: prevalence and perspectives. Int J Sport Nutr Exerc Metab 28:126–138
22. Ventura Comes A, Sánchez-Oliver A, Martínez-Sanz J, Domínguez R (2018) Analysis of nutritional supplements consumption by squash players. Nutrients 10:1341
23. Jäger R, Mohr AE, Carpenter KC, Kerksick CM, Purpura M, Moussa A, Townsend JR, Lamprecht M, West NP, Black K et al (2019) International society of sports nutrition position stand: probiotics. J Int Soc Sports Nutr 16:62
24. Nicolov M, Cocora M, Buda V, Danciu C, Duse AO, Watz C, Borcan F (2021) Hydrosoluble and liposoluble vitamins: new perspectives through ADMET analysis. Medicina 57:1204
25. Peeling P, Binnie MJ, Goods PSR, Sim M, Burke LM (2018) Evidence-based supplements for the enhancement of athletic performance. Int J Sport Nutr Exerc Metab 28:178–187
26. Fernandes H (2021) Dietary and ergogenic supplementation to improve elite soccer players' performance. Ann Nutr Metab 77:197–203

27. Özen AE, Bibiloni MDM, Pons A, Tur JA (2014) Consumo De Alimentos Funcionales En Europa; UnaRevisiónSistemática. Nutr Hosp 470–478
28. Moradi F, Yazdani A, Nematolahi F, Hosseini-Roknabadi SM, Sharifi N (2024) Prevalence of supplement usage and related attitudes and reasons among fitness athletes in the gyms of Kashan and its relationship with feeding behavior: a cross-sectional study. BMC Sports Sci Med Rehabil 16:150.https://doi.org/10.1186/s13102-024-00940-3
29. Reed JL, De Souza MJ, Kindler JM, Williams NI (2014) Nutritional practices associated with low energy availability in Division I female soccer players. J Sports Sci 32:1499–1509
30. Wardenaar F, Brinkmans N, Ceelen I, Van Rooij B, Mensink M, Witkamp R, De Vries J (2017) Macronutrient intakes in 553 dutch elite and sub-elite endurance, team, and strength athletes: does intake differ between sport disciplines? Nutrients 9:119
31. Anderson L, Orme P, Naughton RJ, Close GL, Milsom J, Rydings D, O'Boyle A, Di Michele R, Louis J, Hambly C et al (2017) Energy intake and expenditure of professional soccer players of the english premier league: evidence of carbohydrate periodization. Int J Sport Nutr Exerc Metab 27:228–238
32. Ebine N, Feng J-Y, Homma M, Saitoh S, Jones PJH (2000) Total energy expenditure of elite synchronized swimmers measured by the doubly labeled water method. Eur J Appl Physiol 83:1–6
33. Jenner SL, Trakman G, Coutts A, Kempton T, Ryan S, Forsyth A, Belski R (2018) Dietary intake of professional australian football athletes surrounding body composition assessment. J Int Soc Sports Nutr 15:43
34. Książek A, Zagrodna A, Słowińska-Lisowska M (2020) Assessment of the dietary intake of high-rank professional male football players during a preseason training week. Int J Environ Res Public Health 17:8567
35. Levine BD, Baggish AL, Kovacs RJ, Link MS, Maron MS, Mitchell JH (2015) Eligibility and disqualification recommendations for competitive athletes with cardiovascular abnormalities: task force 1: classification of sports: dynamic, static, and impact: a scientific statement from the American Heart Association and American College of Cardiology. Circulation 132
36. Kerksick CM, Arent S, Schoenfeld BJ, Stout JR, Campbell B, Wilborn CD, Taylor L, Kalman D, Smith-Ryan AE, Kreider RB et al (2017) International society of sports nutrition position stand: nutrient timing. J Int Soc Sports Nutr 14:33
37. Williams C, Rollo I (2015) Carbohydrate nutrition and team sport performance. Sports Med 45:13–22
38. Huecker M, Sarav M, Pearlman M, Laster J (2019) Protein supplementation in sport: source, timing, and intended benefits. Curr Nutr Rep 8:382–396
39. Pérez-Castillo ÍM, Williams JA, López-Chicharro J, Mihic N, Rueda R, Bouzamondo H, Horswill CA (2024) Compositional aspects of beverages designed to promote hydration before, during, and after exercise: concepts revisited. Nutrients 16:17. https://doi.org/10.3390/nu16010017
40. Cerit M, Dalip M, Yildirim DS (2020) Genetics and athletic performance. Res Phys Educ Sport Health 9(2)
41. Hietaranta-Luoma HL, Tahvonen R, Iso-Touru T, Puolijoki H, Hopia A (2014) An intervention study of individual, apoE genotype-based dietary and physicalactivity advice: impact on health behavior. J Nutrigenet Nutrigenomics 7:161–174. https://doi.org/10.1159/000371743
42. Guest NS, Horne J, Vanderhout SM, El-Sohemy A (2019) Sport nutrigenomics: personalized nutrition for athletic performance. Front Nutrition 6. https://doi.org/10.3389/fnut.2019.00008
43. Guest N, Corey P, Vescovi J, El-Sohemy A (2018) Caffeine, CYP1A2 genotype, and endurance performance in athletes. Med Sci Sports Exerc 50:1570–1578. https://doi.org/10.1249/MSS.0000000000001596
44. Womack CJ, Saunders MJ, Bechtel MK, Bolton DJ, Martin M, Luden ND et al (2012) The influence of a CYP1A2 polymorphism on the ergogenic effects of caffeine. J IntSoc Sports Nutr 9:7. https://doi.org/10.1186/1550-2783-9-7
45. Pataky MW, Womack CJ, Saunders MJ, Goffe JL, D'Lugos AC, El-Sohemy A et al (2016) Caffeine and 3-km cycling performance: effects of mouth rinsing, genotype, and time of day. Scand J Med Sci Sports 26:613–9. https://doi.org/10.1111/sms.12501

46. Lietz G, Lange J, Rimbach G (2010) Molecular and dietary regulation of beta, betacarotene 15,15'-monooxygenase 1 (BCMO1). Arch Biochem Biophys 502:8–16. https://doi.org/10.1016/j.abb.2010.06.032
47. Ferrucci L, Perry JR, Matteini A, Perola M, Tanaka T, Silander K et al (2009) Common variation in the beta-carotene 15,15′-monooxygenase gene affectscirculatinglevelsofcarotenoids: agenome-wideassociationstudy. Am J Hum Genet 84:123–133. https://doi.org/10.1016/j.ajhg.2008.12.019
48. Lietz G, Oxley A, Leung W, Hesketh J (2012) Single nucleotide polymorphisms upstream from the beta-carotene 15,15′-monoxygenase gene influence provitamin A conversion efficiency in female volunteers. J Nutr 142:161S-165S. https://doi.org/10.3945/jn.111.140756
49. Garvican LA, Saunders PU, Cardoso T, Macdougall IC, Lobigs LM, Fazakerley R et al (2014) Intravenous iron supplementation in distance runners with low or suboptimal ferritin. Med Sci Sports Exerc 46:376–85. https://doi.org/10.1249/MSS.0b013e3182a53594
50. DellaValle DM, Haas JD (2014) Iron supplementation improves energetic efficiency in iron-depleted female rowers. Med Sci Sports Exerc 46:1204–1215. https://doi.org/10.1249/MSS.0000000000000208
51. Sacirovic S, Asotic J, Maksimovic R, Radevic B, Muric B, Mekic H et al (2013) Monitoring and prevention of anemia relying on nutrition and environmental conditions in sports. Mater Sociomed 25:136–139. https://doi.org/10.5455/msm.2013.25.136-139
52. Myhre KE, Webber BJ, Cropper TL, Tchandja JN, Ahrendt DM, Dillon CA et al (2015) Prevalence and impact of anemia on basic trainees in the impact of anemia on basic trainees in the sports. Med Open 2:23.https://doi.org/10.1186/s40798-0160047
53. Martinsson A, Andersson C, Andell P, Koul S, Engstrom G, Smith JG (2014) Anemia in the general population: prevalence, clinical correlates and prognostic impact. Eur J Epidemiol 29:489–98. https://doi.org/10.1007/s10654-014-9929-9
54. Latunde-Dada GO (2013) Iron metabolism in athletes–achieving a gold standard. Eur J Haematol 90:10–15. https://doi.org/10.1111/ejh.12026
55. Dubnov G, Foldes AJ, Mann G, Magazanik A, Siderer M, Constantini N (2006) High prevalence of iron deficiency and anemia in female military recruits. Mil Med 171:866–869. https://doi.org/10.7205/MILMED.171.9.866
56. Pedlar CR, Brugnara C, Bruinvels G, Burden R (2018) Iron balance and iron supplementation for the female athlete: a practical approach. Eur J Sport Sci 18:295–305. https://doi.org/10.1080/17461391.2017.1416178
57. Antonio J, Pereira F, Curtis J, Rojas J, Evans C (2024) The top5 can't-miss sport supplements. Nutrients 16:3247. https://doi.org/10.3390/nu16193247
58. Antonio J, Newmire DE, Stout JR, Antonio B, Gibbons M, Lowery LM, Harper J, Willoughby D, Evans C, Anderson D et al (2024) Common questions and misconceptions about caffeine supplementation: what does the scientific evidence really show? J Int Soc Sports Nutr 21:2323919
59. Tarnopolsky MA (2008) Effect of caffeine on the neuromuscular system–potential as an ergogenic aid. Appl Physiol Nutr Metab 33:1284–1289
60. Guo W, Wang S (2024) Physiological and performance adaptations to beta alanine supplementation and short sprint interval training in volleyball players. Sci Rep 14:16833
61. Ojeda ÁH, Barahona-Fuentes G, Galdames Maliqueo S, Guzmán Solis M, Cabrera MMY, Jorquera-Aguilera C (2023) Acute supplementation with beta-alanine improves performance in aerobic-anaerobic transition zones in endurance athletes. J Am Nutr Assoc 42:187–194
62. Sas-Nowosielski K, Wycis´lik J, Kaczka P (2021) Beta-alanine supplementation and sport climbing performance. Int J Environ Res Public Health 18:5370
63. Shbib S, Rashidlamir A, Hakak Dokht E (2021) The effects of plyometrictraining and β-alanine supplementation on anaerobic power and serum level of carnosine in handball players. Sport Sci Health 17:569–576
64. Jaques M, Glick D, Greco-Henderson D, Ives SJ (2019) The effects of short term β-alanine supplementation on performance in division iii male and female rowers. J Exerc Nutr 2:15

65. Smith CR, Harty PS, Stecker RA, Kerksick CM (2019) A pilot study to examine the impact of beta-alanine supplementation on anaerobic exercise performance in collegiate rugby athletes. Sports 7:231
66. Glenn JM, Gray M, Stewart RWJ, Moyen NE, Kavouras SA, DiBrezzo R, Turner R, Baum JI, Stone MS (2016) Effects of 28-day beta-alanine supplementation on isokinetic exercise performance and body composition in female masters athletes. J Strength Cond Res 30:200–207
67. Bellinger PM, Minahan CL (2016) The effect of β-alanine supplementation on cycling time trials of different length. Eur J Sport Sci 16:829–836
68. Gross M, Bieri K, Hoppeler H, Norman B, Vogt M (2014) Beta-alanine supplementation improves jumping power and affects severe-intensity performance in professional alpine skiers. Int J SportNutr Exerc Metab 24:665–673
69. Kern BD, Robinson TL (2011) Effects of β-alanine supplementation on performance and body composition in collegiate wrestlers and football players. J Strength Cond Res 25:1804–1815
70. Baguet A, Bourgois J, Vanhee L, Achten E, Derave W (2010) Important role of muscle carnosine in rowing performance. J Appl Physiol 109:1096–1101
71. Derave W, Ozdemir MS, Harris RC, Pottier A, Reyngoudt H, Koppo K, Wise JA, Achten E (2007) Β-alanine supplementation augments muscle carnosine content and attenuates fatigue during repeated isokinetic contraction bouts in trained sprinters. J Appl Physiol 103:1736–1743
72. Van Thienen R, Van Proeyen K, Puype J, Lefere T, Hespel P (2009) Beta-alanine improves sprint performance in endurance cycling. Med Sci Sports Exerc 41:898–903
73. Hodgson AB, Randell RK, Jeukendrup AE (2013) The metabolic and performance effects of caffeine compared to coffee during endurance exercise. PLoS ONE 8:e59561
74. Wu S, Jiang H (2024) Examining the impact of differing caffeine dosages in conjunction with plyometric training on physiological adaptations in basketball players. Sci Rep 14:15571
75. Tamilio RA, Clarke ND, Duncan MJ, Morris RO, Tallis J (2022) How repeatable is the ergogenic effect of caffeine? Limited reproducibility of acute caffeine (3 mg.Kg(-1)) ingestion on muscular strength, power, and muscular endurance. Nutrients 14:4416
76. Berjisian E, Naderi A, Mojtahedi S, Grgic J, Ghahramani MH, Karayigit R, Forbes JL, Amaro-Gahete FJ, Forbes SC (2022) Are caffeine's effects on resistance exercise and jumping performance moderated by training status? Nutrients 14:4840
77. Dittrich N, Serpa MC, Lemos EC, De Lucas RD, Guglielmo LGA (2021) Effects of caffeine chewing gum on exercise tolerance and neuromuscular responses in well-trained runners. J Strength Cond Res 35:1671–1676
78. Candow DG, Chilibeck PD, Burke DG, Mueller KD, Lewis JD (2011) Effect of different frequencies of creatine supplementation on muscle size and strength in young adults. J Strength Cond Res 1831–1838
79. Kreider RB, Wilborn CD, Taylor L, Campbell B, Almada AL, Collins R, Cooke M, Earnest CP, Greenwood M, Kalman DS (2018) Issn exercise and sport nutrition review: research and recommendations. J Int Soc Sports Nutr 15:38
80. Desai I, Wewege MA, Jones MD, Clifford BK, Pandit A, Kaakoush NO, Simar D, Hagstrom AD (2024) The effect of creatine supplementation on resistance training-based changes to body composition: a systematic review and meta-analysis. J Strength Cond Res 38:1813–1821
81. Yamaguchi S, Inami T, Ishida H, Morito A, Yamada S, Nagata N, Murayama M (2024) The effect of prior creatine intake for 28 days on accelerated recovery from exercise-induced muscle damage: a double-blind, randomized, placebo controlled trial. Nutrients 16:896
82. Mills S, Candow DG, Forbes SC, Neary JP, Ormsbee MJ, Antonio J (2020) Effects of creatine supplementation during resistance training sessions in physically active young adults. Nutrients 12:1880
83. Chilibeck PD, Magnus C, Anderson M (2007) Effect of in-season creatine supplementation on body composition and performance in rugby union football players. Appl Physiol Nutr Metab 32:1052–1057

84. Hoffman J, Ratamess N, Kang J, Mangine G, Faigenbaum A, Stout J (2006) Effect of creatine and ß-alanine supplementation on performance and endocrine responses in strength/power athletes. Int J Sport Nutr Exerc Metab 16:430–446
85. Bemben MG, Bemben DA, Loftiss DD, Knehans AW (2001) Creatine supplementation during resistance training in college football athletes. Med Sci Sports Exerc 33:1667–1673
86. Stone MH, Sanborn K, Smith LL, O'Bryant HS, Hoke T, Utter AC, Johnson RL, Boros R, Hruby J, Pierce KC (1999) Effects of in-season (5 weeks) creatine and pyruvate supplementation on anaerobic performance and body composition in American football players. Int J Sport Nutr Exerc Metab 9:146–165
87. Silva KVC, Costa BD, Gomes AC, Saunders B, Mota JF (2022) Factors that moderate the effect of nitrate ingestion on exercise performance in adults: a systematic review with meta-analyses and meta-regressions. Adv Nutr 13:1866–1881
88. Yuschen X, Choi J-H, Seo J, Sun Y, Lee E, Kim S-W, Park H-Y (2024) Effects of acute beetroot juice supplementation and exercise on cardiovascular function in healthy men in preliminary study: a randomized, double-blinded, placebo-controlled, and crossover trial. Healthcare 12:1240
89. Giv V, Aminaei M, Nikoei R (2024) The effect of eight weeks beetroot juice supplement on aerobic, anaerobic power, and field performance of soccer players. Res Sports Med 32:132–144
90. Mosher SL, Sparks SA, Williams EL, Bentley DJ, Mc Naughton LR (2016) Ingestion of a nitric oxide enhancing supplement improves resistance exercise performance. J Strength Cond Res 30:3520–3524
91. Vanhatalo A, Bailey SJ, Blackwell JR, DiMenna FJ, Pavey TG, Wilkerson DP, Benjamin N, Winyard PG, Jones AM (2010) Acute and chronic effects of dietary nitrate supplementation on blood pressure and the physiological responses to moderate intensity and incremental exercise. Am J Physiol Regul Integr Comp Physiol 299:R1121–R1131
92. Bailey SJ, Winyard P, Vanhatalo A, Blackwell JR, Dimenna FJ, Wilkerson DP, Tarr J, Benjamin N, Jones AM (2009) Dietary nitrate supplementation reduces the o2 cost of low-intensity exercise and enhances tolerance to high-intensity exercise in humans. J Appl Physiol 107:1144–1155
93. Larsen FJ, Weitzberg E, Lundberg JO, Ekblom B (2007) Effects of dietary nitrate on oxygen cost during exercise. Acta Physiol 191:59–66
94. Jäger R, Kerksick CM, Campbell BI, Cribb PJ, Wells SD, Skwiat TM, Purpura M, Ziegenfuss TN, Ferrando AA, Arent SM (2017) International society of sports nutrition position stand: protein and exercise. J Int Soc Sports Nutr 14:20
95. Mhamed M, Zarrouk F, Mrad M, Methnani J, Bahlous A, Zaouali M, Lindinger M, Bigard X, Bouhlel E (2024) Effects of whey protein on body composition, biochemical profile, and high intensity physical performances in well-trained endurance runners. Sci Sports 39:588–598
96. Seo JW, Jiang S, Ahn S, Kang YS, Sung Y, Li X, Jamrasi P, Sun EM, Yoo J, Kim BY, et al. (2024) Effect of mixed protein supplementation on golf performance and muscle function: a randomized, double-blind, placebo-controlled study. J Int Soc Sports Nutr 21:2393368
97. Kim CB, Park JH, Park HS, Kim HJ, Park JJ (2023) Effects of whey protein supplement on 4-week resistance exercise-induced improvements in muscle mass and isokinetic muscular function under dietary control. Nutrients 15:1003
98. Jacinto JL, Nunes JP, Gorissen SH, Capel DM, Bernardes AG, Ribeiro AS, Cyrino ES, Phillips SM, Aguiar AF (2022) Whey protein supplementation is superior to leucine-matched collagen peptides to increase muscle thickness during a 10-week resistance training program in untrained young adults. Int J Sport Nutr Exerc Metab 32:133–143
99. Naclerio F, Seijo M (2019) Whey protein supplementation and muscle mass: current perspectives. Nutr Diet Suppl 11:37–48
100. Morgan PT, Breen L (2021) The role of protein hydrolysates for exercise-induced skeletal muscle recovery and adaptation: a current perspective. Nutr Metab 18:44
101. Mazzulla M, Abou Sawan S, Williamson E, Hannaian SJ, Volterman KA, West DW, Moore DR (2020) Protein intake to maximize whole-body anabolism during post exercise recovery

in resistance trained men with high habitual intakes is several fold greater than the current recommended dietary allowance. J Nutr 150:505–511
102. Antonio J, Ellerbroek A, Silver T, Orris S, Scheiner M, Gonzalez A, Peacock CA (2015) A high protein diet (3.4g/kg/d) combined with a heavy resistance training program improves body composition in healthy trained men and women—a follow-up investigation. J Int Soc Sports Nutr 12:39
103. Devries MC, Phillips SM (2015) Supplemental protein in support of muscle mass and health: advantage whey. J Food Sci 80:A8–A15
104. Lima MR, Moreira BJ, Bertuzzi R, Lima-Silva AE (2024) Could nanotechnology improve exercise performance? Evidence from animal studies. Braz J Med Biol Res 57:e13360. https://doi.org/10.1590/1414-431X2024e13360ISSN1414-431X
105. Ramírez-Garza SL, Laveriano-Santos EP, MarhuendaMuñoz M, Storniolo CE, Tresserra-Rimbau A, VallverdúQueralt A et al (2018) Health effects of resveratrol: results from human intervention trials. Nutrients 10:1892. https://doi.org/10.3390/nu10121892
106. Dolinsky VW, Jones KE, Sidhu RS, Haykowsky M, Czubryt MP, Gordon T et al (2012) Improvements in skeletal muscle strength and cardiac function induced by resveratrol during exercise training contribute to enhanced exercise performance in rats. J Physiol 590:2783–2799. https://doi.org/10.1113/jphysiol.2012.230490
107. Jain KK (2020) Role of nanobiotechnology in drug delivery. Methods Mol Biol 2059:55–73
108. Sun J, Zhou Y, Su Y, Li S, Dong J, He Q, et al. (2019) Resveratrol loaded solid lipid nanoparticle supplementation ameliorates physical fatigue by improving mitochondrial quality control. Crystals 9:559. https://doi.org/10.3390/cryst9110559
109. Xu Z, Liu S, Wang H, Gao G, Yu P, Chang Y (2014) Encapsulation of iron in liposomes significantly improved the efficiency of iron supplementation in strenuously exercised rats. Biol Trace Elem Res 162:181–188. https://doi.org/10.1007/s12011014-0143-0
110. Toriumi T, Kim A, Komine S, Miura I, Nagayama S, Ohmori H, et al. (2021) An antioxidant nanoparticle enhances exercise performance in rat high-intensity running models. Adv Healthc Mater 10:e2100067. https://doi.org/10.1002/adhm.202100067
111. Qin L, Wang W, You S, Dong J, Zhou Y, Wang J (2014) In vitro antioxidant activity and in vivo antifatigue effect of layered double hydroxide nanoparticles as delivery vehicles for folic acid. Int J Nanomed 9:5701–5710. https://doi.org/10.2147/IJN.S74306
112. Gao Y, Chen L, Yang SN, Wang H, Yao J, Dai Q et al (2015) Contributions of visuooculomotor abilities to interceptive skills in sports. Optom Vis Sci 92:679–89. https://doi.org/10.1097/OPX.0000000000000599
113. Qin L, Lu T, Qin Y, He Y, Cui N, Du A, et al. (2020) In vivo effect of resveratrol-loaded solid lipid nanoparticles to relieve physical fatigue for sports nutrition supplements. Molecules 25:5302. https://doi.org/10.3390/molecules25225302
114. Yen CC, Chang CW, Hsu MC, Wu YT (2017) Self-nanoemulsifying drug delivery system for resveratrol: enhanced oral bioavailability and reduced physical fatigue in rats. Int J Mol Sci 18:1853. https://doi.org/10.3390/ijms18091853

Index

A. Chatterjee et al. (eds.), *Nanofuel: The Future of Sports Nutrition*, Smart Nanomaterials Technology, https://doi.org/10.1007/978-981-96-5471-0

W

The manufacturer's authorised representative in the EU is Springer Nature Customer Service Centre GmbH, Europaplatz 3, 69115 Heidelberg, Germany. If you have any concerns regarding our products, please contact ProductSafety@springernature.com

Printed and bound by CPI Group (UK) Ltd, Croydon, CR0 4YY
01/10/2026
02225936-0018